Tuncer Cebeci

An Engineering Approach
to the Calculation of Aerodynamic Flows

HORIZONS PUBLISHING

Long Beach, California
Heidelberg

Springer

Berlin *London*
Heidelberg *Milan*
New York *Paris*
Barcelona *Singapore*
Hong Kong *Tokyo*

Tuncer Cebeci

An Engineering Approach to the Calculation of Aerodynamic Flows

With 176 Figures, 11 Tables
and 28 Computer Programs

HORIZONS
PUBLISHING

Springer

Tuncer Cebeci
Aerospace Engineering Department
California State University
1250 Bellflower Blvd.
Long Beach, CA 90840-8304, USA
email: cebeci@engr.csulb.edu

Library of Congress Cataloging-in-Publication Data
Cebeci, Tuncer
An engineering approach to the calculation of aerodynamic flows / Tuncer Cebeci. p. cm.
Includes bibliographical references and index.
1. Air flow – Mathematical models. 2. Aerodynamics – Data processing. I. Title.
TL574.F5 C3 1999 629.132'32'0285–dc21 99-35756 CIP

ISBN 0-9668461-2-5 Horizons Publishing Inc., Long Beach
ISBN 3-540-66181-6 Springer-Verlag Berlin Heidelberg New York

Printed in Germany

Typeset by the author using MS Word and reedited using LaTeX by Kurt Mattes, Heidelberg, Germany
Cover design: Erich Kirchner, Heidelberg, Germany

Printed on acid-free paper 5 4 3 2 1 0

Preface

To Boeing, formerly McDonnell Douglas,
where I found an environment in which
I was encouraged to learn, teach and grow.

It is generally accepted, with a great deal of posterior justification but nevertheless partly as an act of faith, that the unsteady Navier-Stokes equations for a compressible fluid, together with a suitable equation of state, often that for a perfect gas, are sufficient to describe the flow over an airplane. For many years, however, the full numerical solution of these equations was not considered to be a reasonable goal of computational effort, mostly due to the lack of computing power and partly due to the difficulty of resolving all the scales of the motion, especially in regions of turbulent flow. For example, until 1960 the principal effort for predicting flowfields in aerodynamic flows was devoted to the solution of Laplace's equation for calculating two-dimensional inviscid flows; the prediction capability with even inviscid flow approximations was not deemed possible for three-dimensional flows.

With the advent of high-speed computers, however, considerable progress has been made in the calculation of aerodynamic flows, typical examples being airfoils, wings, fuselages and high-lift devices. This progress has led to the development of many powerful design algorithms. In general, two approaches have been pursued in these studies. The first approach is based on the numerical solutions of the Reynolds-averaged Navier-Stokes equations, and the second one, also known as the interactive boundary-layer theory, is based on the numerical solutions of inviscid and viscous flow equations in which an interaction law is used to couple both solutions. The former approach involves the numerical solution of elliptic equations in which information travels instantaneously in all directions through pressure, velocity and viscous and turbulent stress gradients. As a result, the solution method requires simultaneous calculation of the pressure, velocity components, and stress tensor throughout the flowfield, and this, in turn, implies a trade-off between accuracy and cost. This limitation is a function of current capabilities of computers and programming methods; these are likely to improve with time so that solutions of the Navier-Stokes equations are likely to continue to be used for aerodynamic flow problems.

The interactive boundary-layer theory is based on the solutions of the reduced Navier-Stokes equations and involve Euler and boundary-layer equations. While the more complete Navier-Stokes equations are likely to better predict flow variables in which cross-stream convection and diffusion are important, as in fully stalled flow conditions, the interactive approach offers the possibility of accurate numerical solutions for restricted flows with considerable savings of computer time and storage.

Regardless of which approach is used to solve the conservation equations, one must calculate the onset of transition in the development of the design algorithm so that the effects of wind tunnel and flight Reynolds numbers can be properly identified. For example, in the development of algorithms for high-lift configurations, the components of multielement airfoils have relatively lower Reynolds numbers than that of the main airfoil. At chord Reynolds numbers less than 500,000, the components can have large separation bubbles, with the onset of transition occurring inside the separation bubble. As a result, their behavior can be significantly different than the behavior of the main airfoil at higher Reynolds numbers.

Thus, an accurate procedure for calculating aerodynamic flows not only requires the solution of the conservation equations by either of the two approaches discussed above, but also requires the prediction of transition and modeling of the transitional turbulent flow, which is different from the flow at high Reynolds numbers.

This book addresses the calculation of aerodynamic flows with the interactive-boundary-layer stability-transition approach developed by the author. The contents include two-dimensional and three-dimensional steady and unsteady flows with and without compressibility effects. It is prepared in three parts, with Part 1 (Chapters 2 to 6) covering two-dimensional airfoil flows for both incompressible and compressible steady flows, Part 2 (Chapters 7 to 9) covering two-dimensional, incompressible, unsteady airfoil flows and Part 3 (Chapters 10 to 13) covering three-dimensional flows for incompressible and compressible flows. Following the description of the interactive boundary-layer approach for incompressible flows over airfoils in Chapter 1, Chapters 2 to 4 describe the calculation method for airfoils in low speed flows, with Chapter 2 describing the inviscid method and a Fortran program for a panel method, Chapter 3 the boundary-layer method and a Fortran program, and Chapter 4 the transition method and a Fortran program. The applications of the calculation method are described in Chapter 5 for a variety of airfoil flows at low and high Reynolds numbers, including iced airfoils and multielement airfoils.

The extension of the calculation method of Chapters 2 to 4 to compressible flows is discussed in Chapter 6 by describing the boundary-layer method for compressible flows. Unlike in Chapter 2, the discussion of compressible inviscid flow methods with and without viscous effects is not presented in this chapter.

This is partly because of the wide choice of methods that can be used for this purpose and partly because the basic ideas of introducing viscous effects into compressible inviscid flow methods are essentially the same as for panel methods. There are some differences and these are discussed in Chapter 13.

In Chapter 6, the extension of the transition method of Chapter 4 for incompressible flows to compressible flows is also not discussed. This is partly because the equations and the solution procedure for transition prediction are presented in Chapter 12 for three-dimensional compressible flows and partly because the basic ideas for computing transition for compressible flows are identical to those for incompressible flows discussed in Chapter 4.

Unsteady, incompressible airfoil flows are discussed in Chapters 7 and 8, with Chapter 7 describing the inviscid flow and Chapter 8 the boundary-layer method. The prediction of transition in unsteady flows is more involved than that in steady flows, and the calculation method described in Chapter 4 for steady flows does not apply for unsteady flows. As a result, the prediction of transition is not addressed in Chapter 9, which presents the applications of the calculation method of Chapters 7 and 8 to unsteady flows for specified transition locations.

The extension of the calculation method of Chapters 2 to 6 to three-dimensional subsonic and transonic flows is discussed in Chapters 10 to 13, with a procedure similar to that used for two-dimensional steady flows. In each case, the computational strategy is essentially the same: the flowfield is computed by an inviscid method either for subsonic flows by a panel method (Chapter 10) or for transonic flows by Euler methods briefly discussed in Section 13.6, and the viscous flowfield is computed by a boundary-layer method (Chapter 11) formulated in the inverse mode. An interface program is placed between the inviscid method and the three-dimensional boundary-layer method to process the geometry and inviscid velocity data for input to the boundary-layer program. The location of the onset of transition for three-dimensional flows is obtained (Chapter 12) with an extension of the e^n-method discussed in detail in Chapter 3 for two-dimensional flows. The application of the calculation method of Chapters 10 to 12 is described in Chapter 13 for both subsonic and transonic three-dimensional flows.

The computer programs described in this book, with additional details and sample calculations, are available from the author.

The author is indebted to many friends and colleagues for the help they provided in the development of the interactive boundary layer and stability transition approach described here. Special thanks are due to the late A.M.O. Smith of Douglas Aircraft Company, Herb Keller of the California Institute of Technology, the late Keith Stewartson of University College, and Kalle Kaups and John Hess of the Douglas Aircraft Company.

The author is also indebted to Hsun H. Chen, K.C. Chang, J.P. Shao, A.A. Khattab and Eric Besnard for performing many of the calculations reported here and for proof-reading the manuscript and making helpful suggestions. In addition, the author would like to acknowledge the help of Laju Tejwani for typing many drafts of the manuscript and to Kurt Mattes for the excellent job in typing the final manuscript.

Indian Wells, May 1999 *Tuncer Cebeci*

Contents

1 Interactive-Boundary-Layer Approach for Airfoils

1.0 Introduction

Inviscid methods are the most practical methods for calculating flowfields over airfoils in incompressible flows. However, air is undoubtedly a viscous fluid and calculation capabilities based on inviscid flows can only be of limited value to the aircraft designer interested in practical use. Fortunately, this is not really the case and methods in which an external inviscid flow calculation forms a major component part, can be developed for application to a wide variety of important practical aerodynamic flow problems.

All real flows are viscous, but the effects of viscosity can vary greatly in importance from case to case. At one extreme, for example, the flow over a thin symmetrical airfoil at a small angle of attack, for which one can start by calculating the inviscid pressure distribution, use this to calculate the boundary layer and wake, for laminar and turbulent flows, and continue, if necessary, with an improved inviscid calculation taking into account the displacement effect of the viscous layers followed by the final boundary-layer calculation. In this situation – sometimes known as a "weak interaction" problem – this simple procedure, in which separate inviscid and viscous calculations are involved almost independently, leads to estimates of satisfactory accuracy for both the overall forces and detailed pressure distribution. At the other extreme, there are many important problems, involving large regions of flow separation, in which a classical "attached flow" solution can be completely unsatisfactory.

In between these two extremes lies a wide range of practical important flows in which separate inviscid and boundary-layer solutions are still relevant and can provide high overall accuracy provided that they are combined with sufficient attention to the mathematical modeling of the interaction between them. It is essential to arrange the calculation in an iterative manner, so that each successive inviscid solution provides the pressure distribution for the next boundary-layer

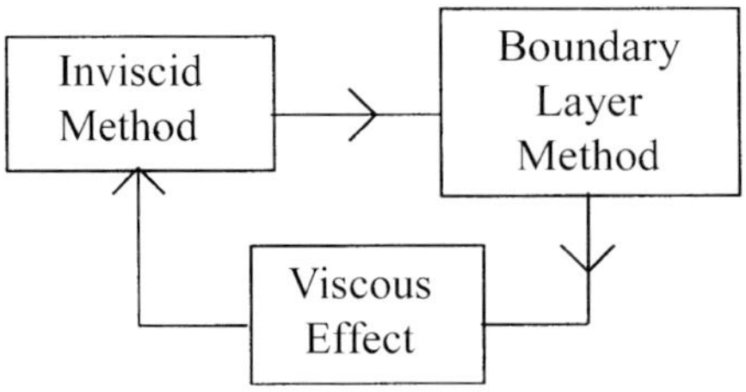

Fig. 1.1. Interactive boundary-layer scheme.

solution. The displacement (viscous) effect is then used to modify the inner boundary conditions for a near inviscid calculation; this "cycle" must then be repeated until convergence is obtained, see Fig. 1.1.

As discussed, for example in [1–4], methods for solving problems of this "strong interaction" type have been under development since the latter part of 1970 and have now reached a high level of sophistication in applications to aerodynamic flows, as is discussed in this book for the approach developed by Cebeci and his associates. They are capable of dealing satisfactorily with low-speed flows about airfoils with high-lift devices (multielement airfoils) as well as transonic flows in which the interaction between a shock wave and a boundary layer is an important phenomenon. In most examples in which the airfoil is generating lift it has been found important to consider the interactive effect of the near wake, since this can appreciably influence the pressure distribution over the whole airfoil, not just near the trailing edge, as perhaps may have been expected.

This chapter outlines the general features of viscous interactions for airfoils to pave the way for the calculation method described later for calculating two- and three-dimensional aerodynamic flows. Following the discussion on the general features of viscous interactions on airfoils in Section 1.1, basic mathematical models for the interaction process for inviscid flows are discussed in Section 1.2. The mathematical models for the interaction process for viscous flows are discussed in Section 1.3. A summary of the interactive procedure for airfoils, including the prediction of transition is presented in Section 1.4.

1.0.1 General Features of Viscous Interactions on Airfoils

The general nature of the viscous flow over an airfoil can vary substantially depending on the Reynolds number and the Mach number. For example, at low chord Reynolds numbers (say 50×10^3 to 500×10^3) and at subsonic speeds, rather large separation bubbles can occur some way downstream of the leading edge (see Fig. 1.2) with the onset of transition taking place within the bubble prior to reattachment. The length of the bubble increases with decreasing Reynolds number and strongly influences the airfoil performance characteristics.

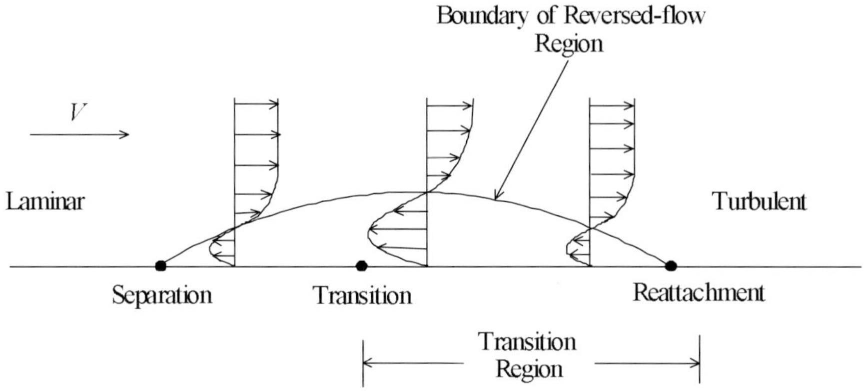

Fig. 1.2. The region of separation bubble on an airfoil at low Reynolds numbers.

At high Reynolds numbers (say 3×10^6 to 60×10^6), the onset of transition occurs before or at the flow separation point and the extent of the transition region is confined to a relatively small region. As will be discussed later in Chapter 5, the transition location is much easier to calculate in this case than its counterpart at low Reynolds numbers. In some cases, especially when empirical methods are used, it is acceptable to assume the onset of transition to correspond to laminar separation if it occurs before the transition location can be computed.

At either high or low Reynolds number, because the mean adverse pressure gradient on the upper surface will be higher than on the lower, the upper surface boundary layer will be the thicker of the two at the trailing edge (see Fig. 1.3), and the net displacement effect will therefore simulate a negative camber over the rear of the airfoil, which is mainly responsible for the overall reduction in circulation, and hence in lift, caused by viscosity. A significant result of the displacement effect is the reduction in pressure at and near the trailing edge from the stagnation value of inviscid flow to the value, just above that of the freestream, observed in experiment; this is caused by the smoothing of the effective shape, produced by the boundary layers and near wake.

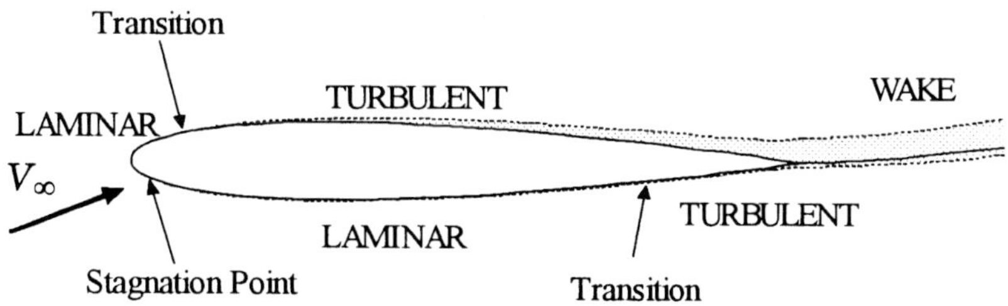

Fig. 1.3. Subsonic flow over an airfoil at incidence.

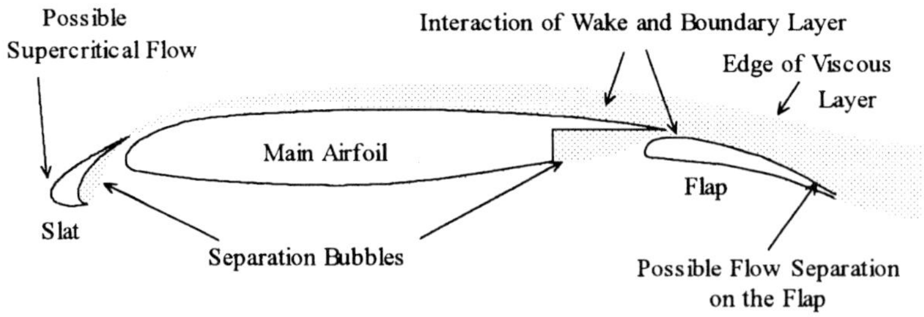

Fig. 1.4. Viscous flow over a multielement airfoil.

In the case of an airfoil under high-lift conditions, with leading-edge slat and trailing-edge flaps deployed, additional viscous effects may further complicate the situation. The chief culprit is the interaction of the wake and the boundary layer. The general nature of the flow over a three-element airfoil is sketched in Fig. 1.4. The wake from the trailing edge of the slat is initially separate from the boundary layer on the main airfoil growing beneath it, but the presence of each will affect the development of the other until eventually they merge to form a single boundary layer. Usually this will happen before the trailing edge of the main airfoil is reached, when the same process starts again as the flow develops over the flap. Practical design requirements usually lead to shapes for the under surfaces of both slat and shroud, which give rise to local separation bubbles in the cavities beneath them, and these must be included in the mathematical model used to represent the flow. In general, by the time the flap is reached, the viscous layers will be relatively thick and normal pressure gradients are likely to be appreciable. The flow will frequently separate from the upper surface of a rear flap well before the final stall takes place.

For transonic flow, a shock wave of appreciable strength – normally on the upper surface of a lifting airfoil – may appear in the supersonic portion of the flow, see Fig. 1.5, and its interaction with the boundary layer and at its foot will be clearly of importance. Again, the boundary layer has a smoothing effect which transforms the discrete pressure jump of the inviscid flow into a rapid, but

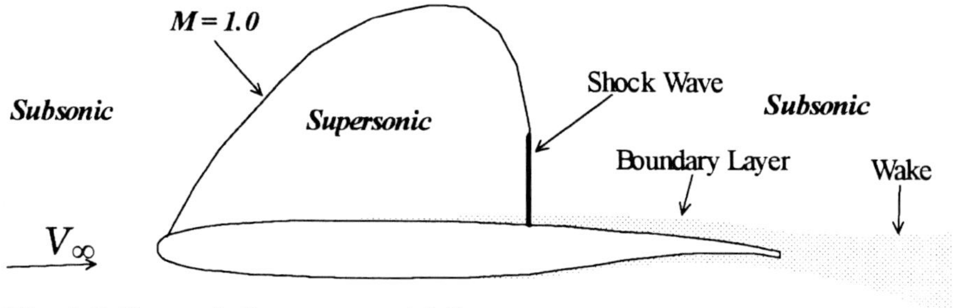

Fig. 1.5. Transonic flow over an airfoil.

a continuous, pressure rise. As the shock increases its strength, it will eventually cause the boundary layer to separate, and the consequent separation bubble will then spread towards the trailing edge. Before this happens, the shock wave will thicken the boundary layer at its foot by an increment which will be amplified by the additional adverse pressure gradient as the trailing edge is approached, thus enhancing the negative camber mentioned before and increasing the overall loss of lift so that viscous effects may be expected to be particularly severe at transonic speeds. The pressure jump of a shock wave (with or without a local separation bubble at its foot) will exacerbate the tendency of the boundary layer to separate before it reaches the trailing edge, a condition which signals the onset of buffeting and limits the usable lift on the airfoil. It is the eventual aim in aerodynamic flows to predict all these phenomenon with reasonable accuracy.

1.1 Mathematical Models for the Interaction Process: Inviscid Flow

Viscous effects can be incorporated into an inviscid method by replacing the real flow with an equivalent fictitious inviscid flow such that the velocity components at the edge of the boundary layer are equal in both cases. We assume that in the fictitious flow, $u = u_e$ for $y \leq \delta$. This fictitious flow can be modeled by using either the "solid" displacement surface concept or the "transpiration" model concept proposed by Lighthill [5]. Figure 1.6a shows the real flow and Figs, 1.6b and 1.6c the equivalent fictitious flow using the displacement surface and blowing velocity concepts, respectively for two-dimensional flows. Figure 11.8 in Chapter 11 shows the same concepts for three-dimensional flows.

Both displacement surface and transpiration models are based on first-order boundary-layer theory, that is; the shear layer thickness is assumed to be thin and the transverse variation in pressure across the layer is assumed to be negligible. When this is not the case, it is necessary to use second-order boundary-layer theory, as discussed in [6].

To describe the mathematical models for each concept, it is convenient, following the discussion in [6], to work in terms of the difference between the two

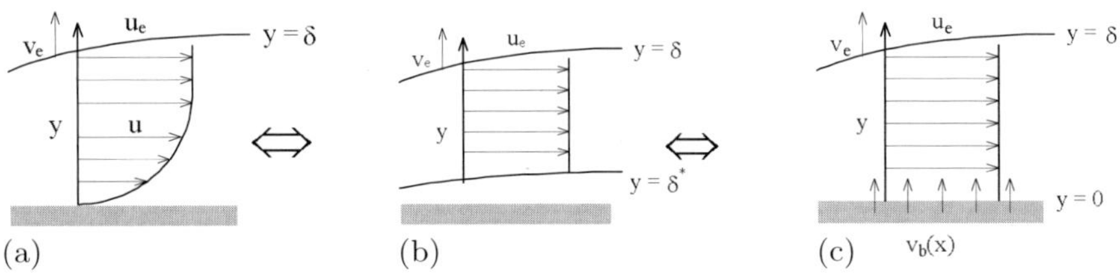

Fig. 1.6. Model of viscous effects in inviscid method in two dimensions (a) real flow, (b), (c) equivalent fictitious flow.

flows corresponding to the real viscous flow and an equivalent inviscid flow defined as analytical continuation, or smooth extrapolation, of the inviscid flow outside the shear layers in the direction normal to the interface between them. For a two-dimensional compressible flow with ρu representing the real viscous flow and $\rho_i u_i$ the equivalent inviscid flow the difference of the continuity equations yields

$$\frac{\partial}{\partial x}(\rho_i u_i - \rho u) + \frac{\partial}{\partial y}(\rho_i v_i - \rho v) = 0 \tag{1.1.1}$$

Integrating across the boundary layer from $y = 0$ to δ and representing the values at the wall by w, we get

$$\rho_{iw} v_{iw} = \frac{d}{dx}\int_0^\delta (\rho_i u_i - \rho u) dy \tag{1.1.2}$$

since both $(\rho_i v_i - \rho v)$ and $(\rho_i u_i - \rho u)$ vanish at $y = \delta$ because of the matching condition, while $v = 0$ at the wall $(y = 0)$. If a displacement thickness, δ_A^* is defined by

$$\delta_A^* = \frac{1}{\rho_{iw} u_{iw}}\int_0^\delta (\rho_i u_i - \rho u) dy \tag{1.1.3}$$

then we obtain the blowing velocity defined by Lighthill,

$$v_{iw} = \frac{1}{\rho_{iw}}\frac{d}{dx}(\rho_{iw} u_{iw} \delta_A^*) \tag{1.1.4}$$

The "classical" displacement thickness, δ_B^*, is defined by the condition that the total mass flow in the equivalent inviscid flow will equal that in the real viscous flow, so that

$$\int_{\delta_B^*}^\delta \rho_i u_i dy = \int_0^\delta \rho u dy \tag{1.1.5}$$

It is shown in [6] that $y = \delta_B^*$ is indeed a streamline of the equivalent inviscid flow. Unlike Eq. (1.2.3), this is only an implicit definition of δ_B^*.

The relation between these two alternative definitions of displacement thickness can be obtained by adding

$$\int_0^{\delta_B^*} \rho_i u_i dy$$

to both sides of Eq. (1.2.5) we obtain

$$\int_0^{\delta_B^*} \rho_i u_i dy = \rho_{iw} u_{iw} \delta_A^* \tag{1.1.6}$$

Thus the transpiration condition given by Eq. (1.2.4) is equivalent to

$$\rho_{iw} v_{iw} = \frac{d}{dx} \int_0^{\delta_B^*} \rho_i u_i dy \tag{1.1.7}$$

which provides a physical explanation of it; that the mass flow injected normal to the wall in the equivalent inviscid flow, in order to displace the dividing streamline outwards the required distance δ_B^* from the surface, must equal to the streamwise rate of change of the mass flow deficit in the boundary layer.

In summary we see that there are two alternative inner boundary conditions which can be used in the calculation of the equivalent inviscid flow, namely, (1) a transpiration condition of nonzero normal velocity at the surface, Eq. (1.2.4), or (2) the condition that the displacement surface is a streamline, Eq. (1.2.5). Both mathematical models are equally valid and general representation of the displacement effect of the boundary layer on the external flow. In practice, however, the transpiration model is more convenient to use. For example, if a panel method is used to calculate the inviscid flow, the matrix of influence coefficients (Chapter 2) can be set up and inverted once and for all, so that subsequent inviscid calculations become trivial, while with a finite difference or finite volume method (Chapter 13) a fixed computational grid can be used throughout. For this reason, the transpiration model will be used throughout this book.

Matching Conditions in the Wake

With the assumptions inherent in first-order boundary-layer theory, the conditions that are needed to effect a matching between the viscous and inviscid calculations to simulate a turbulent wake with the transpiration model are similar to those discussed for the conditions on the airfoil surface. A dividing streamline is chosen in the wake to separate the upper and lower parts of the inviscid flow, and on this line discontinuities are required in the normal components of velocity, so that it can be thought of as a source sheet.

At points C and D on the upper and lower sides of the dividing streamline (Fig. 1.7), the components of transpiration velocity, v_{iu} and v_{il} are, respectively, see Eq. (1.2.4),

$$v_{iu} = \frac{1}{\rho_{iu}} \frac{d}{dx} (\rho_{iu} u_{iu} \delta_u^*) \tag{1.1.8}$$

and

$$v_{il} = -\frac{1}{\rho_{il}} \frac{d}{dx} (\rho_{il} u_{il} \delta_l^*) \tag{1.1.9}$$

Here the sign convention has been used that v is measured positive in the direction of the upward normal to the wake. Hence a jump Δv in the component of velocity normal to the wake is required; it is given by

$$\Delta v_i \equiv v_{iu} - v_{il} = \frac{1}{\rho_{iu}}\frac{d}{dx}(\rho_{iu}u_{iu}\delta_u^*) + \frac{1}{\rho_{il}}\frac{d}{dx}(\rho_{il}u_{il}\delta_l^*) \tag{1.1.10}$$

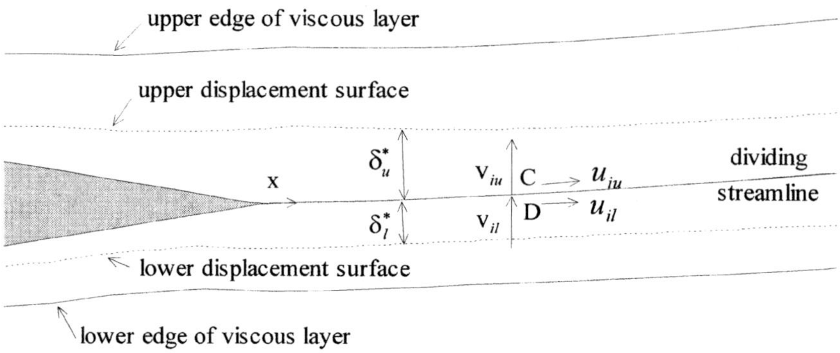

Fig. 1.7. Notation for the airfoil trailing-edge region.

1.2 Mathematical Models for the Interaction Process: Viscous Flow

The blowing velocity needed in the transpiration model can be obtained by solving the boundary-layer equations either in the standard mode (weak interaction problem) or in the inverse mode (strong interaction problem). In the former case, equations are solved for a prescribed external velocity distribution computed by the inviscid method. This procedure, however, is restricted to attached flows since the solutions of the boundary-layer equations become singular at the location where the wall shear vanishes and leads to the breakdown of the solutions. The external velocity distribution in the vicinity of the separation point cannot be assigned arbitrarily and must satisfy conditions associated with the existence of reverse flow downstream of separation. The solutions of the boundary-layer equations, however, are not singular at separation when the external velocity or pressure is computed as part of the solution. This procedure is known as the inverse mode. Catherall and Mangler, [7], were the first to show that modification of the external velocity distribution near the region of flow separation leads to solutions free of numerical difficulties. Prescribing the displacement thickness or wall shear stress as a boundary condition in addition to the usual boundary conditions, the boundary-layer equations can be integrated through the separation location and into the region of reverse flow without any evidence of singularity at the separation point (Chapters 3,5).

A problem associated with using an inverse mode to solve the boundary-layer equations is the lack of a priori knowledge of the required displacement thickness or the wall shear. The appropriate value must be obtained as part of the overall problem from the interaction between the boundary layer and the inviscid flow. In the case of internal flow, the problem is somewhat easier because the conservation of mass in integral form can be used to relate pressure $p(x)$ to velocity $u(x, y)$ in terms of mass balance in the duct.

For two-dimensional external flows, two procedures have been developed to couple the solutions of the inviscid and viscous equations. In the first procedure, developed by LeBalleur [8] and Carter and Wornom [9], the solution of the boundary-layer equations is obtained in the standard mode, and a displacement-thickness, δ^{*°, distribution is determined. If this initial calculation encounters separation, $\delta^{*^\circ}(x)$ is extrapolated to the trailing edge, and one complete cycle of the viscous and inviscid calculation is performed with the boundary-layer equations now solved in the inverse mode. This will, in general, lead to two different external velocity distributions, $u_{ev}(x)$ derived from the inverse boundary-layer solution, and $u_{ei}(x)$ derived from the updated approximation to the inviscid velocity past the airfoil with the added displacement thickness. A relaxation formula is introduced to define an updated displacement-thickness distribution,

$$\delta^* = \delta^{*^\circ}(x)\left\{1 + \omega\left[\frac{u_{ev}(x)}{u_{ei}(x)} - 1\right]\right\} \tag{1.2.1}$$

where ω is a relaxation parameter, and the procedure is repeated with this updated mass flux.

In the second approach, developed by Veldman [10], the external velocity $u_e(x)$ and the displacement thickness $\delta^*(x)$ are treated as unknown quantities, and the equations are solved in the inverse mode simultaneously in successive sweeps over the airfoil surface. For each sweep, the external boundary condition for the boundary-layer equation in dimensionless form, with $u_e(x)$ normalized with u_∞, is written as

$$u_e(x) = u_e^0(x) + \delta u_e(x) \tag{1.2.2a}$$

where $u_e^0(x)$ denotes the inviscid velocity and δu_e the perturbation due to the displacement thickness, which is calculated from the Hilbert integral

$$\delta u_e(x) = \frac{1}{\pi}\int_{x_a}^{x_b} \frac{d}{d\sigma}(u_e\delta^*)\frac{d\sigma}{x - \sigma} \tag{1.2.2b}$$

The above expression is based on the thin airfoil approximation with the term $d/d\sigma(u_e\delta^*)$ denoting the blowing velocity used to simulate the boundary layer in the interaction region (x_a, x_b). This approach is more general and will be used in all the external flow problems considered in this book

1.3 Transition and Its Prediction

In the solution of the boundary-layer equations, as in the Navier-Stokes equations, it is necessary to predict the onset of transition and the region of transition from laminar to turbulent flow. In commercial aircraft the flow is usually turbulent from the leading edge but in wind tunnel tests, which are necessary to aid the development of calculation methods for airfoils and wings, the chord Reynolds numbers are much lower and transition becomes important. Also, specialized aircraft, such as remotely piloted vehicles, make use of wings with dimensions which lead to a significant region of transitional flow and, on occasion, to long separation bubbles.

The onset of transition is associated with the instability of the laminar boundary layer, a nonlinear phenomenon which is not yet well understood. Reviews of the process may be found in articles by Tani [11], Reshotko [12], Kleiser and Zhang [13]. It is clear that freestream turbulence, surface roughness, mass transfer, heat transfer, surface curvature and three-dimensional effects all play a significant role in the determination of the onset of transition. For incompressible flows, correlations have been proposed by Granville [14] and Michel [15]. The former makes use of linear stability theory, and the latter relates the Reynolds number based on momentum thickness, R_θ, and distance from the airfoil leading edge, R_x, by the following expression given in Cebeci and Smith [16],

$$R_\theta = 1.174 \left[1 + \frac{22{,}400}{R_x}\right] R_x^{0.46} \tag{1.3.1}$$

which is now in common use for attached flows on airfoils for chord Reynolds numbers, R_c, greater than, say, 2×10^6.

A more fundamental approach to predict transition is to make use of the theory of small disturbances in the boundary layer, correlating their growth with the onset of transition. This leads to the well-known e^n-method, due to A.M.O. Smith [17] and Van Ingen [18], which is still empirical since it ignores the nonlinearity of the transition process but which has been found to be effective for a wide variety of flows, including separating flows and three-dimensional flows with and without compressibility effects, see Chapters 4, 5, 12 and 13.

In the e^n-method, the value of n is obtained from the solutions of the linear stability equations. When n is between 8 and 10, transition is assumed to occur as we shall discuss in detail in Chapter 6. The effect of freestream turbulence level on transition may partially be accounted for by the following expression given by Mack [19],

$$n = -8.43 - 2.4 \ln T \tag{1.3.2}$$

where T represents the freestream turbulence given by

$$T = \frac{\sqrt{\overline{u'^2}}}{u_\infty} \tag{1.3.3}$$

In some flows, especially those at lower Reynolds numbers, it is necessary to account for the transition region between a laminar and turbulent boundary layer. In zero-equation models (subsection 3.1.4) this is accomplished by an intermittency factor γ_{tr} multiplying the turbulent part of the eddy viscosity. Chen and Thyson [20] have suggested the formula

$$\gamma_{tr} = 1 - \exp\left[-G(x - x_{tr}) \int_{x_{tr}}^{x} \frac{dx}{u_e}\right] \tag{1.3.4}$$

where x_{tr} is the location of the start of the transition and the factor G is given empirically by

$$G = \frac{3}{C^2}\frac{u_e^3}{\nu^2} R_{x_{tr}}^{-1.34} \tag{1.3.5}$$

where the transition Reynolds number $R_{x_{tr}} = (u_e x/\nu)_{tr}$, and C is constant with a recommended value of 60.

We note that the length of the transition region is also susceptible to the degree of freestream turbulence, especially of large scale, flow separation, and surface roughness, decreasing rapidly as these features of the flow become more pronounced. While these features are difficult to incorporate into Eq. (1.4.4) or to correlate with data, in the case of two-dimensional low Reynolds number flows, Eq. (1.4.4) was extended by Cebeci [21] to model the transition region in separation bubbles. The parameter C in Eq. (1.4.5) was expressed in terms of $R_{x_{tr}}$ and its variation with $R_{x_{tr}}$, with the onset of transition obtained from the e^n-method, is shown in Fig. 1.8, together with the experimental data obtained for airfoils NACA 66_3-028, ONERA-D, NACA 65-218, and LNV109A. The data encompass a typical low Reynolds number range from $R_c = 2.4 \times 10^5$ to $2 \times$

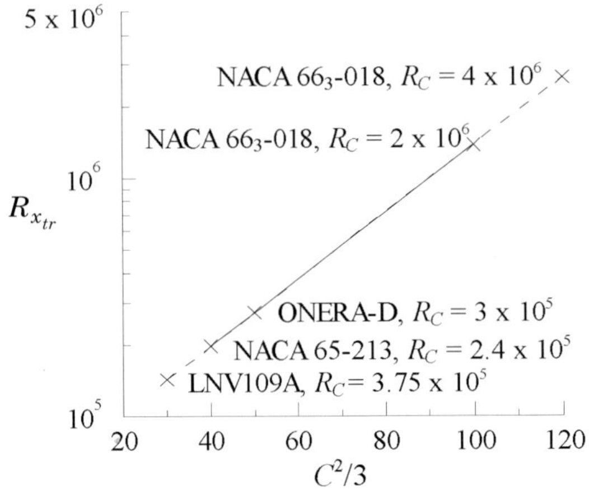

Fig. 1.8. Variation of $C^2/3$ with $R_{x_{tr}}$.

10^6. They fall conveniently on a straight line on a semilog scale and can be represented by the equation

$$C^2 = 213(\log R_{x_{tr}} - 4.7323) \tag{1.3.6}$$

References

[1] Computation of Viscous-Inviscid Interactions, AGARD CP-291, 1981.
[2] Numerical and Physical Aspects of Aerodynamic Flows, II, (Ed. T. Cebeci), Springer-Verlag, N.Y., 1982.
[3] Numerical and Physical Aspects of Aerodynamic Flows, III, (Ed. T. Cebeci), Springer-Verlag, N.Y., 1985.
[4] Numerical and Physical Aspects of Aerodynamic Flows, IV, (Ed. T. Cebeci), Springer-Verlag, N.Y., 1989.
[5] Lighthill, M.J., "On Displacement Thickness," J. Fluid Mech., Vol. 4, p. 383, 1958.
[6] Lock, R.C. and Firmin, M.C.P., "Survey of Techniques for Estimating Viscous Effects in External Aerodynamics," Royal Aircraft Establishment Tech Memo, AERO 1900, Apr 1981.
[7] Catherall, D. and Mangler K.W., "The Integration of the Two-Dimensional Laminar Boundary-Layer Equations Past the Point of Vanishing Skin Friction," Journal of Fluid Mechanics, Vol. 26, 1966, p. 163.
[8] LeBalleur, J.C., "Couplage Visqueux-non Visqueux: Analyse du Probleme Incluant Decollements et Ondes de Choc," La Rech. Aerosp., n. 1977-6, 349. English translation in ESA TT476, 1977.
[9] Carter, J. and Wornom, S.F., "Solutions for Incompressible Separated Boundary Layers Including Viscous-Inviscid Interaction," in Aerodynamic Analysis Requiring Advanced Computers, NASA SP-347, p. 125, 1975.
[10] Veldman, A.E.P., "New Quasi-Simultaneous Method to Calculate Interacting Boundary Layers," AIAA Journal, Vol. 19, p. 769, 1981.
[11] Tani, I., "Boundary-Layer Transition," Annual Review of Fluid Mechanics, Vol. 1, pp. 169–196, 1969.
[12] Reshotko, E., "Boundary-Layer Stability and Transition," Annual Review of Fluid Mechanics, Vol. 8, pp. 311–350, 1976.
[13] Kleiser, L. and Zang, T.A., "Numerical Simulation of Transition in Wall Bounded Shear Flows," Annual Review of Fluid Mechanics, Vol. 23, pp. 495–537, 1991.
[14] Granville, P.S., "The Calculation of the Viscous Drag of Bodies of Revolution," David W. Taylor Model Basin Rept. 849, 1953.
[15] Michel, R., "Etude de la Transition sur les Profiles d'Aile; Establissement d'un Critere de Determination de Point de Transition et Calcul de la Trainee de Profile Incompressible," ONERA Rept. 1/1578A, 1951.
[16] Cebeci, T. and Smith, A.M.O, Analysis of Turbulent Boundary Layers, Academic Press, NY, 1974.
[17] Smith, A.M.O., "Transition, Pressure Gradient, and Stability Theory," Proceedings of the IX International Congress of Applied Mechanics, Vol. 4, pp. 234–244, 1956.
[18] Van Ingen, J.L., "A Suggested Semi-Empirical Method for the Calculation of the Boundary-Layer Region," Delft, Holland, Rept. No. VTH71, VTH74, 1956.
[19] Mack, L.M., "Transition and Laminar Instability," Jet Propoulsion Lab. Pub. 77-15, Pasadena, CA, 1977.
[20] Chen, K.K. and Thyson, N.A., "Extension of Emmons' Spot Theory to Flows on Blunt Bodies," AIAA Journal, Vol. 9, pp. 821–825, 1971.
[21] Cebeci, T., "Essential Ingredients of a Method for Low Reynolds-Number Airfoil," AIAA Journal, Vol. 27, p. 1680, 1989.

2 Two-Dimensional Incompressible Flows: Inviscid Method

2.0 Introduction

Without any question, the panel method is the most efficient and practical method for obtaining flowfield solutions in incompressible flows in which the viscous effects are neglected and the flow is assumed to be irrotational. This method takes account of the linearity of the Laplace equation and avoids the need to generate the grid needed in finite-difference or finite-volume methods for obtaining inviscid flow solutions. It is based on the superposition of solutions: functions that individually satisfy the Laplace equation independent of boundary conditions are added together to describe the desired flowfield. A popular approach is to express the flowfield in terms of the velocity potential based on two or more elemental flows in the presence of an onset flow. The two elemental flows (sometimes called singularities) often used in this approach, referred to as the surface-singularity method, include a source, sink and a vortex, see Bertin and Smith [1] and Anderson [2]. The first is the radially symmetric solution of the Laplace equation, and the second is its complement, the solution independent of radius. A source flows radially outward such that the continuity equation is satisfied everywhere but at the singularity that exists at the source's center. The potential function for a two-dimensional source centered at the origin is

$$\phi_s = \frac{q}{2\pi} \ln r \tag{2.0.1}$$

where r is the radial coordinate from the center of the source and q is the source strength.

A sink is a negative source; that is, fluid flows *into* a sink along radial streamlines. The potential function for a sink centered at the origin is the same as Eq. (2.0.1) except for the minus sign (−) on the right-hand side.

A potential vortex is defined as a singularity about which fluid flows along concentric streamlines. The potential function for a vortex centered at the origin is

$$\phi_v = -\frac{\Gamma}{2\pi}\theta \tag{2.0.2}$$

where Γ is the strength of the vortex. The minus sign on the right-hand side of Eq. (2.0.2) represents a vortex with clockwise circulation.

In the following section we shall describe a panel method due to Hess and Smith [3] to compute the inviscid flowfield around an airfoil at an angle of attack α as well as the modifications needed to include viscous effects in this method computed by the boundary-layer method described in Chapter 3. Its extension to three-dimensional flows is described in Chapter 10. For a more detailed description of panel methods for two-dimensional flows, the reader is referred to Anderson [2] and Moran [4], and for three-dimensional flows to Hess [3].

2.1 Hess-Smith Panel Method

We consider an airfoil at rest in an onset flow of velocity V_∞. We assume that the airfoil is at an angle of attack, α (the angle between its chord line and the onset velocity), and that the upper and lower surfaces are given by functions $Y_u(x)$ and $Y_l(x)$, respectively. These functions can be defined analytically, or (as often is the case) by a set of (x, y) values of the airfoil coordinates. We denote the distance of any field point (x, y) measured from an arbitrary point, b, on the airfoil surface by r, as shown in Fig. 2.1. Let $\vec{n}$ also denote the unit vector normal to the airfoil surface and directed from the body into the fluid and $\vec{t}$, a unit vector tangential to the surface, and assume that its inclination to the x-axis is given by θ. It follows from Fig. 2.1 that with $\vec{i}$ and $\vec{j}$ denoting unit vectors in the x- and y-directions, respectively,

$$\begin{aligned} \vec{n} &= -\sin\theta\,\vec{i} + \cos\theta\,\vec{j} \\ \vec{t} &= \cos\theta\,\vec{i} + \sin\theta\,\vec{j} \end{aligned} \tag{2.1.1}$$

If the airfoil contour is divided into a large number of small segments, ds, then we can write

$$\begin{aligned} dx &= \cos\theta\, ds \\ dy &= \sin\theta\, ds \end{aligned} \tag{2.1.2}$$

We next assume that the airfoil geometry is represented by a finite number (N) of short straight-line elements called panels, defined by $(N+1)(x_j, y_j)$ pairs called boundary points. It is customary to input the (x, y) coordinates starting at the lower surface trailing edge, proceeding clockwise around the airfoil, and ending back at the upper surface trailing edge. If we denote the boundary points by

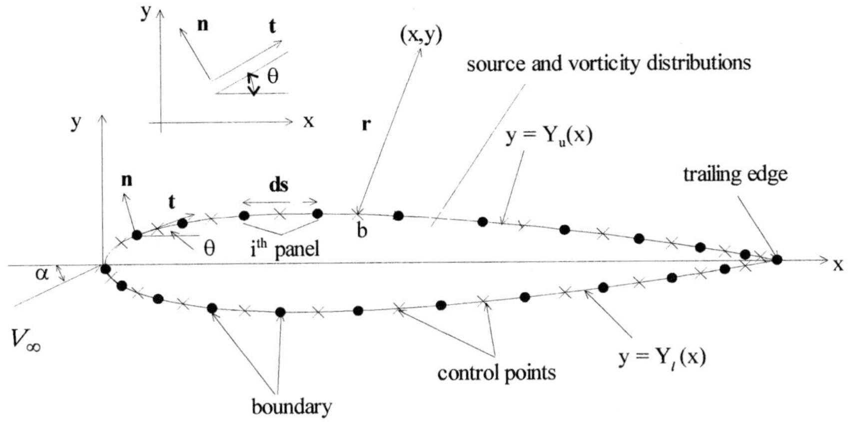

Fig. 2.1. Panel representation of airfoil surface and notation for an airfoil at incidence α.

$$(x_1, y_1), (x_2, y_2), \ldots, (x_N, y_N), (x_{N+1}, y_{N+1}) \tag{2.1.3}$$

then the pairs (x_1, y_1) and (x_{N+1}, y_{N+1}) are identical and represent the trailing edge. It is customary to refer to the element between (x_j, y_j) and (x_{j+1}, y_{j+1}) as the j-th panel, and to the midpoints of the panels as the *control points*. Note from Fig. 2.1 that as one traverses from the i-th boundary point to the $(i+1)$-th boundary point, the airfoil body is on the right-hand side. This numbering sequence is consistent with the common definition of the unit normal vector $\vec{n}_i$ and unit tangential vector $\vec{t}_i$ for all panel surfaces, i.e., $\vec{n}_i$ is directed from the body into the fluid and $\vec{t}_i$ from the i-th boundary point to the $(i+1)$-th boundary point with its inclination to the x-axis given by θ_i.

In the Hess-Smith panel method, the velocity $\vec{V}$ at any point (x, y) is represented by

$$\vec{V} = \vec{U} + \vec{v} \tag{2.1.4}$$

where $\vec{U}$ is the velocity of the uniform flow at infinity

$$\vec{U} = V_\infty(\cos\alpha\,\vec{i} + \sin\alpha\,\vec{j}\,) \tag{2.1.5}$$

and $\vec{v}$ is the disturbance field due to the body which is represented by two elementary flows corresponding to source and vortex flows. A source or vortex on the j-th panel causes an induced source velocity $\vec{v}_s$ at (x, y) or an induced vortex velocity $\vec{v}_v$ at (x, y), respectively, which are obtained by taking gradients of corresponding potentials given by Eq. (2.0.1) and (2.0.2) so that, with integrals applied to the airfoil surface,

$$\vec{v}(x, y) = \int \vec{v}_s q_j(s) ds_j + \int \vec{v}_v \tau_j(s) ds_j \tag{2.1.6}$$

Here $q_j ds_j$ is the source strength for the element ds_j on the j-th panel. Similarly, $\tau_j ds_j$ is the vorticity strength for the element ds_j on the same panel.

Each of the N panels are represented by similar sources and vortices distributed on the airfoil surface. The induced velocities in Eq. (2.1.6) satisfy the irrotationality condition and the boundary condition at infinity

$$u = \frac{\partial \phi}{\partial x} = \frac{\partial \psi}{\partial y} = V_\infty \cos \alpha \tag{2.1.7a}$$

$$v = \frac{\partial \phi}{\partial y} = -\frac{\partial \psi}{\partial x} = V_\infty \sin \alpha \tag{2.1.7b}$$

For uniqueness of the solutions, it is also necessary to specify the magnitude of the circulation around the body. To satisfy the boundary conditions on the body, which correspond to the requirement that the surface of the body is a streamline of the flow, that is,

$$\psi = \text{constant} \quad \text{or} \quad \frac{\partial \phi}{\partial n} = 0 \tag{2.1.8}$$

at the surface on which n is the direction of the normal, the sum of the source-induced and vorticity-induced velocities and freestream velocity is set to zero in the direction normal to the surface of each of the N panels. It is customary to choose the *control points* to numerically satisfy the requirement that the resultant flow is tangent to the surface. If the tangential and normal components of the total velocity at the control point of the i-th panel are denoted by $(V^t)_i$ and $(V^n)_i$, respectively, the flow tangency conditions are then satisfied at panel control points by requiring that the resultant velocity at each control point has only $(V^t)_i$, and

$$(V^n)_i = 0 \quad i = 1, 2, \ldots, N \tag{2.1.9}$$

Thus, to solve the Laplace equation with this approach, at the i-th panel control point we compute the normal $(V^n)_i$ and tangential $(V^t)_i$, $(i = 1, 2, \ldots, N)$ velocity components induced by the source and vorticity distributions on all panels, j $(j = 1, 2, \ldots, N)$, including the i-th panel itself, and separately sum all the induced velocities for the normal and tangential components together with the freestream velocity components. The resulting expressions, which satisfy the irrotationality condition, must also satisfy the boundary conditions discussed above. Before discussing this aspect of the problem, it is convenient to write Eq. (2.1.4) expressed in terms of its velocity components $(V^n)_i$ and $(V^t)_i$ by

$$(V^n)_i = \sum_{j=1}^{N} A^n_{ij} q_j + \sum_{j=1}^{N} B^n_{ij} \tau_j + V_\infty \sin(\alpha - \theta_i) \tag{2.1.10a}$$

$$(V^t)_i = \sum_{j=1}^{N} A^t_{ij} q_j + \sum_{j=1}^{N} B^t_{ij} \tau_j + V_\infty \cos(\alpha - \theta_i) \tag{2.1.10b}$$

where A^n_{ij}, B^n_{ij}, A^t_{ij}, B^t_{ij} are known as influence coefficients defined as the velocities induced at a control point (x_{m_i}, y_{m_i}); more specifically, A^n_{ij} and A^t_{ij}

denote the normal and tangential velocity components, respectively, induced at the i-th panel control point by a unit strength source distribution on the j-th panel, and B_{ij}^n and B_{ij}^t are those induced by unit strength vorticity distribution on the j-th panel. The influence coefficients are related to the airfoil geometry and the panel arrangement; they are given by the following expressions:

$$A_{ij}^n = \begin{cases} \dfrac{1}{2\pi}\left[\sin(\theta_i - \theta_j)\ln\dfrac{r_{i,j+1}}{r_{i,j}} + \cos(\theta_i - \theta_j)\beta_{ij}\right] & i \neq j \\ \dfrac{1}{2} & i = j \end{cases} \tag{2.1.11}$$

$$A_{ij}^t = \begin{cases} \dfrac{1}{2\pi}\left[\sin(\theta_i - \theta_j)\beta_{ij} - \cos(\theta_i - \theta_j)\ln\dfrac{r_{i,j+1}}{r_{i,j}}\right] & i \neq j \\ 0 & i = j \end{cases} \tag{2.1.12}$$

$$B_{ij}^n = -A_{ij}^t, \quad B_{ij}^t = -A_{ij}^n \tag{2.1.13}$$

Here

$$\begin{aligned} r_{i,j+1} &= \left[(x_{m_i} - x_{j+1})^2 + (y_{m_i} - y_{j+1})^2\right]^{1/2} \\ r_{i,j} &= \left[(x_{m_i} - x_j)^2 + (y_{m_i} - y_j)^2\right]^{1/2} \\ x_{m_i} &= \frac{1}{2}(x_i + x_{i+1}), \quad y_{m_i} = \frac{1}{2}(y_i + y_{j+1}) \\ \theta_i &= \tan^{-1}\left(\frac{y_{i+1} - y_i}{x_{i+1} - x_i}\right), \quad \theta_j = \tan^{-1}\left(\frac{y_{j+1} - y_j}{x_{j+1} - x_j}\right) \\ \beta_{ij} &= \tan^{-1}\left(\frac{y_{m_i} - y_{j+1}}{x_{m_i} - x_{j+1}}\right) - \tan^{-1}\left(\frac{y_{m_i} - y_j}{x_{m_i} - x_j}\right) \end{aligned} \tag{2.1.14}$$

Regardless of the nature of $q_j(s)$ and $\tau_j(s)$, Eq. (2.1.10) satisfies the irrotationality condition and the infinity boundary condition, Eq. (2.1.7). To satisfy the requirements given by Eq. (2.1.9) and the condition related to the circulation, it is necessary to adjust these functions. In the approach adopted by Hess and Smith [3], the source strength $q_j(s)$ is assumed to be constant over the j-th panel and is adjusted to give zero normal velocity over the airfoil, and the vorticity strength τ_j is taken to be constant on all panels ($\tau_j = \tau$) and its single value is adjusted to satisfy the condition associated with the specification of circulation. Since the specification of the circulation renders the solution to be unique, a rational way to determine the solution is required.

The best approach is to adjust the circulation to give the correct force on the body as determined by experiment. However, this requires advance knowledge of that force, and one of the principal aims of a flow calculation method is to calculate the force and not to take it as given. Thus, another criterion for determining circulation is needed.

For smooth bodies such as ellipses, the problem of rationally determining the circulation has yet to be solved. Such bodies have circulation associated with them and resulting lift forces, but there is no rule for calculating these forces. If, on the other hand, we deal with an airfoil having a sharp trailing edge, we can apply the Kutta condition [2,4]. It turns out that for every value of circulation except one, the inviscid velocity is infinite at the trailing edge. The Kutta condition states that the particular value of circulation that gives a finite velocity at the trailing edge is the proper one to choose. This condition does not include bodies with nonsharp trailing edges and bodies on which the viscous effects have been simulated by, for example, surface blowing, as shall be discussed in Section 2.2. Thus, the classical Kutta condition is of strictly limited validity. It is customary to apply a "Kutta condition" to bodies outside its narrow definition, but this is an approximation; nevertheless the calculations are often in close accord with experiment.

In the panel method, the Kutta condition is indirectly applied by deducing another property of the flow at the trailing edge that is a direct consequence of the finiteness of velocity; this property is used as "the Kutta condition." Properties that have been used in lieu of "the Kutta condition" in panel methods include the following:

(a) A streamline of the flow leaves the trailing edge along the bisector of the trailing-edge angle.
(b) Upper and lower surface total velocities approach a common limit at the trailing edge. The limiting value is zero if the trailing-edge angle is nonzero.
(c) Source and/or vorticity strengths at the trailing edge must satisfy conditions to allow finite velocity.

Of the above, property (b) is more widely used. At first it may be thought that this property requires setting both the upper and lower surface velocities equal to zero. This gives two conditions, which cannot be satisfied by adjusting a single parameter. The most reasonable choice is to make these two total velocities in the downstream direction at the 1st and N-th panel control points equal so that the flow leaves the trailing edge smoothly. Since the normal velocity on the surface is zero according to Eq. (2.1.9), the magnitude of the two tangential velocities at the trailing edge must be equal to each other, that is,

$$(V^t)_N = -(V^t)_1 \tag{2.1.15}$$

Introducing the flow tangency condition, Eq. (2.1.9), into Eq. (2.1.10a) and noting that $\tau_j = \tau$, we get

$$\sum_{j=1}^{N} A_{ij}^n q_j + \tau \sum_{j=1}^{N} B_{ij}^n + V_\infty \sin(\alpha - \theta_i) = 0, \quad i = 1, 2, \dots, N \tag{2.1.16}$$

In terms of the unknowns, q_j ($j = 1, 2, \dots, N$) and τ, the Kutta condition of Eq. (2.1.15) and Eq. (2.1.16) for a system of algebraic equations can be written in the following form,

$$A\vec{x} = \vec{b} \tag{2.1.17}$$

Here A is a square matrix of order $(N+1)$, that is

$$A \equiv \begin{vmatrix} a_{11} & a_{12} & \cdots & a_{1j} & \cdots & a_{1N} & a_{1,N+1} \\ a_{21} & a_{22} & \cdots & a_{2j} & \cdots & a_{2N} & a_{2,N+1} \\ \vdots & \vdots & \vdots & \vdots & \vdots & \vdots & \vdots \\ a_{i1} & a_{i2} & \cdots & a_{ij} & \cdots & a_{iN} & a_{i,N+1} \\ \vdots & \vdots & \vdots & \vdots & \vdots & \vdots & \vdots \\ a_{N1} & a_{N2} & \cdots & a_{Nj} & \cdots & a_{NN} & a_{N,N+1} \\ a_{N+1,1} & a_{N+1,2} & \cdots & a_{N+1,j} & \cdots & a_{N+1,N} & a_{N+1,N+1} \end{vmatrix} \tag{2.1.18}$$

and $\vec{x} = (q_1, \dots, q_i, \dots, q_N, \tau)^T$ and $\vec{b} = (b_1, \dots, b_i, \dots, b_N, b_{N+1})^T$ with T denoting the transpose. The elements of the coefficient matrix A follow from Eq. (2.1.16)

$$a_{ij} = A^n_{ij}, \quad \begin{array}{l} i = 1, 2, \dots, N \\ j = 1, 2, \dots, N \end{array} \tag{2.1.19a}$$

$$a_{i,N+1} = \sum_{j=1}^{N} B^n_{ij}, \quad i = 1, 2, \dots, N \tag{2.1.19b}$$

A^n_{ij} are given by Eq. (2.1.11) and B^n_{ij} by Eq. (2.1.13). The relation in Eq. (2.1.18) follows from the definition of $\vec{x}$ where τ is essentially x_{N+1}.

To find $a_{N+1,j}$ $(j = 1, \dots, N)$ and $a_{N+1,N+1}$ in the coefficient matrix A, we use the Kutta condition and apply Eq. (2.1.15) to Eq. (2.1.10b) and, with τ as a constant, we write the resulting expression as

$$\begin{aligned} &\sum_{j=1}^{N} A^t_{1j} q_j + \tau \sum_{j=1}^{N} B^t_{1j} + V_\infty \cos(\alpha - \theta_1) \\ &\quad = - \left[\sum_{j=1}^{N} A^t_{Nj} q_j + \tau \sum_{j=1}^{N} B^t_{Nj} + V_\infty \cos(\alpha - \theta_N) \right] \end{aligned}$$

or as

$$\begin{aligned} &\sum_{j=1}^{N} (A^t_{1j} + A^t_{Nj}) q_j + \tau \sum_{j=1}^{N} (B^t_{1j} + B^t_{Nj}) \\ &\quad = -V_\infty \cos(\alpha - \theta_1) - V_\infty \cos(\alpha - \theta_N) \end{aligned} \tag{2.1.20b}$$

so that,

$$a_{N+1,j} = A^t_{1j} + A^t_{Nj}, \quad j = 1, 2, \dots, N \tag{2.1.21a}$$

$$a_{N+1,N+1} = \sum_{j=1}^{N} (B^t_{1j} + B^t_{Nj}) \tag{2.1.21b}$$

where now A^t_{1j} and A^t_{Nj} are computed from Eq. (2.1.12) and B^t_{1j} and B^t_{Nj} from Eq. (2.1.13).

The components of $\vec{b}$ again follow from Eqs. (2.1.16) and (2.1.19). From Eq. (2.1.16),

$$b_i = -V_\infty \sin(\alpha - \theta_i), \quad i = 1, \ldots, N \tag{2.1.22a}$$

and from Eq. (2.1.20),

$$b_{N+1} = -V_\infty \cos(\alpha - \theta_1) - V_\infty \cos(\alpha - \theta_N) \tag{2.1.22b}$$

With all the elements of a_{ij} determined from Eqs. (2.1.19) and (2.1.21) and the elements of $\vec{b}$ from Eq. (2.1.22), the solution of Eq. (2.1.17) can be obtained with the Gaussian elimination method [5]. The elements of $\vec{x}$ are given by

$$x_i = \frac{1}{a_{ii}^{(i-1)}} \left[b_i^{(i-1)} - \sum_{j=i+1}^{N+1} a_{ij}^{(i-1)} x_j \right] \quad i = N+1, \ldots, 1 \tag{2.1.23}$$

where

$$a_{ij}^{(k)} = a_{ij}^{(k-1)} - \frac{a_{ik}^{(k-1)}}{a_{kk}^{(k-1)}} a_{kj}^{(k-1)}, \quad \begin{array}{l} k = 1, \ldots, N \\ j = k+1, \ldots, N+1 \\ i = k+1, \ldots, N+1 \\ a_{ij}^{(0)} = a_{ij} \end{array} \tag{2.1.24a}$$

$$b_i^{(k)} = b_i^{(k-1)} - \frac{a_{ik}^{(k-1)}}{a_{kk}^{(k-1)}} b_k^{(k-1)}, \quad \begin{array}{l} k = 1, \ldots, N \\ i = k+1, \ldots, N+1 \\ b_i^{(0)} = b_i \end{array} \tag{2.1.24b}$$

2.2 Viscous Effects

The viscous effects can be introduced into the panel method by (1) replacing the zero normal velocity condition, Eq. (2.1.9), by a nonzero normal velocity condition $v_{iw}(x)$ and by (2) satisfying the Kutta condition, Eq. (2.1.15), not on the surface of the airfoil trailing edge but at some distance away from the surface.

Here it will be assumed that the nonzero normal velocity distribution $v_{iw}(x)$ along the surface of the airfoil and in its wake is known, together with the distance from the surface, say displacement thickness δ^*, where the Kutta condition is to be satisfied. We now describe how these two new conditions can be incorporated into the panel method.

To include the nonzero normal velocity condition into the solution procedure, we write Eq. (2.1.16) as

$$\sum_{j=1}^{N} A_{ij}^n q_j + \tau \sum_{j=1}^{N} B_{ij}^n q_j + V_\infty \sin(\alpha - \theta_i) = v_{iw}(x_{m_i}) \tag{2.2.1}$$

To satisfy the Kutta condition at the normal distance δ^* from the surface of the trailing edge, called the "off-body" Kutta condition, the total velocities at the N-th and 1-st off-body control points are again required to be equal. Since the normal velocity component is not zero, we write the off-body Kutta condition at distance δ^* as

$$(V)_N = -(V)_1 \tag{2.2.2}$$

where V is the total velocity at the two control points. The off-body total velocities are computed from

$$V = \frac{(V^n)^2 + (V^t)^2}{V} = V^n \frac{V^n}{V} + V^t \frac{V^t}{V} = V^n \sin\phi + V^t \cos\phi \tag{2.2.3}$$

where V^n and V^t are computed by expressions identical to those given by Eqs. (2.1.10) at the two off-body control points, $I = 1$, $I = N$, that is,

$$(V^n)_I = \sum_{j=1}^{N} A^n_{Ij} q_j + \tau \sum_{j=1}^{N} B^n_{Ij} + V_\infty \sin(\alpha - \theta_I) \tag{2.2.4a}$$

$$(V^t)_I = \sum_{j=1}^{N} A^t_{Ij} q_j + \tau \sum_{j=1}^{N} B^t_{Ij} + V_\infty \cos(\alpha - \theta_I) \tag{2.2.4b}$$

and where

$$\phi = \tan^{-1}[(V^n)_I/(V^t)_I] \tag{2.2.5}$$

With Eqs. (2.2.4), the expression for the total velocity given by Eq. (2.2.3) can be written as

$$\begin{aligned} V = & \sum_{j=1}^{N} (A^n_{Ij} \sin\phi + A^t_{Ij} \cos\phi) q_j \\ & + \tau \sum_{j=1}^{N} (B^n_{Ij} \sin\phi + B^t_{Ij} \cos\phi) \\ & + V_\infty \sin(\alpha - \theta_I) \sin\phi + V_\infty \cos(\alpha - \theta_I) \cos\phi \end{aligned} \tag{2.2.6a}$$

or as

$$V = \sum_{j=1}^{N} A'_{Ij} q_j + \tau \sum_{j=1}^{N} B'_{Ij} + V_\infty \cos(\alpha - \theta_I - \phi) \tag{2.2.6b}$$

where

$$A'_{Ij} = A^n_{Ij} \sin\phi + A^t_{Ij} \cos\phi, \quad B'_{Ij} = B^n_{Ij} \sin\phi + B^t_{Ij} \cos\phi \tag{2.2.7a}$$

$$A^n_{Ij} = \frac{1}{2\pi} \left[\sin(\theta_I - \theta_j) \ln \frac{r_{I,j+1}}{r_{I,j}} + \cos(\theta_I - \theta_j) \beta_{Ij} \right] \tag{2.2.7b}$$

$$A^t_{Ij} = \frac{1}{2\pi} \left[\sin(\theta_I - \theta_j) \beta_{Ij} - \cos(\theta_I - \theta_j) \ln \frac{r_{I,j+1}}{r_{I,j}} \right] \tag{2.2.7c}$$

$$B^n_{Ij} = -A^t_{Ij}, \quad B^t_{Ij} = A^n_{Ij} \tag{2.2.7d}$$

If we define

$$\theta'_I = \theta_I + \phi \tag{2.2.8}$$

then it can be shown that Eq. (2.2.6b) can be written as

$$V = \sum_{j=1}^{N} A'_{Ij} q_j + \tau \sum_{j=1}^{N} B'_{Ij} + V_\infty \cos(\alpha - \theta'_I) \tag{2.2.9}$$

where

$$A'_{Ij} = \frac{1}{2\pi} \left[\sin(\theta'_I - \theta_j)\beta_{Ij} - \cos(\theta'_I - \theta_j) \ln \frac{r_{I,j+1}}{r_{I,j}} \right] \tag{2.2.10a}$$

$$B'_{Ij} = \frac{1}{2\pi} \left[\sin(\theta'_I - \theta_j) \ln \frac{r_{I,j+1}}{r_{I,j}} + \cos(\theta'_I - \theta_j)\beta_{Ij} \right] \tag{2.2.10b}$$

The off-body Kutta condition can now be expressed in a form similar to that of Eq. (2.1.20). Applying Eq. (2.2.2) to Eq. (2.2.9), we write the resulting expression as

$$\begin{aligned} &\sum_{j=1}^{N} A'_{Nj} q_j + \tau \sum_{j=1}^{N} B'_{Nj} + V_\infty \cos(\alpha - \theta'_N) \\ &\quad = - \left[\sum_{j=1}^{N} A'_{1j} q_j + \tau \sum_{j=1}^{N} B'_{1j} + V_\infty \cos(\alpha - \theta'_1) \right] \end{aligned} \tag{2.2.11a}$$

or as

$$\begin{aligned} &\sum_{j=1}^{N} (A'_{1j} + A'_{Nj}) q_j + \tau \sum_{j=1}^{N} (B'_{1j} + B'_{Nj}) \\ &\quad + V_\infty \cos(\alpha - \theta'_1) + V_\infty \cos(\alpha - \theta'_N) = 0 \end{aligned} \tag{2.2.11b}$$

2.3 Flowfield Calculation in the Wake

As discussed in Section 5.1, the calculation of airfoils in incompressible viscous flows can be accomplished without taking into account the wake effect; that is, the viscous flow calculations are performed up to the trailing edge only and are not extended into the wake. This procedure, while it may be sufficient at low to moderate angles of attack without flow separation, is not sufficient at higher angles of attack, including post-stall flows. Additional changes are required in the panel method and (in the boundary-layer method), as discussed in this subsection.

The viscous wake calculations usually include a streamline issuing from the trailing edge of the airfoil. The computation of the location of this streamline is relatively simple if conformal mapping methods are used to determine the velocity field. In this case, the stream function ψ is usually known, and because the airfoil surface is represented by $\psi(x, y) = \text{const}$, the calculation of the wake streamline amounts to tracing the curve after it leaves the airfoil. When the flowfield is computed by a panel method or by a finite-difference method, however, the results are known only at discrete points in the field in terms of the velocity components. In this case, the wake streamline is determined from the numerical integration of

$$\frac{dy}{dx} = \frac{v}{u} \tag{2.3.1}$$

aft of the trailing edge with known initial conditions. However, some care is necessary in selecting the initial conditions, especially when the trailing edge is blunt. As a general rule, the initial direction of the streamline is given to a good approximation by the bisector of the trailing-edge angle of the airfoil.

The panel method, which was modified only for an airfoil flow, now requires similar modifications to include the viscous effects in the wake which behaves as a distribution of sinks. It is divided into nwp panels along the dividing streamline with suction velocities or sink strengths $q_i = \Delta v_i$ $(N + 1 \le i \le N + nwp)$, distributed on the wake panels and determined from boundary-layer solutions in the wake by Eq. (1.2.10). As before, off-body boundary points and "control" points are introduced at the intersections of the δ^* surface with the normals through panel boundary points and panel control points, respectively. Summation of all the induced velocities, separately for the normal and tangential components and together with the freestream velocity components, produces $(V^n)_I$ and $(V^t)_I$ at $I = 1, 2, \ldots, N + nwp$. The wake velocity distribution, as the airfoil velocity distribution, is computed on the δ^*-surface, rather than on the dividing streamline.

The total velocities are again computed from Eq. (2.2.3) with $(V^n)_I$ and $(V^t)_I$ from Eqs. (2.2.4) except that now

$$(V^n)_I = \sum_{j=1}^{N+nwp} A^n_{Ij} q_j + \tau \sum_{j=1}^{N} B^n_{Ij} + V_\infty \sin(\alpha - \theta_I) \tag{2.3.2a}$$

$$(V^t)_I = \sum_{j=1}^{N+nwp} A^t_{Ij} q_j + \tau \sum_{j=1}^{N} B^t_{Ij} + V_\infty \cos(\alpha - \theta_I) \tag{2.3.2b}$$

As before, the expression for the total velocity is written in the same form as Eq. (2.2.6a), except that now

$$\begin{aligned} V = & \sum_{j=1}^{N+nwp} (A^n_{Ij} \sin\phi + A^t_{Ij} \cos\phi) q_j \\ & + \tau \sum_{j=1}^{N} (B^n_{Ij} \sin\phi + B^t_{Ij} \cos\phi) \\ & + V_\infty \sin(\alpha - \theta_I) \sin\phi + V_\infty \cos(\alpha - \theta_I) \cos\phi \end{aligned} \tag{2.3.3}$$

where A^n_{Ij}, A^t_{Ij}, B^n_{Ij} and B^t_{Ij}, are identical to those given by Eq. (2.2.7).

Similarly, Eq. (2.2.9) with A_{Ij} and B_{Ij} given by Eq. (2.2.10) is

$$V = \sum_{j=1}^{N+nwp} A'_{Ij} q_j + \tau \sum_{j=1}^{N} B'_{Ij} + V_\infty \cos(\alpha - \theta'_I) \tag{2.3.4}$$

and the Kutta condition given by Eqs. (2.1.33) becomes

$$\begin{aligned} & \sum_{j=1}^{N+nwp} A'_{Nj} q_j + \tau \sum_{j=1}^{N} B'_{Nj} + V_\infty \cos(\alpha - \theta'_N) \\ & = - \left[\sum_{j=1}^{N+nwp} A'_{1j} q_j + \tau \sum_{j=1}^{N} B'_{1j} + V_\infty \cos(\alpha - \theta'_1) \right] \end{aligned} \tag{2.3.5a}$$

or

$$\begin{aligned} & \sum_{j=1}^{N+nwp} (A'_{1j} + A'_{Nj}) q_j + \tau \sum_{j=1}^{N} (B'_{1j} + B'_{Nj}) \\ & + V_\infty \cos(\alpha - \theta'_1) + V_\infty \cos(\alpha - \theta'_N) = 0 \end{aligned} \tag{2.3.5b}$$

In computing the wake velocity distribution at distances δ^* from the wake dividing streamline, the velocities in the upper wake are equal to those on the lower wake for a symmetrical airfoil at zero angle of attack. This is, however, not the case if the airfoil is asymmetric or if the airfoil is at an angle of incidence. While the external velocities on the upper and lower surfaces of the trailing edge are equal to each other, they are not equal to each other in the wake region since the δ^*-distribution in the upper wake is different from the δ^*-distribution in the lower wake.

2.4 FORTRAN Program

In this section we present a Fortran program for the panel method discussed in the previous sections. This program can be used interactively with the boundary-layer program of Section 3.4 so that, as we shall see in Chapter 5, more accurate solutions of inviscid and viscous flow equations can be obtained by including the viscous effects in the panel method of Section 2.1.

The Fortran program of the panel method has five subroutines and MAIN, as described below.

2.4.1 MAIN

MAIN contains the input information which comprises (1) the number of panels along the surface of the airfoil, NODTOT, and the number of panels in the wake, NW. The code is arranged so that it can be used for inviscid flows with and without viscous effects. For inviscid flows, NW is equal to zero. (2) The next input data also comprises airfoil coordinates normalized with respect to its chord c, x/c, y/c, [$\equiv$ (X(I),Y(I))]. If NW $= 0$, then it is necessary to specify the dimensionless displacement thickness δ^*/c ($\equiv$ DLSP(I)), dimensionless blowing velocity u_w/u_∞ ($\equiv$ VNP(I)) distributions on the airfoil, as well as the wake coordinates XW(I), YW(I) of the dividing streamline, the dimensionless displacement thickness distribution on the upper wake DELW(I,1) and lower wake DELW(I,2) and velocity jump QW(I). It should be noted that all input data for wake includes values at the trailing edge. The input also includes angle of attack α ($\equiv$ ALPHA) and Mach number M_∞ ($\equiv$ FMACH).

The panel slopes are calculated from Eq. (2.1.2). The subroutine COEF is called to compute A and $\vec{b}$ in Eq. (2.1.17) subroutine OBKUTA to calculate the off-body Kutta condition, subroutine GAUSS to compute $\vec{x}$, subroutine VPDIS to compute the velocity and pressure distributions, and subroutine CLCM to compute the airfoil characteristics corresponding to lift (CL) and pitching moment (CM) coefficients.

```
      PROGRAM MAIN
      COMMON /BOD/ NODTOT,X(201),Y(201),
     +             XMID(200),YMID(200),COSTHE(200),SINTHE(200)
      COMMON /NUM/ PI,PI2INV
      COMMON/COUPL/DLSP(201),VNP(201),QW(51),DELW(51,2)
      COMMON /OFFBOD/ XOFF(201),YOFF(201),SOFF(200),COFF(200)
      COMMON /OBKU/ STE(2),CTE(2),VNTE(2),VTTE(2),VTOT(2)
      COMMON /COMP / FMACH,C2,C6,C7,C8
      COMMON /WWAKE/ NW,XW(51),YW(51),SINW(51),COSW(51),QWM(51)
      DIMENSION TITLE(20),DBODY(200),SBGEO(201),CBGEO(201)
C - - - - - - - - - - - - - - - - - - - - - - - - - - - - - - - - - - -
      PI      = 3.1415926585
      PI2INV  = .5/PI
      READ (5,*)
      READ (5,501) NODTOT,NW
  501 FORMAT(2I5)
      READ (5,*)
      READ (5,504)(X(I),I=1,NODTOT+1)
      READ (5,*)
      READ (5,504)(Y(I),I=1,NODTOT+1)
      IF(NW.LE.0) THEN
         DO I = 1,NODTOT+1
            VNP(I) = 0.0
            DLSP(I)=0.0
         ENDDO
```

```
         NW   = 0
      ELSE
         READ (5,*)
         READ (5,504)(DLSP(I),I=1,NODTOT+1)
         READ (5,*)
         READ (5,504)(VNP(I),I=1,NODTOT+1)
         READ (5,*)
         READ (5,504)(XW(I),I=1,NW+1)
         READ (5,*)
         READ (5,504)(YW(I),I=1,NW+1)
         READ (5,*)
         READ (5,504)(DELW(I,1),I=1,NW+1)
         READ (5,*)
         READ (5,504)(DELW(I,2),I=1,NW+1)
         READ (5,*)
         READ (5,504)(QW(I),I=1,NW+1)
      ENDIF
C
 504  FORMAT(6F10.0)
C
      READ (5,*)
      READ (5,504) ALPHA
      COSALF  = COS(ALPHA*PI/180.)
      SINALF  = SIN(ALPHA*PI/180.)
      READ (5,*)
      READ (5,504) FMACH
      WRITE (6,1030) ALPHA,FMACH
 1030 FORMAT (//,'SOLUTION AT ALPHA = ',F10.5,'  & MACH = ',f10.5)
      GAM     = 1.4
      C2      = 0.5 * (GAM-1.0)
      C6      = C2 * FMACH**2
      C7      = GAM/(GAM-1.0)
      C8      = 0.5 * GAM * FMACH**2
      DO 100 I=1,NODTOT
      XMID(I) = .5*(X(I) + X(I+1))
      YMID(I) = .5*(Y(I) + Y(I+1))
      DX=X(I+1)-X(I)
      DY=Y(I+1)-Y(I)
      DIST=SQRT(DX*DX+DY*DY)
      SINTHE(I)=DY/DIST
      COSTHE(I)=DX/DIST
      DBODY(I)=DIST
100   CONTINUE
      DO 150 I=1,NW
      DX=XW(I+1)-XW(I)
      DY=YW(I+1)-YW(I)
      DIST=SQRT(DX*DX+DY*DY)
      SINW(I)=DY/DIST
      COSW(I)=DX/DIST
      QWM(I)=0.5*(QW(I)+QW(I+1))
150   CONTINUE
```

```
      DO 200 I=2,NODTOT
      SBGEO(I)=(SINTHE(I)*DBODY(I-1)+SINTHE(I-1)*DBODY(I))/(DBODY(I-1)+
     +         DBODY(I))
200   CBGEO(I)=(COSTHE(I)*DBODY(I-1)+COSTHE(I-1)*DBODY(I))/(DBODY(I-1)+
     +         DBODY(I))
      SBGEO(1)=2.0*SINTHE(1)-SBGEO(2)
      SBGEO(NODTOT+1)=2.0*SINTHE(NODTOT)-SBGEO(NODTOT)
      CBGEO(1)=2.0*COSTHE(1)-CBGEO(2)
      CBGEO(NODTOT+1)=2.0*COSTHE(NODTOT)-CBGEO(NODTOT)
      DO 220 J=1,NODTOT+1
      XOFF(J)=X(J)-SBGEO(J)*DLSP(J)
      YOFF(J)=Y(J)+CBGEO(J)*DLSP(J)
220   CONTINUE
      DO 210 I=1,NODTOT
      DX=XOFF(I+1)-XOFF(I)
      DY=YOFF(I+1)-YOFF(I)
      DIST=SQRT(DX*DX+DY*DY)
      SOFF(I)=DY/DIST
      COFF(I)=DX/DIST
210   CONTINUE
      STE(1)=SOFF(1)
      STE(2)=SOFF(NODTOT)
      CTE(1)=COFF(1)
      CTE(2)=COFF(NODTOT)
      IT=1
 240  CONTINUE
      CALL COEF(SINALF,COSALF)
      CALL GAUSS(1)
      CALL OBKUTA(SINALF,COSALF,INDEX)
      IF(INDEX.EQ.1) GO TO 230
      IF(IT.GT.10) STOP 10
      IT=IT+1
      DO 250 II=1,2
      STEO=STE(II)
      CTEO=CTE(II)
      STE(II)=(STEO*VTTE(II)+CTEO*VNTE(II))/VTOT(II)
      CTE(II)=(CTEO*VTTE(II)-STEO*VNTE(II))/VTOT(II)
250   CONTINUE
      STE(1)=-STE(1)
      CTE(1)=-CTE(1)
      GO TO 240
230   CONTINUE
      CALL VPDIS(SINALF,COSALF)
      CALL VPDWK(SINALF,COSALF)
      CALL CLCM(SINALF,COSALF)
C
      CALL QWIK(ALPHA)
      STOP
      END
```

2.4.2 Subroutine COEF

This subroutine calculates the elements a_{ij} of the coefficient matrix A from Eqs. (2.1.19) and (2.1.21) and the elements of $\vec{b}$ from Eq. (2.1.22). We note that $N+1$ corresponds to KUTTA, and N to NODTOT.

```
      SUBROUTINE COEF(SINALF,COSALF)
      COMMON /BOD/ NODTOT,X(201),Y(201),
     +             XMID(200),YMID(200),COSTHE(200),SINTHE(200)
      COMMON /COF/ A(201,201),BV(201),KUTTA
      COMMON /NUM/ PI,PI2INV
      COMMON/COUPL/DLSP(201),VNP(201),QW(51),DELW(51,2)
      COMMON /OFFBOD/ XOFF(201),YOFF(201),SOFF(200),COFF(200)
      COMMON /OBKU/ STE(2),CTE(2),VNTE(2),VTTE(2),VTOT(2)
      COMMON /WWAKE/ NW,XW(51),YW(51),SINW(51),COSW(51),QWM(51)
C - - - - - - - - - - - - - - - - - - - - - - - - - - - - - - - - - - - -
      KUTTA    = NODTOT + 1
      DO 120  I = 1,NODTOT
      A(I,KUTTA) = 0.0
      DO 110  J = 1,NODTOT
      FLOG     = 0.0
      FTAN     = PI
      IF (J .EQ. I)      GO TO 100
      DXJ      = XMID(I) - X(J)
      DXJP     = XMID(I) - X(J+1)
      DYJ      = YMID(I) - Y(J)
      DYJP     = YMID(I) - Y(J+1)
      FLOG     = .5*ALOG((DXJP*DXJP+DYJP*DYJP)/(DXJ*DXJ+DYJ*DYJ))
      FTAN     = ATAN2(DYJP*DXJ-DXJP*DYJ,DXJP*DXJ+DYJP*DYJ)
  100 CTIMTJ   = COSTHE(I)*COSTHE(J) + SINTHE(I)*SINTHE(J)
      STIMTJ   = SINTHE(I)*COSTHE(J) - COSTHE(I)*SINTHE(J)
      A(I,J)   = PI2INV*(FTAN*CTIMTJ + FLOG*STIMTJ)
      B        = PI2INV*(FLOG*CTIMTJ - FTAN*STIMTJ)
      A(I,KUTTA) = A(I,KUTTA) + B
  110 CONTINUE
      WAQ=0.0
      DO 111 J=1,NW
      DXJ      = XMID(I) - XW(J)
      DXJP     = XMID(I) - XW(J+1)
      DYJ      = YMID(I) - YW(J)
      DYJP     = YMID(I) - YW(J+1)
      FLOG     =.5*ALOG((DXJP*DXJP+DYJP*DYJP)/(DXJ*DXJ+DYJ*DYJ))
      FTAN     = ATAN2(DYJP*DXJ-DXJP*DYJ,DXJP*DXJ+DYJP*DYJ)
      CTIMTJ   = COSTHE(I)*COSW(J) + SINTHE(I)*SINW(J)
      STIMTJ   = SINTHE(I)*COSW(J) - COSTHE(I)*SINW(J)
      AA       = PI2INV*(FTAN*CTIMTJ + FLOG*STIMTJ)
  111 WAQ      = WAQ+AA*QWM(J)
      BV(I) = SINTHE(I)*COSALF - COSTHE(I)*SINALF
     +         +0.5*(VNP(I)+VNP(I+1))-WAQ
  120 CONTINUE
 200  DO 90   J = 1,KUTTA
```

```
 90   A(KUTTA,J)    = 0.0
      WAQ=0.0
      I=1
      II=1
230   CONTINUE
      XMIDO    = .5*(XOFF(I) + XOFF(I+1))
      YMIDO    = .5*(YOFF(I) + YOFF(I+1))
      DLSH=0.5*(DLSP(I)+DLSP(I+1))
      DO 210  J = 1,NODTOT
      FLOG=0.0
      FTAN=PI
      IF(J.EQ.I .AND. DLSH.LT.0.001)GO TO 209
      DXJ      = XMIDO - X(J)
      DXJP     = XMIDO - X(J+1)
      DYJ      = YMIDO - Y(J)
      DYJP     = YMIDO - Y(J+1)
      FLOG     = .5*ALOG((DXJP*DXJP+DYJP*DYJP)/(DXJ*DXJ+DYJ*DYJ))
      FTAN     = ATAN2(DYJP*DXJ-DXJP*DYJ,DXJP*DXJ+DYJP*DYJ)
 209  CTIMTJ   = CTE(II)*COSTHE(J) + STE(II)*SINTHE(J)
      STIMTJ   = STE(II)*COSTHE(J) - CTE(II)*SINTHE(J)
      AA       = PI2INV*(FTAN*CTIMTJ + FLOG*STIMTJ)
      B        = PI2INV*(FLOG*CTIMTJ - FTAN*STIMTJ)
      A(KUTTA,J) = A(KUTTA,J) - B
      A(KUTTA,KUTTA) = A(KUTTA,KUTTA) +AA
210   CONTINUE
      DO 211 J=1,NW
      DXJ      = XMIDO - XW(J)
      DXJP     = XMIDO - XW(J+1)
      DYJ      = YMIDO - YW(J)
      DYJP     = YMIDO - YW(J+1)
      FLOG     = .5*ALOG((DXJP*DXJP+DYJP*DYJP)/(DXJ*DXJ+DYJ*DYJ))
      FTAN     = ATAN2(DYJP*DXJ-DXJP*DYJ,DXJP*DXJ+DYJP*DYJ)
      CTIMTJ   = CTE(II)*COSW(J) + STE(II)*SINW(J)
      STIMTJ   = STE(II)*COSW(J) - CTE(II)*SINW(J)
      B        = PI2INV*(FLOG*CTIMTJ - FTAN*STIMTJ)
      WAQ      = WAQ-B*QWM(J)
211   CONTINUE
      IF(I.EQ.NODTOT) GO TO 220
      I=NODTOT
      II=2
      GO TO 230
220   CONTINUE
      BV(KUTTA) = - (CTE(1) + CTE(2))*COSALF
     +            - (STE(1) + STE(2))*SINALF - WAQ
      RETURN
      END
```

2.4.3 Subroutine OBKUTA

This subroutine is used to calculate the off-body Kutta condition.

```
      SUBROUTINE OBKUTA(SINALF,COSALF,INDEX)
      COMMON /BOD/ NODTOT,X(201),Y(201),
     +             XMID(200),YMID(200),COSTHE(200),SINTHE(200)
      COMMON /COF/ A(201,201),BV(201),KUTTA
      COMMON /NUM/ PI,PI2INV
      COMMON/COUPL/DLSP(201),VNP(201),QW(51),DELW(51,2)
      COMMON /OFFBOD/ XOFF(201),YOFF(201),SOFF(200),COFF(200)
      COMMON /OBKU/ STE(2),CTE(2),VNTE(2),VTTE(2),VTOT(2)
      COMMON /WWAKE/ NW,XW(51),YW(51),SINW(51),COSW(51),QWM(51)
      DIMENSION Q(200)
C - - - - - - - - - - - - - - - - - - - - - - - - - - - - - - - - - - -
      DO 50   I = 1,NODTOT
 50   Q(I)    = BV(I)
      GAMMA   = BV(KUTTA)
      I=1
      II=1
   55 CONTINUE
      XMIDO   = .5*(XOFF(I) + XOFF(I+1))
      YMIDO   = .5*(YOFF(I) + YOFF(I+1))
      VTANG   = COSALF*CTE(II) + SINALF*STE(II)
      VNOFF   = SINALF*CTE(II) - COSALF*STE(II)
      DLSH=0.5*(DLSP(I)+DLSP(I+1))
      DO 120  J = 1,NODTOT
        FLOG    = 0.0
        FTAN    = PI
        IF(J.EQ.I.AND.DLSH.LT.0.001)GO TO 100
 300    DXJ     = XMIDO - X(J)
        DXJP    = XMIDO - X(J+1)
        DYJ     = YMIDO - Y(J)
        DYJP    = YMIDO - Y(J+1)
        FLOG    = .5*ALOG((DXJP*DXJP+DYJP*DYJP)/(DXJ*DXJ+DYJ*DYJ))
        FTAN    = ATAN2(DYJP*DXJ-DXJP*DYJ,DXJP*DXJ+DYJP*DYJ)
 100    CTIMTJ  = CTE(II)*COSTHE(J) + STE(II)*SINTHE(J)
        STIMTJ  = STE(II)*COSTHE(J) - CTE(II)*SINTHE(J)
        AA      = PI2INV*(FTAN*CTIMTJ + FLOG*STIMTJ)
        B       = PI2INV*(FLOG*CTIMTJ - FTAN*STIMTJ)
        VTANG   = VTANG - B*Q(J) + GAMMA*AA
        VNOFF   = VNOFF + GAMMA*B + AA*Q(J)
 120  CONTINUE
      DO 121 J=1,NW
        DXJ     = XMIDO - XW(J)
        DXJP    = XMIDO - XW(J+1)
        DYJ     = YMIDO - YW(J)
        DYJP    = YMIDO - YW(J+1)
        FLOG    = .5*ALOG((DXJP*DXJP+DYJP*DYJP)/(DXJ*DXJ+DYJ*DYJ))
        FTAN    = ATAN2(DYJP*DXJ-DXJP*DYJ,DXJP*DXJ+DYJP*DYJ)
        CTIMTJ  = CTE(II)*COSW(J) + STE(II)*SINW(J)
```

```
          STIMTJ  = STE(II)*COSW(J) - CTE(II)*SINW(J)
          AA      = PI2INV*(FTAN*CTIMTJ + FLOG*STIMTJ)
          B       = PI2INV*(FLOG*CTIMTJ - FTAN*STIMTJ)
          VTANG   = VTANG-B*QWM(J)
          VNOFF   = VNOFF+AA*QWM(J)
121     CONTINUE
        VTTE(II)= VTANG
        VNTE(II)= VNOFF
        VTOT(II)= SQRT(VTANG*VTANG+VNOFF*VNOFF)
        IF(II.EQ.2) GO TO 130
        I=NODTOT
        II=2
        GO TO 55
130     INDEX=0
        IF(ABS(VTOT(1)-VTOT(2)).LT.0.00001)INDEX=1
        RETURN
        END
```

2.4.4 Subroutine GAUSS

The solution of Eq. (2.1.17) is obtained with the Gauss elimination method described in Section 2.1.

```
      SUBROUTINE GAUSS(M)
      COMMON /COF/ A(201,201),B(201,1),N
C - - - - - - - - - - - - - - - - - - - - - - - - - - - - - - - - - - - -
      DO 100  K = 1,N-1
      KP        = K + 1
      DO 100  I = KP,N
      R         = A(I,K)/A(K,K)
      DO 200  J = KP,N
 200  A(I,J)    = A(I,J) - R*A(K,J)
      DO 100  J = 1,M
 100  B(I,J)     = B(I,J) - R*B(K,J)
      DO 300  K = 1,M
      B(N,K) = B(N,K)/A(N,N)
      DO 300  I = N-1,1,-1
      IP        = I + 1
      DO 400  J = IP,N
 400  B(I,K)     = B(I,K) - A(I,J)*B(J,K)
 300  B(I,K)     = B(I,K)/A(I,I)
      RETURN
      END
```

2.4.5 Subroutine VPDIS

Once $\vec{x}$ is determined by subroutine GAUSS so that source strengths q_i ($i = 1, 2, \ldots, N$) and vorticity τ on the airfoil surface are known, the tangential velocity component (V^t) at each control point can be calculated. Denoting q_i

with Q(I) and τ with GAMMA, the tangential velocities $(V^t)_i$ are obtained with the help of Eq. (2.1.10b). This subroutine also determines the distributions of the dimensionless pressure coefficient C_p ($\equiv CP$) defined by

$$C_p = \frac{p - p_\infty}{(1/2)\rho V_\infty^2} \tag{2.4.1a}$$

which in terms of velocities can be written as

$$C_p = 1 - \left(\frac{V^t}{V_\infty}\right)^2 \tag{2.4.1b}$$

It is common to use panel methods for low Mach number flows by introducing compressibility corrections which depend upon the linearized form of the compressibility velocity potential equation and are based on the assumption of small perturbations and thin airfoils [2]. A simple correction formula for this purpose is the Karman-Tsien formula which uses the tangent gas approximation to simplify the compressible potential-flow equations. According to this formula, the effect of Mach number on the pressure coefficient is estimated from

$$C_p = \frac{C_{p_i}}{\beta + [M_\infty^2/(1+\beta)](C_{p_i}/2)} \tag{2.4.2}$$

and the corresponding velocities are computed from

$$\frac{V}{V_\infty} = \frac{(1-\lambda)(V/V_\infty)_i}{1 - \lambda(V/V_\infty)_i^2} \tag{2.4.3}$$

Here C_{p_i} denotes the incompressible pressure coefficient, M_∞ the freestream Mach number and

$$\lambda = \frac{M_\infty^2}{1+\beta^2}, \quad \beta = \sqrt{1 - M_\infty^2} \tag{2.4.4}$$

In this subroutine we also include this capability into the Hess-Smith panel method.

```
      SUBROUTINE VPDIS(SINALF,COSALF)
      COMMON /BOD/ NODTOT,X(201),Y(201),
     +             XMID(200),YMID(200),COSTHE(200),SINTHE(200)
      COMMON /COF/ A(201,201),BV(201),KUTTA
      COMMON /CPD/ UE(250),CP(250),UEW(50,2),CPW(50,2)
      COMMON /COMP / FMACH,C2,C6,C7,C8
      COMMON /NUM/ PI,PI2INV
      COMMON/COUPL/DLSP(201),VNP(201),QW(51),DELW(51,2)
      COMMON /OFFBOD/ XOFF(201),YOFF(201),SOFF(200),COFF(200)
      COMMON /WWAKE/ NW,XW(51),YW(51),SINW(51),COSW(51),QWM(51)
      DIMENSION Q(200)
C - - - - - - - - - - - - - - - - - - - - - - - - - - - - - - - - - - -
      WRITE (6,1000)
      DO 50   I = 1,NODTOT
```

```
 50   Q(I)     = BV(I)
      GAMMA    = BV(KUTTA)
      DO 130  I = 1,NODTOT
      XMIDO    = .5*(XOFF(I) + XOFF(I+1))
      YMIDO    = .5*(YOFF(I) + YOFF(I+1))
      VTANG    = COSALF*COFF(I) + SINALF*SOFF(I)
      VNOFF    = SINALF*COFF(I) - COSALF*SOFF(I)
      DLSH=0.5*(DLSP(I)+DLSP(I+1))
      DO 120  J = 1,NODTOT
      FLOG     = 0.0
      FTAN     = PI
      IF (J .EQ. I .AND. DLSH. LT. 0.001) GO TO 100
 300  DXJ      = XMIDO - X(J)
      DXJP     = XMIDO - X(J+1)
      DYJ      = YMIDO - Y(J)
      DYJP     = YMIDO - Y(J+1)
      FLOG     = .5*ALOG((DXJP*DXJP+DYJP*DYJP)/(DXJ*DXJ+DYJ*DYJ))
      FTAN     = ATAN2(DYJP*DXJ-DXJP*DYJ,DXJP*DXJ+DYJP*DYJ)
 100  CTIMTJ   = COFF(I)*COSTHE(J) + SOFF(I)*SINTHE(J)
      STIMTJ   = SOFF(I)*COSTHE(J) - COFF(I)*SINTHE(J)
      AA       = PI2INV*(FTAN*CTIMTJ + FLOG*STIMTJ)
      B        = PI2INV*(FLOG*CTIMTJ - FTAN*STIMTJ)
      VTANG    = VTANG - B*Q(J) + GAMMA*AA
      VNOFF    = VNOFF + GAMMA*B + AA*Q(J)
 120  CONTINUE
      DO 121 J=1,NW
      DXJ      = XMIDO - XW(J)
      DXJP     = XMIDO - XW(J+1)
      DYJ      = YMIDO - YW(J)
      DYJP     = YMIDO - YW(J+1)
      FLOG     =.5*ALOG((DXJP*DXJP+DYJP*DYJP)/(DXJ*DXJ+DYJ*DYJ))
      FTAN     = ATAN2(DYJP*DXJ-DXJP*DYJ,DXJP*DXJ+DYJP*DYJ)
      CTIMTJ   = COFF(I)*COSW(J) + SOFF(I)*SINW(J)
      STIMTJ   = SOFF(I)*COSW(J) - COFF(I)*SINW(J)
      AA       = PI2INV*(FTAN*CTIMTJ + FLOG*STIMTJ)
      B        = PI2INV*(FLOG*CTIMTJ - FTAN*STIMTJ)
      VTANG    = VTANG-B*QWM(J)
      VNOFF    = VNOFF+AA*QWM(J)
 121  CONTINUE
      UE(I)=SQRT(VNOFF*VNOFF+VTANG*VTANG)
      IF(VTANG.LT.0.0) UE(I)=-UE(I)
      CP(I)    = 1.0 - UE(I)*UE(I)
 130  CONTINUE
C     COMPRESSIBILITY CORRECTION BASED ON KARMAN-TSIEN FORMULA
      BETA     = SQRT(1.-FMACH**2)
      CLAMDA   = FMACH**2 / (1. + BETA**2)
      DO 150 I =1,NODTOT
      CP(I)    = CP(I)/(BETA+(FMACH**2/(1.+BETA))*CP(I)*0.5)
      UE(I)    = (1. - CLAMDA)* UE(I) / (1. - CLAMDA*UE(I)**2)
      WRITE (6,1050) I,XMID(I),YMID(I),CP(I),UE(I)
  150 CONTINUE
```

```
 1000 FORMAT(/,4X,'J',4X,'X(J)',6X,'Y(J)',6X,'CP(J)',5X,'V(J)',/)
 1050 FORMAT(I5,4F10.5)
      RETURN
      END
```

2.4.6 Subroutine CLCM

The dimensionless pressure in the appropriate directions is integrated to compute the aerodynamic force and the coefficients for lift (CL) and pitching moment (CM) about the leading edge of the airfoil.

```
      SUBROUTINE CLCM(SINALF,COSALF)
      COMMON /BOD/ NODTOT,X(201),Y(201),
     +             XMID(200),YMID(200),COSTHE(200),SINTHE(200)
      COMMON /CPD/ UE(250),CP(250),UEW(50,2),CPW(50,2)
      COMMON /COMP / FMACH,C2,C6,C7,C8
C - - - - - - - - - - - - - - - - - - - - - - - - - - - - - - - - - - -
      CFX    = 0.0
      CFY    = 0.0
      CM     = 0.0
      DO 100  I = 1,NODTOT
        DX      = X(I+1) - X(I)
        DY      = Y(I+1) - Y(I)
        CFX     = CFX + CP(I)*DY
        CFY     = CFY - CP(I)*DX
        CM      = CM + CP(I)*(DX*XMID(I) + DY*YMID(I))
  100 CONTINUE
      CL       = CFY*COSALF - CFX*SINALF
      WRITE (6,1000) CL,CM
 1000 FORMAT(//,'    CL =',F10.5,'    CM =',F10.5)
      RETURN
      END
```

2.4.7 Subroutine VPDWK

This subroutine calculates the total velocity and pressure coefficient at each control point along the upper and lower wakes separately. The normal and tangential components of the total velocities are computed from Eqs. (2.3.2a) and (2.3.2b).

```
      SUBROUTINE VPDWK(SINALF,COSALF)
      COMMON /BOD/ NODTOT,X(201),Y(201),
     +             XMID(200),YMID(200),COSTHE(200),SINTHE(200)
      COMMON /COF/ A(201,201),BV(201),KUTTA
      COMMON /CPD/ UE(250),CP(250),UEW(50,2),CPW(50,2)
      COMMON /NUM/ PI,PI2INV
      COMMON/COUPL/DLSP(201),VNP(201),QW(51),DELW(51,2)
      COMMON /OFFBOD/ XOFF(201),YOFF(201),SOFF(200),COFF(200)
      COMMON /WWAKE/ NW,XW(51),YW(51),SINW(51),COSW(51),QWM(51)
```

```
      COMMON /COMP / FMACH,C2,C6,C7,C8
      DIMENSION Q(200)
C - - - - - - - - - - - - - - - - - - - - - - - - - - - - - - - - - - -
      DO 50   I = 1,NODTOT
 50   Q(I)    = BV(I)
      GAMMA   = BV(KUTTA)
      DO 130  I = 1,NW
         DO 125  K = 1,2
            XMIDO   = .5*(XW(I) + XW(I+1))
            YMIDO   = .5*(YW(I) + YW(I+1))
            VTANG   = COSALF*COSW(I) + SINALF*SINW(I)
            VNOFF   = SINALF*COSW(I) - COSALF*SINW(I)
            DLSH=0.5*(DELW(I,K)+DELW(I+1,K))
            IF (K.EQ.1) THEN
               XMIDO   = XMIDO+SINW(I)*DLSH
               YMIDO   = YMIDO-COSW(I)*DLSH
            ELSE
               XMIDO   = XMIDO-SINW(I)*DLSH
               YMIDO   = YMIDO+COSW(I)*DLSH
            ENDIF
            DO 120  J = 1,NODTOT
               DXJ      = XMIDO - X(J)
               DXJP     = XMIDO - X(J+1)
               DYJ      = YMIDO - Y(J)
               DYJP     = YMIDO - Y(J+1)
               FLOG =.5*ALOG((DXJP*DXJP+DYJP*DYJP)/(DXJ*DXJ+DYJ*DYJ))
               FTAN     = ATAN2(DYJP*DXJ-DXJP*DYJ,DXJP*DXJ+DYJP*DYJ)
               CTIMTJ   = COSW(I)*COSTHE(J) + SINW(I)*SINTHE(J)
               STIMTJ   = SINW(I)*COSTHE(J) - COSW(I)*SINTHE(J)
               AA       = PI2INV*(FTAN*CTIMTJ + FLOG*STIMTJ)
               B        = PI2INV*(FLOG*CTIMTJ - FTAN*STIMTJ)
               VTANG    = VTANG-B*Q(J)+GAMMA*AA
               VNOFF    = VNOFF + GAMMA*B + AA*Q(J)
 120        CONTINUE
            DO 121 J=1,NW
               FLOG     = 0.0
               FTAN     = PI
               IF (J .EQ. I .AND. DLSH. LT. 0.001) GO TO 100
               DXJ      = XMIDO - XW(J)
               DXJP     = XMIDO - XW(J+1)
               DYJ      = YMIDO - YW(J)
               DYJP     = YMIDO - YW(J+1)
               FLOG     = .5*ALOG((DXJP*DXJP+DYJP*DYJP)/(DXJ*DXJ+
     1                    DYJ*DYJ))
               FTAN     = ATAN2(DYJP*DXJ-DXJP*DYJ,DXJP*DXJ+DYJP*DYJ)
 100           CTIMTJ   = COSW(I)*COSW(J) + SINW(I)*SINW(J)
               STIMTJ   = SINW(I)*COSW(J) - COSW(I)*SINW(J)
               AA       = PI2INV*(FTAN*CTIMTJ + FLOG*STIMTJ)
               B        = PI2INV*(FLOG*CTIMTJ - FTAN*STIMTJ)
               VTANG    = VTANG-B*QWM(J)
               VNOFF    = VNOFF+AA*QWM(J)
```

```
 121        CONTINUE
            UEW(I,K)   = SQRT(VNOFF*VNOFF+VTANG*VTANG)
            CPW(I,K)   = 1.0 - UEW(I,K)*UEW(I,K)
 125     CONTINUE
 130  CONTINUE
C  COMPRESSIBILITY CORRECTION BASED ON KARMAN-TSIEN FORMULA
      BETA    = SQRT(1.-FMACH**2)
      CLAMDA  = FMACH**2 / (1. + BETA**2)
      DO 157 K = 1,2
      DO 157 I = 1,NW
         CPW(I,K) = CPW(I,K)/(BETA+(FMACH**2/(1.+BETA))*CPW(I,K)*0.5)
         UEW(I,K) = (1. - CLAMDA)* UEW(I,K) / (1.-CLAMDA*UEW(I,K)**2)
  157 CONTINUE
      RETURN
      END
```

References

[1] Bertin, J.J. and Smith, M.L., Aerodynamics for Engineers, Prentice Hall, Englewood Cliffs, 2nd Ed., 1989.

[2] Anderson, J., Aerodynamics, McGraw-Hill, NY, 1988.

[3] Hess, J.L. and Smith, A.M.O., "Calculation of Potential Flow About Arbitrary Bodies," Progress in Aerospace Sciences, Vol. 8, Pergamon Press, N.Y., 1966.

[4] Moran, J., An Introduction to Theoretical and Computational Aerodynamics, John Wiley, NY, 1984.

[5] Isaacson, E. and Keller, H.B., Analysis of Numerical Methods, John Wiley, NY, 1966.

3 Two-Dimensional Incompressible Flows: Boundary-Layer Method

3.0 Introduction

Unlike the Laplace's equation, the boundary-layer equations are nonlinear and cannot be solved by the superposition of elemental flows. Their solution procedure, however, is considerably simpler than the procedure for the Navier-Stokes equations. While the boundary-layer equations are parabolic, the Navier-Stokes equations for incompressible flows are elliptic and as such require much larger computer resources than the boundary-layer equations.

The solution of the boundary-layer equations can be obtained by integral or differential methods [1]. For example, in an integral method the velocity field in an incompressible flow is obtained from the solution of the momentum integral equation

$$\frac{d\theta}{dx} + (H+2)\frac{\theta}{u_e}\frac{du_e}{dx} = \frac{c_f}{2} \tag{3.0.1}$$

which contains three unknowns: momentum thickness θ, shape factor H, which represents the ratio of displacement thickness δ^* to θ, and local skin friction coefficient $c_f[\equiv \tau_w/(1/2)\rho u_e^2]$. While Eq. (3.0.1) is an exact equation and is applicable to both laminar and turbulent flows, its solution requires auxiliary equations which are often empirical; as a result, in an integral method the computed results are not as accurate as a differential method in which the continuity and the momentum equations,

$$\frac{\partial u}{\partial x} + \frac{\partial v}{\partial y} = 0 \tag{3.0.2}$$

$$u\frac{\partial u}{\partial x} + v\frac{\partial u}{\partial y} = u_e\frac{du_e}{dx} + \frac{\partial}{\partial y}\left(\nu\frac{\partial u}{\partial y} - \overline{u'v'}\right) \tag{3.0.3}$$

and their boundary conditions

$$y = 0: \quad u = 0 \quad v = v_w(x) \tag{3.0.4a}$$

$$y = \delta: \quad u = u_e \tag{3.0.4b}$$

and with a suitable term for the Reynolds shear stress, $-\rho\overline{u'v'}$, are solved in partial differential equation form and not as an ordinary differential equation form, as in an integral method. In both cases, it is assumed that the external velocity distribution $u_e(x)$ is given and the solutions are obtained in the standard mode. For further details, see [1].

The solution procedure in an inverse mode, as discussed in Section 1.3, treats $u_e(x)$ as an unknown and requires that it is obtained as part of the solution. The procedures used in integral methods for this purpose are discussed in references 1 to 4 of Chapter 1. Those most common and convenient in differential methods are discussed in Section 1.3. In the Hilbert integral approach, the inviscid velocity distribution, $u_e^0(x)$, in Eq. (1.3.2a) is assumed to be given and the perturbation velocity $\delta u_e(x)$ due to the displacement effect is computed from Eq. (1.3.2b). In this case, the boundary condition in Eq. (3.0.4b) is replaced by

$$y = \delta, \quad u = u_e = u_e^0 + \frac{1}{\pi}\int_{x_a}^{x_b} \frac{d}{d\sigma}(u_e\delta^*)\frac{d\sigma}{x - \sigma} \tag{3.0.4c}$$

The boundary conditions given by Eqs. (3.0.4) are for flow on the airfoil. If the calculations are to be performed for the wake, then with $y = 0$ denoting the dividing streamline, the centerline boundary conditions are

$$y = 0, \quad v = 0, \quad \frac{\partial u}{\partial y} = 0 \tag{3.0.5}$$

The edge boundary condition is identical to Eq. (3.0.4b) in the standard mode and to (3.0.4c) in the inverse mode.

In Section 3.3, the numerical solution of the system given by Eqs. (3.0.2) and (3.0.3) is discussed subject to the boundary conditions imposed by Eqs. (3.0.4) for both laminar and turbulent flows. Their solution subject to the wake boundary conditions is not considered since their incorporation into the solution procedure is straightforward as discussed in subsection 3.3.3.

The Reynolds shear stress, $-\rho\overline{u'v'}$, represents an additional unknown and requires an additional equation or equations. This can be accomplished by using several turbulence models. In the following section we present a brief description of three popular turbulence models, including the one used in this book. This is followed by Section 3.2 where the boundary-layer equations are placed into a more convenient form by using transformed variables for the numerical method described in Section 3.3. A Fortran program for solving the boundary-layer equations in both standard and inverse modes is given in Section 3.4.

$$2\nu\nu_t\left(\frac{\partial^2 u}{\partial y^2}\right)^2$$

is added as a source term to Eq. (3.1.7). With ν_t now defined by

$$\nu_t = c_\mu f_\mu \frac{k^2}{\varepsilon} \tag{3.1.10}$$

the momentum equation (3.0.3) and the transport equations (3.1.6) and (3.1.7) are written as

$$u\frac{\partial u}{\partial x} + v\frac{\partial u}{\partial y} = u_e\frac{du_e}{dx} + \nu\frac{\partial}{\partial y}\left(b\frac{\partial u}{\partial y}\right) \tag{3.1.11}$$

$$u\frac{\partial k}{\partial x} + v\frac{\partial k}{\partial y} = \frac{\partial}{\partial y}\left(\frac{\nu_t}{\sigma_k}\frac{\partial k}{\partial y}\right) + \nu_t\left(\frac{\partial u}{\partial y}\right)^2 - \varepsilon - 2\nu\left(\frac{\partial k^{1/2}}{\partial y}\right)^2 \tag{3.1.12}$$

$$u\frac{\partial \varepsilon}{\partial x} + v\frac{\partial \varepsilon}{\partial y} = \frac{\partial}{\partial y}\left(\frac{\nu_t}{\sigma_\varepsilon}\frac{\partial \varepsilon}{\partial y}\right) + c_{\varepsilon 1}\frac{\varepsilon}{k}\nu_t\left(\frac{\partial u}{\partial y}\right)^2 - f_2 c_{\varepsilon 2}\frac{\varepsilon^2}{k} + 2\nu\nu_t\left(\frac{\partial^2 u}{\partial y^2}\right)^2 \tag{3.1.13}$$

where

$$b = 1 + \frac{\nu_t}{\nu} \tag{3.1.14}$$

$$f_\mu = \exp\left[-\frac{3.4}{(1+R_t/50)^2}\right], \quad R_t = \frac{k^2}{\nu\varepsilon} \tag{3.1.15a}$$

$$f_2 = 1 - 0.3\exp(-R_t^2) \tag{3.1.15b}$$

The boundary conditions are

$$y = 0 \quad k = \varepsilon = 0 \tag{3.1.16a}$$

$$y \to \delta, \quad k \to k_e, \quad \varepsilon \to \varepsilon_e \tag{3.1.16b}$$

To avoid numerical problems, k_e and ε_e should not be zero. In addition, k_e and ε_e can not be prescribed arbitrarily because their development is governed by the transport equations (3.1.11) and (3.1.12) written at the boundary-layer edge,

$$u_e\frac{dk_e}{dx} = -\varepsilon_e \tag{3.1.17a}$$

$$u_e\frac{d\varepsilon_e}{dx} = -c_{\varepsilon 2}\frac{\varepsilon_e^2}{k_e} \tag{3.1.17b}$$

The above equations can be integrated with respect to x with initial conditions corresponding to k_{e0} and ε_{e0} at x_0. The solution provides the evolutions of $k(x)$ and $\varepsilon(x)$ as boundary conditions for the k- and ε-equations. Equations (3.1.11) to (3.1.13) are then solved subject to the boundary conditions given by Eqs. (3.0.4), (3.1.16a) and the conditions given by Eq. (3.1.17).

Another approach, sometimes referred to as the *two-layer method*, employs a simple model near the wall (a mixing length model [7] or a one equation model [8] which is valid only near the wall region and a transport equation model in the outer region of the boundary-layer; the two solutions are matched at a certain point in the boundary-layer as discussed by Arnal et al. [7].

3.1.2 Reynolds Stress Models

The Reynolds shear stress $-\rho\overline{u'v'}$ appearing in the momentum equation (3.0.3) can also be modelled by using the Reynolds transport equation. In tensor form, this equation can be written as

$$\frac{D}{Dt}(\overline{u_i'u_j'}) = P_{ij} - D_{ij} + \Phi_{ij} + \frac{\partial J_{ij\ell}}{\partial x_\ell} \tag{3.1.18}$$

where

$$P_{ij} = -(\overline{u_i'u_\ell'})\frac{\partial u_j}{\partial x_\ell} - (\overline{u_j'u_\ell'})\frac{\partial u_i}{\partial x_\ell} \tag{3.1.19a}$$

$$D_{ij} = 2\nu\overline{\frac{\partial u_i'}{\partial x_\ell}\frac{\partial u_j'}{\partial x_\ell}} \tag{3.1.19b}$$

$$\Phi_{ij} = \overline{\frac{p'}{\rho}\left(\frac{\partial u_i'}{\partial x_j} + \frac{\partial u_j'}{\partial x_i}\right)} \tag{3.1.19c}$$

$$J_{ij\ell} = -\overline{u_i'u_j'u_\ell'} - \overline{\frac{p'}{\rho}(u_i'\delta_{j\ell} + u_j'\delta_{i\ell})} + \nu\frac{\partial\overline{u_i'u_j'}}{\partial x_\ell} \tag{3.1.19d}$$

The meaning of various terms in the above equation are

- P_{ij} is the production term due to the interaction between Reynolds stresses and velocity gradients.
- D_{ij} is the destruction term due to the action of viscosity
- Φ_{ij} is called the pressure-strain term which distributes kinetic energy k of turbulence among its components, $\overline{u_i'^2}$ since Φ_{ij} is equal to zero by the continuity equation for the fluctuating velocities.
- $J_{ij\ell}$ is a diffusion term due to viscosity and the fluctuations of velocity and pressure.

As in the momentum equation, closure assumptions are needed for Eq. (3.1.18) before it can be solved. There are several models developed for this purpose, most of which are applicable to turbulent flows where the effects of viscosity and the presence of the wall are neglected.

A popular model is due to Launder, Reece and Rodi [9]. According to this model, with P_{ij} given by Eq. (3.1.19a), the terms D_{ij}, Φ_{ij} and $J_{ij\ell}$ are given by

$$D_{ij} = \frac{2}{3}\delta_{ij}\varepsilon \tag{3.1.20a}$$

$$\Phi_{ij} + \Phi_{ji} = -C_1\frac{\varepsilon}{k}\left(\overline{u_i'u_j'} - \frac{2}{3}\delta_{ij}k\right) - C_2'\left(P_{ij} - \frac{2}{3}\delta_{ij}P\right) \tag{3.1.20b}$$

$$k = \frac{\overline{u_i'u_i'}}{2}; \quad P = -\overline{u_m'u_\ell'}\frac{\partial u_m}{\partial x_\ell} \tag{3.1.20c}$$

$$J_{ij\ell} = C_8' \frac{k}{\varepsilon} \overline{u_\ell' u_m'} \frac{\partial}{\partial x_m} \overline{u_i' u_j'} \tag{3.1.20d}$$

The model is completed by a transport equation for the rate of dissipation ε,

$$\frac{D\varepsilon}{Dt} = -C_{\varepsilon 1} \frac{\varepsilon}{k} \overline{u_m' u_l'} \frac{\partial u_m}{\partial x_\ell} - C_{\varepsilon 2} \frac{\varepsilon^2}{k} + \frac{\partial}{\partial x_\ell} \left(C_\varepsilon' \frac{k}{\varepsilon} \overline{u_m' u_\ell'} \frac{\partial \varepsilon}{\partial x_m} \right) \tag{3.1.21}$$

where

$$\begin{aligned} C_1 &= 1.8 & C_2' &= 0.6 \\ C_{\varepsilon 1} &= 1.44 & C_{\varepsilon 2} &= 1.92 \\ C_8' &= 0.22 & C_\varepsilon' &= 0.15 \end{aligned}$$

To include the wall effects in this model, Gibson and Launder [10] modified the pressure-strain term by adding the following terms to Eq. (8.3.26b)

$$\phi_{ij,w} + \phi_{ji,w} = f \left(f_{km} n_k n_m \delta_{ij} - \frac{3}{2} f_{ki} n_i n_j - \frac{3}{2} f_{kj} n_k n_i \right) \tag{3.1.22}$$

where, with x_m representing distance from the wall and n_i, the components of the unit vector normal to the wall,

$$\begin{gathered} f = \frac{k^{3/2}}{2.5 \varepsilon x_n} \\ f_{\alpha\beta} = C_{1w} \frac{\varepsilon}{k} \overline{u_\alpha' u_\beta'} + C_{2w} (\phi_{\alpha\beta,2} + \phi_{\beta\alpha,2}) \\ C_{1w} = 0.5 \quad C_{2w} = 0.3 \end{gathered} \tag{3.1.23}$$

$$\frac{D}{Dt} \overline{u'v'} = -A_{12} \overline{v'^2} \frac{\partial u}{\partial y} - B_{12} \frac{\varepsilon}{k} \overline{u'v'} + \frac{\partial}{\partial y} \left(C_s' \frac{k}{\varepsilon} \overline{v'^2} \frac{\partial \overline{u'v'}}{\partial y} \right) \tag{3.1.24}$$

$$\frac{D}{Dt} \overline{u'^2} = -A_{11} \overline{u'v'} \frac{\partial u}{\partial y} - B_{11} \frac{\varepsilon}{k} \overline{u'^2} + C_{11} \varepsilon + D_{11} \frac{\varepsilon}{k} \overline{v'^2} + \frac{\partial}{\partial y} \left(C_s' \frac{k}{\varepsilon} \overline{v'^2} \frac{\partial \overline{u'^2}}{\partial y} \right) \tag{3.1.25}$$

$$\frac{D}{Dt} \overline{v'^2} = -A_{22} \overline{u'v'} \frac{\partial u}{\partial y} - B_{22} \frac{\varepsilon}{k} \overline{v'^2} + C_{22} \varepsilon + \frac{\partial}{\partial y} \left(C_s' \frac{k}{\varepsilon} \overline{v'^2} \frac{\partial}{\partial y} \overline{v'^2} \right) \tag{3.1.26}$$

$$\frac{D}{Dt} \overline{w'^2} = -A_{33} \overline{u'v'} \frac{\partial u}{\partial y} - B_{33} \frac{\varepsilon}{k} \overline{w'^2} + C_{33} \varepsilon + D_{33} \frac{\varepsilon}{k} \overline{v'^2} + \frac{\partial}{\partial y} \left(C_s' \frac{k}{\varepsilon} \overline{v'^2} \frac{\partial \overline{w'^2}}{\partial y} \right) \tag{3.1.27}$$

$$\frac{D\varepsilon}{Dt} = -C_{\varepsilon 1} \frac{\varepsilon}{k} \overline{u'v'} \frac{\partial u}{\partial y} - C_{\varepsilon 2} \frac{\varepsilon^2}{k} + \frac{\partial}{\partial y} \left(C_\varepsilon' \frac{k}{\varepsilon} \overline{v'^2} \frac{\partial \varepsilon}{\partial y} \right) \tag{3.1.28}$$

where with f given by

$$f = \frac{k^{3/2}}{2.5 \varepsilon y} \tag{3.1.29}$$

the various terms in the above equations are

$$
\begin{aligned}
A_{12} &= 1 - C_2' + \frac{3}{2} f C_2' C_{2w}, & B_{12} &= C_1 + \tfrac{3}{2} f C_{1w} \\
A_{11} &= 2 - \frac{4}{3} C_2' + \frac{2}{3} f C_2' C_{2w}, & B_{11} &= C_1 \\
C_{11} &= \frac{2}{3} C_1 - \frac{2}{3}, & D_{11} &= C_{1w} f
\end{aligned}
\tag{3.1.30}
$$

$$
A_{22} = \frac{2}{3} C_2' - \frac{4}{3} f C_2' C_{2w}, \quad B_{22} = C_1 + 2 f C_{1w}, \quad C_{22} = \frac{2}{3} C_1 - \frac{2}{3}
$$

$$
A_{33} = \frac{2}{3} C_2' + \frac{2}{3} f C_2' C_{2w}, \quad B_{33} = C_1, \quad C_{33} = \frac{2}{3} C_1 - \frac{2}{3}, \quad D_{33} = f C_{1w}
$$

The presence of the wall is neglected merely by setting $f = 0$.

It should be noted that the above model is applicable only to high Reynolds number flows at some distance y_0 away from the wall where the flow is fully turbulent. Modifications are required to extend the model and the calculations all the way to the wall.

3.1.3 Comments on k-ε and Reynolds Stress Models

The two turbulence models discussed in the previous two subsections contain many important features of turbulent flows needed in understanding and calculating turbulent flows. For example, in the k-ε model, the turbulence energy is known to influence mean flow properties, and the turbulent dissipation rate equation can represent several characteristic scales which change with location in many flow configurations. Extensively separated flows, in particular, should be represented by two-equation models. Transport equation models are usually applicable over a wider range of flows than zero-equation models, although the latter can be expected to be more precise over the range of data upon which they are based.

Reynolds stress models involve solving larger numbers of equations than two equation models. They are insufficiently tested or uneconomical for use with the Navier-Stokes equations and have not yet been shown to be advantageous for wall boundary-layer flows. In addition, the numerical solution of these equations for boundary-layer flows can involve thirteen first-order equations and can cause numerical difficulties in adverse pressure gradient flows as discussed in [11].

Problems associated with the specification of boundary conditions for transport models should not be overlooked. They tend to affect the results much less than the mean velocities, but can give rise to error. In a boundary-layer flow, for example, values of each dependent variable or its gradient are required on three sides of the solution domain. Measurement can provide guidance for the initial and external boundary conditions for k and ε although they are frequently evaluated from a mixing-length assumption. Since their influence is usually considerably less than that of the mean-velocity conditions, approximations of this type are unlikely to limit the accuracy of the calculation of the mean flow. For a detailed discussion, the reader is referred to Wilcox [2].

3.1.4 Zero-Equation Models

Zero-equation models used to model the Reynolds stress term in the momentum equations employ algebraic eddy viscosity or algebraic mixing-length formulations. For a two-dimensional incompressible laminar and turbulent flow, the continuity and momentum equations, (3.0.2) and (3.0.3), are solved subject to given initial conditions and boundary conditions with $-\overline{u'v'}$ represented either by Eq. (3.1.1) or (3.1.2). For a limited range of flows, it is possible to specify the turbulent viscosity of Eq. (3.1.1) as the mixing length of Eq. (3.1.3) in closed algebraic expressions. In approaches based on zero-equations models, no differential equations are solved for turbulence quantities, and turbulence transport is, consequently, not represented.

For external flows, a popular and widely used algebraic eddy viscosity formulation is due to Cebeci and Smith [5]. This two-layer method assumes that a turbulent boundary-layer can be considered in terms of eddy viscosities for inner and outer layers. The corresponding functions are empirical and based on limited ranges of experimental data; the amount of data is, however, extensive so that the full algebraic formulation is the most general available at present.

In the inner region of a boundary-layer on a smooth surface, with or without mass transfer, the eddy viscosity ν_t is written as

$$(\nu_t)_i = \ell^2 \left|\frac{\partial u}{\partial y}\right| \gamma_{\text{tr}}, \quad 0 \le y \le y_c \tag{3.1.31}$$

Here the mixing length ℓ is given by

$$\ell = \kappa y \left[1 - \exp\left(-\frac{y}{A}\right)\right] \tag{3.1.32a}$$

where $\kappa = 0.40$ and A is a damping-length constant, which may be represented by

$$A = 26\frac{\nu}{N}u_\tau^{-1}, \quad N = \left\{\frac{p^+}{v_w^+}[1 - \exp(11.8v_w^+)] + \exp(11.8v_w^+)\right\}^{1/2} \tag{3.1.32b}$$

$$p^+ = \frac{\nu u_e}{u_\tau^3}\frac{du_e}{dx}, \quad v_w^+ = \frac{v_w}{u_\tau} \tag{3.1.32c}$$

For flows with no mass transfer ($v_w^+ = 0$), Eq. (3.1.32b) reduces to

$$N = (1 - 11.8p^+)^{1/2} \tag{3.1.32d}$$

and for a flow with no mass transfer and pressure gradient, $N = 1$.

In Eq. (3.1.31) γ_{tr} is an intermittency factor which represents the streamwise region from the onset of transition to turbulent flow. It is defined by the empirical correlation given by Eqs. (1.4.4) and (1.4.5). For high Reynolds numbers, the parameter C in Eq. (1.4.5) is a constant with recommended value of 60. For lower Reynolds numbers, it is given by Eq. (1.4.7).

In the outer region, the eddy viscosity is given by

$$(\nu_t)_0 = \alpha u_e \delta^* \gamma_{tr} \gamma, \quad y_c \leq y \leq \delta \tag{3.1.33}$$

Here γ accounts for the intermittency of the outer region and is represented by

$$\gamma = \left[1 + 5.5\left(\frac{y}{y_0}\right)^6\right]^{-1} \tag{3.1.34}$$

where y_0 is defined as the y-location where $u/u_e = 0.995$. Continuity of the expressions for the eddy viscosities in the inner and outer regions, Eqs. (3.1.31) and (3.1.33), defines these regions.

In Eq. (3.1.33) the parameter α is equal to 0.0168 for values of R_θ greater than 5000 and, for R_θ less than 5000, α varies according to the formula

$$\alpha = 0.0168 \frac{1.55}{1 + \Pi} \tag{3.1.35a}$$

where the variation of Π with R_θ can be correlated empirically by

$$\Pi = 0.55[1 - \exp(-0.243 z_1^{1/2} - 0.298 z_1)] \tag{3.1.35b}$$

where $z_1 = R_\theta/425 - 1$ for $R_\theta > 425$.

Extensive studies, mostly employing boundary-layer equations, show that while many external turbulent flow problems can satisfactorily be calculated with the above Cebeci-Smith eddy-viscosity formulation, improvements are needed for regions of strong adverse pressure gradient and flow separation. Comparison of calculated eddy-viscosity distributions with experimental data show that the parameter α decreases with increasing adverse pressure gradient. This is expected since the law of the wall region given by Eq. (3.1.31) becomes smaller with strong adverse pressure gradient and flow separation.

Currently there are two approaches that have been pursued to improve the CS model in strong adverse pressure-gradient flows. In one approach, pursued by Johnson and King [12], and Johnson and Coakley [13], the parameter α is taken not to be equal to 0.0168 at higher Reynolds number but computed from an ordinary differential equation derived from the turbulence kinetic energy equation. Another approach assumes α to be computed not from an ordinary differential equation but from an algebraic formula [14],

$$\alpha = \frac{0.0168}{[1 - \beta(\partial u/\partial x)/(\partial u/\partial y)]^{1.5}} \tag{3.1.36}$$

The parameter β is computed from

$$\beta = \begin{cases} \dfrac{6}{1 + 2R_t(2 - R_t)} & R_t \le 1.0 \\ \dfrac{2R_t}{1 + R_t} & R_t \ge 1.0 \end{cases} \tag{3.1.37}$$

where R_t denotes the ratio of wall shear to maximum Reynolds shear stress, $-(\rho\overline{u'v'})_m$,

$$R_t = \frac{\tau_w}{(-\rho\overline{u'v'})_m} \tag{3.1.38}$$

Another improvement made in the CS model to improve the predictions in strong adverse pressure prediction flows is to replace the intermittency parameter γ in Eq. (3.1.33) by another intermittency expression recommended in [15], According to the experiments conducted by Fiedler and Head, it was found that the pressure gradient has a marked effect on the distribution of intermittency defined as the ratio of time the flow was turbulent to total time at any point so that it measures the probability of finding turbulent flow at any instant at the point considered. Their experiments indicated that in the boundary-layer proceeding to separation, the intermittent zone decreases in width and moved further from the surface as shape factor H increases. The reverse trend is observed with decreasing H in a favorable pressure gradient.

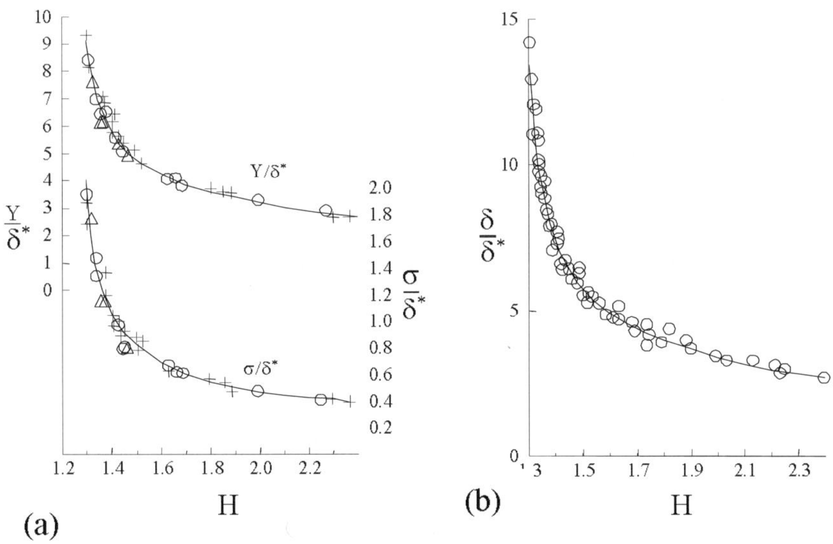

Fig. 3.1. Variation of (a) Y/δ^* and σ/δ^* and (b) δ/δ^* with H according to the data of [15].

In the improved CS model the intermittency expression of [15] was written in the form

$$\gamma = \frac{1}{2}\left[1 - \operatorname{erf}\frac{[y - Y]}{\sqrt{2}\sigma}\right] \tag{3.1.39}$$

where Y and σ are general intermittency parameters with Y denoting value of y for which $\gamma = 0.5$ and σ, the standard deviation. The dimensionless intermittency parameters Y/δ^* and σ/δ^* were expressed as functions of H as shown in Fig. 3.1a. The variation of the ratio of boundary-layer thickness δ to δ^* with H is shown in Fig. 3.1b.

3.2 Transformed Boundary-Layer Equations

Since the boundary-layer equations are parabolic, their solution procedure employs a marching scheme discussed in Section 3.3 and in more detail in [1]. It begins with initial conditions at $x = x_0$ and proceeds in the positive x-direction with dependent variables determined in sequence normal to the flow at each x-station subject to the boundary conditions. In general, the boundary-layer thickness $\delta(x)$ increases with increasing downstream distance for both laminar and turbulent flows, and small steps must be taken in the streamwise direction to maintain computational accuracy.

Transformed coordinates employing similarity variables are advantageous in the solution of boundary-layer equations since they can reduce and even eliminate the growth of transformed boundary-layer thickness, $\eta_e(x)$, and thus allow larger steps to be taken in the streamwise direction. Furthermore, and as shown below, they can also be used to generate initial conditions needed for the solution of the boundary-layer equations. In this chapter and throughout the book, the use of similarity variables is advocated; their form depends on whether the flow is laminar or turbulent, and whether it is wall boundary-layer flow or free-shear layer flow. There are several transformations that can be used for this purpose, and even though most of them have been developed for laminar flows, they can also be used for turbulent flows.

In this section we employ two separate but closely related transformations. The first one is the Falkner-Skan transformation in which the dimensionless similarity variable η and a dimensionless stream function $f(x, \eta)$ are defined by

$$\eta = \sqrt{u_e/\nu x}\, y, \quad \psi(x, y) = \sqrt{u_e \nu x} f(x, \eta) \tag{3.2.1}$$

This is a convenient and useful transformation to generate the initial conditions needed at the airfoil leading edge. It is also useful to perform the boundary-layer calculations subject to the boundary conditions in Eq. (3.0.4), which correspond to the standard mode. The resulting equations from this transformation are given in subsection 3.2.1.

In the inverse mode, since $u_e(x)$ is also an unknown, slight changes are made to the transformation given by Eq. (3.2.1), replacing $u_e(x)$ by u_∞ and redefining new variables Y and F by

$$Y = \sqrt{u_\infty/\nu x}\, y, \quad \psi(x,y) = \sqrt{u_\infty \nu x} F(\xi, Y) \tag{3.2.2}$$

The resulting equations from this transformation are given in subsection 3.2.2.

3.2.1 Transformed Equations: Standard Mode

Noting the chain rule relations, it can be shown that, with a prime denoting differentiation with respect to η, and with

$$\xi = \frac{x}{L}, \quad w = \frac{u_e}{u_\infty}$$

the continuity equation, (3.0.2), and the momentum equation, employing the eddy-viscosity concept, Eq. (3.1.1), can be written as

$$(bf'')' + \frac{m+1}{2} ff'' + m[1-(f')^2] = \xi \left(f' \frac{\partial f'}{\partial \xi} - f'' \frac{\partial f}{\partial \xi} \right) \tag{3.2.3}$$

Here m is a dimensionless pressure-gradient parameter defined by

$$m = \frac{\xi}{w} \frac{dw}{d\xi} \tag{3.2.4}$$

The velocity components u and v are related to the dimensionless stream function $f(\xi, \eta)$ by

$$u = u_e f', \quad v = -\sqrt{u_e \nu x} \left[\frac{f}{\sqrt{u_e x}} \frac{d}{dx} \sqrt{u_e x} + \frac{\partial f}{\partial x} + f' \frac{\partial \eta}{\partial x} \right] \tag{3.2.5}$$

In terms of the transformed variables defined by Eq. (3.2.1), the boundary conditions given by Eq. (3.0.4a,b) become

$$\eta = 0, \quad f' = 0 \quad f(\xi, 0) \equiv f_w(\xi) = -\frac{\sqrt{R_L}}{\sqrt{w\xi}} \int_0^\xi \frac{v_w}{u_\infty} d\xi \tag{3.2.6a}$$

$$\eta = \eta_e, \quad f' = 1 \tag{3.2.6b}$$

Here η_e corresponds to a transformed boundary-layer thickness, $\sqrt{u_e/\nu x}\,\delta$, or $\sqrt{wR_L/\xi}\,\delta/L$; R_L is a Reynolds number based on reference velocity u_∞ and length L.

3.2.2 Transformed Equations: Inverse Mode

Using a similar procedure and the transformation given by Eq. (3.2.2), the continuity and momentum equations can again be expressed as a third-order partial-differential equation in a form analogous to that given by Eq. (3.2.5)

$$(bF'')' + \frac{1}{2}FF'' = \xi\left(F'\frac{\partial F'}{\partial \xi} - F''\frac{\partial F}{\partial \xi}\right) - \xi w\frac{dw}{d\xi} \tag{3.2.7}$$

Here primes denote differentiation with respect to Y.

In terms of the transformed variables defined by Eq. (3.2.2), the wall boundary conditions given by Eq. (3.0.4a) and with mass transfer starting at $\xi = \xi_0$ become

$$Y = 0, \quad F' = 0 \quad F_w = -\frac{\sqrt{R_L}}{\sqrt{\xi}}\int_{\xi_0}^{\xi}\frac{v_w}{u_\infty}d\xi \tag{3.2.8}$$

The boundary condition corresponding to Eq. (3.0.4c) is obtained by applying a discretization approximation to the Hilbert integral

$$u_e(x_i) = u_e^0(x_i) + C_{ii}D_i + \sum_{j=1}^{i-1}C_{ij}D_j + \sum_{j=i+1}^{N}C_{ij}D_j \tag{3.2.9}$$

where the subscript i denotes the ξ-station where the inverse calculations are to be performed, C_{ij} is a matrix of interaction coefficients obtained by the procedure described in subsection 3.4.4, and D is given by $D = u_e\delta^*$. In terms of transformed variables, the parameter D becomes

$$\overline{D} = \frac{D}{Lu_\infty} = \sqrt{\frac{\xi}{R_L}}(Y_e w - F_e) \tag{3.2.10}$$

and the relation between the external velocity u_e and displacement thickness δ^* provided by the Hilbert integral can then be written in dimensionless form as

$$Y = Y_e, \quad F_e'(\xi^i) - \lambda[Y_e F_e'(\xi^i) - F_e(\xi^i)] = g_i \tag{3.2.11}$$

where

$$\lambda = C_{ii}\sqrt{\xi^i/R_L} \tag{3.2.12a}$$

$$g_i = \overline{u_e^0}(\xi^i) + \sum_{j=1}^{i-1}C_{ij}\overline{D}_j + \sum_{j=i+1}^{N}C_{ij}\overline{D}_j \tag{3.2.12b}$$

3.2.3 Transformed Equations: General Form

It is convenient to write the transformed equations for standard and inverse modes in one general form so that the numerical method described in the following section can be discussed for both cases. For this purpose, we write Eqs. (3.2.3) and (3.2.7) in the following form

$$\begin{aligned}(bf'')' + \lambda_1 \left\{\frac{m}{2} ff'' + m[1-(f')^2]\right\} + \frac{1}{2} ff'' \\ = \xi \left\{\frac{\partial}{\partial \xi}\left[\frac{(f')^2}{2}\right] - f'' \frac{\partial f}{\partial \xi} - \lambda_2 \frac{d}{d\xi}\left(\frac{w^2}{2}\right)\right\}\end{aligned} \tag{3.2.13}$$

where for $\lambda_1 = 0$ and $\lambda_2 = 0$ we recover Eq. (3.2.3) which is subject to the boundary conditions given by Eqs. (3.2.6). For $\lambda_1 = 0$ and $\lambda_2 = 1$, we recover Eq. (3.2.7) provided that we note that F now corresponds to f, primes denote differentiation with respect to Y and not η. For consistency, we also rewrite Eqs. (3.2.8) and (3.2.11) in the adopted notation of Eq. (3.2.13),

$$\eta = 0; \quad f' = 0 \quad f_w = -\frac{\sqrt{R_L}}{\sqrt{\xi}} \int_{\xi_0}^{\xi} \frac{v_w}{u_\infty} d\xi \tag{3.2.14a}$$

$$\eta = \eta_e; \quad f' - \lambda(\eta f' - f) = g_i \tag{3.2.14b}$$

3.3 Numerical Method

There are several numerical methods for solving the boundary-layer equations. Finite-difference methods offer the greatest flexibility, and those of Crank-Nicolson [16] and Keller [17] have been widely used as discussed in detail in [6]. The latter method provides significant advantages over the former method and will be used to solve the boundary-layer equations for steady two-dimensional flows, as described here, unsteady boundary-layer equations in Chapter 8, three-dimensional flows in Chapter 11, as well as the stability equations in Chapters 4 and 12.

One of the requirements of Keller's box method is that the governing equations are written as a first-order system. As described in subsection 3.3.1, first, derivatives of f with respect to η are introduced as new functions, and the resulting first-order equations are approximated on an arbitrary rectangular net, Figure 3.2. The simple "centered-difference" derivatives and averages at the midpoints of net rectangles and net segments are then used to get finite-difference equations with a truncation error of second order. The resulting difference equations, which are implicit and nonlinear, are linearized by Newton's method described in subsection 3.3.2, and are solved by the algorithms discussed in subsections 3.3.3 and 3.3.4.

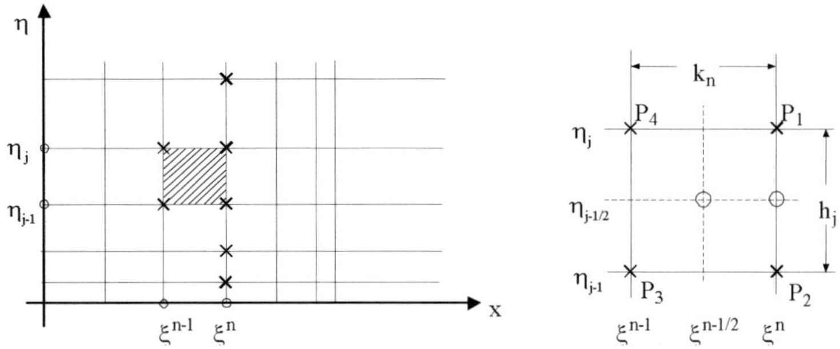

Fig. 3.2. Net rectangle for difference approximations.

3.3.1 Numerical Formulation

Equation (3.2.13) can be expressed as a first-order system by introducing new dependent variables $u(\xi, \eta), v(\xi, \eta)$ and by defining

$$f' = u \tag{3.3.1a}$$

$$u' = v \tag{3.3.1b}$$

Equation (3.2.13) becomes

$$\begin{aligned}(bv)' + \lambda_1 \left[\frac{m}{2} fv + m(1-u^2)\right] + \frac{1}{2} fv \\ = \xi \left[\frac{\partial}{\partial \xi}\left(\frac{u^2}{2}\right) - v\frac{\partial f}{\partial \xi} - \lambda_2 \frac{d}{d\xi}\left(\frac{w^2}{2}\right)\right]\end{aligned} \tag{3.3.1c}$$

The boundary conditions for the standard mode follow from Eqs. (3.2.6),

$$\eta = 0; \quad u = 0, \quad f = f_w \tag{3.3.2a}$$

$$\eta = \eta_e; \quad u = 1 \tag{3.3.2b}$$

Similarly, for the inverse mode the boundary conditions are

$$\eta = 0; \quad u = 0, \quad f = f_w \tag{3.3.3a}$$

$$\eta = \eta_e; \quad \lambda f + (1 - \lambda \eta_e) w = g_i \tag{3.3.3b}$$

Next, we denote the net points of the net rectangle shown in Fig. 3.2 by

$$\begin{aligned}\xi^0 = 0, \quad \xi^n = \xi^{n-1} + k_n, \quad n = 1, 2, \ldots, N \\ \eta_0 = 0, \quad \eta_j = n_{j-1} + h_j, \quad j = 1, 2, \ldots, J \\ \eta_J = \eta_e\end{aligned} \tag{3.3.4}$$

and write the difference equations that are to approximate Eqs. (3.3.1) by considering one mesh rectangle, as shown in Fig. 3.2. We start by writing the finite-difference approximations of the ordinary differential equations (3.3.1a,b) for the midpoint $(\xi^n, \eta_{j-1/2})$ of the segment P_1P_2 using centered-difference derivatives; this process is called "centering about $(\xi^n, \eta_{j-1/2})$". We get

$$\frac{f_j^n - f_{j-1}^n}{h_j} = \frac{u_j^n + u_{j-1/2}^n}{2} \equiv u_{j-1/2}^n \tag{3.3.5a}$$

$$\frac{u_j^n - u_{j-1}^n}{h_j} = \frac{v_j^n + v_{j-1}^n}{2} \equiv v_{j-1/2}^n \tag{3.3.5b}$$

Similarly, the partial differential equation (3.3.1c) is approximated by centering about the midpoint $(\xi^{n-1/2}, \eta_{j-1/2})$ of the rectangle $P_1P_2P_3P_4$,

$$\begin{aligned}
&\frac{1}{2}\left[\frac{(bv)_j^n - (bv)_{j-1}^n}{h_j} + \frac{(bv)_j^{n-1} - (bv)_{j-1}^{n-1}}{h_j}\right] \\
&\quad + \frac{\lambda_1}{2}\left[\frac{m^{n-1/2}}{2}\{(fv)_{j-1/2}^n + (fv)_{j-1/2}^{n-1}\}\right. \\
&\quad \left. + 2m^{n-1/2}\left\{1 - \frac{(u^2)_{j-1/2}^n + (u^2)_{j-1/2}^{n-1}}{2}\right\}\right] + \frac{1}{2}[(fv)_{j-1/2}^n + (fv)_{j-1/2}^{n-1}] \\
&= \xi^{n-1/2}\left[FLARE\frac{(u^2)_{j-1/2}^n - (u^2)_{j-1/2}^{n-1}}{2k_n}\right. \\
&\quad \left. - \left(\frac{v_{j-1/2}^n + v_{j-1/2}^{n-1}}{2}\right)\left(\frac{f_{j-1/2}^n - f_{j-1/2}^{n-1}}{k_n}\right) - \lambda_2\left\{\frac{(u_J^2)^n - (u_J^2)^{n-1}}{2k_n}\right\}\right]
\end{aligned}$$

In the above equation, the parameter *FLARE* refers to the Flügge-Lotz Reyhner approximation [18] used to set $u\,\partial u/\partial x$ equal to zero in the momentum equation wherever $u < 0$. As a result, the numerical instabilities that plague attempts to integrate the boundary-layer equations against the local direction of flow are avoided. In regions of positive streamwise velocity $(u_j > 0)$, it is taken as unity and as zero in regions of negative streamwise velocity $(u_j \leq 0)$

After some algebra and rearranging, the above expression can be written as

$$\begin{aligned}
&h_j^{-1}[(bv)_j^n - (bv)_{j-1}^n] + \alpha_1(fv)_{j-1/2}^n - \alpha_2(u^2)_{j-1/2}^n \\
&+ \alpha^n[v_{j-1/2}^{n-1}f_{j-1/2}^n - f_{j-1/2}^{n-1}v_{j-1/2}^n + \lambda_2(u_J^2)^n] = R_{j-1/2}^{n-1}
\end{aligned} \tag{3.3.5c}$$

where

$$\alpha^n = \frac{\xi^{n-1/2}}{k_n} \tag{3.3.6a}$$

$$\alpha_1 = 1 + \frac{\lambda_1}{2} m^{n-1/2} + \alpha^n \tag{3.3.6b}$$

$$\alpha_2 = \lambda_1 m^{n-1/2} + \alpha^n \tag{3.3.6c}$$

$$\begin{aligned} R^{n-1}_{j-1/2} = & -L^{n-1}_{j-1/2} + \alpha^n [-FLARE(u^2)^{n-1}_{j-1/2} \\ & + v^{n-1}_{j-1/2} f^{n-1}_{j-1/2} + \lambda_2 (u^2_J)^{n-1}] \end{aligned} \tag{3.3.6d}$$

$$\begin{aligned} L^{n-1}_{j-1/2} = & h_j^{-1}[(bv)^{n-1}_j - (bv)^{n-1}_{j-1}] + \left(1 + \frac{\lambda_1}{2} m^{n-1/2}\right)(fv)^{n-1}_{j-1/2} \\ & + 2\lambda_1 m^{n-1/2}\left[1 - \frac{(u^2)^{n-1}_{j-1/2}}{2}\right] \end{aligned} \tag{3.3.6e}$$

Equations (3.3.5), which follow from Eqs. (3.3.1), are imposed for $j = 1, 2, \ldots, J$ at given n. They are accompanied with the boundary conditions given by Eqs. (3.3.2) for the standard mode,

$$f_0^n = f_w^n \quad u_0^n = 0 \tag{3.3.7a}$$

$$u_J^n = 1 \tag{3.3.7b}$$

and by Eqs. (3.3.3) for the inverse mode,

$$f_0^n = f_w^n, \quad u_0^n = 0 \tag{3.3.8a}$$

$$\lambda f_J^n + (1 - \lambda \eta_J) u_J^n = g_i \tag{3.3.8b}$$

The next step is to solve the algebraic equations represented by Eqs. (3.3.5) and (3.3.7) or by Eqs. (3.3.5) and (3.3.8). Since these equations are nonlinear, the linearization procedure, using Newton's method described in the following section, is introduced before they are solved by the algorithms described in subsections 3.3.3 and 3.3.4.

Studies with separating flows (inverse mode) indicate that the use of central differences in the x-direction can cause oscillations. They can, however, be reduced and in most cases, avoided by representing streamwise derivatives by backward differences which can be first-order or second order. For example, the $\frac{\partial u}{\partial x}$ term can be written as

$$\frac{\partial u}{\partial x} = A_1 u^n + A_2 u^{n-1} \qquad \text{(first-order)}$$

$$\frac{\partial u}{\partial x} = A_1 u^n + A_2 u^{n-1} + A_3 u^{n-2} \qquad \text{(second-order)}$$

where for first-order

$$A_1 = \frac{1}{x_n - x_{n-1}}, \quad A_2 = -\frac{1}{x_n - x_{n-1}}$$

and second-order

$$A_1 = \frac{2x_n - x_{n-1} - x_{n-2}}{(x_n - x_{n-1})(x_n - x_{n-2})}$$

$$A_2 = \frac{(x_n - x_{n-2})}{(x_{n-1} - x_{n-2})(x_{n-1} - x_n)}$$

$$A_3 = \frac{x_n - x_{n-1}}{(x_{n-2} - x_n)(x_{n-2} - x_{n-1})}$$

Further details can be obtained from the author.

3.3.2 Newton's Method

Assume f_j^{n-1}, u_j^{n-1}, and v_j^{n-1} to be known for $0 \le j \le J$; then Eqs. (3.3.5) and (3.3.7) form a system of $3J+3$ equations for the solution of $3J+3$ unknowns (f_j^n, u_j^n, v_j^n), $j = 0, 1, \ldots, J$. We use Newton's method to solve this nonlinear system. Thus, the iterates $[f_j^{(\nu)}, u_j^{(\nu)}, v_j^{(\nu)}]$, $\nu = 0, 1, 2, \ldots$, are introduced with initial values ($\nu = 0$) equal to those at the previous ξ-station, ξ^{n-1}. For the higher iterates we set

$$f_j^{(\nu+1)} = f_j^{(\nu)} + \delta f_j^{(\nu)}, \quad u_j^{(\nu+1)} = u_j^{(\nu)} + \delta u_j^{(\nu)}, \quad v_j^{(\nu+1)} = v_j^{(\nu)} + \delta v_j^{(\nu)} \quad (3.3.9)$$

We insert the right-hand sides of these expressions in place of f_j^n, u_j^n and v_j^n in Eqs. (3.3.5) and drop the terms that are quadratic in $\delta f_j^{(\nu)}$, $\delta u_j^{(\nu)}$ and $\delta v_j^{(\nu)}$. This procedure yields the following *linear* system (the superscript n is dropped from f_j, u_j, v_j and ν from δ quantities for simplicity).

$$\delta f_j - \delta f_{j-1} - \frac{h_j}{2}(\delta u_j + \delta u_{j-1}) = (r_1)_j \quad (3.3.10a)$$

$$\delta u_j - \delta u_{j-1} - \frac{h_j}{2}(\delta v_j + \delta v_{j-1}) = (r_3)_{j-1} \quad (3.3.10b)$$

$$\begin{aligned}(s_1)_j \delta v_j &+ (s_2)_j \delta v_{j-1} + (s_3)_j \delta f_j + (s_4)_j \delta f_{j-1} + (s_5)_j \delta u_j \\ &+ (s_6)_j \delta u_{j-1} + (s_7)_j \delta u_J = (r_2)_j\end{aligned} \quad (3.3.10c)$$

where for $1 \le j \le J$

$$(r_1)_j = f_{j-1}^{(\nu)} - f_j^{(\nu)} + h_j u_{j-1/2}^{(\nu)} \quad (3.3.11a)$$

$$(r_3)_{j-1} = u_{j-1}^{(\nu)} - u_j^{(\nu)} + h_j v_{j-1/2}^{(\nu)} \quad (3.3.11b)$$

$$\begin{aligned}(r_2)_j = R_{j-1/2}^{n-1} &- \Big[h_j^{-1}(b_j^{(\nu)} v_j^{(\nu)} - b_{j-1}^{(\nu)} v_{j-1}^{(\nu)}) + \alpha_1 (fv)_{j-1/2}^{(\nu)} \\ &- \alpha_2 (u^2)_{j-1/2}^{(\nu)} \\ &+ \alpha^n (v_{j-1/2}^{n-1} f_{j-1/2}^{(\nu)} - f_{j-1/2}^{n-1} v_{j-1/2}^{(\nu)}) + \lambda_2 (u_J^{(\nu)})^2\Big]\end{aligned} \quad (3.3.11c)$$

As will be shown later, in writing the system given by Eqs. (3.3.10), a certain order has been chosen to ensure that the A_0-matrix in Eq. (3.3.19a) is not singular.

The coefficients of the linearized momentum equation, where A_2 is zero for the standard mode and is unity for the inverse mode, are

$$(s_1)_j = h_j^{-1} b_j^{(\nu)} + \frac{\alpha_1}{2} f_j^{(\nu)} - \frac{\alpha^n}{2} f_{j-1/2}^{n-1} \tag{3.3.12a}$$

$$(s_2)_j = -h_j^{-1} b_{j-1}^{(\nu)} + \frac{\alpha_1}{2} f_{j-1}^{(\nu)} - \frac{\alpha^n}{2} f_{j-1/2}^{n-1} \tag{3.3.12b}$$

$$(s_3)_j = \frac{\alpha_1}{2} v_j^{(\nu)} + \frac{\alpha^n}{2} v_{j-1/2}^{n-1} \tag{3.3.12c}$$

$$(s_4)_j = \frac{\alpha_1}{2} v_{j-1}^{(\nu)} + \frac{\alpha^n}{2} v_{j-1/2}^{n-1} \tag{3.3.12d}$$

$$(s_5)_j = -\alpha_2 u_j^{(\nu)} \tag{3.3.12e}$$

$$(s_6)_j = -\alpha_2 u_{j-1}^{(\nu)} \tag{3.3.12f}$$

$$(s_7)_J = 2\alpha^n \lambda_2 u_J^{(\nu)} \tag{3.3.12g}$$

The boundary conditions become

Standard Mode

$$\delta f_0 = (r_1)_0 = 0, \quad \delta u_0 = (r_2)_0 = 0, \tag{3.3.13a}$$

$$\delta u_J = (r_3)_J = 0 \tag{3.3.13b}$$

Inverse Mode

$$\delta f_0 = (r_1)_0 = 0, \quad \delta u_0 = (r_2)_0 = 0, \tag{3.3.14a}$$

$$\lambda \delta f_J + (1 - \lambda \eta_J)\delta u_J = (r_3)_J \tag{3.3.14b}$$

where

$$(r_3)_J = g_i - [\lambda f_J + (1 - \lambda \eta_J) u_J] \tag{3.3.15}$$

3.3.3 Block-Elimination Method

The linear system given by Eqs. (3.3.10) and (3.3.13) with $\lambda_2 = 0$ has a block tridiagonal structure and can be expressed in matrix- vector form

$$A\vec{\vec{\delta}} = \vec{\vec{r}} \tag{3.3.16}$$

with the coefficient matrix A and the compound vectors $\vec{\vec{\delta}}$ and $\vec{\vec{r}}$ given by

$$\vec{\vec{\delta}} = \begin{vmatrix} \vec{\delta}_0 \\ \vec{\delta}_1 \\ \cdot \\ \cdot \\ \vec{\delta}_j \\ \cdot \\ \cdot \\ \vec{\delta}_J \end{vmatrix}, \quad \vec{\vec{r}} = \begin{vmatrix} \vec{r}_0 \\ \vec{r}_1 \\ \cdot \\ \cdot \\ \vec{r}_j \\ \cdot \\ \cdot \\ \vec{r}_J \end{vmatrix}, \quad A = \begin{vmatrix} A_0 & C_0 & & & & & \\ B_1 & A_1 & C_1 & & & & \\ & \cdot & \cdot & \cdot & & & \\ & & \cdot & \cdot & \cdot & & \\ & & B_j & A_j & C_j & & \\ & & & \cdot & \cdot & \cdot & \\ & & & B_{J-1} & A_{J-1} & C_{J-1} \\ & & & & B_J & A_J \end{vmatrix} \tag{3.3.17}$$

The three-dimensional vectors, $\vec{\delta}_j$ and $\vec{r}_j$, for each value of j, are

$$\vec{\delta}_j = \begin{vmatrix} \delta f_j \\ \delta u_j \\ \delta v_j \end{vmatrix}, \quad \vec{r}_j = \begin{vmatrix} (r_1)_j \\ (r_2)_j \\ (r_3)_j \end{vmatrix}, \quad 0 \le j \le J \tag{3.3.18}$$

and the 3×3 matrices A_j, B_j, and C_j are given by

$$A_0 = \begin{vmatrix} 1 & 0 & 0 \\ 0 & 1 & 0 \\ 0 & -1 & -h_1/2 \end{vmatrix},$$
$$A_j \equiv \begin{vmatrix} 1 & -h_j/2 & 0 \\ (s_3)_j & (s_5)_j & (s_1)_j \\ 0 & -1 & -h_{j+1}/2 \end{vmatrix} \quad 1 \le j \le J-1 \tag{3.3.19a}$$

$$A_J \equiv \begin{vmatrix} -1 & -h_J/2 & 0 \\ (s_3)_J & (s_5)_J & (s_1)_J \\ 0 & 1 & 0 \end{vmatrix}$$
$$B_j \equiv \begin{vmatrix} -1 & -h_j/2 & 0 \\ (s_4)_j & (s_6)_j & (s_2)_j \\ 0 & 0 & 0 \end{vmatrix} \quad 1 \le j \le J \tag{3.3.19b}$$

$$C_j \equiv \begin{vmatrix} 0 & 0 & 0 \\ 0 & 0 & 0 \\ 0 & 1 & -h_{j+1}/2 \end{vmatrix} \quad 0 \le j \le J-1 \tag{3.3.19c}$$

Note that the first two rows of A_0 and C_0 and the last row of B_J and A_J correspond to the boundary conditions. To solve the boundary-layer equations for different boundary conditions, only the matrix rows mentioned above need altering. For example, to include the wake boundary conditions given by Eq. (3.0.5), it is only sufficient to change the second row of A_0-matrix from 0 1 0 to 0 0 1.

As a result of the boundary conditions given by Eq. (3.3.13), we note that $(r_1)_0 = (r_2)_0 = 0$ and $(r_3)_J = 0$ so that

$$\vec{r}_0 = \begin{vmatrix} 0 \\ 0 \\ (r_3)_0 \end{vmatrix}, \quad \vec{r}_J = \begin{vmatrix} (r_1)_J \\ (r_2)_J \\ 0 \end{vmatrix} \tag{3.3.20}$$

Equation (3.3.16) can be solved efficiently and effectively by using the block-elimination method which involves two steps. In the first part of the forward sweep, the matrices Γ_j and Δ_j, with Γ_j having the same structure as B_j, are computed from the recursion formulas given by

$$\Delta_0 = A_0 \tag{3.3.21a}$$

$$\Gamma_j \Delta_{j-1} = B_j \quad j = 1, 2, \ldots, J \tag{3.3.21b}$$

$$\Delta_j = A_j - \Gamma_j C_{j-1} \quad j = 1, 2, \ldots, J \tag{3.3.21c}$$

In the second part of the forward sweep, the vector $\vec{w}_j$ defined by

$$\vec{w}_j = \begin{vmatrix} (w_1)_j \\ (w_2)_j \\ (w_3)_j \end{vmatrix} \quad 0 \le j \le J \tag{3.3.22}$$

are computed from

$$\vec{w}_0 = \vec{r}_0 \tag{3.3.23a}$$

$$\vec{w}_j = \vec{r}_j - r_j \vec{w}_{j-1} \quad 1 \le j \le J \tag{3.3.23b}$$

In the backward sweep, the vectors $\vec{\delta}_j$ are computed from

$$\Delta_J \vec{\delta}_J = \vec{w}_J \tag{3.3.24a}$$

$$\Delta_j \vec{\delta}_j = \vec{w}_j - C_j \vec{\delta}_{j+1} \quad j = J-1, J-2, \ldots, 0 \tag{3.3.24b}$$

For further details of the block elimination method, the reader is referred to [1,19].

3.3.4 Bordering Algorithm

The linear system given by Eqs. (3.3.10) and (3.3.14) with $\lambda_2 = 1$ can again be written in matrix vector form as

$$G\vec{\vec{\delta}} = \vec{\vec{r}} \tag{3.3.25}$$

where $\vec{\vec{\delta}}$ and $\vec{\vec{r}}$ are identical to those defined by Eq. (3.3.17), and

$$G = \left|\begin{array}{c|c} A & D \\ \hline E^T & A_J \end{array}\right| \tag{3.3.26}$$

The coefficient matrix A is also identical to that defined in Eq., (3.3.17) except that the index of A_j, B_j and C_j matrices run up to $j = J - 1$, that is, the last row of A is now B_{J-1}, A_{J-1}. For this reason, for convenience we rewrite $\vec{\vec{\delta}}$ and $\vec{\vec{r}}$ as

$$\vec{\vec{\delta}} = \left|\begin{array}{c} \vec{\Delta} \\ \vec{\delta}_J \end{array}\right|, \quad \vec{\vec{r}} = \left|\begin{array}{c} \vec{R} \\ \vec{r}_J \end{array}\right| \tag{3.3.27}$$

and define $\vec{\Delta}$ and $\vec{R}$ by

$$\vec{\Delta} = \left|\begin{array}{c} \vec{\delta}_0 \\ \cdot \\ \cdot \\ \vec{\delta}_j \\ \cdot \\ \cdot \\ \vec{\delta}_{J-1} \end{array}\right|, \quad \vec{R} = \left|\begin{array}{c} \vec{r}_0 \\ \cdot \\ \cdot \\ \vec{r}_j \\ \cdot \\ \cdot \\ \vec{r}_{J-1} \end{array}\right| \tag{3.3.28}$$

with $\vec{\delta}_j$ and $\vec{r}_j$ as defined by Eqs. (3.3.18).

The matrices D and E, with B_j given by Eq. (3.3.19b), are

$$D = \left|\begin{array}{c} D_0 \\ \cdot \\ \cdot \\ D_j \\ \cdot \\ \cdot \\ D_{J-1} \end{array}\right|, \quad E = \left|\begin{array}{c} 0 \\ \cdot \\ \cdot \\ 0 \\ \cdot \\ \cdot \\ B_J^T \end{array}\right| \tag{3.3.29}$$

$$D_0 = \left|\begin{array}{ccc} 0 & 0 & 0 \\ 0 & 0 & 0 \\ 0 & 0 & 0 \end{array}\right|, \quad D_j = \left|\begin{array}{ccc} 0 & 0 & 0 \\ 0 & (s_7)_j & 0 \\ 0 & 0 & 0 \end{array}\right| \quad 1 \le j \le J-2$$
$$D_{J-1} = \left|\begin{array}{ccc} 0 & 0 & 0 \\ 0 & (s_7)_{J-1} & 0 \\ 0 & 1 & -h_J/2 \end{array}\right| \tag{3.3.30}$$

For convenience, we write the D matrix as

$$D = [\vec{D}_1, \vec{D}_2, \vec{D}_3] \tag{3.3.31}$$

where

$$\vec{D}_1 = \begin{vmatrix} 0 \\ 0 \\ 0 \\ \cdot \\ \cdot \\ \cdot \\ 0 \\ 0 \\ 0 \\ \cdot \\ \cdot \\ \cdot \\ 0 \\ 0 \\ 0 \end{vmatrix} \quad \vec{D}_2 = \begin{vmatrix} 0 \\ 0 \\ 0 \\ 0 \\ (s_7)_1 \\ 0 \\ \cdot \\ \cdot \\ \cdot \\ 0 \\ (s_7)_j \\ 0 \\ \cdot \\ \cdot \\ 0 \\ (s_7)_{J-1} \\ 1 \end{vmatrix} \quad \vec{D}_3 = \begin{vmatrix} 0 \\ 0 \\ 0 \\ \cdot \\ \cdot \\ \cdot \\ 0 \\ 0 \\ 0 \\ \cdot \\ \cdot \\ \cdot \\ \cdot \\ \cdot \\ 0 \\ 0 \\ -\frac{h_j}{2} \end{vmatrix} \tag{3.3.32}$$

The term $(s_7)_J \delta u_j$ in Eq. (3.3.10c) causes the coefficient matrix G to have a form other than tridiagonal, as indicated in Eq. (3.3.23), and this prevents the use of the efficient block-elimination method discussed in subsection 3.3.3. A consequence was the development and the use of the approach discussed in [20,21], where $w_j(\equiv u_J)$ was an extra unknown and the coefficient matrix G became tridiagonal with 4×4 blocks rather than the more convenient 3×3 arrangement associated with the standard mode. An immediate consequence was the slight reduction of the efficiency of the numerical method. As the number of equations increases, for example to represent compressibility, heat transfer, three-dimensional flows, and transport turbulence models, this deviation from the tridiagonal arrangement increasingly makes the solution of the linear system more difficult and less efficient.

To overcome this deficiency and to return to a 3×3 tridiagonal arrangement, we use Keller's bordering algorithm [20] discussed in detail in [21], and write Eq. (3.3.25) as

$$A\vec{\Delta} + D\vec{\delta}_J = \vec{R} \tag{3.3.33a}$$

$$E^T\vec{\Delta} + A_J\vec{\delta}_J = \vec{r}_J \tag{3.3.33b}$$

If F is introduced such that

$$AF = D \tag{3.3.34}$$

then by substituting Eq. (3.3.34) into Eq. (3.3.33a), we get

$$A\vec{w} = \vec{R} \tag{3.3.35}$$

where, unlike F and D, which are matrices, $\vec{w}$ and $\vec{R}$ are vectors with $\vec{w}$ given by

$$\vec{w} = \vec{\Delta} + F\vec{\delta}_J \tag{3.3.36}$$

With Eq. (3.3.36), Eq. (3.3.33b) can be written as

$$(A_J - E^T F)\vec{\delta}_J = \vec{r}_J - E^T \vec{w} \tag{3.3.37}$$

According to this bordering algorithm, the solution of Eq. (3.3.22) is obtained in the following three steps:

(1) Solve Eqs. (3.3.34) and (3.3.35). Since the coefficient matrix A has a block tridiagonal structure, the solution of Eq. (3.3.35) can be obtained efficiently with the block elimination method discussed in subsection 3.3.3. With $F = [\vec{F}_1, \vec{F}_2, \vec{F}_3]$, Eq. (3.3.34) can be written as vector equations with the definition given by Eq. (3.3.31). $\vec{D}_1$ is zero, and therefore it is necessary to solve only two vector equations,

$$A\vec{F}_2 = \vec{D}_2, \quad A\vec{F}_3 = \vec{D}_3 \tag{3.3.38}$$

Since A_j, B_j, C_j $(j \le J-1)$ are identical to those given in Eq. (3.3.19), Γ_j and Δ_j calculated in the first part of the forward sweep are identical to those calculated in the standard mode. The second part of the forward sweep is also the same. For example, to solve

$$A\vec{w} = \vec{R}$$

we use the same procedure that was used to solve Eq. (3.3.16). Denoting the elements of the vector $\vec{w}$ by Eq. (3.3.22), we use the recursion formulas given by Eq. (3.3.23), and with $\vec{X}_j$ replacing $\vec{w}_j$ in Eq. (3.3.23) and with $\vec{X}_j$ defined by

$$\vec{X}_j = \begin{vmatrix} (X_1)_j \\ (X_2)_j \\ (X_3)_j \end{vmatrix} \tag{3.3.39}$$

we obtain expressions identical to those calculated in the standard mode for $j = 0$ and $1 \le j \le J-1$.

In the backward sweep, $\vec{w}_j$ is computed from Eqs. (3.3.24) with j now starting from $J-1$ so that

$$\Delta_{J-1}\vec{w}_{J-1} = \vec{X}_{J-1} \tag{3.3.40a}$$

$$\Delta_j \vec{w}_j = \vec{X}_j - C_j \vec{w}_{j+1}, \quad j = J-2, J-3, \ldots, 0 \tag{3.3.40b}$$

The components of $\vec{w}_j$ computed from Eq. (3.3.37b) are identical to those in the standard mode with $[\delta f_j, \delta u_j, \delta v_j]$ corresponding to $[(w_1)_j, (w_2)_j, (w_3)_j]$ with $\vec{w}_j$'s then replaced by $\vec{X}_j$.

(2) With $\vec{w}_j$ and F known from step 1, we compute

$$A'_J = A_J - E^T F \tag{3.3.41}$$

$$\vec{r}'_J = \vec{r}_J - E^T \vec{w} \tag{3.3.42}$$

then solve

$$A'_J \vec{\delta}_J = \vec{r}'_J \tag{3.3.43}$$

This solution is easily found since A'_J is a 3×3 matrix, and one solves a 3×3 system to obtain $\vec{\delta}_J$.

To compute $\vec{\delta}_J$ from Eq. (3.3.43), we make use of Eqs. (3.3.41) and (3.3.42), which define A'_J and r'_J by

$$A'_J = \begin{vmatrix} 1 & \alpha'_{12} & \alpha'_{13} \\ (s_3)_J & \alpha'_{22} & \alpha'_{23} \\ \gamma_1 & \gamma_2 & 0 \end{vmatrix}, \quad \vec{r}'_J = \begin{vmatrix} (r_1)'_J \\ (r_2)'_J \\ (r_3)'_J \end{vmatrix}$$

where

$$\alpha'_{12} = -\frac{h_J}{2} + (F_{21})_{J-1} + \frac{h_J}{2}(F_{22})_{J-1}$$

$$\alpha'_{22} = (s_5 + s_7)_J - (s_4)_J(F_{21})_{J-1} - (s_6)_J(F_{22})_{J-1} - (s_2)_J(F_{23})_{J-1}$$

$$\alpha'_{13} = (F_{31})_{J-1} + \frac{h_J}{2}(F_{32})_{J-1}$$

$$\alpha'_{23} = (s_1)_J - (s_4)_J(F_{31})_{J-1} - (s_6)_J(F_{32})_{J-1} - (s_2)_J(F_{33})_{J-1}$$

$$\gamma_1 = \lambda$$

$$\gamma_2 = 1 - \lambda\eta_J$$

$$(r_1)'_J = (r_1)_J + (w_1)_{J-1} + \frac{h_J}{2}(w_2)_{J-1}$$

$$(r_2)'_J = (r_2)_J - (s_4)_J(w_1)_{J-1} - (s_6)_J(w_2)_{J-1} - (s_2)_J(w_3)_{J-1}$$

$$(r_3)'_J = (r_3)_J$$

The solution of Eq. (P6.23) is then obtained by

$$\delta f_J = \left\{(r'_3)_J(\alpha'_{12}\alpha'_{23} - \alpha'_{22}\alpha'_{13}) - \gamma_2[(r'_1)_J\alpha'_{23} - (r'_2)_J\alpha'_{13}]\right\}/\Delta$$

$$\delta u_J = \left\{\gamma_1[(r'_1)_J\alpha'_{23} - (r'_2)_J\alpha'_{13}] - (r'_3)_J[\alpha'_{23} - (s_3)_J\alpha'_{13}]\right\}/\Delta$$

$$\delta v_J = [(r'_2)_J - (s_3)_J\delta f_J - \alpha'_{22}\delta u_J]/\alpha'_{23}$$

where $\Delta = \gamma_1(\alpha'_{12}\alpha'_{23} - \alpha'_{22}\alpha'_{13}) - \gamma_2[\alpha'_{23} - (s_3)_J\alpha'_{13}]$

(3) Now compute

$$\vec{\Delta} = \vec{w} - F\vec{\delta}_J \tag{3.3.44}$$

where $\vec{\Delta}$ is defined by Eq. (3.3.25) with its index running up to $J-1$, rather than J, to get

$$\begin{aligned} \delta f_j &= (w_1)_j - \delta u_J(F_{21})_j - \delta v_J(F_{31})_j \\ \delta u_j &= (w_2)_j - \delta u_J(F_{22})_j - \delta v_J(F_{32})_j \\ \delta v_j &= (w_3)_j - \delta u_J(F_{23})_j - \delta v_J(F_{33})_j \end{aligned} \tag{3.3.45}$$

3.4 FORTRAN PROGRAM

In this section, we describe and present an inverse boundary-layer program, IBLP, for two-dimensional incompressible flows. It is applicable to laminar and turbulent flows for which the external velocity, u_e/u_∞ is given as a function of dimensionless surface distance ξ with the location of transition $\xi_{\rm tr}$ and freestream Reynolds number $R_L(\equiv u_\infty L/\nu)$. It is a simplified version of the general boundary-layer program used to obtain the results discussed in Chapter 5. The general code takes the input from the inviscid method of Chapter 2, establishes a boundary-layer grid and the stagnation point, performs calculations on the upper airfoil surface and its wake, then performs calculations on the lower airfoil surface and its wake. The computed blowing velocity and displacement thickness distributions are then transformed to the panel method to obtain a modified external velocity distribution. The simplified code, IBLP, is designed to compute the flow over one surface of the airfoil and its wake which is separated into an upper wake flow region and lower wake flow region by the dividing streamline. The calculations start near the leading edge of the airfoil, $\xi = 0$, where the flow is laminar; at $\xi > 0$, it becomes turbulent at any ξ-location by specifying or computing the transition location. Initially the calculations are performed in the standard mode; later from a specified ξ-station, they are continued in the inverse mode.

IBLP also contains a simplified Cebeci-Smith eddy viscosity formulation than the one discussed in subsection 3.1.4 and is used to obtain the results in Chapter 5. It can, however, be changed to the formulation of subsection 3.1.4 or to any other algebraic eddy-viscosity or mixing-length formulation.

IBLP consists of a MAIN routine (subsection 3.4.1) which contains the logic of the computations and 14 subroutines: INPUT, IVPL, GROWTH, HIC, EDDY, COEF, SWTCH, WAKEPR, INTEG, DIFF1, LNTP, AMEAN, BORALG, and OUTPUT.

3.4.1 MAIN

MAIN contains the logic of the boundary-layer calculations, checks the convergence of the solutions at each ξ-station, checks whether the calculations are for laminar and/or turbulent flows in the standard or inverse modes and prints the profiles f_j, u_j, v_j and b_j as a function of j and η_j.

Since the nonlinear equations are linearized, one must use an iteration procedure in which the solutions of Eq. (3.3.16) or (3.3.25) are obtained for successive estimates of velocity profiles, and the convergence of the solutions are checked. When the calculations are performed for the airfoil it is common to use a criterion based on the wall shear parameter $v_0(\equiv f''_w)$ for this purpose. The iterations, which are generally quadratic for laminar flows, are stopped when

$$|\delta v_0| < \varepsilon_1 \tag{3.4.1a}$$

For turbulent flows, due to the approximate linearization procedure used for the turbulent diffusion term, the rate of convergence is not quadratic and solutions are usually acceptable when the ratio of $|\delta v_0/v_0|$ is less than 0.01. With proper linearization, as discussed in [22,23], quadratic convergence of the solutions can be obtained.

When the calculations are performed for the wake, the convergence criterion is based on the centerline velocity by requiring

$$|\delta u_0| < \varepsilon_2 \tag{3.4.1b}$$

with ε_2 taken as 0.0025.

The program restricts the number of iterations per ξ-station, for example by setting ITMAX = 15. Laminar flows usually require 3 to 4 iterations, and depending on the problem, turbulent flow calculations may require a few more.

When the calculations are performed in the inverse mode with the Hilbert integral formulation, several sweeps on the body are required for a prescribed external velocity distribution. In this subroutine, the calculations are performed and controlled for the number of sweeps (ISWPT). This value is usually 3 but may be larger for some problems.

The transformed boundary-layer thickness, $\eta_e(\xi)$, is almost constant for most laminar boundary-layer flows and generally increases with increasing ξ for turbulent boundary-layer flows. Since the calculations assume $\eta_e = 8$ initially, the value of $f''(\eta_e)[\equiv V(NP, 2)]$ is checked to see whether it is sufficiently small (such as being less than 0.001) for $\xi > 0$ at the convergence of the solutions at $\xi = \xi^n$. If this test is satisfied, $\eta_e(\xi^{n+1})$ is set equal to $\eta_e(\xi^n)$, and if it is not, J_{new} is set equal to $J_{\text{old}} + t$, where t is a number of points, taken to be two in the present code. The values for new η_j points $(f_j^n, u_j^n, v_j^n, b_j^n)$ are defined by the values of $u_j^n = u_J^n$, $v_j^n = 0$, $f_j^n = (\eta_j - \eta_J)u_J^n + f_J^n$, $b_j^n = b_J^n$ at ξ^n, and f_j^{n-1}, u_j^{n-1}, v_j^{n-1} and b_j^{n-1} at ξ^{n-1}.

MAIN also computes the g_i-parameter (GI) given by Eq. (3.2.12b); the term

$$\sum_{j=1}^{i-1} C_{ij}\overline{D}_j \tag{3.4.2a}$$

named SUM1 represents the upstream effect and the term

$$\sum_{j=i+1}^{N} C_{ij}\overline{D}_j \tag{3.4.2b}$$

named SUM2 represents the downstream effects in the Hilbert integral relation.

```
C     MAIN
      COMMON /BLC001/ NX,NP,NXTE,NXT,NPT,NTR,NXS,IT,ITV,ISWP,ISWPT
      COMMON /BLPRC1/ F(101,2),U(101,2),V(101,2),W(101,2),B(101,2)
      COMMON /BLC2SO/ DELF(101),DELU(101),DELV(101),DELW(101)
      COMMON /BLGRD1/ X(201),ETA(101),A(101),DETA(101)
      COMMON /BLGTYY/ P1(201),P2(201)
      COMMON /BLHLBS/ CC(201,201),DD(201),DB(201),GI
      COMMON /BLINP3/ RL, RX, SQRL, SQRX
      COMMON /BLRC01/ XBL(201),UBL(201),CFS(201),THETA(201),
     1                DLS(201),UEVS(201),NPSAV(201),ITSAV(201)
      DIMENSION       D1(401)
C     LAGRANGE INTERPOLATION FORMULA
      GRANG(X1,X2,X3,Y1,Y2,Y3,X0)= (X0-X2)*(X0-X3)/(X1-X2)/(X1-X3)*Y1
     1      +(X0-X1)*(X0-X3)/(X2-X1)/(X2-X3)*Y2+(X0-X1)*(X0-X2)
     2      /(X3-X1)/(X3-X2)*Y3
C - - - - - - - - - - - - - - - - - - - - - - - - - - - - - - - - - - -
      CALL INPUT
      CALL IVPL
      X(NXT+1)  = 3.0 * X(NXT) -3.0 * X(NXT-1) + X(NXT-2)
      CALL HIC(NXT,RL,X(1))
      DD(NXT+1) =GRANG(X(NXT-2),X(NXT-1),X(NXT),
     1             DD(NXT-2),DD(NXT-1),DD(NXT),X(NXT+1))
      ISWP    = 1
      ITV     = 1
      NX      = 0
  100 NX      = NX + 1
      RX      = UEVS(NX)*X(NX)*RL
      SQRX    = SQRT(RX)
      WRITE(6,6000) NX,XBL(NX),X(NX)
      IF(NX.GT.1) THEN
        DO 102 J=1,NP
           F(J,1) = F(J,2)
           U(J,1) = U(J,2)
           V(J,1) = V(J,2)
           B(J,1) = B(J,2)
  102   CONTINUE
      END IF
      IF(ITV .EQ. 0) THEN
        SUM2      = 0.
        DO 104 K = NX+1,NXT+1
           SUM2      = SUM2 + CC(K,NX)*DD(K)
  104   CONTINUE
        SUM1      = 0.
        DO 108 K = 3,NX-1
           SUM1      = SUM1 + CC(K,NX)*DD(K)
  108   CONTINUE
        GI        = UBL(NX) + SUM1 + SUM2
      END IF
  110 IT      = 0
      IGROW   = 0
  120 IT      = IT+1
```

```
      IF(IT .GE. 15) THEN
         WRITE(6,9999) NX
         GO TO 999
      ENDIF
      IF(NX .GE. NTR) CALL EDDY
      CALL COEF
      CALL BORALG
      IF(ITV .EQ. 0) UEVS(NX) = U(NP,2)
      IF (V (1,2) .LE. 0.0 .AND. ITV .EQ. 1) THEN
         WRITE(6,201) NX,X(NX),UBL(NX),P1(NX),P2(NX),V(1,2)
         STOP
      ENDIF
      IF(NX .LE. NXTE) THEN
C  CONVERGENCE OF THE SOLUTIONS - AIRFOIL
         IF(NX .LE. NTR ) THEN
           IF(ABS(DELV(1)).GT. 0.0001) GOTO 120
         ELSE
           IF(ABS(DELV(1)/V(1,2)).GT. 0.01) GOTO 120
         END IF
      ELSE
C  CONVERGENCE OF THE SOLUTIONS - WAKE
         IF(ABS(DELU(1)).GT. 0.0025) GOTO 120
      ENDIF
      IF(NP .LT. NPT .AND. IGROW .LT. 3) THEN
         IF( ABS(V(NP,2)) .GT. 0.0010) THEN
            IGROW = IGROW + 1
C  BOUNDARY LAYER GROWTH
            NPO   = NP
            NP    = MINO(NPT,NP+2)
            DO 125 J=NPO+1,NP
               DO 122 I = 1,2
                  V(J,I)= 0.0
                  U(J,I)= U(NPO,I)
                  F(J,I)= F(J-1,I)+U(J,I)*DETA(J-1)
                  B(J,I)= B(NPO,I)
  122          CONTINUE
  125       CONTINUE
            IT    = 1
            GO TO 120
         ENDIF
      ENDIF
  999 CONTINUE
      CALL OUTPUT
      WRITE (6,9000) (J,ETA(J),F(J,2),U(J,2),V(J,2),B(J,2),J=1,NP)
      IF(NX .LT. NXS ) GO TO 100
      IF(NX .EQ. NXS) THEN
        ITV     = 0
        CALL SWTCH(0)
        GO TO 100
      END IF
      IF(NX .LT. NXT  ) THEN
```

```
        IF(NX.EQ.NXTE) THEN
            CALL WAKEPR
        ENDIF
        GOTO 100
     ENDIF
     IF(ISWP .LT. ISWPT) THEN
       CALL AMEAN(NXS,NXT,X,DD,D1,1)
       DD(NXT+1)=GRANG(X(NXT-2),X(NXT-1),X(NXT),
    1            DD(NXT-2), DD(NXT-1),DD(NXT),X(NXT+1))
       ISWP   = ISWP + 1
       NX     = NXS
       CALL SWTCH(1)
       GOTO 100
     END IF
     CALL AMEAN(NXS,NXTE,X,CFS,D1,1)
     WRITE(6,9100)  NXS,X(NTR)
     WRITE (6,9120) (I,XBL(I),THETA(I),CFS(I),DLS(I),I=2,NXT)
     PRINT *, '   '
     PRINT *, 'NORMAL TERMINATION IN MAIN '
     STOP
 201 FORMAT (' FLOW SEPARATES IN THE STANDARD MODE; PRINT DATA',
    1        ' FOR PROGRAM CHECK '/1H ,2X,'NX = ',I3,2X,'X = ',
    2        F10.5,2X,'UBL = ',F10.5,2X,'P1 = ',F10.5,2X,'P2 = ',
    3        F10.5,2X,'VW = ',F10.5,/)
6000 FORMAT('NX=',I3,2X,'X/C=',F7.3,3X,'S =',F7.3)
9000 FORMAT('  J ',4X,'ETA',9X,'F',13X,'U',13X,'V',13X,'B'/
    &       (I3,F8.3,4E14.5))
9120 FORMAT (3H NX,3X,2HXC,6X,5HTHETA,8X,2HCF,7X,5HDELST/
    &       (I3,F8.4,3E11.3))
9100 FORMAT(//'** SUMMARY OF BOUNDARY LAYER SOLUTIONS '/
    &        'NXS =',I3,4X,'** TRANSITION LOCATION : XCTR =   ',F10.4)
9999 FORMAT('ITERATION EXCEEDS MAX  AT  NX =',I3/)
     END
```

3.4.2 Subroutine INPUT

This subroutine contains the input to IBLP, generates the grid in the η-direction, computes the $\gamma_{\rm tr}$-term in the Cebeci-Smith model which accounts for the transition region between a laminar and turbulent flow and the D-term appearing in the Hilbert integral. Before we describe the input, it is useful to discuss the generation of the streamwise and η-grids.

The streamwise grid is generated by reading the values of ξ^i. In general the ξ-grid distribution depends on the variation of u_e with ξ so that rapid variations in external velocity distribution and the approach to separation require small $\Delta\xi$-steps (k_n). For laminar flows, it is often sufficient to use a uniform grid in the η-direction. A choice of transformed boundary-layer thickness η_e equal to 8 often ensures that the dimensionless slope of the velocity profile at the edge, $f''(\eta_e)$, is sufficiently small $(< 10^{-3})$ and that approximately 41 grid points are adequate for most flows. For turbulent flows, however, a uniform grid is not satisfactory because the boundary-layer thickness η_e and the dimensionless wall shear parameter $v_w(\equiv f''_w)$ are much larger in turbulent flows than laminar flows. Due to the rapid variation of the velocity profile close to the wall, it is necessary to make much smaller steps in η close to the wall.

This subroutine generates an η-grid which has the property that the ratios of lengths of any two adjacent intervals is a constant, that is, $h_j = Kh_{j-1}$ and the distance to the j-th line is given by

$$\eta_j = h_1 \frac{K_j - 1}{K - 1} \quad j = 1, 2, \ldots, J \qquad K > 1 \tag{3.4.3a}$$

There are two parameters: h_1, the length of the first $\Delta\eta$-step, and K, the variable grid parameter. The total number of points, J, is calculated from

$$J = \frac{\ln[1 + (K-1)(\eta_e/h_1)]}{\ln K} + 1 \tag{3.4.3a}$$

In practice, it is common to choose h_1 [DETA(1)] and K(VGP) so that, for an assumed maximum value of η_e(ETAE), the total number of j-points do not exceed the total number J(NP) that is specified in IBLP. Figure 3.3 can help in the selection of K and shows that, for example for $J = 61$ and $h_1 = 0.01$ the value of K must be about 1.1.

In general, it is sufficient to take specified values of h_1 and K corresponding to 0.01 and 1.10, respectively. For turbulent flows at modest Reynolds numbers, say up to 10^7, this choice of h_1 and K is usually reasonable if the transformed boundary-layer thickness η_e does not exceed 50. At higher Reynolds numbers,

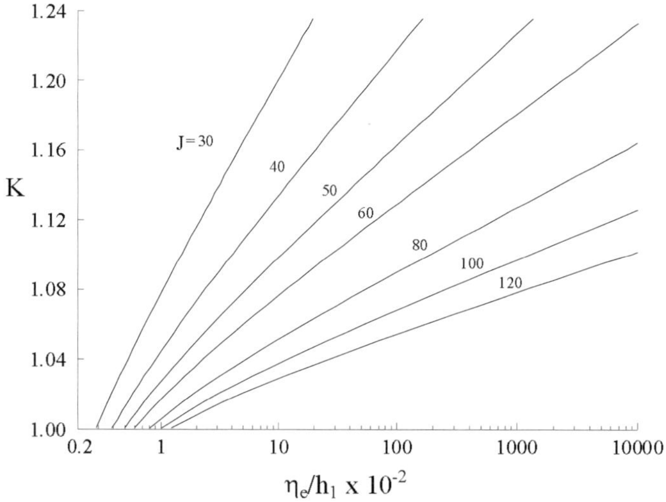

Fig. 3.3. Variation of J with K and η_e/h_1.

it is better to use smaller values of h_1 (around 0.001) and larger values of K (around 1.15 to 1.2) so that large variations in the slope of the profiles near the wall can be accurately represented. If the calculations are performed for laminar flow only, then it is best to choose a uniform grid with $K = 1$ and $h_1 = 0.20$ for $\eta_e = 8$.

The input to IBLP consists of NXT (total number of ξ-stations), location of ξ-station corresponding to trailing edge, NXTE, location of ξ-station where calculations switch from standard mode to inverse mode, NXS, total number of grid points J in the η-direction, NPT, maximum number of sweeps, ISWPT on the airfoil and wake, chord Reynolds number $R_L (\equiv u_\infty c/\nu)$, the ξ-station for transition, XTR, η-grid parameters η_e, K, h_1, defined by ETAE, VGP and DETA(1). In addition, the streamwise grid ξ, X(I), and dimensionless external velocity, UBL(I) are read in for all NXT-stations.

With dimensionless velocities $w(\xi)$ are known at each ξ-station, the pressure gradient parameters $m \equiv (P2)$ and $[(m+1)/2 \equiv (P1)]$ are computed from their definitions for all ξ-station except the first one at which m is specified, after the derivatives of $dw/d\xi = (DWDS)$ needed in the calculations of m is obtained by using three-point Lagrange interpolation formulas given by

$$\begin{aligned}\left(\frac{dw}{d\xi}\right)_n = & -\frac{w^{n-1}}{A_1}(\xi^{n+1} - \xi^n) + \frac{w^n}{A_2}(\xi^{n+1} - 2\xi^n + \xi^{n-1}) \\ & + \frac{w^{n+1}}{A_3}(\xi^n - \xi^{n-1})\end{aligned} \tag{3.4.4a}$$

Here

$$\begin{aligned} A_1 &= (\xi^n - \xi^{n-1})(\xi^{n+1} - \xi^{n-1}) \\ A_2 &= (\xi^n - \xi^{n-1})(\xi^{n+1} - \xi^n) \\ A_3 &= (\xi^{n+1} - \xi^n)(\xi^{n+1} - \xi^{n-1}) \end{aligned} \tag{3.4.4b}$$

The derivatives of $(dw/d\xi)$ at the end point $n = N$ is given by

$$\begin{aligned}\left(\frac{dw}{d\xi}\right)_N = & -\frac{w^{N-2}}{A_1}(\xi^N - \xi^{N-1}) - \frac{w^{N-1}}{A_2}(\xi^N - \xi^{N-2}) \\ & + \frac{w^N}{A_3}(2\xi^N - \xi^{N-2} - \xi^{N-1})\end{aligned} \tag{3.4.5a}$$

where now

$$\begin{aligned} A_1 &= (\xi^{N-1} - \xi^{N-2})(\xi^N - \xi^{N-2}) \\ A_2 &= (\xi^{N-1} - \xi^{N-2})(\xi^N - \xi^{N-1}) \\ A_3 &= (\xi^N - \xi^{N-1})(\xi^N - \xi^{N-2}) \end{aligned} \tag{3.4.5b}$$

If the ξ-station is greater than that corresponding to NXS, say $\xi = \xi^1$, so that the calculations are to be performed in the inverse mode (ITV$\neq$0), then the term in Eqs. (3.4.2a) and (3.4.2b) are calculated. The initial distribution of those terms is also calculated in this subroutine prior to the start of the inverse-flow calculation. This is done by assuming the flow to be turbulent and the displacement thickness distribution on the airfoil is given by the expression for a flat plate flow

$$\delta^*(\xi) = 0.046875\frac{\xi}{R_x^{0.2}} \tag{3.4.6}$$

Its distribution in the wake region is computed from

$$\delta^*(x) = \delta^*_{t.e.} \exp - \left(\frac{\xi_i - \xi_{t.e.}}{\xi_t - \xi_{t.e.}}\right) \tag{3.4.7}$$

Here subscript $t.e.$ denotes the trailing edge and t denotes the total number of ξ-stations.

```
      SUBROUTINE INPUT
      COMMON /BLGTYY/ P1(201),P2(201)
      COMMON /BLC001/ NX,NP,NXTE,NXT,NPT,NTR,NXS,IT,ITV,ISWP,ISWPT
      COMMON /BLGRD1/ X(201),ETA(101),A(101),DETA(101)
      COMMON /BLHLBS/ CC(201,201),DD(201),DB(201),GI
      COMMON /BLINP3/ RL, RX, SQRL, SQRX
      COMMON /BLEDY1/ GAMTR(201),FFS(201),ALFAS(201),EDV(101)
      COMMON /BLRC01/ XBL(201),UBL(201),CFS(201),THETA(201),
     1                DLS(201),UEVS(201),NPSAV(201),ITSAV(201)
      GRANG(X1,X2,X3,Y1,Y2,Y3,X0)= (2.*X0-X2-X3)/(X1-X2)/(X1-X3)*Y1
     +          +(2.*X0-X1-X3)/(X2-X1)/(X2-X3)*Y2+(2.*X0-X1-X2)/(X3-X1)
     +          /(X3-X2)*Y3
C - - - - - - - - - - - - - - - - - - - - - - - - - - - - - - - - - - -
      READ (5,*  ) NXT,NXTE,NXS,NPT,ISWPT
      READ (5,*  ) RL,XTR,ETAE,VGP,DETA(1)
      READ (5,*  ) (XBL(I),I=1,NXT)
      READ (5,*  ) (X(I),I=1,NXT)
      READ (5,*  ) (UBL(I),I=1,NXT)
      WRITE(6,6000) RL,XTR
      WRITE (6,6040) ETAE,VGP,DETA(1)
      IF((VGP-1.0) .GT. 0.001) THEN
        NP = ALOG((ETAE/DETA(1))*(VGP-1.0)+1.0)/ALOG(VGP)+1.0001
      ELSE
        NP = ETAE/DETA(1)+1.0001
      END IF
      IF(NP .GT. NPT) THEN
         WRITE (6,9000)
         STOP
      END IF
      ETA(1) = 0.0
      DO 20 J=2,NPT
        DETA(J)  = VGP*DETA(J-1)
        ETA(J)   = ETA(J-1)+DETA(J-1)
        A(J)     = 0.5*DETA(J-1)
   20 CONTINUE
C     DETERMINE TRANSITION LOCATION
```

```
      DO 30 I=1,NXT
         IF (XBL(I) .GE. XTR) GO TO 40
   30 CONTINUE
   40 NTR = I
C     CALCULATE GAMTR
      RXNTR   = X(NTR-1)* UBL(NTR-1) * RL
      GGFT    = RL**2/RXNTR**1.34*UBL(NTR-1)**3
      UEINTG  = 0.0
      U1      = 0.5/UBL(NTR-1)/ 1200.0
      DO 60 I = NTR,NXT
         U2      = 0.5/UBL(I)/1200.0
         UEINTG  = UEINTG+(U1+U2)*(X(I)-X(I-1))
         U1      = U2
         GG      = GGFT*UEINTG*(X(I)-X(NTR-1))
         IF(GG .GT. 10.0) THEN
            GAMTR(I)=1.0
         ELSE
            GAMTR(I)= 1.0-EXP(-GG)
         ENDIF
   60 CONTINUE
      DO 100 I=1,NXT
  100 UEVS(I) = UBL(I)
C  INITIAL DELTASTAR DISTRIBUTION
      SQRL = SQRT(RL)
      DO 200 I = 2,NXTE
            RX        = RL*X(I)
            DLS(I) = 0.046875*X(I)/RX**0.2
            DD(I)  = DLS(I)*SQRT(RL)
  200 CONTINUE
      DO 210 I=NXTE+1,NXT
      DD(I) = DD(NXTE)/EXP( (X(I)-X(NXTE))/(X(NXT)-X(NXTE)) )
  210 CONTINUE
      P2(1) = 1.0
      P1(1) = 1.0
      DO 240 I=2,NXTE
      DUDX = GRANG(X(I-1),X(I),X(I+1),UBL(I-1),UBL(I),UBL(I+1),X(I))
      P2(I) = X(I)/UBL(I)*DUDX
      P1(I) = 0.5*(1.0+P2(I))
  240 CONTINUE
      RETURN
 5000 FORMAT(A)
 6000 FORMAT(//5X,'RL =',E10.3,5X,'XCTR =',F7.3)
 6040 FORMAT('ETAE=',F8.3,3X,'VGP=',F7.3,3X,'DETA(1)=',F7.3//)
 9000 FORMAT('NP EXCEEDED NPT - PROGRAM TERMINATED')
      END
```

3.4.3 Subroutine IVPL

At $\xi = 0$, Eq. (3.2.3) reduces to the Falkner-Skan equation, which can be solved subject to the boundary conditions of Eq. (3.2.6). Since the equations are solved in linearized form, initial estimates of f_j, u_j and v_j are needed in order to obtain the solutions of the nonlinear Falkner-Skan equation. Various expressions can be used for this purpose. Since Newton's method is used, however, it is useful to provide as good an estimate as is possible and an expression of the form.

$$u_j = \frac{3}{2}\frac{\eta_j}{\eta_e} - \frac{1}{2}\left(\frac{\eta_j}{\eta_e}\right)^3 \tag{3.4.8}$$

usually satisfies this requirement. The above equation is obtained by assuming a third-order polynomial of the form

$$f' = a + b\eta + c\eta^3$$

and by determining constants a, b, c from the boundary conditions given by Eq. (3.2.6) for the zero-mass transfer case and from one of the properties of momentum equation which requires that $f'' = 0$ at $\eta = \eta_e$.

The other profiles f_j, v_j follow from Eq. (3.4.8) and can be written as

$$f_j = \frac{\eta_e}{4}\left(\frac{\eta_j}{\eta_e}\right)^2\left[3 - \frac{1}{2}\left(\frac{\eta_j}{\eta_e}\right)^2\right] \tag{3.4.9a}$$

$$v_j = \frac{3}{2}\frac{1}{\eta_e}\left[1 - \left(\frac{\eta_j}{\eta_e}\right)^2\right] \tag{3.4.9b}$$

```
      SUBROUTINE IVPL
      COMMON /BLC7/ ETA(101),DETA(101),A(101)
      COMMON /BLC2/ NX,NXT,NP,NPT,NTR,NVRS,IT,INVRS,ISWPT
      COMMON /BLC8/ F(101,2),U(101,2),V(101,2),W(101,2),B(101,2)
      DO 20  J=1,NP
         ETAB = ETA(J)/ETA(NP)
         ETAB2 = ETAB**2
         F(J,2) = 0.25*ETA(NP)*ETAB2*(3.0 - 0.5*ETAB2)
         U(J,2) = 0.5*ETAB*(3.0 - ETAB2)
         V(J,2) = 1.5*(1.0 - ETAB2)/ETA(NP)
         B(J,2) = 1.0
   20 CONTINUE
      RETURN
      END
```

3.4.4 Subroutine GROWTH

For most laminar-boundary-layer flows the transformed boundary-layer thickness $\eta_e(\xi)$ is almost constant. A value of $\eta_e = 8$ is sufficient. However, for turbulent boundary-layers, $\eta_e(\xi)$ generally increases with increasing ξ. An estimate of $\eta_e(\xi)$ is determined by the following procedure.

We always require that $\eta_e(\xi^n) \geq \eta_e(\xi^{n-1})$, and in fact the calculations start with $\eta_e(0) = \eta_e(\xi_1)$. When the computations on $\xi = \xi^n$ (for any $n \geq 1$) have been completed, we test to see if $v_J^n \le \varepsilon_v$ at $\eta_e(\xi^n)$ where, say $\varepsilon_\nu = 5 \times 10^{-4}$. This test is done in MAIN. If this test is satisfied, we set $\eta_e(\xi^{n+1}) = \eta_e(\xi^n)$. Otherwise, we call GROWTH and set $J_{\text{new}} = J_{\text{old}} + t$, where t is a number of points, say $t = 2$. In this case we also specify values of $(f_j^n, u_j^n, v_j^n, b_j^n)$ for the new n_j points. We take the values of $u_j = 1$, $v_j^n = 0$, $f_j^n = (\eta_j - \eta_e)u_J^n + f_J^n$, and $b_j^n = b_J^n$. This is also done for the values of f_j^{n-1}, v_j^{n-1} and b_j^{n-1}.

```
      SUBROUTINE GROWTH(INDEX)
      COMMON /BLC7/ ETA(101),DETA(101),A(101)
      COMMON /BLC2/ NX,NXT,NP,NPT,NTR,NVRS,IT,INVRS,ISWPT
      COMMON /BLC8/ F(101,2),U(101,2),V(101,2),W(101,2),B(101,2)
      NP1 = NP + 1
      IF (INDEX.EQ.1) THEN
      NP = MIN0(NPT,NP+2)
      NPEND = NP
      ELSE
      NPEND = NPT
      ENDIF
      DO 30 J=NP1,NPEND
      F(J,2) = F(J-1,2) + DETA(J-1)*U(J-1,2)
      U(J,2) = U(J-1,2)
      V(J,2) = 0.0
      B(J,2) = B(J-1,2)
   30 CONTINUE
      RETURN
      END
```

3.4.5 Subroutine HIC

This subroutine calculates the coefficients of the Hilbert integral denoted by C_{ij}. While they can be generated from any suitable integration procedure, we use the following procedure which is appropriate with the box method.

We evaluate

$$H_i = \int_{\xi^i}^{\xi^L} G(\sigma) \frac{d\sigma}{\xi^i - \sigma} \tag{3.4.10}$$

where

$$G(\sigma) = \frac{dF}{d\sigma}$$

with F denoting any function, so that over each subinterval (ξ^{n-1}, ξ^n), except the two enclosing the point $\xi = \xi^i$, replace $G(\sigma)$ by its midpoint value:

$$\int_{\xi^{n-1}}^{\xi^n} \frac{G(s)d\sigma}{\xi^i - \sigma} = G_{n-1/2} \int_{\xi^{n-1}}^{\xi^n} \frac{d\sigma}{\xi^i - \sigma} = G_{n-1/2} \ln \left| \frac{\xi^i - \xi^{n-1}}{\xi^i - \xi^n} \right| \tag{3.4.11}$$

Making the further approximation,

$$G_{n-1/2} = \frac{F_n - F_{n-1}}{\xi^n - \xi^{n-1}}$$

we can write

$$\int_{\xi^{n-1}}^{\xi^n} \frac{dF}{d\sigma}\frac{d\sigma}{\xi^i - \sigma} = E_n^i(F_n - F_{n-1}) \tag{3.4.12}$$

where for $n \neq i$ or $i+1$

$$E_n^i = (\xi^n - \xi^{n-1})^{-1} \ln\left|\frac{\xi^i - \xi^{n-1}}{\xi^i - \xi^n}\right| \tag{3.4.13}$$

for the two subintervals ξ^{i-1} to ξ^i and ξ^I to ξ^{I-1}, because of the cancellation with the constant term, account should be taken of the linear variation of G from one interval to the next. Thus, we take the linear interpolation

$$G = \frac{G_{i-1/2}(\xi^{i+1} - \xi^i) + G_{i+1/2}(\xi^i - \xi^{i-1}) + 2(G_{i+1/2} - G_{i-1/2})(\sigma - \xi^i)}{\xi^{i+1} - \xi^{i-1}}$$

so that

$$\int_{\xi^{i-1}}^{\xi^{i+1}} \frac{G d\sigma}{\xi^i - \sigma} = \frac{G_{i-1/2}(\xi^{i+1} - \xi^i) + G_{i+1/2}(\xi^i - \xi^{i-1})}{\xi^{i+1} - \xi^{i-1}} \ln\left|\frac{\xi^i - \xi^{i-1}}{\xi^i - \xi^{i+1}}\right| - 2(G_{i+1/2} - G_{i-1/2}) \tag{3.4.14}$$

Replacing the midpoint derivative values by difference quotients

$$\int_{\xi^{i-1}}^{\xi^{i+1}} \frac{dF}{d\sigma}\frac{d\sigma}{\xi^i - \sigma} = E_1^i(F_i - F_{i-1}) + E_{i+1}^i(F_{i+1} - F_i) \tag{3.4.15}$$

where

$$E_1^i = \frac{\dfrac{\xi^{i+1} - \xi^i}{\xi^{i+1} - \xi^{i-1}} \ln\left|\dfrac{\xi^i - \xi^{i-1}}{\xi^i - \xi^{i+1}}\right| + 2}{\xi^i - \xi^{i-1}} \tag{3.4.16a}$$

$$E_{i+1}^i = \frac{\dfrac{\xi^i - \xi^{i-1}}{\xi^{i+1} - \xi^{i-1}} \ln\left|\dfrac{\xi^i - \xi^{i-1}}{\xi^i - \xi^{i+1}}\right| - 2}{\xi^{i+1} - \xi^i} \tag{3.4.16b}$$

Thus

$$\begin{aligned} H_i &= E_2^i(F_2 - F_1) + E_3^i(F_3 - F_2) + \ldots + E_{L-1}^i(F_{L-1} - F_{L-2}) + E_L^i(F_L - F) \\ &= -E_2^i F_1 + (E_2^i - E_3^i)F_2 + \ldots + (E_{L-1}^i - E_L^i)F_{L-1} + E_L^i F_L \end{aligned} \tag{3.4.17}$$

so, finally the C_{ij} of Eq. (3.2.9) are given by

$$C_{ij} = \frac{1}{\pi}(E_j^i - E_{j+1}^i) \tag{3.4.18}$$

and the E^i are given by Eqs. (3.4.11) and (3.4.14) with $E_1^i = E_{L+1}^i = 0$.

```
      SUBROUTINE HIC(NXT,RL,X)
      COMMON /BLHLBS/ CC(201,201),DD(201),DB(201),GI
      DIMENSION       X(*),E(201)
C - - - - - - - - - - - - - - - - - - - - - - - - - - - - - - - - - - -
      PI      = 4. * ATAN(1.0 )
      PISQRL  = PI * SQRT( RL )
      DO 65 I = 2, NXT
         E (1)    = 0.
         DO 60 J= 2, I-1
            E(J)= ALOG(ABS((X(I)-X(J-1))/(X(I)-X(J))))
     1            /(X(J)-X(J-1))
   60    CONTINUE
         J      = I
         R1     = (X(J+1) -X(J)) /(X(J+1) - X(J-1))
         E(J)   = (R1*ALOG(ABS((X(I)-X(J-1))/(X(I)-X(J+1))))+2.0)
     1            /(X(J) - X(J-1))
         J      = I + 1
         R1     = (X(J-1) - X(J-2)) / (X(J) - X(J-2))
         E(J  ) = (R1*ALOG(ABS((X(I)-X(J-2))/(X(I)-X(J))))-2.0)
     1            /(X(J) - X(J-1))
         DO 61 J= I+2,NXT+1
            E(J)= ALOG(ABS((X(I)-X(J-1))/(X(I)-X(J))))
     1            /(X(J) - X(J-1))
   61    CONTINUE
         DO 62 J =2,NXT+1
            CC(J-1,I) = ( E(J-1) - E(J) ) / PISQRL
   62    CONTINUE
         CC(NXT+1,I) = E(NXT+1) / PISQRL
   65 CONTINUE
      RETURN
      END
```

3.4.6 Subroutine EDDY

A simplified form of the eddy-viscosity formulation of Cebeci and Smith (subsection 3.1.4) is used to define the turbulence model in the momentum equation for the airfoil flow

$$\nu_t = \begin{cases} (\nu_t)_i = \ell^2 \left| \dfrac{\partial u}{\partial y} \right| \gamma_{\text{tr}} \\ (\nu_t)_0 = 0.0168 u_e \delta^* \gamma_{\text{tr}} \gamma \end{cases} \tag{3.4.19}$$

In terms of the Falkner-Skan variables defined by Eq. (3.2.1) or those defined by Eq. (3.2.2), the transformed eddy viscosity formulas can be written as

$$\nu_t^+ = \begin{cases} (\nu_t^+)_i = 0.16 R_x^{1/2} \left[1 - \exp\left(-\dfrac{y}{A} \right) \right]^2 \eta^2 v \gamma_{\text{tr}} \\ (\nu_t^+)_0 = 0.168 R_x^{1/2} (\eta_e C - f_e) \gamma_{\text{tr}} \gamma \end{cases} \tag{3.4.20}$$

where $C = 1$, for the standard mode (ITV = 1) and $C = w$ ($\equiv u_e/u_\equiv$) for the inverse mode with y/A and R_x defined by

$$\frac{y}{A} = \frac{R_x^{1/2}}{26}, \quad R_x = \frac{Vx}{\nu} \tag{3.4.21}$$

Here $V = u_e$ for ITV = 1 and $V = u_\infty$ for ITV$\neq$1.

Note that the expression for the transition region between a laminar and turbulent flow, γ_{tr}, is defined in subroutine INPUT since it is a function of ξ only. The parameter γ denotes the intermittency expression defined by Eq. (3.1.34).

For the wake region, the turbulence model is

$$\nu_t = (\nu_t)_w + [(\nu_t)_{te} - (\nu_t)_w] \exp\left[-\frac{x - x_{te}}{20\delta_{te}}\right] \tag{3.4.22}$$

Here $(\nu_t)_{te}$ denotes the eddy viscosity at the trailing edge computed from (Eq. 3.4.17) and $(\nu_t)_w$ is the eddy viscosity in the far wake given by

$$(\nu_t)_w = 0.064 \int_0^\infty (u_e - u)dy \tag{3.4.23}$$

In terms of dimensionless quantities, Eqs. (3.4.22) and (3.4.23) become

$$\nu_t^+ = (\nu_t^+)_w + [(\nu_t^+)_{te} - (\nu_t^+)_w] \exp\left[-\frac{x - x_{te}}{20\delta_{te}}\right] \tag{3.4.24}$$

$$(\nu_t^+)_w = 0.064\sqrt{\xi R_L}(\eta_e C - f_e) \tag{3.4.25}$$

```
      SUBROUTINE EDDY
      COMMON /BLC001/ NX,NP,NXTE,NXT,NPT,NTR,NXS,IT,ITV,ISWP,ISWPT
      COMMON /BLPRC1/ F(101,2),U(101,2),V(101,2),W(101,2),B(101,2)
      COMMON /BLEDY1/ GAMTR(201),FFS(201),ALFAS(201),EDV(101)
      COMMON /BLGRD1/ X(201),ETA(101),A(101),DETA(101)
      COMMON /BLINP3/ RL, RX, SQRL, SQRX
      COMMON /BLCWAK/ EDSAVE(101),YTEDGE(101),WSCALE,YDELWK
      DIMENSION        FINT(101),YY(101),EDINTP(101)
C     ------------------------------------------------------
      UDEL    = 0.995 * U(NP,2)
      DO 15  J = 2 , NP
         JJ    = J
         IF (U(J,2) .GT.UDEL) GOTO 20
   15 CONTINUE
   20 EDEL    = ETA(JJ-1)+(ETA(JJ)-ETA(JJ-1))/(U(JJ,2)-U(JJ-1,2))*
     1           (UDEL-U(JJ-1,2))
      DO 25 J=1,NP
      FINT(J)  = 1./(1.0+5.5*(ETA(J)/EDEL)**6)
   25 CONTINUE
      IF(NX .GT. NXTE) GO TO 260
      IF(ITV .NE. 0) THEN
```

```
         SQREY = SQRX
      ELSE
         SQXC  = SQRT(X(NX))
         SQREY = SQXC * SQRL
      ENDIF
      CYA   = SQRT(SQREY)/26.0
      EDVIIC= 0.16 * SQREY * GAMTR(NX)
      CEDV  = SQREY*GAMTR(NX) * U(NP,2)
      EDVOO = 0.0168*(ETA(NP)-F(NP,2)/U(NP,2))*CEDV
      DO 168 J = 1,NP
         YOA   = CYA*SQRT(ABS(V(1,2)))*ETA(J)
         IF(YOA .LT. 10.0) EL    = 1.0 - EXP(-YOA)
         EDV(J)= EDVIIC*ABS(V(J,2))*(ETA(J)*EL)**2
  168 CONTINUE
      DO 175 J = 1,NP
         JJ    = J
         IF(EDV(J).GT.EDVOO) GOTO 190
  175 CONTINUE
  190 DO 195 J = JJ,NP
         EDV(J)= EDVOO
  195 CONTINUE
      DO 210 J=1,NP
         EDV(J) = EDV(J) * FINT(J)
         B(J,2) = 1.0+EDV(J)
  210 CONTINUE
      RETURN
  260 CONTINUE
      SQXC  = SQRT(X(NX))
      SQREY = SQXC * SQRL
      EDVOO  = 0.064 * (ETA(NP)-F(NP,2)/U(NP,2)) * SQREY * U(NP,2)
      CEXPT  = AMIN1(20.,(X(NX)-X(NXTE))/(WSCALE*YDELWK))
      EXPT   = 1.0 - EXP(-CEXPT)
      YY(1) = 0.0
      YOETA = X(NX) / SQREY
      DO 330 J = 2, NP
         YY(J)  = YOETA * ETA(J)
  330 CONTINUE
      CALL LNTP(NPT,YTEDGE,EDSAVE,NP,YY,EDINTP)
      DO 340 J = 1,NP
         EDVWJ = EDVOO * FINT(J)
         EDV(J)= AMAX1(0.0,EDINTP(J)+(EDVWJ-EDINTP(J)) * EXPT)
         B(J,2)= 1.0+EDV(J)
  340 CONTINUE
      RETURN
      END
```

3.4.7 Subroutine COEF

The main purpose of this subroutine is to calculate the coefficients of the linearized momentum equation expressed in Falkner-Skan or in modified Falkner-

Skan variables with the coefficients and the boundary conditions corresponding to those either in the standard (ITV = 1) or inverse modes.

```
      SUBROUTINE COEF
      COMMON /BLC001/ NX,NP,NXTE,NXT,NPT,NTR,NXS,IT,ITV,ISWP,ISWPT
      COMMON /BLPRC1/ F(101,2),U(101,2),V(101,2),W(101,2),B(101,2)
      COMMON /BLCS30/ S1(101),S2(101),S3(101),S4(101),S5(101),S6(101),
     1                S7(101),S8(101),R1(101),R2(101),R3(101)
      COMMON /BLCS31/ GAMMA1,GAMMA2
      COMMON /BLGRD1/ X(201),ETA(101),A(101),DETA(101)
      COMMON /BLGTYY/ P1(201),P2(201)
      COMMON /BLHLBS/ CC(201,201),DD(201),DB(201),GI
C - - - - - - - - - - - - - - - - - - - - - - - - - - - - - - - - - - -
      IF(IT .GT. 1) GO TO 5
      CEL = 0.0
      IF(NX .GT. 1) CEL = 0.5*(X(NX)+X(NX-1))/(X(NX)-X(NX-1))
      L1  = ITV
      L2  = 1 - ITV
      P1P = L1*P1(NX) + CEL + 0.5*L2
5     CONTINUE
      DO 100 J=2,NP
C  PRESENT STATION
        USB   = 0.5*(U(J,2)**2+U(J-1,2)**2)
        UB    = 0.5*(U(J,2)+U(J-1,2))
        VB    = 0.5*(V(J,2)+V(J-1,2))
        FB    = 0.5*(F(J,2)+F(J-1,2))
        FVB   = 0.5*(F(J,2)*V(J,2)+F(J-1,2)*V(J-1,2))
        DERBV = (B(J,2)*V(J,2)-B(J-1,2)*V(J-1,2))/DETA(J-1)
        FLARE = 1.0
        IF(UB .LT. 0.0) FLARE = 0.0
        P2P = L1*P2(NX) + CEL*FLARE
        IF(NX.EQ.1) THEN
          CRB = -P2(NX)
        ELSE
C   PREVIOUS STATION
          CFB   = 0.5*(F(J,1)+F(J-1,1))
          CVB   = 0.5*(V(J,1)+V(J-1,1))
          CUSB  = 0.5*(U(J,1)**2+U(J-1,1)**2)
          CFVB  = 0.5*(F(J,1)*V(J,1)+F(J-1,1)*V(J-1,1))
          CDERBV= (B(J,1)*V(J,1)-B(J-1,1)*V(J-1,1))/DETA(J-1)
          CLB = CDERBV + L1*(P1(NX-1)*CFVB + P2(NX-1)*(1.0-CUSB)) +
     &          0.5*L2*CFVB
          CRB = -CLB - L1*P2(NX) - CEL*FLARE*CUSB + CEL*CFVB
        ENDIF
C  DEFINITIONS OF Sj
        S1(J) = B(J,2)/DETA(J-1) + 0.5*P1P*F(J,2) -0.5*CEL*CFB
        S2(J) = -B(J-1,2)/DETA(J-1) + 0.5*P1P*F(J-1,2) - 0.5*CEL*CFB
        S3(J) = 0.5*(P1P*V(J,2) + CEL*CVB)
        S4(J) = 0.5*(P1P*V(J-1,2) + CEL*CVB)
        S5(J) = -P2P*U(J,2)
        S6(J) = -P2P*U(J-1,2)
```

```
         S7(J) = L2*CEL*U(NP,2)
         S8(J) = L2*CEL*U(NP,2)
C  DEFINITIONS OF RJ
         R1(J) = CRB - (DERBV + P1P*FVB - P2P*USB + L2*CEL*
     &                (U(NP,2)**2-U(NP,1)**2) + CEL*(FB*CVB-VB*CFB))
         R2(J-1)= F(J-1,2) - F(J,2) + UB*DETA(J-1)
         R3(J-1)= U(J-1,2) - U(J,2) + VB*DETA(J-1)
  100 CONTINUE
C  BOUNDARY CONDITIONS
       R1(1)  = 0.0
       R2(NP) = 0.0
       IF (ITV .EQ. 1 ) THEN
C  STANDARD MODE
          GAMMA1 = 0.0
          GAMMA2 = 1.0
          R3(NP) = 0.0
       ELSE
C  INVERSE MODE
          GAMMA1 = CC(NX,NX) * SQRT(X(NX))
          GAMMA2 = 1.0 - ETA(NP) * GAMMA1
          R3(NP) = GI - (GAMMA2*U(NP,2) +GAMMA1*F(NP,2))
       ENDIF
       RETURN
       END
```

3.4.8 Subroutine SWTCH

This subroutine switches the variables f, f', f'' which are based on the F.S. transformation

$$\eta = y\sqrt{\frac{u_e}{\nu x}}, \quad f' = \frac{u}{u_e} \tag{3.4.26}$$

to the variables F, F', F'' which are based on

$$\eta = y\sqrt{\frac{u_0}{\nu x}}, \quad F' = \frac{u}{u_0} \tag{3.4.27}$$

The latter is used for the inverse problem.

```
      SUBROUTINE SWTCH(INDEX)
      COMMON /BLC001/ NX,NP,NXTE,NXT,NPT,NTR,NXS,IT,ITV,ISWP,ISWPT
      COMMON /BLPRC1/ F(101,2),U(101,2),V(101,2),W(101,2),B(101,2)
      COMMON /BLRC01/ XBL(201),UBL(201),CFS(201),THETA(201),
     1                DLS(201),UEVS(201),NPSAV(201),ITSAV(201)
      COMMON /BLGRD1/ X(201),ETA(101),A(101),DETA(101)
      DIMENSION       FSAVE(101),USAVE(101),VSAVE(101),BSAVE(101)
C     ------------------------------------------------------------
      IF(INDEX.GT.0) GOTO 40
      WNP   = UBL(NX)
      AK    = SQRT(1.0/WNP)
      DO 20 J=1,NP
```

```
         U(J,2) = WNP*U(J,2)
         V(J,2) = WNP*V(J,2)/AK
         F(J,2) = WNP*F(J,2)*AK
         FSAVE(J)  = F(J,2)
         USAVE(J)  = U(J,2)
         VSAVE(J)  = V(J,2)
         BSAVE(J)  = B(J,2)
   20 CONTINUE
      DO 30 J= 1,NPT
         ETA(J) = AK*ETA(J)
   30 CONTINUE
      DO 35 J=2,NPT
         DETA(J-1)= ETA(J)-ETA(J-1)
         A(J)  = 0.5*DETA(J-1)
   35 CONTINUE
      NPSAVE = NP
      RETURN
   40 NP      = NPSAVE
      DO 50 J=1,NP
         F(J,2) =  FSAVE(J)
         U(J,2) =  USAVE(J)
         V(J,2) =  VSAVE(J)
         B(J,2) = BSAVE(J)
   50 CONTINUE
      RETURN
      END
```

3.4.9 Subroutine WAKEPR

The profile at the airfoil trailing edge is used as the initial profile for the wake by setting $f''(0) = 0$. The profiles f, f' near the airfoil surface (or wake centerline) are modified to match the new condition $f''(0) = 0$. At first, the f' profile is modified in a small region near the wake centerline so that its value at the centerline is very small, but not zero. To ensure that the f and f'' profiles are smooth, f is recalculated by integrating f', and f'' is recalculated by differentiating f'.

The boundary-layer thickness δ is calculated and saved for later use in the eddy viscosity for the wake region as defined by Eq. (3.4.22).

```
      SUBROUTINE WAKEPR
      COMMON /BLC001/ NX,NP,NXTE,NXT,NPT,NTR,NXS,IT,ITV,ISWP,ISWPT
      COMMON /BLPRC1/ F(101,2),U(101,2),V(101,2),W(101,2),B(101,2)
      COMMON /BLEDY1/ GAMTR(201),FFS(201),ALFAS(201),EDV(101)
      COMMON /BLINP3/ RL, RX, SQRL, SQRX
      COMMON /BLINP8/ D1(401),D2(401),D3(401),D4(401)
      COMMON /BLGRD1/ X(201),ETA(101),A(101),DETA(101)
      COMMON /BLCWAK/ EDSAVE(101),YTEDGE(101),WSCALE,YDELWK
      DIMENSION       Y(201),UU(201),VV(201)
C - - - - - - - - - - - - - - - - - - - - - - - - - - - - - - - - - - -
```

```
      IF (NX.GE.NTR) THEN
        DO 20 J=1,NP
          EDV(J) = EDV(J) /GAMTR(NX)
 20     CONTINUE
      ELSE
        GAMTR(NX) = 1.0
        NTR       = NX
        IT        = 1
        CALL EDDY
        DO 30 J =1,NP
          EDSAVE(J) = EDV(J)
          B(J,2) = 1.0
 30     CONTINUE
        GAMTR(NX) = 0.0
        NTR       = NX + 1
      ENDIF
      DO 40 J=1,NP
         JJ     = NP-1+J
         UU(JJ)= U (J,2)
         Y(JJ) = ETA(J)
 40   CONTINUE
      DO 50 J = 2,NP
         JJ     = NP-J+1
         UU(JJ) = U (J,2)
         Y(JJ)  = -ETA(J)
 50   CONTINUE
      JN     = 2*NP-1
      CALL AMEAN(NP-3,NP+3,Y,UU,D1,2)
      CALL DIFF1(JN,Y,UU,VV,D1)
      DO 60 JJ= NP,JN
         J        = JJ - NP + 1
         U (J ,2)= UU(JJ)
         V (J ,2)= VV(JJ)
         EDSAVE(J)= EDV(J)
 60   CONTINUE
      EDSANP  = EDSAVE(NP)
      DO 65 J = NP,NPT
         EDSAVE(J)= EDSANP
 65   CONTINUE
      V(1,2)  = 0.0
      CALL INTEG(NP ,ETA, U(1,2), V(1,2), F(1,2))
      YTEDGE(1 )  = 0.0
      CY     = (1.0/SQRL) *  SQRT(X(NX) )
      UDEL   = 0.995 * U(NP,2)
      DO 230 J = 2,NP
         YTEDGE(J)  = CY * ETA(J)
         IF ( U(J,2) .LT. UDEL ) JJ  = J
230   CONTINUE
      DO 240 J = NP+1,NPT
         YTEDGE(J) = CY * ETA(J)
240   CONTINUE
```

```
      YDELWK = YTEDGE(JJ)+(YTEDGE(JJ+1)-YTEDGE(JJ))/(U(JJ+1,2)-U(JJ,2))*
     1         (UDEL-U(JJ,2))
      WSCALE  = 20.0
      RETURN
      END
```

3.4.10 Subroutine INTEG

This is a general integration routine.

```
      SUBROUTINE INTEG (N,X,F,FP,H)
      DIMENSION X(N),F(N),FP(N),H(N)
C - - - - - - - - - - - - - - - - - - - - - - - - - - - - - - - - - - -
      H(1)=0.0
      DO 10 I=2,N
         I1  = I-1
         DX  = X(I)-X(I1)
         DF  = F(I)-F(I1)
         DFP = FP(I)-FP(I1)
         FPF =FP(I)+FP(I1)
         DX2 =DX*DX/2.0
         DX3 =DX2*DX/3.0
         DX4 =DX3*DX/4.0
         FPPP=(-2.*DF+DX*FPF)/DX3
         FPP =DFP/DX - DX*FPPP/2.0
         H(I)=H(I1) + F(I1)*DX + FP(I1)*DX2 + FPP*DX3 + FPPP*DX4
   10 CONTINUE
      RETURN
      END
```

3.4.11 Subroutine DIFF1

This is a general differentiation routine.

```
      SUBROUTINE DIFF1 (N,X,F,FP,FPP)
      DIMENSION X(N),F(N),FP(N),FPP(N)
C - - - - - - - - - - - - - - - - - - - - - - - - - - - - - - - - - - -
      DO 5 I = 1,N-1
         DX     = X(I+1)-X(I)
         DF     = F(I+1)-F(I)
         FPP(I)= ATAN2(DF,DX)
    5 CONTINUE
      DO 10 I  = 2,N-1
         DX1   = X(I) - X(I-1)
         DX2   = X(I+1) - X(I)
         ANG   = (DX2*FPP(I-1)+DX1*FPP(I))/(DX1+ DX2)
         FP(I)  = TAN(ANG)
   10 CONTINUE
      FP(1)   = 2.*TAN(FPP(1)) - FP(2)
      FP(N)   = 2.*TAN(FPP(N-1))- FP(N-1)
```

```
      RETURN
      END
```

3.4.12 Subroutine LNTP

This is a general linear interpolation routine.

```
      SUBROUTINE LNTP(NN,XN,YN,NO,XO,YO)
      DIMENSION XN(NN),YN(NN),XO(NO),YO(NO)
C - - - - - - - - - - - - - - - - - - - - - - - - - - - - - - - - - - -
      IF(NN.LE.1) RETURN
      JS    = 2
      JJ    = 2
      DO 100 I=1,NO
         DO 50  J=JS,NN
            JJ      = J
            IF(XN(J).GE.XO(I)) GOTO 75
 50      CONTINUE
 75      YO(I)    = YN(JJ-1) + (YN(JJ)-YN(JJ-1))/(XN(JJ)-XN(JJ-1))*
     1              (XO(I)-XN(JJ-1))
         JS       = JJ
 100  CONTINUE
      RETURN
      END
```

3.4.13 Subroutine AMEAN

This is a general smoothing routine using average values.

```
      SUBROUTINE AMEAN(NS,ND,X,Y,D,IT)
      DIMENSION X(ND),Y(ND),D(ND)
C - - - - - - - - - - - - - - - - - - - - - - - - - - - - - - - - - - -
      IF((ND-NS).GT.2 .AND. IT.GT.0) THEN
         DO 20 K=1,IT
            DO 10 I = NS,ND
               D(I) = Y(I)
 10         CONTINUE
            DO 15 I=NS+1,ND-1
               F1    = (X(I + 1) -X(I))/(X(I+1) - X(I-1))
               Y(I)  = 0.5*(F1*D(I-1)+D(I)+(1.0-F1)*D(I+1))
 15         CONTINUE
 20      CONTINUE
      ENDIF
      RETURN
      END
```

3.4.14 Subroutine BORALG

```
      SUBROUTINE BORALG
C     BORDERING ALGORITHM
      COMMON /BLCS31/ GAMMA1,GAMMA2
      COMMON /BLC001/ NX,NP,NXTE,NXT,NPT,NTR,NXS,IT,ITV,ISWP,ISWPT
      COMMON /BLGRD1/ X(201),ETA(101),A(101),DETA(101)
      COMMON /BLC2SO/ DELF(101),DELU(101),DELV(101),DELW(101)
      COMMON /BLPRC1/ F(101,2),U(101,2),V(101,2),W(101,2),B(101,2)
      COMMON /BLCS30/ S1(101),S2(101),S3(101),S4(101),S5(101),S6(101),
     1                S7(101),S8(101),R1(101),R2(101),R3(101)
      COMMON /BLCAA4/ A11(101),A12(101),A13(101),A14(101),
     1                A21(101),A22(101),A23(101),A24(101),
     1                A31(101),A32(101),A33(101),A34(101)
      DIMENSION W1(101,4),W2(101,4),W3(101,4),V1(101,4),V2(101,4),
     1          V3(101,4),G11(101),G12(101),G13(101)
C - - - - - - - - - - - - - - - - - - - - - - - - - - - - - - - - - - -
      NPM1=NP-1
      NPM2=NP-2
      A11(1)=1.0
      A12(1)=0.0
      A13(1)=0.0
      DO 10 J=1,NPM1
        DO 20 K=2,4
          V1(J,K)=0.
          V2(J,K)=0.
          V3(J,K)=0.
   20   CONTINUE
        V1(J,1)=R1(J)
        V2(J,1)=R2(J)
        V3(J,1)=R3(J)
        V1(J,3)=S7(J)+S8(J)
   10 CONTINUE
      V1(1,3)=0.0
      DO 100 J=2,NPM1
        DEN=A(J)*A(J)*A11(J-1)-A(J)*A12(J-1)+A13(J-1)
        G11(J)=(A(J)*A(J)*S4(J)-A(J)*S6(J)+S2(J))/DEN
        G12(J)=A11(J-1)*G11(J)-S4(J)
        G13(J)=A12(J-1)*G11(J)-A(J)*G12(J)-S6(J)
        A11(J)=S3(J)-G12(J)
        A12(J)=S5(J)+A(J)*G12(J)-G13(J)
        A13(J)=S1(J)+A(J)*G13(J)
        V1(J,1)=V1(J,1)-G11(J)*V1(J-1,1)-G12(J)*V2(J-1,1)
     1                 -G13(J)*V3(J-1,1)
        V1(J,3)=V1(J,3)-G11(J)*V1(J-1,3)-G12(J)*V2(J-1,3)
     1                 -G13(J)*V3(J-1,3)
  100 CONTINUE
      V2(NPM1,2)=1.
      V2(NPM1,3)=-A(NP)
      V3(NPM1,3)=1.
      V3(NPM1,4)=-A(NP)
C  BACKWARD SWEEP
```

```
      DO 150 K=1,4
        J=NPM1
        DEN=A(J+1)*A(J+1)*A11(J)-A(J+1)*A12(J)+A13(J)
        W2(J,K)=-(A(J+1)*A11(J)*V2(J,K)+A(J+1)*V1(J,K)+A13(J)*V3(J,K))
     1         /DEN
        W1(J,K)=-(V2(J,K)+A(J+1)*W2(J,K))
        W3(J,K)=-(V3(J,K)+W2(J,K))/A(J+1)
        DO 150 JJ=1,NPM2
          J=NPM1-JJ
          V2(J,K)=V2(J,K)-W1(J+1,K)+A(J+1)*W2(J+1,K)
          V3(J,K)=V3(J,K)-W2(J+1,K)+A(J+1)*W3(J+1,K)
          DEN=A(J+1)*A(J+1)*A11(J)-A(J+1)*A12(J)+A13(J)
          W2(J,K)=-     (A(J+1)*A11(J)*V2(J,K)+A(J+1)*V1(J,K)+A13(J)*V3(J,K))
     1         /DEN
          W1(J,K)=-(V2(J,K)+A(J+1)*W2(J,K))
          W3(J,K)=-(V3(J,K)+W2(J,K))/A(J+1)
  150 CONTINUE
      A11(NP)=S3(NP)-S4(NP)*W1(NPM1,2)-S6(NP)*W2(NPM1,2)-S2(NP)*
     1         W3(NPM1,2)
      A12(NP)=S5(NP)+S7(NP)+S8(NP)-S4(NP)*W1(NPM1,3)-S6(NP)*W2(NPM1,3)-
     1         S2(NP)*W3(NPM1,3)
      A13(NP)=S1(NP)-S4(NP)*W1(NPM1,4)-S6(NP)*W2(NPM1,4)-S2(NP)*
     1         W3(NPM1,4)
      R1(NP)=R1(NP)-S4(NP)*W1(NPM1,1)-S6(NP)*W2(NPM1,1)-S2(NP)*
     1         W3(NPM1,1)
      IF (NX .LE. NXTE) THEN
        R2(NP)=R2(NP)-W2(1,1)
        DEN=A13(NP)*(GAMMA1*W2(1,3)-GAMMA2*W2(1,2))
     1     +W2(1,4)*(GAMMA2*A11(NP)-GAMMA1*A12(NP))
        DELF(NP)=(A13(NP)*(GAMMA2*R2(NP)+R3(NP)*W2(1,3))
     1           +W2(1,4)*(GAMMA2*R1(NP)-A12(NP)*R3(NP)))/DEN
      ELSE
        R2(NP)=R2(NP)-W3(1,1)
        DEN=A13(NP)*(GAMMA1*W3(1,3)-GAMMA2*W3(1,2))
     1     +W3(1,4)*(GAMMA2*A11(NP)-GAMMA1*A12(NP))
        DELF(NP)=(A13(NP)*(GAMMA2*R2(NP)+R3(NP)*W3(1,3))
     1           +W3(1,4)*(GAMMA2*R1(NP)-A12(NP)*R3(NP)))/DEN
      END IF
      DELU(NP)=(R3(NP)-GAMMA1*DELF(NP))/GAMMA2
      DELV(NP)=(R1(NP)-A11(NP)*DELF(NP)-A12(NP)*DELU(NP))/A13(NP)
      DO 160 J=1,NPM1
      DELF(J)=W1(J,1)-DELF(NP)*W1(J,2)-DELU(NP)*W1(J,3)-DELV(NP)*W1(J,4)
      DELU(J)=W2(J,1)-DELF(NP)*W2(J,2)-DELU(NP)*W2(J,3)-DELV(NP)*W2(J,4)
      DELV(J)=W3(J,1)-DELF(NP)*W3(J,2)-DELU(NP)*W3(J,3)-DELV(NP)*W3(J,4)
  160 CONTINUE
      DELF(1)=0.
      UMIN=U(2,2)
      IF(IT.LT.5.AND.UMIN.GT.0.0) THEN
         DO 190 J = 1,NP
            F(J,2) = F(J,2)+DELF(J)
            U(J,2) = U(J,2)+DELU(J)
```

```
            V(J,2) = V(J,2)+DELV(J)
190      CONTINUE
      ELSE
         DO 200 J=1,NP
            F(J,2)=F(J,2)+0.5*DELF(J)
            U(J,2)=U(J,2)+0.5*DELU(J)
            V(J,2)=V(J,2)+0.5*DELV(J)
200      CONTINUE
      ENDIF
      RETURN
      END
```

3.4.15 Subroutine OUTPUT

This subroutine computes the boundary-layer parameters R_θ, δ^*, θ, R_{δ^*}, and c_f in standard (ITV$\neq$0) and inverse (ITV $=$ 0) modes together with $\vec{D}_j$ used in SUM1 and SUM2.

```
      SUBROUTINE OUTPUT
      COMMON /BLCOO1/ NX,NP,NXTE,NXT,NPT,NTR,NXS,IT,ITV,ISWP,ISWPT
      COMMON /BLPRC1/ F(101,2),U(101,2),V(101,2),W(101,2),B(101,2)
      COMMON /BLGRD1/ X(201),ETA(101),A(101),DETA(101)
      COMMON /BLHLBS/ CC(201,201),DD(201),DB(201),GI
      COMMON /BLINP3/ RL, RX, SQRL, SQRX
      COMMON /BLRCO1/ XBL(201),UBL(201),CFS(201),THETA(201),
     1                DLS(201),UEVS(201),NPSAV(201),ITSAV(201)
      DIMENSION       D1(101)
C - - - - - - - - - - - - - - - - - - - - - - - - - - - - - - - - - - - -
      NPSAV(NX)  = NP
      ITSAV(NX)  = IT
      IF(NX .EQ. 1 ) THEN
         CFS(NX) = 0.0
         RETURN
      ENDIF
      SUM  = 0.0
      DO 5 J =1,NP
         D1(J) = U(J,2) * (1.0-U(J,2)/U(NP,2))
    5 CONTINUE
      DO 10 J=2,NP
         SUM  = SUM  + A(J)*(D1(J)  + D1(J-1))
   10 CONTINUE
      IF(ITV .NE. 0) THEN
         RTH       = SQRX * SUM
         DLS(NX)= X(NX)/SQRX*(ETA(NP)- F(NP,2))
         THETA(NX)= X(NX)/SQRX*SUM
         RDS       = SQRX * (ETA(NP) - (F(NP,2)-F(1,2)))
         CFS(NX)= 2.0 * V (1,2) / SQRX
      ELSE
         SQXC      = SQRT(X(NX))
         DLS(NX)   = SQXC/SQRL*(ETA(NP)- F(NP,2)/U(NP,2))
```

```
      THETA(NX)= SQRT(X(NX)/RL)*SUM/U(NP,2)
      RTH      = SQRL * SUM * SQXC
      RDS      = RX * (DLS(NX)/X(NX))
      IF(NX.LE.NXTE) THEN
         CFS(NX)= 2.0*V(1,2)/(SQRL*SQXC *U(NP,2) ** 2)
      ELSE
         CFS(NX)= 0.0
      ENDIF
      UEVS(NX) =  U(NP,2)
   ENDIF
   DD(NX)   =  DLS(NX) * UEVS(NX) * SQRL
   RETURN
   END
```

References

[1] Cebeci, T. and Cousteix, J., Modeling and Computation of Boundary-Layer Flows, Horizons and Springer-Verlag, 1998.

[2] Wilcox, D.C., Turbulence Modeling for CFD, 2nd DCW Industries, La Canada, CA

[3] Jones, W.P. and Launder, B.E., "The Prediction of Laminization with a Two-Equation Model of Turbulence," Int. J. Heat and Mass Transfer, Vol. 15, p. 301, 1972.

[4] Marvin, J.G., "Turbulence Modeling for Computational Aerodynamics," AIAA J., Vol. 21, p. 941, 1983.

[5] Cebeci, T. and Smith, A.M.O., Analysis of Turbulent Boundary Layers, Academic Press, N.Y., 1974.

[6] Launder, B.E. and Sharma, B.I., "Application of the Energy-Dissipation Model of Turbulence to the Calculation of Flow Near a Spinning Disc," Letters in Heat and Mass Transfer, Vol. 1, pp. 131–138, 1974.

[7] Arnal, D., Cousteix, J. and Michel, R., "Couche limite se développant avec gradient de pression positif dans un écoulement turbulent," La Rech. Aérosp., 1976-1, 1976.

[8] Norris, L.H. and Reynolds, W.C., "Turbulent Channel Flow with a Moving Wavy Boundary," Report FM-10, Department of Mechanical Engineering, Stanford University, Stanford, California, 1975.

[9] Launder, B.E., Reece, G.J. and Rodi, W., "Progress in the Development of a Reynolds Stress Turbulence Closure," J. Fluid Mech., Vol. 68, No., 3, p. 537, 1975.

[10] Gibson, M.M. and Launder, B.E., "Ground Effects on Pressure Fluctuation in the Atmospheric Boundary Layer," J. Fluid Mech., Vol. 86, pp. 491–511, 1978.

[11] Chang, K.C. and Cebeci, T., "Numerical Aspects of Turbulence Models," 1st International Symposium on Computational Fluid Dynamics, Tokyo, Japan, Sept. 1985.

[12] Johnson, D.A. and King, L.S., "Mathematically Simple Turbulence Closure Model for Attached and Separated Turbulent Boundary Layers," AIAA J., Vol. 23, No. 11, pp. 1684–1692, 1985.

[13] Johnson, D.A. and Coakley, T.J., "Improvement to a Nonequilibrium Algebraic Turbulence Model," AIAA J., Vol. 28, No. 11, pp. 2000–2003, 1990.

[14] Cebeci, T. and Change, K.C., "An Improved Cebeci-Smith Turbulence Model for Boundary-Layer and Navier-Stokes Methods," 20th Congress of the International Council of the Aeronautical Sciences, paper No. ICAS-96-1.7.3, Sorrento, Italy, Sept. 8–13, 1996.

[15] Fiedler, H. and Head, M.R., "Intermittency Measurements in the Turbulent Boundary Layer," J. Fluid Mech., Vol. 25, No. 4, pp. 719–735, 1966.

[16] Crank, J. and Nicolson, P., "A Practical Method for Numerical Evaluation of Solutions of Partial-Differential Equations of the heat-conduction type," Proc. Camb. Phil. Soc., Vol. 43, No. 50, 1947.

[17] Keller, H.B., "A New Difference Scheme for Parabolic Problems," In: Numerical Solution of Partial Differential Equations, (J. Bramble, ed.), Vol. II, Academic Press, NY, 1970.
[18] Reyhner, T.A. and Flügge-Lotz, I., "The Interaction of a Shock Wave with a Laminar Boundary Layer," Int. J. Non-Linear Mech., Vol. 3, No. 2, p. 173, 1968.
[19] Cebeci, T. and Bradshaw, P., Physical and Computational Aspects of Convective Heat Transfer, Springer, Heidelberg, 1988.
[20] Keller, H.B., "The Bordering Algorithm and Path Following Near Singular Points of Higher Nullity," SIAM J. on Scientific and Statistical Computing, Vol. 4, No. 4, pp. 573–582, 1983.
[21] Chen, H.H. and Cebeci, T., "Bordering Algorithm for Solution of Boundary-Layer Equations in Inverse Mode," AIAA J., Vol. 29, no. 12, pp. 2257–2259, Dec. 1991.
[22] Cebeci, T., Chang, K.C., and Mack, "Linearization of Turbulent Boundary-Layer Equations," AIAA J., Vol. 22, no. 12, pp. 1819–1821, Dec. 1984.
[23] Jang, H.M., Cebeci, T. and Keller, H.B., "A Preferred Approach to the Linearization of Turbulent Boundary-Layer Equations," Computers and Fluids, Vol. 17, pp. 571–578, 1989.

4 Two-Dimensional Incompressible Flows: Transition Method

4.0 Introduction

Current methods available for predicting transition, excluding correlation formulas such as those discussed in Section 1.4, are based on the solutions of the unsteady Navier-Stokes equations (DNS) discussed in [1], on the solution of the parabolized stability equations (PSE) discussed for example, in [2] and on the solutions of the linear stability equations discussed in this chapter for two-dimensional incompressible flows and in Chapter 12 for two-dimensional and three-dimensional compressible flows. Despite the progress being made with the PSE approach, this method currently is some time away from being used as an engineering tool; it is still under development. The direct numerical simulation (DNS) approach offers exciting possibilities; the excellent review of Kleiser and Zang [1] describes the rapid progress with this approach and shows that the prediction of transition can already be achieved in some simple flows with this method. The computer requirements of DNS, however, are large, and it is unlikely that this approach can be used for transition calculations on complex bodies in the immediate future. The only engineering calculation method for predicting transition at this time is the e^n-method based on the solutions of the linear stability equations, which for two-dimensional incompressible flows are given by the following fourth-order ordinary differential equation

$$\nu(\phi^{iv} - 2\alpha^2\phi'' + \alpha^4\phi) = i[(\alpha u - \omega)(\phi'' - \alpha^2\phi) - \alpha u''\phi] \tag{4.0.1}$$

known as the Orr-Sommerfeld equation [3]. Here primes denote differentiation with respect to y, u denotes the streamwise velocity profile and $\phi(y)$ $(\equiv \phi_r + i\phi_i)$ denotes the complex amplitude of the disturbance stream function ψ' defined by the fluctuating velocity components u' and v'

$$u' = \frac{\partial \psi'}{\partial y}, \quad v' = -\frac{\partial \psi'}{\partial x} \tag{4.0.2}$$

The parameter α is the wave number of the disturbance related to the wave-length of the disturbance, λ, by $\lambda = 2\pi/\alpha$ and ω is the circular frequency. If α is complex ($\equiv \alpha_r + i\alpha_i$), then the amplitude of the disturbance will vary with x as $\exp(-\alpha_i x)$. The analysis for this case is referred to as the *spatial-amplification theory*. If ω is complex, ($\equiv \omega_r + i\omega_i$), the amplitude will vary with time as $\exp(-\omega_i t)$. The analysis for this case is referred to as the *temporal-amplification theory*.

It is convenient to introduce dimensionless variables by dividing all velocities by a reference velocity u_0, and all lengths by a reference length l. It is also convenient to introduce a dimensionless time $\tau(\equiv tu_0/l)$ so that Eq. (4.0.1) can be written in dimensionless form as

$$\phi^{iv} - 2\alpha_l^2\phi'' + \alpha_l^4\phi = iR[(\alpha_l\bar{u} - \bar{\omega})(\phi'' - \alpha_l^2\phi) - \alpha_l\bar{u}''\phi] \tag{4.0.3}$$

Now the primes denote differentiation with respect to $\bar{y}$ ($\equiv y/l$). The parameter $\phi(\bar{y})$ is the complex amplitude of the disturbance stream function $\psi'(\bar{x}, \bar{y}, \tau)$ and is defined by

$$\psi'(\bar{x}, \bar{y}, \tau) = \phi(\bar{y}) \exp[i(\alpha_l\bar{x} - \bar{\omega}\tau)]$$

The dimensionless parameters $\alpha_l, \bar{u}, \bar{\omega}$ and R are given by

$$\alpha_l = \alpha l, \quad \bar{u} = \frac{u}{u_0}, \quad \bar{\omega} = \frac{u_0\omega}{l}, \quad R = \frac{u_0 l}{\nu} \tag{4.0.4}$$

The solution of Eq. (4.0.3) requires boundary conditions and, for boundary-layer flows, these are satisfied by requiring that at the wall ($y = 0$) and away from the wall ($y \to \infty$), $u' = v' = 0$. In terms of the dimensionless disturbance stream function, ϕ, the conditions at the wall can be written as

$$\bar{y} = 0, \quad \phi = 0, \quad \phi' = 0 \tag{4.0.5a}$$

As discussed in [3], it is convenient to write the "infinity" boundary conditions by taking into account that perturbation velocities decay as the edge of the boundary-layer δ is approached. In this case, the "infinity" boundary conditions can be satisfied at $y = \delta$ by

$$\begin{aligned} \bar{y} = \delta, \ &(D^2 - \xi_1^2)\phi + (\xi_1 + \xi_2)(D + \xi_1)\phi = 0 \\ &(D + \xi_2)(D^2 - \xi_1^2)\phi = 0 \end{aligned} \tag{4.0.5b}$$

where $D \equiv d/d\bar{y}$ and

$$\xi_1^2 = \alpha_l^2, \quad \xi_2^2 = \xi_1^2 + iR(\alpha_l\bar{u} - \bar{\omega}) \tag{4.0.6}$$

The onset of transition location is obtained by solving Eqs. (4.0.3) and (4.0.5) and following the e^n-procedure described in Section 4.1. The solution procedure described in Section 4.2 may be based on either temporal or spatial amplification theories. While both procedures have advantages, in this book we will use the spatial amplification theory since the amplitude change of disturbance with distance can be measured in a steady mean flow. The amplitude at a fixed point is independent of time and spatial theory gives the amplitude change in a more direct manner than does the temporal theory.

4.1 e^n-Procedure

The e^n-procedure requires the calculation of the amplification rates $(-\alpha_i)$ as a function of x (or R) for a range of dimensional frequencies ω^* defined by

$$\omega^* = \bar{\omega}\frac{u_0}{l} \tag{4.1.1}$$

The stability calculations are preceded by boundary-layer calculations and, for a given external velocity distribution $u_e(x)$ and freestream Reynolds number, the laminar boundary-layer equations are solved to obtain the streamwise velocity profile u and its second derivative u''. The stability calculations begin at a Reynolds number based on displacement thickness, R_{δ^*},

$$R_{\delta^*} = \frac{u_e\delta^*}{\nu} \tag{4.1.2}$$

slightly larger than the critical Reynolds number, $R_{\delta^*_{\rm cr}}$ (see Fig. 4.1) on the lower branch of the neutral stability curve ($\alpha_i = 0$) at an x-location, say $x = x_1$ (see Fig. 4.2). At this point, since u, u'' and R are known, the eigenvalues α_r and ω are computed by the procedure of subsection 4.2.1 and the dimensional frequency ω^* determined from Eq. (4.2.1) is kept constant along line 1 defined

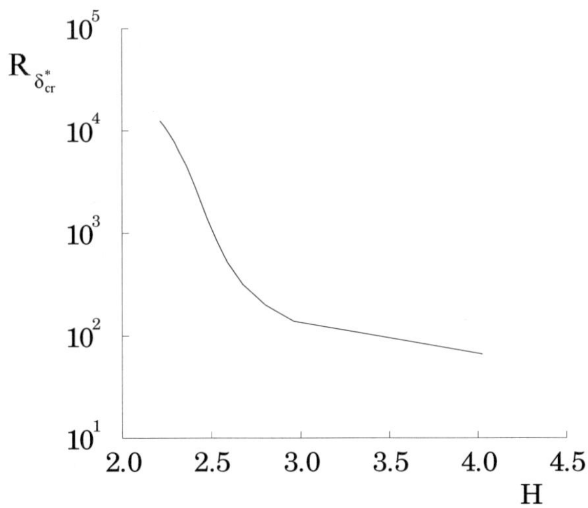

Fig. 4.1. Variation of critical Reynolds number $R_{\delta^*_{\rm cr}}$, with shape factor H [3].

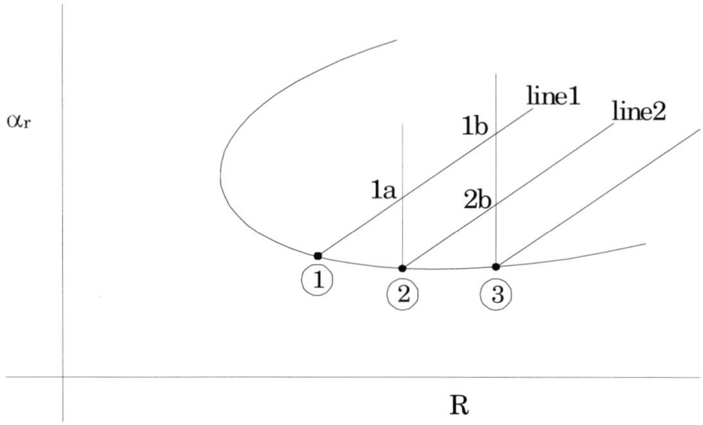

Fig. 4.2. Strategy of calculating transition with the e^n-method.

by this constant dimensional frequency. At the next x-location, x_2, two separate calculations are performed for the newly computed boundary-layer profiles u and u'' and Reynolds number R. In one set of calculations (point 2 in Fig. 4.2), α_r and ω are computed on the neutral curve with the procedure used to obtain α_r and ω at point 1 (subsection 4.2.1) so that a new dimensional frequency can be defined on line 2. In the second set of calculations, point 1a, the dimensionless frequency ω, with l and u_0 known, is first determined from the dimensional frequency ω^* on line 1. With ω known and R defined at point 2, α is obtained with the eigenvalue procedure of subsection 4.2.2. The procedure at point 1a is then repeated at points 2b and 1b, and a dimensional frequency is computed for line 3. For example, at point 1b, values of (α_r, α_i) are computed with the known dimensional frequency on line 1 and the specified Reynolds number at point 3; at point 2b, they are computed with the known dimensional frequency on line 2 and the specified Reynolds number at point 3. This procedure is repeated for several lines and the variation of the integrated amplification rate, defined by

$$n = -\int_{x_0}^{x} \alpha_i dx \tag{4.1.3}$$

with x_0 corresponding to each value of x on the neutral stability curve, is computed for each line. This procedure leads to the curves of constant frequency, Fig. 4.3, and their envelope corresponds to the maximum amplification factors from which transition is computed with a value for n, commonly assumed between 8 and 9.

This envelope procedure is useful and convenient for two- dimensional attached flows but, as pointed out by Cebeci and Egan [4], has limitations for more complicated flows such as those with separation, as discussed in Section 5.2. A more general method would require the determination of the first dimensional frequency, referred to as the critical frequency, f_{cr} ($\equiv \omega^*_{\text{cr}}/2\pi$), which

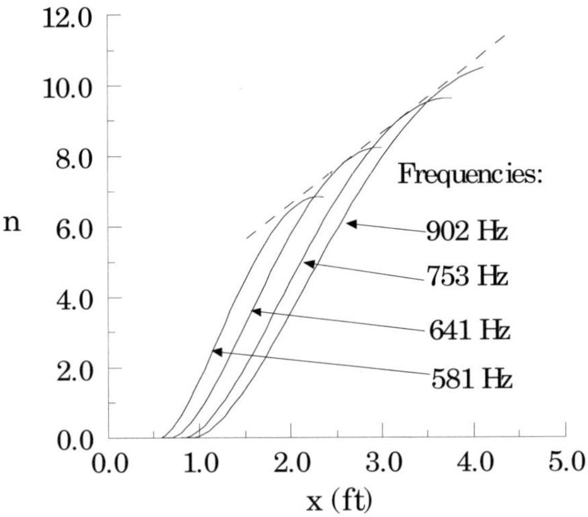

Fig. 4.3. Variation of the integrated amplification factors with distance and frequency for zero-pressure gradient flow.

achieves the predetermined amplification factor corresponding to transition. For example, Fig. 4.3 shows that an amplification factor of 9 has not been achieved with $f_{\text{cr}} = 885$ Hz and that the appropriate value f_{cr} is between 662 and 759.

4.2 Numerical Method

The solution of the Orr-Sommerfeld equation and its boundary conditions, Eqs. (4.0.3) and (4.0.5), exists only for certain combinations of R, α, ω since all the boundary conditions are homogeneous. Thus the problem is an eigenvalue problem in which values of R, α and ω (subscript l in α and bar on ω are dropped for convenience) are the eigenvalues and the corresponding amplitude functions are eigenfunctions. Hence, in general, no nontrivial solution of this system exists. Only if α, ω, R satisfy one or more relations of the form

$$F(\alpha, \omega, R) = 0 \tag{4.2.1}$$

can such a solution be found.

Until 1964 the numerical methods used to solve the Orr-Sommerfeld equation made use of the initial-value schemes and were limited to the lower range of Reynolds numbers. The integration procedure usually started at the edge of the boundary-layer where the asymptotic solutions of the stability equation are of the form (see [3])

$$\phi \sim A_1 e^{-\xi_1 y} + A_2 e^{-\xi_2 y}$$

with ξ_1 and ξ_2 given by Eq. (4.0.6) and with ξ_2^2 evaluated at the boundary-layer edge, that is,

$$\xi_1^2 = \alpha^2, \quad \xi_2^2 = \xi_1^2 + iR(\alpha u_e - \omega) \tag{4.2.2}$$

Since ξ_1 or α and ω are all of order unity and $R \gg 1$, the two rates of change are drastically different with $\xi_2 \gg \xi_1$ so that of the four independent solutions of Eq. (4.0.3), the two associated with $e^{\xi_1 y}$ grow slowly, while the two associated with $e^{\xi_2 y}$ grow rapidly. Even though the solutions are linearly independent at the beginning, with increasing Reynolds number they become linearly dependent as the solutions progress to the wall due to the contamination error growth of the "slower"-growing solutions by the "faster"-growing solutions. Double precision can alleviate this difficulty at moderate Reynolds numbers, though at some cost, but is unable to overcome this parasitic error growth at higher Reynolds numbers. Kaplan [5] devised a scheme where the slowly growing solution was purified of its parasitic error at each integration step so that the solution was not dominated by the faster-growing solution at the end of the integration interval. This "purification scheme" was found successful in solving the Orr-Sommerfeld equation at Reynolds number R_{δ^*} up to 10,000.

Wazzan et al. [6] applied Kaplan's technique to the Orr-Sommerfeld equation and solved it with the Gram-Schmidt orthogonalization procedure with good success and obtained accurate solutions for a wide range of Reynolds numbers. The main shortcoming of this technique is the eigenvalue procedure which is cumbersome, especially in heat transfer problems, and which leads to large computer time requirements as well as to a breakdown of the solutions when the initial estimates of the eigenvalues are not close to the final answer.

In general, finite-difference methods do not have the problems associated with parasitic error growth and have been successfully used in recent years to solve the stability equations, including flows at high Reynolds numbers. A popular method is that based on a finite-difference matrix used by Srokowski and Orszag [7] for three-dimensional incompressible flows and by Malik [8] for compressible flows. Another successful finite-difference method is based on Keller's box scheme described for two-dimensional steady boundary-layer equations in Chapter 3, unsteady boundary-layer flows in Chapter 8 and three-dimensional boundary-layer flows in Chapter 11. It is described here for the solution of the linear stability equations for two-dimensional flows and in Chapter 12 for the solution of three-dimensional flows.

4.2.1 Numerical Formulation

With the definitions in Eq. (4.0.6) and with

$$\xi_3 = iR\alpha u'' \tag{4.2.2}$$

Eq. (4.0.3) can be written in the compact form:

$$\phi^{iv} - \xi_1^2\phi'' - \xi_2^2(\phi'' - \xi_1^2\phi) + \xi_3\phi = 0 \tag{4.2.3}$$

Eq. (4.2.3) and its boundary conditions, Eq. (4.0.5), are written as an equivalent first-order system, as was done for the boundary-layer equations in subsection 3.3.1. Thus, the definitions

$$\phi' = f \tag{4.2.4a}$$

$$f' = s + \xi_1^2 \phi \tag{4.2.4b}$$

$$s' = g \tag{4.2.4c}$$

reduce Eq. (4.2.3) and its boundary conditions to

$$g' = \xi_2^2 s - \xi_3 \phi \tag{4.2.4d}$$

$$y = 0; \quad \phi = 0, \quad f = 0 \tag{4.2.5a}$$

$$y = \delta; \quad s + (\xi_1 + \xi_2) f + \xi_1(\xi_1 + \xi_2)\phi = 0, \quad g + \xi_2 s = 0 \tag{4.2.5b}$$

where ξ_2, defined in Eq. (4.0.6), attains its value at $y = \delta$.

With a nonuniform mesh, and with $y = \delta$ represented by y_J, the quantities (f, s, g, ϕ) are approximated at points y_j by (f_j, s_j, g_j, ϕ_j). As in subsection 3.3.1, we write the finite-difference approximations of Eqs. (4.2.4) for the midpoint $y_{j-1/2}$ using centered-difference derivatives to obtain

$$\phi_j - \phi_{j-1} - (c_3)_j (f_j + f_{j-1}) = (r_1)_j = 0 \tag{4.2.6a}$$

$$- f_j + f_{j-1} + (c_3)_j (s_j + s_{j-1}) + (c_1)_j (\phi_j + \phi_{j-1}) = (r_3)_{j-1} = 0 \tag{4.2.6b}$$

$$s_j - s_{j-1} - (c_3)_j (g_j + g_{j-1}) = (r_2)_j = 0 \tag{4.2.6c}$$

$$- g_j + g_{j-1} + (c_4)_j (s_j + s_{j-1}) + (c_2)_j (\phi_j + \phi_{j-1}) = (r_4)_{j-1} = 0 \tag{4.2.6d}$$

Here, with h_{j-1} denoting $y_j - y_{j-1}$,

$$\begin{aligned} (c_1)_j &= \xi_1^2 (c_3)_j, \quad (c_2)_j = -(\xi_3)_{j-1/2} (c_3)_j, \\ (c_3)_j &= \frac{h_{j-1}}{2}, \quad (c_4)_j = -(\xi_2^2)_{j-1/2} (c_3)_j \end{aligned} \tag{4.2.7}$$

For convenience, in Eqs. (4.2.6b) and (4.2.6d), both sides of the equations are multiplied by -1. As in the procedure for solving the boundary-layer equations, Eqs. (4.2.6) are written in a sequence which ensures that the A-matrix in Eq. (3.3.14) is not singular.

Equations (4.2.6) are imposed for $j = 1, 2, \ldots, J$, and additional conditions at $j = 0$ and $j = J$ are obtained from the boundary conditions given by Eqs. (4.2.5) and can be written as

$$\phi_0 = (r_1)_0 = 0, \quad f_0 = (r_2)_0 = 0, \tag{4.2.8a}$$

$$f_J + (\tilde{c}_3)_J s_J + (\tilde{c}_1)_J d_J = (r_3)_J = 0, \quad g_J + (\tilde{c}_4)_J s_J = (r_4)_J = 0 \tag{4.2.8b}$$

where now

$$(\tilde{c}_3)_J = \frac{1}{\xi_1 + \xi_2}, \quad (\tilde{c}_1)_J = \xi_1, \quad (\tilde{c}_4)_J = \xi_2 \tag{4.2.9}$$

Since the Orr-Sommerfeld equation and its boundary conditions are homogeneous, the trivial solutions $f_j = s_j = \phi_j = g_j = 0$ for all j are valid for all values of α, ω and R. For this reason, the eigenvalues and the eigenfunctions must be calculated using an iteration procedure to find the values of parameters for which nontrivial solutions exist. The boundary condition, $\phi'(0) = 0$ (that is $f_0 = 0$) of Eq. (4.2.8), is replaced by $\phi''(0) = 1$, that is $s_0 = 1$, and the difference equations have a nontrivial solution since $\phi''(0) \neq 0$; parameter values are then determined so that the original boundary condition is satisfied. This is achieved by an iteration scheme, again based on Newton's method, with Eqs. (4.2.6), the wall boundary conditions,

$$\phi_0 = (r_1)_0 = 0, \qquad s_0 = (r_2)_0 = 0 \tag{4.2.10}$$

and the edge boundary conditions of Eq. (4.2.8b) written in matrix- vector form as in Eq. (3.3.16), that is

$$A\vec{\vec{\delta}} = \vec{\vec{r}} \tag{3.3.16}$$

where now

$$\vec{\delta}_j = \begin{vmatrix} \phi_j \\ s_j \\ f_j \\ g_j \end{vmatrix}, \quad \vec{r}_j = \begin{vmatrix} (r_1)_j \\ (r_2)_j \\ (r_3)_j \\ (r_4)_j \end{vmatrix} \tag{4.2.11}$$

and the A_j, B_j, C_j matrices in the coefficient matrix A denote 4×4 matrices given by

$$A_0 \equiv \begin{vmatrix} 1 & 0 & 0 & 0 \\ 0 & 1 & 0 & 0 \\ (c_1)_1 & (c_3)_1 & 1 & 0 \\ (c_2)_1 & (c_4)_1 & 0 & 1 \end{vmatrix}, \quad A_j \equiv \begin{vmatrix} 1 & 0 & -(c_3)_j & 0 \\ 0 & 1 & 0 & -(c_3)_j \\ (c_1)_{j+1} & (c_3)_{j+1} & 1 & 0 \\ (c_2)_{j+1} & (c_4)_{j+1} & 0 & 1 \end{vmatrix} \quad 1 \le j \le J{-}1$$

$$A_J = \begin{vmatrix} 1 & 0 & -(c_3)_J & 0 \\ 0 & 1 & 0 & -(c_3)_J \\ (\tilde{c}_1) & (\tilde{c}_3) & 1 & 0 \\ 0 & (\tilde{c}_4) & 0 & 1 \end{vmatrix}, \quad B_j = \begin{vmatrix} -1 & 0 & -(c_3)_j & 0 \\ 0 & -1 & 0 & -(c_3)_j \\ 0 & 0 & 0 & 0 \\ 0 & 0 & 0 & 0 \end{vmatrix} \quad 1 \le j \le J$$

$$C_j = \begin{vmatrix} 0 & 0 & 0 & 0 \\ 0 & 0 & 0 & 0 \\ (c_1)_{j+1} & (c_3)_{j+1} & -1 & 0 \\ (c_2)_{j+1} & (c_4)_{j+1} & 0 & -1 \end{vmatrix} \quad 0 \le j \le J-1 \tag{4.2.12}$$

The solution of Eq. (3.3.16), with the difference equations defined by Eqs. (4.2.6), (4.2.8b) and (4.2.10), again makes use of the block-elimination method of subsection 3.3.3 and depends upon the three parameters α, ω and R. This dependence

$$\vec{\vec{\delta}} = \vec{\vec{\delta}}(\alpha, \omega, R) \tag{4.2.13}$$

with α complex and ω real in spatial-amplification theory, implies that the solution of Eq. (3.3.16) depends upon four scalars. With any two of these scalars fixed, the remaining scalar can be determined so that the missing boundary condition $\phi'(0) = 0$ is satisfied. In our finite-difference notation, this corresponds to the condition

$$f_0(\alpha, \omega, R) = 0 \tag{4.2.14}$$

4.2.2 Calculation of Dimensional Frequencies: Neutral Stability Curve

The eigenvalues of the Orr-Sommerfeld equation for the spatial-amplification case are often presented in (α, R) and (ω, R) diagrams that describe the three states of a disturbance at a given Reynolds number R as damped, neutral or amplified. For two-dimensional flows the locus $\alpha_i = 0$ called the *curve of neutral* stability, separates the damped (stable) region from the amplified (unstable) region. The point on this curve at which R has its smallest value is of special interest, because at values of R less than this value, all disturbances are stable. This smallest Reynolds number is known as the critical Reynolds number R_{cr}. The neutral curve is the same in both temporal and spatial amplification theories.

The dimensional frequency ω^* needed in the e^n-method can be obtained by solving the eigenvalue problem for α_r, ω with α_i and R specified. Since on the neutral stability curve $\alpha_i = 0$, and ω is real in spatial amplification theory, and since f_0 is complex, Eq. (4.2.14) represents two equations with two unknowns (α_r, ω), and the equations can be solved by Newton's method. Specifically, if $(\alpha_r^\nu, \omega^\nu)$ are the ν-th iterates, the $(\nu+1)$-th iterates are determined by inserting

$$\alpha_r^{\nu+1} = \alpha_r^\nu + \delta\alpha_r^\nu \tag{4.2.15a}$$

$$\omega_r^{\nu+1} = \omega_r^\nu + \delta\omega_r^\nu \tag{4.2.15b}$$

in Eq. (4.2.14), expanding it around α_r^ν and ω^ν by the Taylor series, and retaining only linear terms in the expansion. This gives the linear system

$$f_r^\nu + \left(\frac{\partial f_r}{\partial \alpha_r}\right) \delta\alpha_r^\nu + \left(\frac{\partial f_r}{\partial \omega}\right) \delta\omega^\nu = 0 \tag{4.2.16a}$$

$$f_i^\nu + \left(\frac{\partial f}{\partial \alpha_r}\right)^\nu \delta\alpha_r^\nu + \left(\frac{\partial f}{\partial \omega}\right) \delta\omega^\nu = 0 \tag{4.2.16b}$$

The solution of Eqs. (4.2.16) is given by

$$\delta\alpha_r^\nu = \frac{1}{\Delta_0}\left[f_i^\nu \left(\frac{\partial f_r}{\partial \omega}\right)^\nu - f_r^\nu \left(\frac{\partial f_r}{\partial \omega}\right)^\nu\right] \tag{4.2.17a}$$

$$\delta\omega_r^\nu = \frac{1}{\Delta_0}\left[f_r^\nu \left(\frac{\partial f_i}{\partial \alpha_r}\right)^\nu - f_i^\nu \left(\frac{\partial f_r}{\partial \alpha_r}\right)^\nu\right] \tag{4.2.17b}$$

where

$$\Delta_0 = \left(\frac{\partial f_r}{\partial \alpha_r}\right)^\nu \left(\frac{\partial f_i}{\partial \omega}\right)^\nu - \left(\frac{\partial f_i}{\partial \alpha_r}\right)^\nu \left(\frac{\partial f_r}{\partial \omega}\right)^\nu \tag{4.2.17c}$$

To evaluate the derivatives of f_r and f_i with respect to α_r and ω, we differentiate Eq. (3.3.16) with respect to α_r and ω and, since the vector $\vec{r}$ is independent of α_r and ω,

$$A\left(\frac{\partial \vec{\delta}}{\partial \alpha_r}\right)^\nu = -\left(\frac{\partial A}{\partial \alpha_r}\right)^\nu \vec{\delta}^\nu \tag{4.2.18a}$$

$$A\left(\frac{\partial \vec{\delta}}{\partial \omega}\right)^\nu = -\left(\frac{\partial A}{\partial \omega}\right)^\nu \vec{\delta}^\nu \tag{4.2.18b}$$

The above equations are *the variational equations of Eq. (3.3.16) with respect to* α_r *and* ω, respectively. Thus, to obtain the required derivatives, two linear systems have to be solved with the same coefficient matrix A already computed and factored for Eq. (3.3.16). The vectors on the right-hand side of Eqs. (4.2.18a) are determined from Eqs. (4.2.6), (4.2.8b) and (4.2.10). For Eq. (4.2.18a), $(r_1)_j = (r_2)_j = 0$ for $0 \le j \le J$, but $(r_3)_j$ and $(r_4)_j$ for $1 \le j \le J$, are given by

$$(r_3)_{j-1} = -2\left(\frac{\partial c_1}{\partial \alpha_r}\right)_j \phi_{j-1/2} \tag{4.2.19a}$$

$$(r_4)_{j-1} = 2\left(\frac{\partial c_4}{\partial \alpha_r}\right)_j s_{j-1/2} + 2\left(\frac{\partial c_2}{\partial \alpha_r}\right)_j \phi_{j-1/2} \tag{4.2.19b}$$

$$(r_3)_J = -\left(\frac{\partial \tilde{c}_3}{\partial \alpha_r}\right)_J s_J - \left(\frac{\partial \tilde{c}_1}{\partial \alpha_r}\right)_J \phi_J \tag{4.2.19c}$$

$$(r_4)_J = -\left(\frac{\partial \tilde{c}_4}{\partial \alpha_r}\right)_J s_J \tag{4.2.19d}$$

For Eq. (4.2.18b), again $(r_1)_j = (r_2)_j = 0$ for $0 \le j \le J$, but with $(c_1)_j$, $(c_2)_j$ and $(c_3)_j$, are independent of ω for $j \le J-1$, and the coefficients $(r_3)_j$ and $(r_4)_j$ for $1 \le j \le J$ are given by

$$(r_3)_{j-1} = 0 \tag{4.2.20a}$$

$$(r_4)_{j-1} = -2\left(\frac{\partial c_4}{\partial \omega}\right)_j s_{j-1/2} \tag{4.2.20b}$$

$$(r_3)_J = -\left(\frac{\partial \tilde{c}_3}{\partial \omega}\right)_J s_J \tag{4.2.20c}$$

$$(r_4)_J = -\left(\frac{\partial \tilde{c}_4}{\partial \omega}\right)_J s_J \tag{4.2.20d}$$

To summarize one part of the iteration process for a fixed value of R and for assumed values of α_r and ω, we solve Eq. (3.3.16). If the initial estimates of α_r and ω satisfy Eq. (4.2.14) and if $f_0 = 0$, there is no need to compute new values of α_r and ω. On the other hand, if $f_0 \neq 0$, new estimates of α_r and ω are obtained by Newton's method. With the right-hand side of Eqs. (4.2.18) given by Eqs. (4.2.19) and (4.2.20), Eqs. (4.2.18) are solved in order to compute $\partial f/\partial \alpha_r$ and $\partial f/\partial \omega$. Then $\delta\alpha_r$ and $\delta\omega$ are computed from Eqs. (4.2.17) and are inserted in Eqs. (4.2.15) so that Eq. (3.3.16) can be solved with new estimates of α_r and ω to satisfy Eq. (4.2.14). This process is repeated until the increments $|\delta\omega|$ and $|\alpha_r|$ are less than a specified tolerance parameter.

Once a solution of Eq. (4.2.14) is obtained at a specified Reynolds number R_0, $(\partial\alpha_r/\partial R)_0$ and $(\partial\omega/\partial R)_0$ can be determined at $R = R_0$. For this purpose we take the total derivative of Eq. (4.2.14) with respect to R and after separating the real and imaginary parts of the resulting expression, we get

$$\left(\frac{\partial f_r}{\partial \alpha_r}\right)_0 \left(\frac{\partial \alpha_r}{\partial R}\right)_0 + \left(\frac{\partial f_r}{\partial \omega}\right)_0 \left(\frac{\partial \omega}{\partial R}\right)_0 = -\left(\frac{\partial f_r}{\partial R}\right)_0 \tag{4.2.21a}$$

$$\left(\frac{\partial f_i}{\partial \alpha_r}\right)_0 \left(\frac{\partial \alpha_r}{\partial R}\right)_0 + \left(\frac{\partial f_i}{\partial \omega}\right)_0 \left(\frac{\partial \omega}{\partial R}\right)_0 = -\left(\frac{\partial f_i}{\partial R}\right)_0 \tag{4.2.21b}$$

It follows from Eqs. (4.2.21) that

$$\left(\frac{\partial \alpha_r}{\partial R}\right)_0 = \frac{1}{\Delta_0}\left[\left(\frac{\partial f_i}{\partial R}\right)_0 \left(\frac{\partial f_r}{\partial \omega}\right)_0 - \left(\frac{\partial f_r}{\partial R}\right)_0 \left(\frac{\partial f_i}{\partial \omega}\right)_0\right] \tag{4.2.22a}$$

$$\left(\frac{\partial \omega}{\partial R}\right)_0 = \frac{1}{\Delta_0}\left[\left(\frac{\partial f_r}{\partial R}\right)_0 \left(\frac{\partial f_i}{\partial \alpha_r}\right)_0 - \left(\frac{\partial f_i}{\partial R}\right)_0 \left(\frac{\partial f_r}{\partial \alpha_r}\right)_0\right] \tag{4.2.22b}$$

where

$$\Delta_0 = \left(\frac{\partial f_r}{\partial \alpha_r}\right)_0 \left(\frac{\partial f_i}{\partial \omega}\right)_0 - \left(\frac{\partial f_r}{\partial \omega}\right)_0 \left(\frac{\partial f_i}{\partial \alpha_r}\right)_0 \tag{4.2.22c}$$

To evaluate the derivatives f_r and f_i with respect to R, this time we differentiate Eq. (3.3.16) to obtain the variational equations with respect to R.

$$A\left(\frac{\partial \vec{\delta}^{\nu}}{\partial R}\right) = -\left(\frac{\partial A}{\partial R}\right)^{\nu} \vec{\delta} \tag{4.2.23}$$

The right hand side vectors of the above equation with $(r_1)_j = (r_2)_j = 0$ for $0 \leq j \leq J$, and with $(r_3)_j$ and $(r_4)_j$ for $1 \leq j \leq J$ are:

$$(r_3)_{j-1} = 0 \tag{4.2.24a}$$

$$(r_4)_{j-1} = 2\left(\frac{\partial c_4}{\partial R}\right) s_{j-1/2} + 2\left(\frac{\partial c_2}{\partial R}\right) \phi_{j-1/2} \tag{4.2.24b}$$

$$(r_3)_J = -\left(\frac{\partial \tilde{c}_3}{\partial R}\right)_J s_J \tag{4.2.24c}$$

$$(r_4)_J = -\left(\frac{\partial \tilde{c}_4}{\partial R}\right)_J s_J \tag{4.2.24d}$$

As before, with the coefficient matrix A already known, we now compute the relations in Eqs. (4.2.24) so that $\partial f/\partial R$ can be obtained from the solutions of Eq. (4.2.23). Since $\partial f/\partial \alpha_r$ and $\partial f/\partial \omega$ from the solutions of Eqs. (4.2.18), $\partial \alpha_r/\partial R$ and $\partial \omega/\partial R$ can be computed easily from Eqs. (4.2.22) to decide on the choice of the eigenvalue procedure.

4.2.3 Calculation of Transition

The eigenvalue procedure for computing α for specified values of ω and R is analogous to the procedure described in the previous subsection. In this case, Eq. (4.2.14) is expanded with the Taylor series rather than Eqs. (4.2.16), and linear terms are retained to give

$$f_r^\nu + \left(\frac{\partial f_r}{\partial \alpha_r}\right)^\nu \delta\alpha_r^\nu + \left(\frac{\partial f_r}{\partial \alpha_i}\right)^\nu \delta\alpha_i^\nu = 0 \tag{4.2.25a}$$

$$f_i^\nu + \left(\frac{\partial f_i}{\partial \alpha_r}\right)^\nu \delta\alpha_r^\nu + \left(\frac{\partial f_i}{\partial \alpha_i}\right)^\nu \delta\alpha_i^\nu = 0 \tag{4.2.25b}$$

The solution of these equations is similar to those given by Eq. (4.2.17),

$$\delta\alpha_r^\nu = \frac{1}{\Delta_0}\left[f_i^\nu \left(\frac{\partial f_r}{\partial \alpha_i}\right)^\nu - f_r^\nu \left(\frac{\partial f_i}{\partial \alpha_i}\right)^\nu\right] \tag{4.2.26a}$$

$$\delta\alpha_r^\nu = \frac{1}{\Delta_0}\left[f_r^\nu \left(\frac{\partial f_i}{\partial \alpha_r}\right)^\nu - f_i^\nu \left(\frac{\partial f_r}{\partial \alpha_r}\right)^\nu\right] \tag{4.2.26b}$$

where

$$\Delta_0 = \left(\frac{\partial f_r}{\partial \alpha_r}\right)^\nu \left(\frac{\partial f_i}{\partial \alpha_i}\right)^\nu - \left(\frac{\partial f_i}{\partial \alpha_r}\right)^\nu \left(\frac{\partial f_r}{\partial \alpha_i}\right)^\nu \tag{4.2.26c}$$

Differentiation of Eq. (3.3.16), with respect to α leads to the derivatives of f_r and f_i with respect to α_r and α_i, and an equation identical to Eq. (4.2.18a) with ω replaced by α_i is obtained. Again the vectors on the right-hand side of Eq. (4.2.18a) are determined from Eqs. (4.2.6), (4.2.8) and (4.2.10), and $(r_1)_j = (r_2)_j = 0$ for $0 \le j \le J$, and $(r_3)_j$ and $(r_4)_j$ for $0 \le j \le J$, are given by Eqs. (4.2.19) with c_1, c_2, c_3 and c_4 now differentiated with respect to α.

4.2.4 Estimation of Eigenvalues

The solution of Eq. (4.2.14) requires initial estimates of two eigenvalues (α_r, α_i) and since Newton's method is used to compute the subsequent estimates, good estimates are needed. A convenient procedure advocated by Keller (sometimes referred to as the continuation method) is described below for this purpose.

We define

$$u = u_{\text{ref}} + n(u_f - u_{\text{ref}}) \tag{4.2.27a}$$

$$u'' = u''_{\text{ref}} + n(u''_f - u''_{\text{ref}}) \tag{4.2.27b}$$

$$R = R_{\text{ref}} + n(R_f - R_{\text{ref}}) \tag{4.2.27c}$$

Here u_f corresponds to the velocity profile of the flow under consideration at a Reynolds number R_f, and u_{ref} denotes a reference velocity profile (say that of Blasius) with a Reynolds number R_{ref} and with eigenvalues of α_0 and ω_0. The parameter n is a sequence of specified numbers ranging from 0 to 1. Eqs. (4.2.27) show that the profiles u and u'' correspond to the reference profiles for $n = 0$ at $R = R_{\text{ref}}$, as well as to those for which eigenvalues α and ω are to be determined at the specified R_f for $n = 1$.

According to the continuation method, initial estimates of (α_r, α_i) for the next x-station x_{i+1}, are obtained by varying n in small increments. Studies show that this is a very useful and effective procedure.

4.3 FORTRAN Program

In this section we present a stability-transition program (STP) for calculating transition in two-dimensional external flows with the e^n-method discussed in the previous sections. STP assumes that the boundary-layer velocity profiles are calculated with IBLP in the standard mode so that u and u'' are available at each station as a function of η. For convenience, the velocity u_0 and length l scales in Eq. (4.0.4) are chosen to correspond to

$$u_0 = u_e, \quad l = \sqrt{\frac{\nu x}{u_e}} = \frac{x}{\sqrt{R_x}} \tag{4.3.1}$$

with Reynolds number in the Orr-Sommerfeld equation now given by

$$R = \frac{u_0 l}{\nu} = \sqrt{R_x} \tag{4.3.2}$$

so that the boundary-layer grid of the velocity profiles can be used in the solution of the Orr-Sommerfeld equation. This means that u and u'' in STP are related to the output of IBLP by

$$u = f', \quad u'' = f'' \tag{4.3.3}$$

and that the output of IBLP can be arranged to calculate u'' once the solutions of the boundary-layer equations are known.

The parameter f''' can be obtained either by differentiating f'' with respect to η or from the finite-differenced momentum equation, Eq. (3.3.5c), which for laminar flow can be written as

$$v_j \equiv (f'''_j)^n = -\alpha_1 (fv)^n_j + \alpha_2 (u^2_j)^n - \alpha^n (v^{n-1}_j f^n_j - f^{n-1}_j v^n_j) + R^{n-1}_j$$

where R^{n-1}_j is given by Eq. (3.3.6d).

With boundary-layer profiles defined at each ξ-station, the stability calculations can be started at any ξ-station where the critical Reynolds number, R_{cr}, is less than the Reynolds number used in the boundary-layer calculations. For external flows, an estimate of R_{cr} can be obtained from Fig. 4.1 with R_{δ^*} known and included in the output subroutine of IBLP.

The stability calculations are first performed for a neutral stability curve at the specified ξ-location where R $(= \sqrt{R_x})$ is known. The calculation of ω and α requires initial estimates for the velocity profiles of u and u''. A convenient procedure to achieve this is the continuation method of subsection 4.2.4 but, to retain a comparatively simple program, this is not part of the present description. The initial estimates of ω and α for a given R can be obtained for Blasius flow except where the pressure gradient parameter m is far from zero, and either trial and error or the continuation method is required.

To calculate the location of the onset of transition, it is usually sufficient to perform neutral stability calculations at four ξ-locations so that four frequencies can be computed and amplification rates determined on four curves with constant dimensional frequencies. While STP is general and can be used for external flows with suction and injection as well as free shear flows such as jets and wakes, changes are required to accommodate boundary conditions and provide better estimates of initial eigenvalues where required [3].

STP comprises several blocks, labeled as MAIN, VELPRO, CSAVE, NEWTON and NEWTONI and are described in the following subsections.

4.3.1 MAIN

In MAIN the total number of ξ-stations (NXT) and the ξ-station where the stability calculations begin (NXO) are identified together with the requirement of neutral stability (IXT $= 0$) or transition. In the latter case, it is necessary to specify the number of frequencies, and this is done by setting IXT to 1, 2, or any number other than zero but less than twenty so that, for example, 1 will lead to transition calculations for one frequency and 2 for two frequencies, etc.

To calculate the neutral stability curves, the dimensionless external velocity UE(I) $(\equiv u_e/u_0)$ is required as a function of S(I) $(\equiv s/L)$, together with Reynolds number R_L $(\equiv u_0 L/\nu)$, a reference length L and velocity UINF $(\equiv u_0)$. The stability Reynolds number REY(I) $(\equiv R)$ is then calculated from Eq. (4.3.2) with R_x given by

$$R_x = \bar{u}_e \left(\frac{s}{L}\right) R_L \tag{4.3.5}$$

The calculation of the eigenvalues α and ω at NX $=$ NXO $(s = s_0)$ also requires the specification of their initial estimates at the Reynolds number corresponding to its value at s_0. For Blasius flow, they can be obtained from the already constructed stability diagrams, such as those given by Fig. 4.4 in [3]. When other

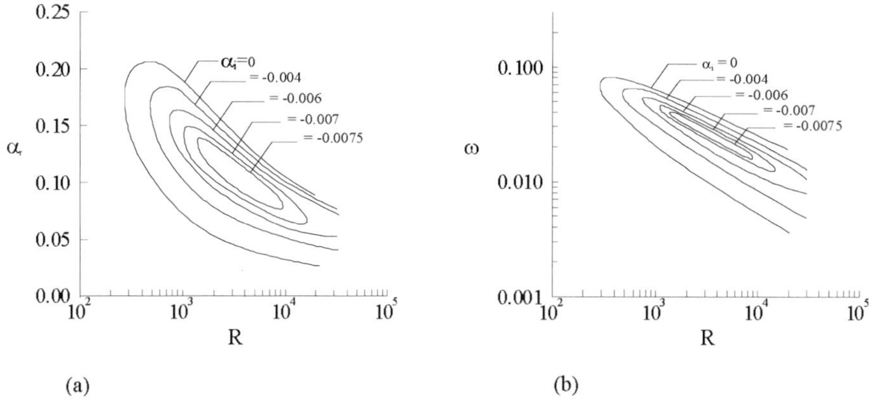

Fig. 4.4. Stability diagrams for Blasius flow. (a) α_r vs R. (b) ω vs R.

figures are used, care should be taken to ensure that the estimated eigenvalues have the same length and velocity scales as those used in STP.

The velocity profiles u and u'' needed in the stability equation are obtained by calling VELPRO, and the calculations for neutral stability are performed by calling NEWTON and for amplification rates α_i and n-factor by calling NEWTONI.

```
C     MAIN PROGRAM
      COMPLEX ALFA,OMEGA,DELA
      COMMON/BLCO/ NP,NX,NXT,NXO,IX,IXT,RL,C,
     +             UINF,ETA(101),DETA(101),A(101)
      COMMON/BLCC/ X(101),S(101),UE(101),RX(101)
      COMMON/BLC5/ ALFA,OMEGA,DELA,DELR,REY(101)
      COMMON/BLC8/ QSUM(101),RLAST(101),LAST(101)
      COMMON/BLCP/ U(101),V(101),UUDP(101),UUB(101),UUDPB(101)
      COMMON/BLCD/ IEND
C - - - - - - - - - - - - - - - - - - - - - - - - - - - - - - - - - - -
      REWIND 1
      READ(5,8000)NXT,NXO,IXT
      READ(5,8101) UINF,C,RL
      READ(5,8101) ALFA,OMEGA
      WRITE(6,9000) NXT,NXO,IXT
      WRITE(6,9100) UINF,RL,C
      READ (5,8101) (X(I),I=1,NXT)
      READ (5,8101) (S(I),I=1,NXT )
      READ (5,8101) (UE(I),I=1,NXT)
C     CALCULATION OF STABILITY REYNOLDS NUMBER
      DO 400 I=1,NXT
      QSUM(I) = 0.0
      RX(I) = UE(I)*S(I)*RL
      REY(I) = SQRT(RX(I))
  400 CONTINUE
```

```
      WRITE(6,9200) (I,X(I),UE(I),RX(I),REY(I),I=1,NXT)
      IX      = 0
      NX      = 1
      IEND    = 0
C     CALCULATION OF VELOCITY PROFILE
 100  CALL VELPRO
      IF ( NX .LT. NXO ) GO TO 500
      IX      = IX + 1
      LAST(IX) = 0
      IF ( IEND .GT. 0 ) LAST(IX) = 1
      IF (IXT.NE.O.AND.IX.GT.IXT) GO TO 350
C     CALCULATION OF NEUTRAL STABILITY CURVE
      IF ( IEND .EQ. 0 ) CALL NEWTON
      IF ( IXT.EQ.0.OR.IX .EQ. 1 ) GO TO 500
C     CALCULATION OF AMPLIFICATION FACTOR
  350 CALL NEWTONI
  500 NX      = NX + 1
      IF ( NX.GT.NXT ) GO TO 1000
      GO TO 100
 8000 FORMAT(20I3)
 8101 FORMAT(7F10.0)
 9000 FORMAT(1H0,5X,10HNXT      = ,I10,2X,10HNXO      = ,I10,2X,
     +     10HIXT      = ,I10 )
 9100 FORMAT(1H0,5X,10HUINF     = ,F10.3,2X,10HRL       = ,F15.1,
     +       2X,10HC        = ,F10.5,///)
 9200 FORMAT(1H0,6X,2HNX,8X,1HX,14X,2HUE,13X,2HRX,12X,3HREY /
     +       (1H ,5X,I3,4E15.6) )
 1000 STOP
      END
```

4.3.2 Subroutine VELPRO

The velocity profiles for STP can be generated by any boundary- layer method or, in some cases such as that of a fully developed duct flow, can be generated analytically. For external flows, we assume that they are provided by the boundary-layer program of Chapter 3. The read-in values of KX and NP in this subroutine refer to the NX- station and the j-points (J) where the profiles are computed. With the specification of U(J) ($\equiv f'_j$) and V(J) ($\equiv f''_j$) as a function of ETA(J) ($\equiv \eta_j$), UUDP(J) ($\equiv u''_j$) are calculated, by differentiating V(J) with respect to η.

```
      SUBROUTINE VELPRO
      COMMON/BLCO/ NP,NX,NXT,NXO,IX,IXT,RL,C,
     +             UINF,ETA(101),DETA(101),A(101)
      COMMON/BLCC/ X(101),S(101),UE(101),RX(101)
      COMMON/BLCP/ U(101),V(101),UUDP(101),UUB(101),UUDPB(101)
      COMMON/BLCD/ IEND
C - - - - - - - - - - - - - - - - - - - - - - - - - - - - - - - - - - -
      READ (1,50) KX,NP
```

```
      IF ( KX .EQ. NX ) GO TO 150
      WRITE ( 6,3000 ) NX,KX,NP
      STOP 2
  150 READ (1,60) ( ETA(J),J=1,NP )
      READ (1,60) ( U(J),J=1,NP )
      READ (1,60) ( V(J),J=1,NP )
      DO 400 J=2,NP
      DETA(J-1)=ETA(J)-ETA(J-1)
      A(J)=0.5*DETA(J-1)
 400  CONTINUE
C     COMPUTE UUDP(J)
      NP1     = NP-1
      DETAS   = ETA(2)-ETA(1)
      DV      = V(2)-V(1)
      ANG2    = ATAN2(DV,DETAS)
      DETAI   = DETAS
      DVI     = DV
      DETA2   = DETAS
      DO 10 I = 2,NP1
      ANG1    = ANG2
      DETA1   = DETA2
      I1      = I+1
      DETAS   = ETA(I1)-ETA(I)
      DV      = V(I1)-V(I)
      ANG2    = ATAN2(DV,DETAS)
      DETA2   = DETAS
      ANG     = (DETA2*ANG1+DETA1*ANG2)/(DETA1+DETA2)
      UUDP(I) = TAN(ANG)
   10 CONTINUE
      DVF     = DV
      DETAF   = DETA2
C     EXTRAPOLATE FOR END VALUES
      UUDP(1) = 2.*DVI/DETAI - UUDP(2)
      UUDP(NP)= 2.*DVF/DETAF - UUDP(NP1)
      DO 180 J=2,NP
      UUB(J)    = 0.5*(U(J)+U(J-1))
      UUDPB(J)  = 0.5*(UUDP(J)+UUDP(J-1))
 180  CONTINUE
 50   FORMAT ( 2I5 )
 60   FORMAT ( 6E14.6 )
 3000 FORMAT(1H0,2X,'INCONSISTENT DATA.  NX = ',I3,2X,'KX = ',I3,2X,
     +       'NP = ',I4 / 3X,'PROGRAM STOPS.' )
      RETURN
      END
```

4.3.3 Subroutine CSAVE

This subroutine is used to obtain the solutions of the Orr-Sommerfeld equation for a given set of α and ω when the neutral stability curve is required or α_r and α_i for the determination of the location of the onset of transition. The standard problem refers to the solution of Eq. (3.3.16) subject to the boundary condition that $\phi''(0)$ ($\equiv s_0$) is equal to 1. The definitions of $(c_1)_j$ to $(c_4)_j$ in the Orr-Sommerfeld equation, as well as the edge definitions, are given by Eqs. (4.2.7) and (4.2.9), respectively, and the $\vec{r}_j$ terms by Eqs. (4.2.8b) and (4.2.10). All values of $\vec{r}_j$ are zero except for $(r_2)_0$ which is equal to 1 because of the requirement that $s_0 \equiv 1$.

This subroutine also contains the coefficients of the variational equations of the standard problem, Eq. (3.3.16), with respect to α, ω and R, together with the right-hand sides of these equations as given by Eq. (4.2.19) for those with respect to α, Eq. (4.2.20) for ω, and Eq. (4.2.24) for R.

The variational equations are written with respect to α, ω and R, but the coefficient matrix A in Eq. (3.3.16) is the same. For this reason, it is only necessary to define those $(\vec{r}_j)$ terms that are not zero.

This subroutine also contains the block-elimination algorithm to solve Eq. (3.3.16) with the procedure described in subsection 3.3.3. Note that AA(1,1,1) and AA(2,2,1) denote $(a_{11})_0$ and $(a_{22})_0$ and correspond to the "wall" boundary conditions.

```
      SUBROUTINE CSAVE
      COMPLEX  ALFA,OMEGA,SQM1,ALSQ,REYI,GAMMA,GAMMAS,AIREY,
     1         SQUIG,SSIV,R,C1,C2,C3,C4,BC1,BC2,BC3,BC4,
     2         SS,FS,WS,GS,WSAR,SSAR,WSOM,SSOM,WSR,SSR,WB,SB,
     3         DELS,DELF,DELW,DELG,AA,AJ,AA1,AA2,AA3,AA4,DET,
     4         GG,W1,W2,W3,W4,BB1,BB2,BB3,BB4,BB5,BB6,DETT,CC1,CC2,DD1,
     5         DD2,DD3,DD4,DD5,DD6,DENO,UM,UMA,UMO,UMR,DELA
      COMMON/BLCO/ NP,NX,NXT,NXO,IX,IXT,RL,C,
     +             UINF,ETA(101),DETA(101),A(101)
      COMMON/BLCC/ X(101),S(101),UE(101),RX(101)
      COMMON/BLC5/ ALFA,OMEGA,DELA,DELR,REY(101)
      COMMON/BLC3/ C1(101),C2(101),C3(101),C4(101)
      COMMON/BLC4/ UM(4,1),UMA(4,1),UMO(4,1),UMR(4,1)
      COMMON/BLCP/ U(101),V(101),UUDP(101),UUB(101),UUDPB(101)
      DIMENSION AA(4,4,101),W1(101),W2(101),W3(101),W4(101),R(4,101),
     1          GG(4,4,101),DELW(101),DELS(101),DELF(101),DELG(101),
     2          SS(101),FS(101),WS(101),GS(101),WSAR(101),
     3          SSAR(101),WSOM(101),SSOM(101),WSR(101),SSR(101)
C - - - - - - - - - - - - - - - - - - - - - - - - - - - - - - - - - - -
      SQM1     = (0.0,1.0)
      GAMMAS   = ALFA**2
      GAMMA    = CSQRT(GAMMAS)
      REYI     = SQM1*REY(NX)
      SQUIG    = CSQRT(GAMMAS+REYI*(ALFA*U(NP)-OMEGA))
      II       = 0
      II       = II + 1
      DO 150 J = 2,NP
      C3(J)    = 0.5*DETA(J-1)
      C1(J)    = C3(J)*GAMMAS
      C2(J)    = -C3(J)*REYI*ALFA*UUDPB(J)
```

```
      C4(J)     = C3(J)*(GAMMAS+REYI*(ALFA*UUB(J)-OMEGA))
C     STANDARD PROBLEM
      DO 150 K = 1,4
  150 R(K,J)    = (0.0,0.0)
      R(2,1)    = (1.0,0.0)
      R(1,1)    = (0.0,0.0)
      R(3,1)    = (0.0,0.0)
      R(4,1)    = (0.0,0.0)
C     EDGE BOUNDARY CONDITIONS
      ALSQ      = GAMMA+SQUIG
      BC1       = GAMMA
      BC2       = (0.0,0.0)
      BC3       = 1.0/ALSQ
      BC4       = SQUIG
      GO TO 2000
C     VARIATIONAL EQUATIONS
  200 II        = II + 1
C     WRT  ALFA
      AIREY     = REYI*ALFA
      DO 250 J = 2,NP
      WB        = 0.5*(WS(J)+WS(J-1))
      SB        = 0.5*(SS(J)+SS(J-1))
      R(3,J-1) = -2.0*WB*DETA(J-1)*ALFA
      R(4,J-1) = -SB*DETA(J-1)*(2.0*ALFA+REYI*UUB(J))+
     1            WB*DETA(J-1)*REYI*UUDPB(J)
  250 CONTINUE
      R(3,NP)   = SS(NP)*(ALFA/GAMMA+ALFA/SQUIG+0.5*REYI*U(NP)/
     1            SQUIG)/(GAMMA+SQUIG)**2-WS(NP)*ALFA/GAMMA
      R(4,NP)   = -SS(NP)*(2.0*ALFA+REYI*U(NP))/SQUIG*0.5
      R(2,1)    = (0.0,0.0)
      GO TO 2000
  400 II        = II + 1
C     WRT OMEGA
      DO 425 J = 2,NP
      SB        = 0.5*(SS(J)+SS(J-1))
      R(3,J-1) = (0.0,0.0)
      R(4,J-1) = SB*DETA(J-1)*REYI
  425 CONTINUE
      R(4,NP)   = SS(NP)*REYI/SQUIG*0.5
      R(3,NP)   = -R(4,NP)/(GAMMA+SQUIG)**2
      R(2,1)    = (0.0,0.0)
      GO TO 2000
  500 II        = II + 1
C     WRT REY
      DO 525 J = 2,NP
      SB        = 0.5*(SS(J)+SS(J-1))
      WB        = 0.5*(WS(J)+WS(J-1))
      R(3,J-1) = (0.0,0.0)
      R(4,J-1) = -SQM1*DETA(J-1)*(SB*(ALFA*UUB(J)-OMEGA)
     1            -WB*ALFA*UUDPB(J))
  525 CONTINUE
```

```
      SSIV     = ALFA*U(NP)-OMEGA
      R(4,NP)  = -SS(NP)*SQM1*SSIV*0.5/SQUIG
      R(3,NP)  = -R(4,NP)/(GAMMA+SQUIG)**2
      R(2,1)   = (0.0,0.0)
C     BLOCK ELIMINATION ALGORITHM
 2000 DO 2050 I = 1,4
      DO 2050 K = 1,4
      AA(I,K,1) = (0.0,0.0)
 2050 CONTINUE
C     WALL BOUNDARY CONDITIONS
      AA(1,1,1) = (1.0,0.0)
      AA(2,2,1) = (1.0,0.0)
      W1(1)    = R(1,1)
      W2(1)    = R(2,1)
      W3(1)    = R(3,1)
      W4(1)    = R(4,1)
C     FORWARD SWEEP
      DO 2100 J = 2,NP
      AJ       = C3(J)
      AA1      = AA(1,1,J-1)-C1(J  )*AA(1,3,J-1)-C2(J  )*AA(1,4,J-1)
      AA2      = AA(2,1,J-1)-C1(J  )*AA(2,3,J-1)-C2(J  )*AA(2,4,J-1)
      AA3      = AA(1,2,J-1)-C3(J  )*AA(1,3,J-1)-C4(J  )*AA(1,4,J-1)
      AA4      = AA(2,2,J-1)-C3(J  )*AA(2,3,J-1)-C4(J  )*AA(2,4,J-1)
      DET      = AA1*AA4-AA2*AA3
      GG(1,1,J) = (AA4*(C1(J  )*AJ-1.0)-AA2*C3(J  )*AJ)/DET
      GG(1,2,J) = (AA1*C3(J  )*AJ-AA3*(C1(J  )*AJ-1.0))/DET
      GG(1,3,J) = -AJ-AA(1,3,J-1)*GG(1,1,J)-AA(2,3,J-1)*GG(1,2,J)
      GG(1,4,J) = -AA(1,4,J-1)*GG(1,1,J)-AA(2,4,J-1)*GG(1,2,J)
      GG(2,1,J) = (C2(J  )*AJ*AA4-AA2*(C4(J  )*AJ-1.0))/DET
      GG(2,2,J) = (AA1*(C4(J  )*AJ-1.0)-AA3*C2(J  )*AJ)/DET
      GG(2,3,J) = -AA(1,3,J-1)*GG(2,1,J)-AA(2,3,J-1)*GG(2,2,J)
      GG(2,4,J) = -AJ-AA(1,4,J-1)*GG(2,1,J)-AA(2,4,J-1)*GG(2,2,J)
      AA(1,1,J) = (1.0,0.0)-C1(J  )*GG(1,3,J)-C2(J  )*GG(1,4,J)
      AA(1,2,J) = -GG(1,3,J)*C3(J  )-GG(1,4,J)*C4(J  )
      AA(1,3,J) = -AJ+GG(1,3,J)
      AA(1,4,J) = GG(1,4,J)
      AA(2,1,J) = -C1(J  )*GG(2,3,J)-C2(J  )*GG(2,4,J)
      AA(2,2,J) = (1.0,0.0)-GG(2,3,J)*C3(J  )-GG(2,4,J)*C4(J  )
      AA(2,3,J) = GG(2,3,J)
      AA(2,4,J) = -AJ+GG(2,4,J)
      W1(J)    = R(1,J)-GG(1,1,J)*W1(J-1)-GG(1,2,J)*W2(J-1)
     1           -GG(1,3,J)*W3(J-1)-GG(1,4,J)*W4(J-1)
      W2(J)    = R(2,J)-GG(2,1,J)*W1(J-1)-GG(2,2,J)*W2(J-1)
     1           -GG(2,3,J)*W3(J-1)-GG(2,4,J)*W4(J-1)
      W3(J)    = R(3,J)
      W4(J)    = R(4,J)
 2100 CONTINUE
C     BACKWARD SWEEP
      J = NP
      BB1      = AA(1,1,J)-AA(1,3,J)*BC1  -AA(1,4,J)*BC2
      BB2      = AA(1,2,J)-AA(1,3,J)*BC3  -AA(1,4,J)*BC4
```

```
      BB3      = AA(2,1,J)-AA(2,3,J)*BC1  -AA(2,4,J)*BC2
      BB4      = AA(2,2,J)-AA(2,3,J)*BC3  -AA(2,4,J)*BC4
      BB5      = W1(J)-AA(1,3,J)*W3(J)-AA(1,4,J)*W4(J)
      BB6      = W2(J)-AA(2,3,J)*W3(J)-AA(2,4,J)*W4(J)
      DETT     = BB1*BB4-BB2*BB3
      DELW(J)  = (BB4*BB5-BB2*BB6)/DETT
      DELS(J)  = (BB1*BB6-BB3*BB5)/DETT
      DELF(J)  = W3(J)-BC1  *DELW(J)-BC3  *DELS(J)
      DELG(J)  = W4(J)-BC2  *DELW(J)-BC4  *DELS(J)
 2150 J        = J-1
      CC1      = W3(J)-C1(J+1)*DELW(J+1)-C3(J+1)*DELS(J+1)+DELF(J+1)
      CC2      = W4(J)-C2(J+1)*DELW(J+1)-C4(J+1)*DELS(J+1)+DELG(J+1)
      DD1      = AA(1,1,J)-AA(1,3,J)*C1(J+1)-AA(1,4,J)*C2(J+1)
      DD2      = AA(1,2,J)-AA(1,3,J)*C3(J+1)-AA(1,4,J)*C4(J+1)
      DD3      = AA(2,1,J)-AA(2,3,J)*C1(J+1)-AA(2,4,J)*C2(J+1)
      DD4      = AA(2,2,J)-AA(2,3,J)*C3(J+1)-AA(2,4,J)*C4(J+1)
      DD5      = W1(J)-AA(1,3,J)*CC1-AA(1,4,J)*CC2
      DD6      = W2(J)-AA(2,3,J)*CC1-AA(2,4,J)*CC2
      DENO     = DD1*DD4-DD2*DD3
      DELW(J)  = (DD5*DD4-DD2*DD6)/DENO
      DELS(J)  = (DD1*DD6-DD3*DD5)/DENO
      DELF(J)  = CC1-C1(J+1)*DELW(J)-C3(J+1)*DELS(J)
      DELG(J)  = CC2-C2(J+1)*DELW(J)-C4(J+1)*DELS(J)
      IF (J .GT. 1) GO TO 2150
      GO TO ( 2500, 2600, 2800,2900 ), II
C     SOLUTION OF THE STANDARD PROBLEM
 2500 DO 2510 J = 1,NP
      WS(J)    = DELW(J)
      SS(J)    = DELS(J)
      FS(J)    = DELF(J)
      GS(J)    = DELG(J)
 2510 CONTINUE
      UM(3,1)  = DELF(1)
      GO TO 200
C     SOLUTION OF THE VARIATIONAL EQUATION -- WRT ALFA
 2600 DO 2610 J = 1,NP
      WSAR(J)  = DELW(J)
      SSAR(J)  = DELS(J)
 2610 CONTINUE
      UMA(3,1) = DELF(1)
      GO TO 400
C     SOLUTION OF THE VARIATIONAL EQUATION --WRT OMEGA
 2800 DO 2810 J = 1,NP
      WSOM(J)  = DELW(J)
      SSOM(J)  = DELS(J)
 2810 CONTINUE
      UMO(3,1) = DELF(1)
      GO TO 500
C     SOLUTION OF THE VARIATIONAL EQUATION --WRT REY
 2900 DO 2910 J=1,NP
      WSR(J)=DELW(J)
```

```
      SSR(J)=DELS(J)
 2910 CONTINUE
      UMR(3,1)=DELF(1)
      RETURN
      END
```

4.3.4 Subroutine NEWTON

For initial estimates of ALFA ($\equiv \alpha_r$) and OMEGA ($\equiv \omega$), this subroutine computes the eigenvalues α and ω according to the procedure described in subsection 4.2.2. The perturbation quantities DALFA ($\equiv \delta\alpha_r$) and DOMEGA ($\equiv \delta\omega$) are computed according to Eq. (4.2.17). Upon convergence of the iterations, the dimensional frequencies WSO(IX) ($\equiv \omega^*$) are calculated and printed for that NX-station, and for each corresponding frequency, IX, together with the values of α and ω which serve as initial estimates for the next NX-station (or R), are stored. Also stored are the values of UM ($\equiv f_0$), UMA ($\equiv \partial f_0/\partial\alpha$), UMO ($\equiv \partial f_0/\partial\omega$), UMR ($\equiv \partial f_0/\partial R$), UE and REY.

```
      SUBROUTINE NEWTON
      COMPLEX  ALSAVE,OMSAVE,FSSAVE,FASAVE,FOSAVE,FRSAVE,ALFA1,OMEGA1,
     1         UM,UMA,UMO,UMR,ALFA,OMEGA,FSTD,FALFA,FOMGA,DELA
      DIMENSION  WSO(101)
      COMMON/BLCO/ NP,NX,NXT,NXO,IX,IXT,RL,C,
     +             UINF,ETA(101),DETA(101),A(101)
      COMMON/BLCC/ X(101),S(101),UE(101),RX(101)
      COMMON/BLCP/ U(101),V(101),UUDP(101),UUB(101),UUDPB(101)
      COMMON/BLC5/ ALFA,OMEGA,DELA,DELR,REY(101)
      COMMON/BLC8/ QSUM(101),RLAST(101),LAST(101)
      COMMON/SAVE/ ALSAVE(2,20),OMSAVE(2,20),FSSAVE(2,20),
     1             FASAVE(2,20),FOSAVE(2,20),FRSAVE(2,20),
     2             RSAVE(2,20),UESAVE(2,20),QSAVE(2,20)
      COMMON/BLC4/ UM(4,1),UMA(4,1),UMO(4,1),UMR(4,1)
      COMMON/BLCD/ IEND
      DATA     EPS / 1.0E-04 /
C - - - - - - - - - - - - - - - - - - - - - - - - - - - - - - - - - - -
      WRITE ( 6, 10 ) NX, X(NX), S(NX), REY(NX), NP
      ISOLV = 0
      ITIME   = 0
      DALPHA  = 0.
      DOMEGA  = 0.
      WRITE ( 6, 20 )
      IF ( IX .EQ. 1 ) GO TO 90
      ALFA    = ALFA1
      OMEGA   = OMEGA1
   90 AR      = REAL ( ALFA )
      OR      = REAL ( OMEGA )
      AI      = AIMAG ( ALFA )
      OI      = AIMAG ( OMEGA )
      WRITE ( 6, 30 ) ITIME,AR,OR,DALPHA,DOMEGA
```

```
      CALL CSAVE
      ITIME    = ITIME + 1
      FSTD     = UM (3,1)
      FALFA    = UMA(3,1)
      FOMGA    = UMO(3,1)
      DELTAO   = REAL(FALFA)*AIMAG(FOMGA) - REAL(FOMGA)*AIMAG(FALFA)
      IF ( ABS(DELTAO) .LE. 1.0E-06 ) GO TO 500
      D1       = 1.0 / DELTAO
      DALPHA   = D1 *(AIMAG(FSTD)*REAL(FOMGA)-REAL(FSTD)*AIMAG(FOMGA) )
      DOMEGA   = D1 *(REAL(FSTD)*AIMAG(FALFA)-AIMAG(FSTD)*REAL(FALFA) )
      GO TO 114
  500 WRITE (6,100) DELTAO
      IEND    = 1
      RETURN
  114 AR1      = AR + DALPHA
      OM1      = OR + DOMEGA
      ALFA     = CMPLX ( AR1,AI )
      OMEGA    = CMPLX ( OM1,OI )
      IF ( (ABS(DALPHA) .LE. EPS) .AND. (ABS(DOMEGA) .LE. EPS) )
     +      GO TO 115
      IF ( ITIME .LE. 10 ) GO TO 90
      WRITE ( 6, 50 )
      STOP 5
C     SAVE DATA FOR DEFAULT VALUE
  115 WRITE ( 6, 30 ) ITIME,AR1,OM1,DALPHA,DOMEGA
      WSO(IX)  = REAL(OMEGA)*UE(NX)*UINF*REY(NX)/S(NX)/C
      IF ( IX .GT. 20 ) GO TO 130
      ALSAVE(1,IX) = ALFA
      OMSAVE(1,IX) = OMEGA
      FSSAVE(1,IX) = UM(3,1)
      FASAVE(1,IX) = UMA(3,1)
      FOSAVE(1,IX) = UMO(3,1)
      FRSAVE(1,IX) = UMR(3,1)
      RSAVE (1,IX) = REY(NX)
      UESAVE(1,IX) = UE(NX)
      QSAVE (1,IX) = 0.0
      WRITE ( 6, 60 ) IX, WSO(IX)
  130 CONTINUE
      ALFA1    = ALFA
      OMEGA1   = OMEGA
   10 FORMAT ( 1H0,5X,5HNX = ,I3,2X, 8HX(NX) = , F12.6, 2X,8HS(NX) = ,
     +         F12.6,2X,6HREY = , F12.3, 2X, 5HNP = , I3 )
   20 FORMAT ( 1H0, 5X, 2HIT, 6X, 4HALFA, 11X,
     +         5HOMEGA,10X, 5HDALFA,  10X, 5HDOMGA / )
   30 FORMAT ( 1H , 5X, I2, 6(2X, E13.6) )
   50 FORMAT ( 1H0,5X,42HITERATION EXCEEDED MAXIMUM. PROGRAM STOPS. )
   60 FORMAT ( 1H0, 5X, 9HLINE NO. , I3, 2X, 21HPHYSICAL FREQUENCY = ,
     +         2E15.6 )
  100 FORMAT ( 1H0, 5X, 15H+++++ DELTAO = , E13.6, 14H TOO SMALL TO ,
     +         25HCALCULATE DALFA AND DOMGA  )
```

```
      RETURN
      END
```

4.3.5 Subroutine NEWTONI

In the calculation of transition, the amplification rates α_i are computed for a dimensional frequency ω^* according to Eq. (4.1.3) which in terms of the dimensionless quantities used in the solution of the Orr-Sommerfeld equation with CLOGA corresponding to the exponent n in the e^n-method, can be written as

$$n = -\int_{\xi_0}^{\xi} \left(\frac{\alpha_i R}{\xi} \right) d\xi \tag{4.3.6}$$

```
      SUBROUTINE NEWTONI
      COMPLEX  ALFA,OMEGA,DELA,UM,UMA,UMO,UMR,FSTD,FALFA,FOMGA,
     1         FREY,DODR,DADR,ALSAVE,OMSAVE,FSSAVE,FASAVE,FOSAVE,
     2         FRSAVE
      COMMON/BLCO/ NP,NX,NXT,NXO,IX,IXT,RL,C,
     +             UINF,ETA(101),DETA(101),A(101)
      COMMON/BLCP/ U(101),V(101),UUDP(101),UUB(101),UUDPB(101)
      COMMON/SAVE/ ALSAVE(2,20),OMSAVE(2,20),FSSAVE(2,20),
     1             FASAVE(2,20),FOSAVE(2,20),FRSAVE(2,20),
     2             RSAVE(2,20),UESAVE(2,20),QSAVE(2,20)
      COMMON/BLCC/ X(101),S(101),UE(101),RX(101)
      COMMON/BLC4/ UM(4,1),UMA(4,1),UMO(4,1),UMR(4,1)
      COMMON/BLC5/ ALFA,OMEGA,DELA,DELR,REY(101)
      COMMON/BLC8/ QSUM(101),RLAST(101),LAST(101)
      DATA     EPS / 1.0E-04 /
C - - - - - - - - - - - - - - - - - - - - - - - - - - - - - - - - - - -
      WRITE ( 6, 10 ) NX, X(NX), S(NX), REY(NX), NP
      IXM      = IX-1
      DO 1000 I = 1,IXM
      DELAR    = 0.
      DELAI    = 0.
      IT       = 0
      K        = IX-I
      IF ( K .GT. IXT ) GO TO 1000
      J        = I+1
      IF ( LAST(K) .GT. 0 ) GO TO 1000
      WRITE ( 6, 83 ) K, J
      AR       = REAL ( ALSAVE(1,K) )
      AI       = AIMAG( ALSAVE(1,K) )
      OR       = REAL ( OMSAVE(1,K) )
      OI       = AIMAG( OMSAVE(1,K) )
      FSTD     = FSSAVE ( 1,K )
      FALFA    = FASAVE ( 1,K )
      FOMGA    = FOSAVE ( 1,K )
      FREY     = FRSAVE ( 1,K )
      UEO      = UESAVE ( 1,K )
```

```
      RO       = RSAVE  ( 1,K )
      DR       = REY(NX) - RO
      OMEGA    = OMSAVE(1,K)*(REY(NX)/RO)*(UEO**2/UE(NX)**2)
      CC       = OR*UEO**2 / RO
      DODR     = (OMEGA-OMSAVE(1,K))/DR
      DADR     = -(1.0/FALFA) * ( FOMGA*DODR+FREY )
      DELAR    = REAL( DADR ) * DR
      DELAI    = AIMAG( DADR ) * DR
      DELA     = CMPLX ( DELAR, DELAI )
      AR1      = AR + DELAR
      AI1      = AI + DELAI
      WRITE (6,84)
  520 ALFA     = CMPLX( AR1,AI1 )
      WRITE (6,85) IT,ALFA,OMEGA,DELA
      CALL CSAVE
      FSTD     = UM(3,1)
      FALFA    = UMA(3,1)
      FOMGA    = UMO(3,1)
      FREY     = UMR(3,1)
      DELA     = -FSTD / FALFA
      IF ((ABS(REAL(DELA)) .LE. EPS) .AND. (ABS(AIMAG(DELA)) .LE. EPS))
     +     GO TO 700
      AR1      = REAL( ALFA ) + REAL( DELA )
      AI1      = AIMAG( ALFA ) + AIMAG( DELA )
      IT       = IT + 1
      IF ( IT .LE. 10 ) GO TO 520
      WRITE ( 6, 70 )
      STOP 7
  700 IT       = IT + 1
      ALFA = ALFA +DELA
      WRITE (6,85) IT,ALFA,OMEGA,DELA
      ALSAVE(2,K) = ALFA
      OMSAVE(2,K) = OMEGA
      FSSAVE(2,K) = UM(3,1)
      FASAVE(2,K) = UMA(3,1)
      FOSAVE(2,K) = UMO(3,1)
      FRSAVE(2,K) = UMR(3,1)
      RSAVE  (2,K) = REY(NX)
      UESAVE(2,K) = UE(NX)
      QSAVE  (2,K) = ( REY(NX) * AIMAG( ALFA ) ) / S(NX)
      QSUM(K) = QSUM(K)+((QSAVE(2,K)+QSAVE(1,K))/2.0)*(S(NX)-S(NX-1))
      CLOGA    = -QSUM(K)
      WRITE ( 6,50 ) QSAVE(2,K),CLOGA
      PROD     = QSAVE(2,K)*QSAVE(1,K)
      IF ( PROD .GE. 0.0 ) GO TO 960
      LAST(K)  = 1
      WRITE (6,88) K, J
      GO TO 1000
  960 IF ( ABS(QSAVE(2,K)).GT.1.0E-05 ) GO TO 1000
      LAST(K) = 1
      WRITE (6,87) K, J
```

```
1000 CONTINUE
     DO 1400 I=1,IXM
     K       = IX-I
     IF ( K .GT. IXT ) GO TO 1400
     ALSAVE(1,K) = ALSAVE(2,K)
     OMSAVE(1,K) = OMSAVE(2,K)
     FSSAVE(1,K) = FSSAVE(2,K)
     FASAVE(1,K) = FASAVE(2,K)
     FOSAVE(1,K) = FOSAVE(2,K)
     FRSAVE(1,K) = FRSAVE(2,K)
     RSAVE (1,K) = RSAVE (2,K)
     UESAVE(1,K) = UESAVE(2,K)
     QSAVE (1,K) = QSAVE (2,K)
1400 CONTINUE
  10 FORMAT ( 1H0,5X,5HNX = ,I3,2X, 8HX(NX) = , F12.6, 2X,8HS(NX) = ,
   +         F12.6,2X,6HREY = , F12.3, 2X, 5HNP = , I3 )
  50 FORMAT ( 1H0, 5X, 8HQSAVE = ,F15.6,8HCLOGA = , F15.6 )
  70 FORMAT ( 1H0,5X,39HITERATION EXCEEDS 10 ( FOR AR AND AI ) , 2X,
   +         13HPROGRAM STOPS  )
  83 FORMAT ( 1H0, 5X, 9HLINE NO. , I3, 2X, 10HPOINT NO. , I3 /
   +         1H , 5X, 27(1H-) / )
  84 FORMAT ( 1H0, 5X, 2HIT, 6X, 5HALFAR, 10X, 5HALFAI, 10X, 5HOMGAR,
   +         10X, 5HOMGAI, 10X, 5HDELAR, 10X, 5HDELAI / )
  85 FORMAT ( 1H , 5X, I2, 6( 2X, E13.6 ) )
  87 FORMAT ( 1H0, 5X, 43H***** ALFA(IMAG) IS LE 1.0E-05 AT LINE NO. ,
   +         I2, 16H  AND POINT NO. , I2 / )
  88 FORMAT ( 1H0, 5X, 36H***** ALFAI IS POSITIVE AT LINE NO. , I2,
   +         16H  AND POINT NO. , I2 / )
     RETURN
     END
```

References

[1] Kleiser, L. and Zang, T.A., "Numerical simulation of transition in wall bounded shear flows," Annual review of Fluid Mech., Vol. 23, pp. 495-537, 1991.

[2] Herbert, T., "Parabolized Stability Equations," Special Course on Progress in Transition Modeling, AGARD Report 793, April 1994.

[3] Cebeci, T. and Cousteix, J., Modeling and Computation of Boundary-Layer Flows, Horizons and Springer-Verlag, 1998.

[4] Cebeci, T. and Egan, D., "Prediction of transition due to isolated roughness," AIAA J., Vol. 27, pp. 870-875, 1989.

[5] Kaplan, R.E., "The stability of laminar incompressible boundary-layers in the presence of compliant boundaries," M.I.T. Aero-Electric and Structures Research Lab Report ASR L-TR-116-1, 1964.

[6] Wazzan, A.R., Okamura, T.T. and Smith, A.M.O., "Spatial and temporal stability charts for the Falkner-Skan Boundary-Layer Profiles," Douglas Aircraft Company, Long Beach, California, Rep. DAC-67086, 1968.

[7] Skorkowski, A. and Orszag, S.A., "Mass flow requirements for LFC wing design," AIAA paper no. 77-1222, 1977.

[8] Malik, M.R., "COSAL-A black-box compressible stability analysis code for transition prediction in three-dimensional boundary-layers," NASA CR 155952, 1982.

5 Applications of the Calculation Method: Airfoils in Incompressible Flows

5.0 Introduction

Before we discuss the applications of the calculation method described in Chapters 2 to 4 to several airfoils in incompressible flows, it is useful to present a brief description of its solution procedure.

The panel method provides an external velocity distribution for the specified airfoil geometry and identifies the airfoil stagnation point. The inverse boundary-layer method obtains solutions on the upper and lower surfaces from the stagnation point. If the onset of the transition location is not specified, the laminar flow solutions are used to calculate the transition location by using either Michel's formula, Eq. (1.4.1), or the e^n-method described in Chapter 4. As discussed in Section 1.4, when the Reynolds number is high and the angles of attack are low to moderate, the onset of transition occurs before the flow separation point, and the extent of the transition region is confined to a relatively small region. In this case, it is satisfactory to determine the onset of transition from Eq. (1.4.1) by calculating local Reynolds numbers based on momentum thickness, R_θ, and surface distance from the leading edge, R_x, at each chordwise station. Initially the computed Reynolds numbers in the coordinates of R_θ vs R_x are always beneath the transition curve, and if transition occurs, then the calculated R_θ vs R_x values will cross over this curve. The onset of transition location is then determined by interpolating between the two sets of R_θ and R_x values on either side of the transition curve. Sometimes, however, especially at higher angles of attack, before transition location can be computed with this procedure, laminar separation takes place. In that case, the separation location may be assumed to correspond to the onset of transition.

When the Reynolds number is low and the angles of attack are low to moderate, it is necessary to use the e^n-method. Of course, this method can also be used when the Reynolds number is high. In either case, the computed laminar velocity profiles are analyzed with the linear stability method and growth rates are determined for different physical frequencies. Transition is assumed to take place when the integrated amplification factor reaches to an n-value between 8 and 10. At higher angles of attack, as in the case of high Reynolds number flows, the prediction of the onset of transition with the e^n-method must be replaced with the assumption that the onset of transition corresponds to the separation location.

Once the onset of transition location is known, the viscous flow solutions are obtained for both laminar and turbulent flows on the airfoil and in the wake, with calculations taking place separately on the upper and lower surfaces of the airfoil and wake. For a given external velocity distribution, these calculations are repeated. Each boundary-layer calculation, starting at the stagnation point and ending at some specified ξ-location in the wake is called a sweep. In sweeping through the boundary-layer, the right-hand side of Eq. (3.2.9) uses the values of δ^* from the previous sweep when $j > i$ and the values from the current sweep when $j < i$. Thus, at each ξ-station the right-hand side of Eq. (3.2.9) provides a prescribed value for the linear combination of $u_e(\xi^i)$ and $\delta^*(\xi^i)$. After convergence of the Newton iterations at each station, the summations of Eq. (3.2.9) are updated for the next ξ-station. Note that the Hilbert integral coefficients C_{ij} discussed in subsection 3.4 have been computed and stored at the start of the boundary-layer calculations.

At the completion of the boundary-layer sweeps on the airfoil and in the wake, boundary-layer solutions are available on the airfoil and in the wake. The blowing velocity on the airfoil v_{iw} [see Eqs. (1.2.8) and (1.2.9)] and a jump in the normal velocity component Δv_i in the wake [see Eq. (1.2.10)], which for an incompressible flow are

$$v_{iw} = \frac{d}{dx}(u_{iw}\delta^*_A) \tag{5.0.1a}$$

$$\Delta v_i = \frac{d}{dx}(u_{iu}\delta^*_u) + \frac{d}{dx}(u_{il}\delta^*_l) \tag{5.0.1b}$$

are calculated and are used to obtain a new distribution of external velocity $u_{ei}(x)$ from the inviscid method. As before, the onset of transition location is determined from the laminar flow solutions and the boundary-layer calculations are performed on the upper and lower surfaces of the airfoil and in the wake by making several specified sweeps. This sequence of calculations is repeated for the whole flowfield until convergence is achieved.

In the following sections, this calculation procedure is first applied to single airfoils and then to multielement airfoils. Section 5.1 describes the applications to clean airfoils, that is, airfoils with smooth surfaces, at high Reynolds numbers. Section 5.3 describes the applications to clean airfoils at low Reynolds numbers following an assessment of the e^n-method in Section 5.2 for predicting the onset of transition in flows with separation. It is followed by Section 5.4, which describes the application of the method to airfoils at low Mach numbers.

Section 5.5 describes a popular method for predicting ice shapes on airfoils and Section 5.6 describes the applications of the calculation method to iced airfoils and to airfoils with leading-edge roughness.

Section 5.7 describes the application of the calculation method to high-lift flows. Again the inviscid flow is computed with a panel method based on the extension of the panel method with viscous corrections described in Chapter 2 for single airfoils. Results are presented for a single airfoil with a flap well as well as for two-element airfoils with and without flap wells.

5.1 Airfoils at High Reynolds Numbers

In this section the calculation method of Chapters 2 and 3 is applied to airfoils at high Reynolds numbers with the onset of transition location computed by Eq. (1.4.1), except where the boundary-layer separates upstream of this location, in which case transition is assumed to correspond to the separation point. Results are presented in a sequence which corresponds to the complexity of the flow. Thus, the first results are presented for a symmetrical airfoil, NACA 0012, and are followed by a cambered airfoil, NACA 23012 airfoil, and a NASA supercritical airfoil. All three airfoils have been investigated for angles of attack beyond stall.

In performing these calculations for airfoils, either at high or low Reynolds numbers, it is usually sufficient to neglect the wake effect and perform the calculations on the airfoil only, provided that there is no or little flow separation on the airfoil. With flow separation, the relative importance of including the wake effect in the calculations depends on the flow separation as shown in Fig. 5.1. Figure 5.1a shows the computed separation location on a NACA 0012 airfoil at a chord Reynolds number, R_c, of 6×10^6. When the wake effect is included, separation is encountered for angles of attack greater than $10°$, and attempts to obtain results without consideration of the wake effect lead to erroneously large regions of recirculation that increases with angle of attack, as discussed in [1]. Figure 5.1b shows that the difference in displacement thickness at the trailing edge is negligible for $\alpha = 10°$ but more than 30% for $\alpha = 16°$. Since the viscous effects are introduced into the panel method through the blowing velocity and off-body Kutta condition, both of which involve δ^*, the accuracy of the inviscid flow depends on the accuracy of determining δ^* everywhere in the flowfield, especially at the airfoil trailing edge and in the wake.

Results for the NACA 0012 airfoil, with calculations including the wake effects, are shown in Figs. 5.2 to 5.4 with Figs. 5.2 and 5.3 corresponding to a chord Reynolds number of 3×10^6 and Fig. 5.4 to Reynolds numbers of 6×10^6 and 9×10^6. Figure 5.2 shows the variation of the lift and drag coefficients of the NACA 0012 airfoil for a chord Reynolds number of 3×10^6. As can be seen from Fig. 5.2a, viscous effects have a considerable effect on $(c_l)_{\max}$ of the airfoil,

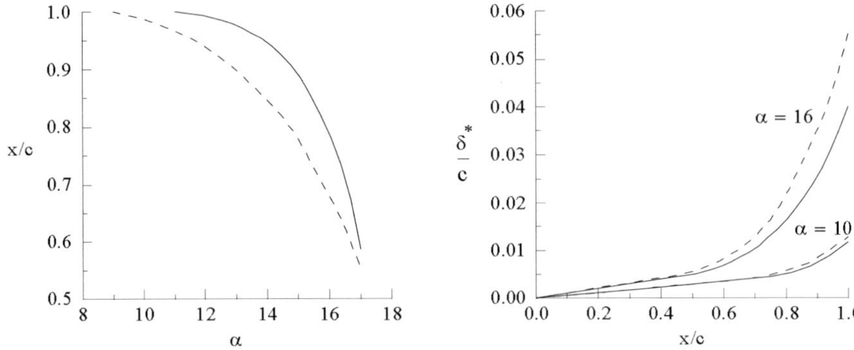

(a) (b)

Fig. 5.1. Wake effect on (a) flow separation and (b) displacement thickness – NACA 0012 airfoil. —, with wake; - - -, without wake.

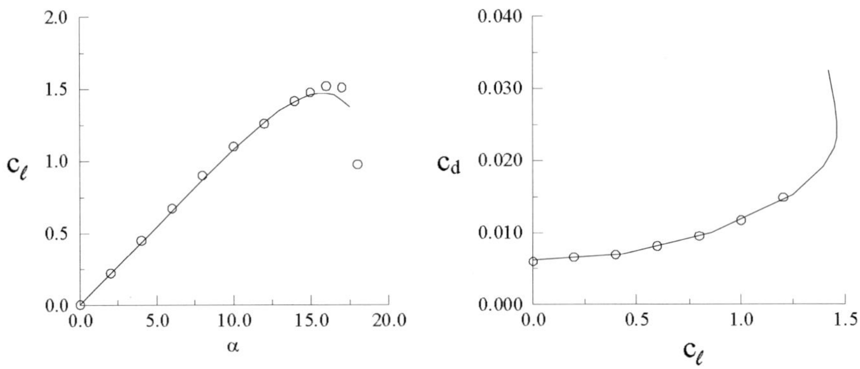

(a) (b)

Fig. 5.2. Comparison between calculated (solid lines) and experimental values (symbols) of: (a) c_l vs α, and (b) c_d vs c_l. NACA 0012 airfoil at $R_c = 3 \times 10^6$.

which occurs at a stall angle of around 16°, and the calculated results agree well with measurements [2].

Figure 5.2b shows the variation of the drag coefficient with lift coefficient. As can be seen, the measurements of drag coefficients do not extend beyond an angle of attack of 12 degrees and at smaller angles agree well with the calculations. The nature of the lift-drag curve is interesting at higher angles of attack with the expected increase in drag coefficient and reduction in lift coefficient for post-stall angles.

Figure 5.3 shows the variation of the local skin-friction coefficient c_f and dimensionless displacement thickness δ^*/c distribution, for the same airfoil at the same Reynolds number. As can be seen from Fig. 5.3a, flow separation occurs around $\alpha = 10°$ and its extent increases with increasing angle of attack.

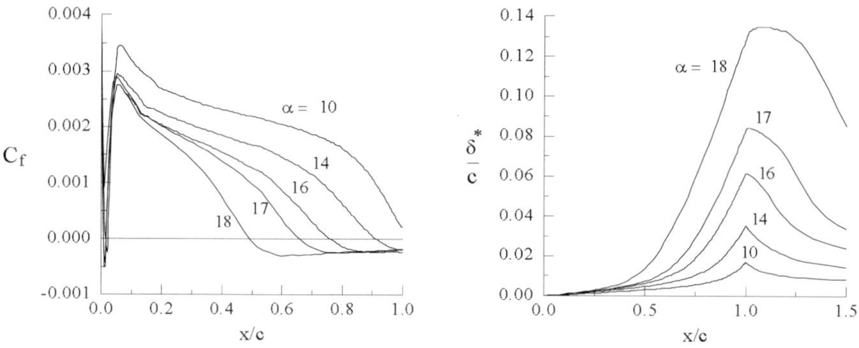

(a) (b)

Fig. 5.3. Variation of (a) local skin-friction coefficient and (b) dimensionless displacement thickness distribution. NACA 0012 airfoil at $R_c = 3 \times 10^6$.

At an angle of attack $\alpha = 18°$, the flow separation on the airfoil is 50% of the chord length.

The variation of dimensionless displacement thickness along the airfoil and wake of the airfoil shown in Fig. 5.3b indicates that, as expected, displacement thickness increases along the airfoil, becoming maximum at the trailing edge, and decreases in the wake. For $\alpha = 10°$, δ^*/c at the trailing edge is around 2% of the chord, becoming 4% at $\alpha = 14°$ and 7% at $\alpha = 16°$. With increase in angle of attack, the trailing-edge displacement thickness increases significantly, becoming 12% of the chord at $\alpha = 17°$ and 16% at $\alpha = 18°$. However, what is quite interesting, aside from this rather sharp increase in displacement thickness, is the behavior of the maximum value of the displacement thickness. While for angles of attack up to and including stall angle, $\alpha = 16°$, its maximum value is at the trailing edge, at higher angles of attack corresponding to post-stall, its maximum is after the trailing edge.

Figure 5.4, together with Fig. 5.2, shows the effect of the Reynolds number on the lift coefficient. In accord with the measurements, the calculation method satisfactorily accounts for the effects of Reynolds number. The results show that the maximum lift coefficient, $(c_l)_{\max}$, increases with an increase in Reynolds number in agreement with measurements.

Figures 5.5 and 5.6 correspond to the NACA 23012 airfoil for which measurements have been reported in [2] with Fig. 5.5 corresponding to Reynolds numbers of 6×10^6 and 8.8×10^6. In common with the results of Fig. 5.2, the results of Fig. 5.5 show a fairly gentle approach to stall, which in this case occurs around 17 degrees. The calculated drag coefficients are in good agreement with the measured values but to a lesser degree than those obtained for the NACA 0012 airfoil.

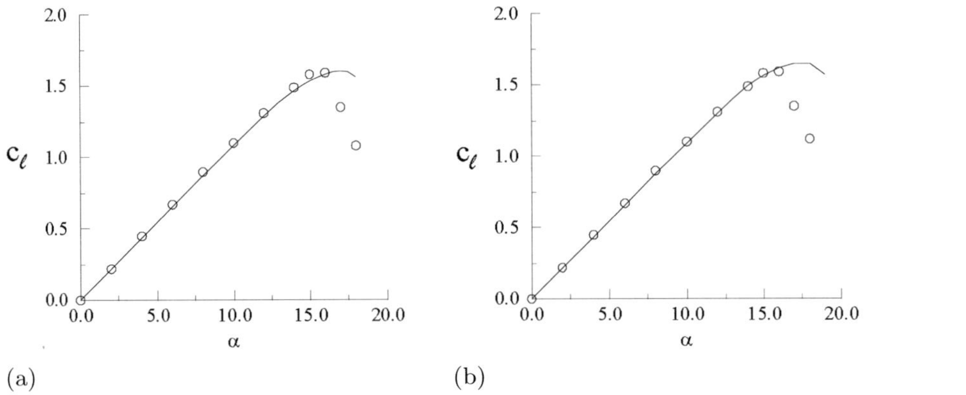

Fig. 5.4. Effect of Reynolds number on the lift coefficient. NACA 0012 airfoil at (a) $R_c = 6 \times 10^6$, and (b) $R_c = 9 \times 10^6$.

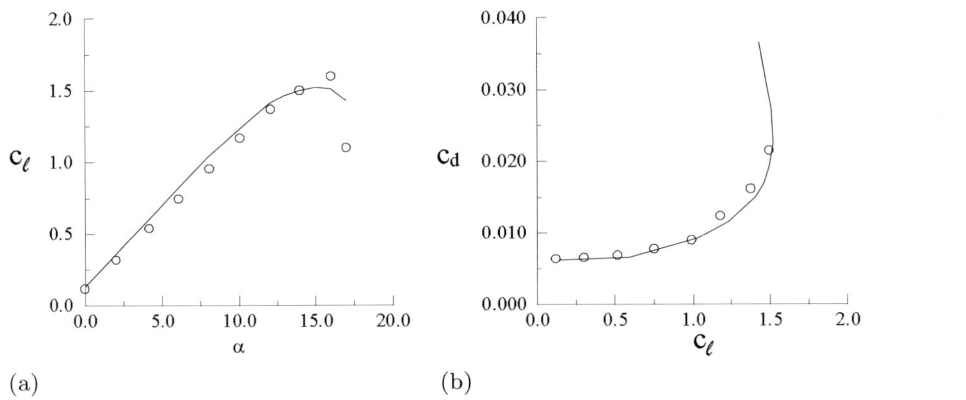

Fig. 5.5. Comparison between calculated (solid lines) and experimental (symbols) values of, (a) c_l vs α, and (b) c_d vs c_l. NACA 23012 airfoil.

Figures 5.5a and 5.6 show the Reynolds number effect on the lift coefficient. As in the case of the NACA 0012 airfoil, again this effect is well reflected in the calculations.

Finally, Fig. 5.7 shows the variation of lift coefficient with angle of attack for a Boeing supercritical airfoil at a chord Reynolds number of 2.83×10^6 and 4.02×10^6. The models and experiments are described in [3].

5.2 Accuracy of the e^n-Method for Flows with Separation

The accuracy of the e^n-method has been studied extensively for attached flows on two-dimensional and axisymmetric bodies, including heat transfer. For ex-

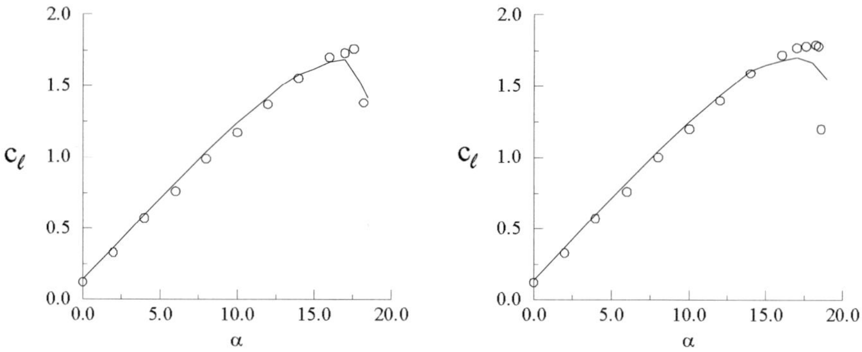

(a) (b)

Fig. 5.6. Effect of Reynolds number on the lift coefficient. NACA 23012 airfoil at (a) $R_c = 6 \times 10^6$, and (b) $R_c = 8.8 \times 10^6$.

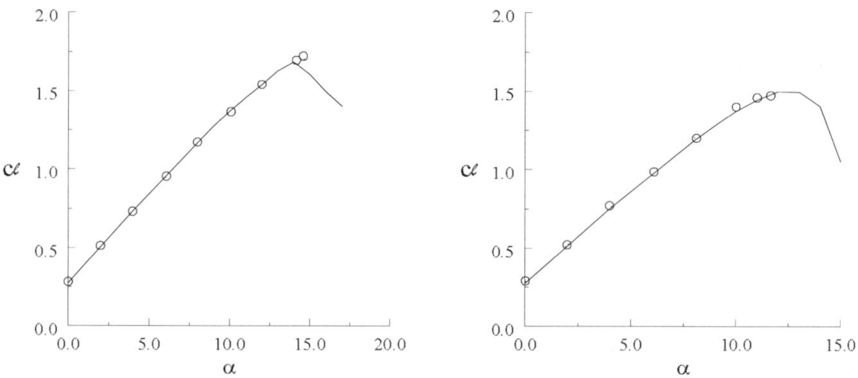

(a) (b)

Fig. 5.7. Comparison between calculated (solid line) and experimental values (symbols) of for a Boeing supercritical airfoil at chord Reynolds numbers of (a) $R_c = 2.83 \times 10^6$ and (b) $R_c = 4.02 \times 10^6$.

ample, Wazzan [4] has presented a substantial review which shows that the method provides values of transition in good agreement with experiment. In flows with separation, however, there has been less activity and Nayfeh, Ragab and Al-Maaitah [5] have reported one of the few attempts to apply the method, which they did in the context of the corrugated-plate flows of Fage [6]. Cebeci and Egan [7] addressed the same problem with detailed differences in the application of the e^n-method and provided results which suggested that the method can represent the effects of separation on the location of transition in a manner which can be used for flows such as those addressed here. In a subsequent study, Cebeci applied the calculation method of Chapters 2 and 3 with the onset of transition computed with the e^n-method to airfoil flows with separation bubbles [8]. A sample of results presented in this section are taken from [8] in order to demonstrate the accuracy of the e^n-method to flows with separation.

The stability-transition calculations are started by first obtaining the solutions of the laminar boundary-layers with the calculation method of Chapters 2 and 3. Since an inverse procedure was being used, the solutions included flows with or without separation. The stability calculations were then initiated at the first x-station where the Reynolds number based on the displacement thickness, R_{δ_*} (Eq. 4.1.2), exceeded the $R_{\delta^*_{cr}}$ established for similar boundary-layers, Fig. 4.1. As discussed in Section 4.1, several dimensional frequencies at different x-locations were then computed on the lower branch of the neutral stability curve in order to determine the amplification rates, $\alpha_i(x)$. In regions of no flow separation, the eigenvalue procedure described in subsection 4.2.3 operated satisfactorily and did not require the continuation method discussed in subsection 4.2.4. In regions of flow separation, however, this was not the case and the velocity profiles u and u'' at R were found to be rather sensitive to the initial estimates of the eigenvalues for α. With the use of the continuation method with reference velocity profiles corresponding to those of a previous x-station and with increments of Δn around 0.1, the sensitivity vanished and the eigenvalues were computed satisfactorily as in attached flows.

As discussed in Section 4.1, the envelope procedure used in the e^n-method which allows the dimensional critical frequency to be determined indirectly and conveniently, is not applicable to flows with separation. It is necessary to search for this frequency in the calculations. The results presented here indicated that in some cases the amplification rates were sensitive to the dimensional frequencies computed on the neutral stability curve and that it was necessary to take relatively small Δx increments on the airfoil.

Once the onset of transition location was computed, the remaining portion of the flow on the airfoil and in the wake, which involved transitional and turbulent flows, was computed with the calculation method of Chapters 2 and 3. To account for the region of transitional flow and separation-induced transition, the modeling of the transition region made use of Eqs. (1.4.4), (1.4.5) with C^2 given by Eq. (1.4.7) in the Cebeci-Smith eddy-viscosity formulation of subsection 3.1.4. Several sweeps were performed for this transition location and for the inviscid pressure distribution until the solutions converged. A new inviscid pressure distribution which made use of the boundary-layer solutions through the use of the blowing velocity was computed in order to again obtain laminar boundary-layer solutions on the airfoil to predict the onset of transition and to enable their subsequent transitional and turbulent flow calculations on the airfoil and in the wake until convergence. The results obtained in this way are presented in the following subsections and have indicated that in general this procedure, which involved the calculation of transition by the e^n-method and the calculation of inviscid and viscous flows, required three to four iterations.

5.2.1 Experimental Data of Gault

The experiments of Gault [9] were performed in the NASA Ames Research Center 7 × 10 ft wind tunnel with two airfoils, Reynolds numbers from 1.5×10^6 to 10×10^6 and angles of attack from 0 to 15 degrees. Emphasis was placed on the NACA 66_3-018 airfoil at angles of attack of 0 and 2 degrees for which detailed velocity information was provided through regions of separated flow which occurred near midchord. A main purpose was to identify locations of separation, transition and reattachment and, with freestream turbulence intensities less than 0.2%, the flow might be expected to be reproducible. Figure 5.8 displays the 66_3-018 airfoil together with measured and calculated mean velocity profiles for zero angle of attack and for a chord Reynolds number of 2×10^6.

Due to the presence of the separation bubble near midchord, the amplification rates were found to grow very fast in that region, see Fig. 5.9a, and required a very fine spacing in the x-direction in order to compute the eigenvalues. As can be seen from Fig. 5.9a, the envelope procedure used in attached flows does not apply to flows with separation. Furthermore, the amplification rates are not overly sensitive to the dimensional frequencies and that with those computed at $s/c = 0.35$ and 0.40 produce similar results due to the rapid growth in the separation region. At first, transition was computed at a location close to the separation point ($s/c = 0.62$), but moved inside the bubble in the second cycle. The experimental transition location was reported to be at $(s/c)_{\text{tr}} = 0.715$. The

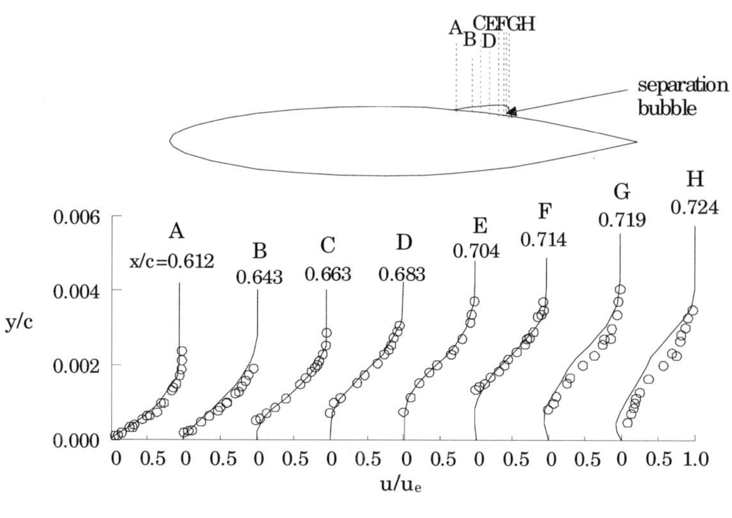

Fig. 5.8. Comparison of calculated (solid lines) and measured (symbols) velocity profiles for the NASA 66_3-018 airfoil for $\alpha = 0°$, $R_c = 2 \times 10^6$.

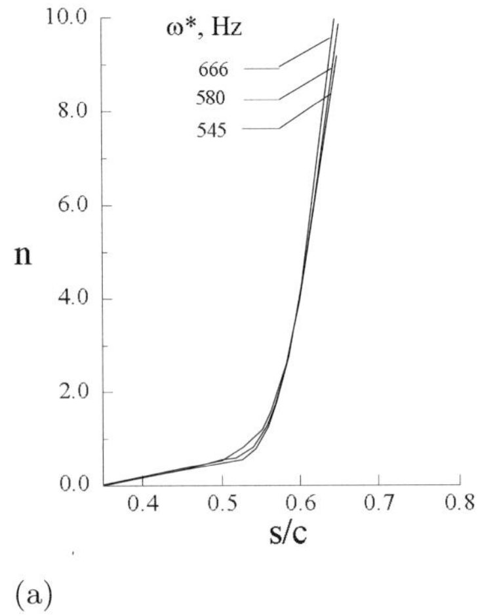

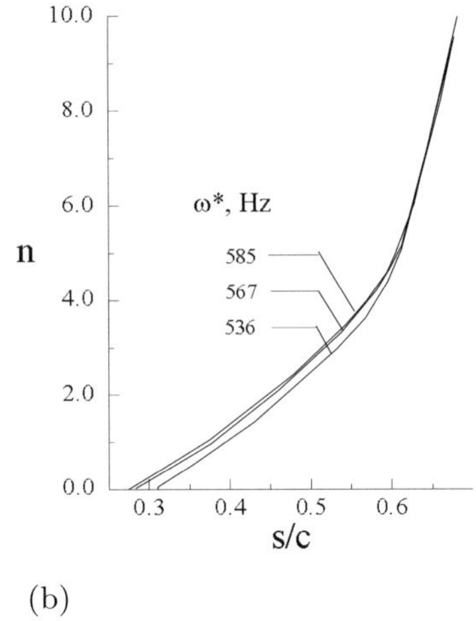

Fig. 5.9. Variation of amplification rates on the upper surface of the NACA 66_3-018 airfoil at different frequencies for (a) $\alpha = 0°$. (b) $= 2°$, $R_c = 2 \times 10^6$.

computed transition location was at $s/c = 0.685$ for $n = 8$ and $s/c = 0.693$ for $n = 9$. A value of $n = 10.6$ was required to match the experimental location. In all the calculations reported here, the value of $n = 8$ was taken as in [8].

Figures 5.9b, 5.10 and 5.11 show the results for the same Reynolds number but for $\alpha = 2°$. As can be seen from Fig. 5.9b, the amplification rates have the same behavior as those for $\alpha = 0°$ in that they vary rapidly in the region of flow reversal and they are not very sensitive to the choice of dimensional frequencies. In this case the experimental transition location was at $s/c = 0.710$ in contrast to the computed values at $s/c = 0.66$ for $n = 8$, $s/c = 0.676$ for $n = 9$, and $n = 12.5$ for $s/c = 0.71$. The computed velocity profiles in Fig. 5.10 indicate good agreement with experiment except near the trailing edge. The pressure coefficient results of Fig. 5.11 also indicate good agreement with data except near the separation bubble region where the calculations do not exhibit the relatively flat behavior in C_p.

5.2.2 Experimental Data of Cousteix and Pailhas

The ONERA-D airfoil was examined by Cousteix and Pailhas [10] in a wind tunnel with a chord Reynolds number of 3×10^5 at zero angle of attack. The airfoil and mean velocity profiles, variation of amplification rates, and distributions of freestream velocity and skin-friction coefficient are shown in Figs. 5.12 to 5.14. In this case, the measured and calculated results are in close agreement with appreciable differences only in the velocity profiles immediately upstream of boundary-layer separation where we may expect cross-stream pressure gradients and normal stresses to have a locally-important role. In this

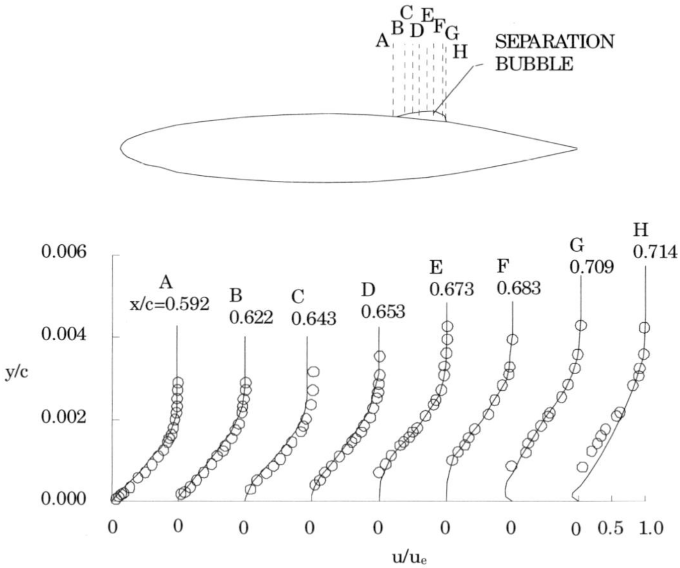

Fig. 5.10. Comparison of calculated (solid lines) and measured (symbols) velocity profiles for the NACA 66_3-018 airfoil for $\alpha = 2°$, $R_c = 2 \times 10^6$.

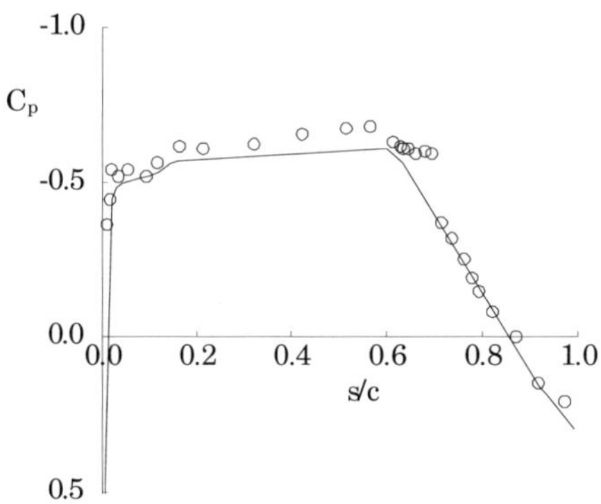

Fig. 5.11. Comparison of calculated (solid lines) and measured (symbols) pressure coefficients for the NACA 66_3-018 airfoil for $\alpha = 2°$, $R_c = 2 \times 10^6$.

case, transition occurred within the separated flow region and caused reattachment shortly thereafter. The calculations revealed transition at $x/c = 0.79$ for $n = 8$, at $x/c = 0.81$ for $n = 9$ in comparison with measurements which revealed transition at $x/c = 0.808$.

5.2.3 Experimental Data of Hoheisel et al.

Hoheisel et al. [11] were also concerned with zero angle of attack, this time with a NACA 65-213 airfoil at a chord Reynolds number of 2.4×10^5 so that

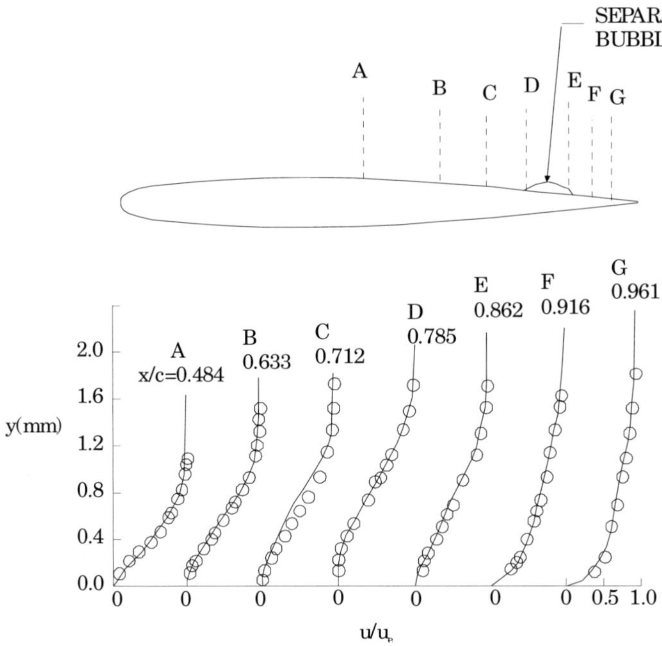

Fig. 5.12. Comparison of calculated (solid lines) and measured (symbols) velocity profiles for the ONERA-D airfoil for $\alpha = 0°$, $R_c = 3.0 \times 10^5$.

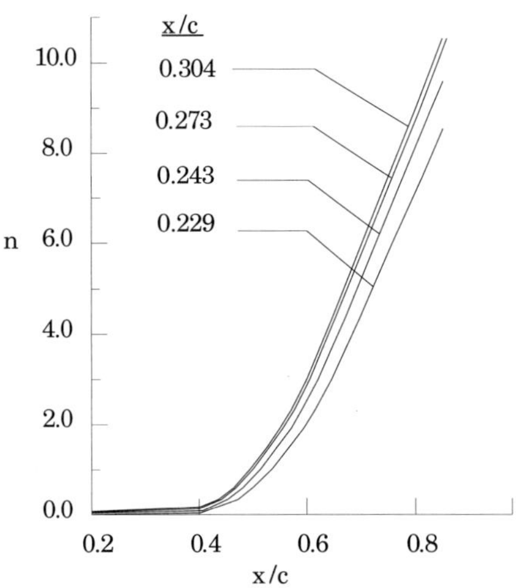

Fig. 5.13. Variation of amplification rates on the upper surface of the ONERA-D airfoil at four different chordwise locations.

the flow was separated from x/c of around 0.6 to 0.8. The experiments were carried out in the wind tunnel of the French-German Institute at St. Louis which had a freestream turbulence intensity of 0.2%. The airfoil and velocity profiles

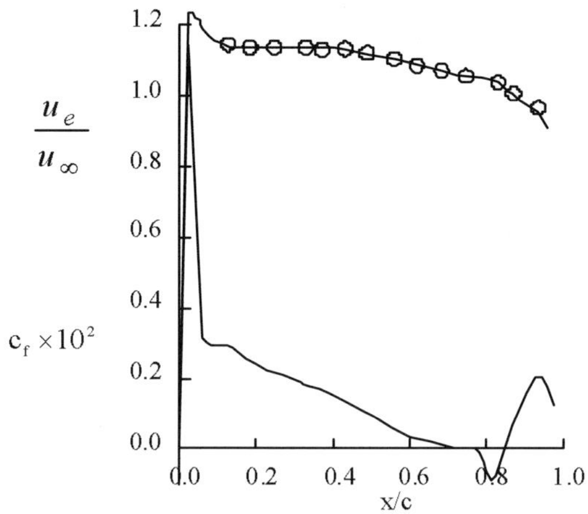

Fig. 5.14. Results for the upper surface of the ONERA-D airfoil. Symbols denote experimental data.

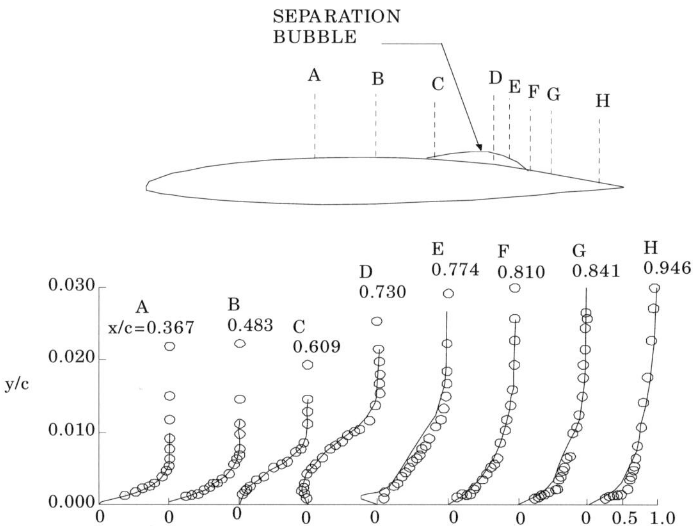

Fig. 5.15. Comparison of calculated (solid lines) and measured (symbols) velocity profiles for the NACA 65-213 airfoil.

and distributions of external velocity and skin-friction coefficient are shown in Figs. 5.15 and 5.16. As in Fig. 5.8, we can see a slightly slower calculated variation in external velocity between the x/c-stations corresponding to laminar separation and turbulent reattachment which were reported to occur at 0.609 and 0.774c, respectively. The velocity profiles imply separation at a similar location with reattachment slightly downstream of the measured value. The location of onset of transition was calculated at x/c of 0.721, which is consistent

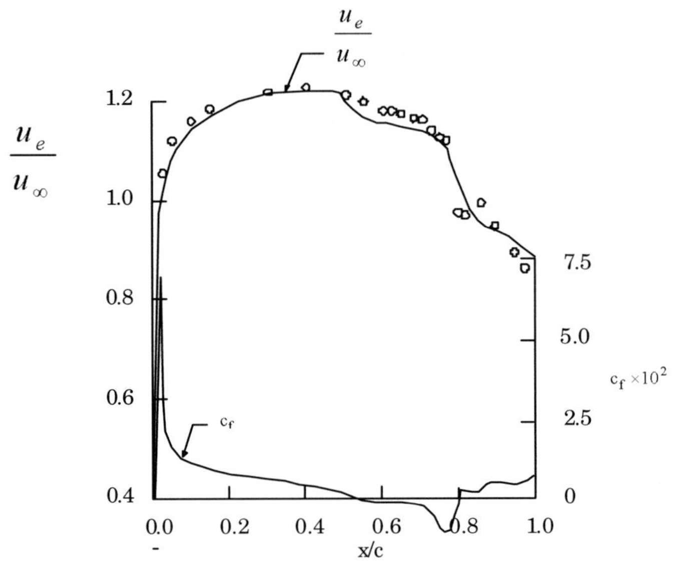

Fig. 5.16. Comparison of calculated (solid lines) and measured (symbols) external velocity and skin-friction distributions for the NACA 65-213 airfoil.

with an experimental measurement of maximum fluctuations in the near-wall region at the reattachment location of around 0.774.

5.3 Airfoils at Low Reynolds Numbers

In recent years the subject of low Reynolds number airfoils has received considerable interest in both civil and military applications including remotely piloted vehicles, propeller and wind turbine aerodynamics, aircraft with high-aspect ratio wings, and ultralight human-powered vehicles, as evidenced in a review article by Lissaman [12], two proceedings volumes edited by Mueller [13, 14], and one by the Royal Aeronautical Society [15]. The behavior of these airfoils differs from those at high Reynolds numbers, discussed in Section 5.1, in that rather large separation bubbles can occur some way downstream of the leading edge with transition taking place within the bubble prior to reattachment. The length of the bubble increases with a decreasing Reynolds number and strongly influences the aerodynamic characteristics of the airfoils.

In this section, the calculation method of Chapters 2 to 4 is applied to airfoil flows at low Reynolds numbers. At low to moderate angles of attack the onset of transition is calculated with the e^n-method as described in the previous section. At higher angles of attack, it is not possible to calculate the onset of transition with the e^n-method as in high Reynolds number flows. For this reason, the onset of transition is assumed to correspond to the location of laminar flow separation.

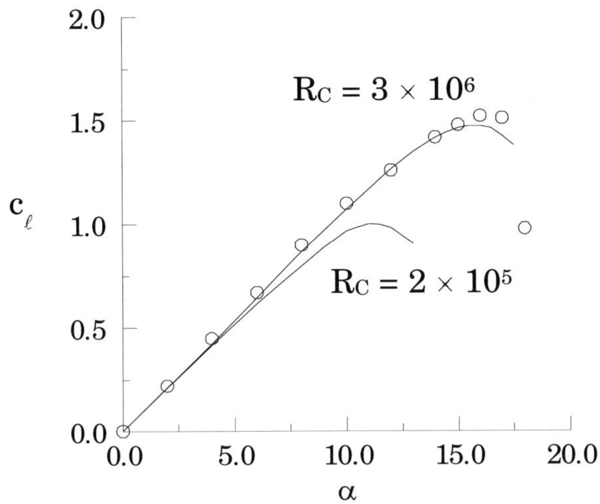

Fig. 5.17. Variation of lift coefficient of the NACA 0012 airfoil at two Reynolds numbers.

Before the results are presented for a sample of low Reynolds number airfoils, and comparisons are made with experimental data in order to demonstrate the accuracy of the calculation method, it is useful to demonstrate the effect of low Reynolds number on the reduction of the maximum lift coefficient.

Figure 5.17 shows the variation of the lift coefficient of the NACA 0012 airfoil for a chord Reynolds number of 2×10^5 together with the results obtained at $R_c = 3 \times 10^6$ (Fig. 5.2a). As can be seen, the Reynolds number has a significant effect, not only on the value of the maximum lift coefficient but also on the stall angle. With this change in Reynolds number, the $(c_l)_{\max}$ is reduced from a value of 1.5 to a value slightly over 1. The corresponding reductions in stall angle of attack are from 16° to around 11°.

5.3.1 Eppler Airfoil

To demonstrate the accuracy of the calculation method for airfoils at low Reynolds numbers, the results are now presented for the Eppler airfoil (Fig. 5.18), which is a low-drag airfoil. The experimental data for this airfoil was obtained by McGhee et al. [16] in the Langley Low-Turbulence Pressure Tunnel (LTPT). The tests were conducted over a Mach number range from 0.03 to 0.13 and a chord Reynolds number range from 60×10^3 to 460×10^3. Lift and pitching-moment data were obtained from airfoil surface pressure measurements

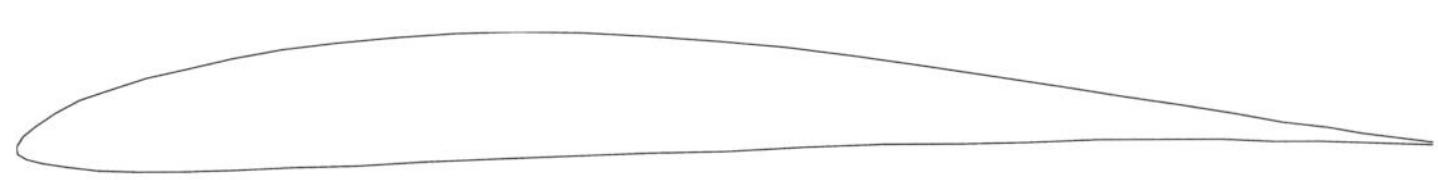

Fig. 5.18. Eppler 387 airfoil.

and drag data from wake surveys. Oil flow visualization was used to determine laminar-separation and turbulent-reattachment locations.

In [8], the calculations for this airfoil were performed for chord Reynolds numbers of 10^5, 2×10^5, 3×10^5, and 4.6×10^5 and for a range of angles of attack up to stall. Since the Reynolds number range of the experimental data was less than that of Fig. 1.6, which led to the expression given by Eq. (1.4.7), studies were first conducted to investigate the influence of the $C^2/3$ parameter at the Reynolds number of 1×10^5. It was found that values of $C^2/3$ from 10 to 30 led to negligible differences in the lift coefficient and differences of less than 3% in the drag coefficient. Thus, the straight-line formulation of Fig. 1.6 appears to be valid at a chord Reynolds number of 1×10^5, and the calculation of the transition-region flow is adequate.

Figures 5.19 and 5.20 show a comparison between calculated and measured results for a chord Reynolds number of 2×10^5. In the experimental data, the measured turbulence level in the LTPT was 0.06% for a total pressure $P_t = 15$ psi, which according to Eq. (1.4.2), corresponds to a value of $n = 9.4$ and was used in all the calculations presented below.

Figure 5.19 presents a comparison between measured and calculated distributions of pressure and local skin-friction coefficients for angles of attack of 0°, 4° and 8°. As can be seen, at lower angles of attack, the separation bubble is long and located away from the leading edge. It becomes smaller with increasing angle of attack and moves towards the leading edge.

Figure 5.20 shows a similar comparison for the lift and drag coefficients up to and beyond the stall angle. In general, the calculated results agree remarkably well with the measured ones.

Further details of the results shown in Figs. 5.19 and 5.20 are presented in Table 5.1. The calculated values of the chordwise location of laminar separation (LS), turbulent reattachment (TR), and the onset of transition are given for several angles of attack. The experimental results of this table are subject to some uncertainty because of difficulties associated with the surface visualization technique. With this proviso, comparison between measured and calculated values must be considered outstanding. It should be noted that when there is a separation bubble, the transition location obtained from the e^n-method occurs within the bubble in all cases, and, in accord with experimental observation, leads to reattachment some distance downstream.

5.3.2 Liebeck Airfoils

The theory used in the design of Liebeck airfoils for low Reynolds numbers is described in [17]. The basic approach is to first define an "optimized" airfoil pressure distribution for a given Reynolds number, flow condition (i.e. laminar or turbulent boundary-layer on the rooftop region), and design lift coefficient. This pressure distribution comes from a family of pressure distributions which

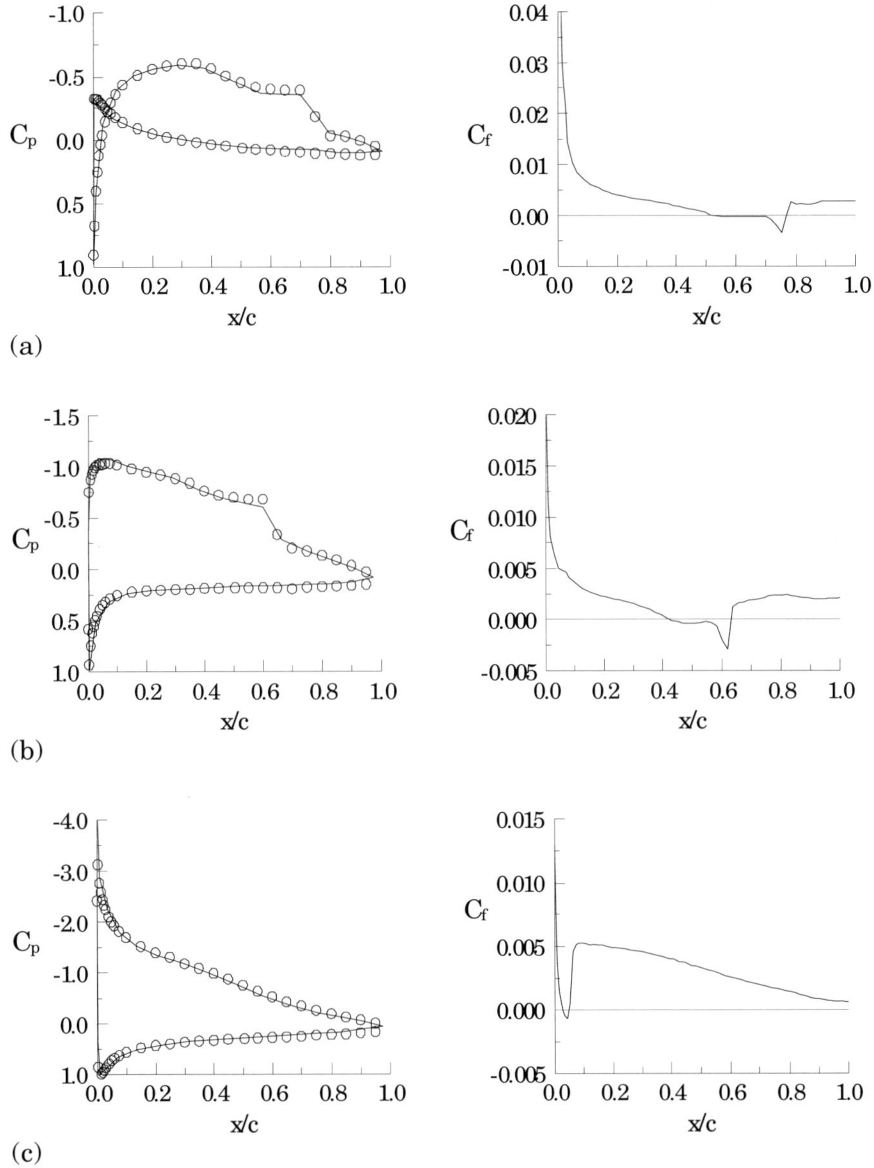

Fig. 5.19. Comparison of calculated (solid lines) and measured (symbols) pressure-coefficient and local skin-friction-coefficient distributions for the Eppler airfoil at (a) $\alpha = 0°$, (b) $\alpha = 4°$, and (c) $\alpha = 8°$ for $R_c = 2 \times 10^5$.

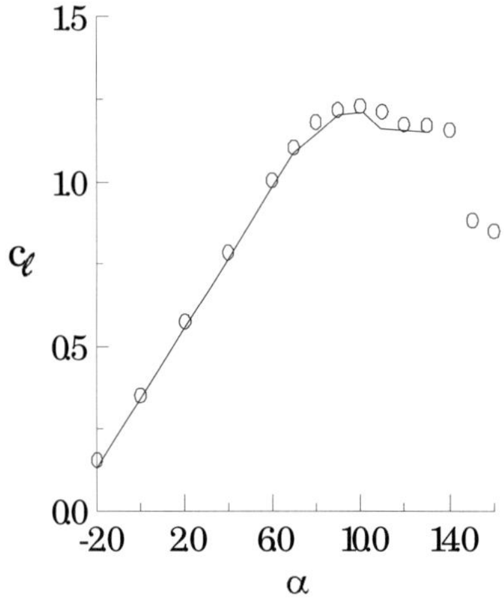

(a)

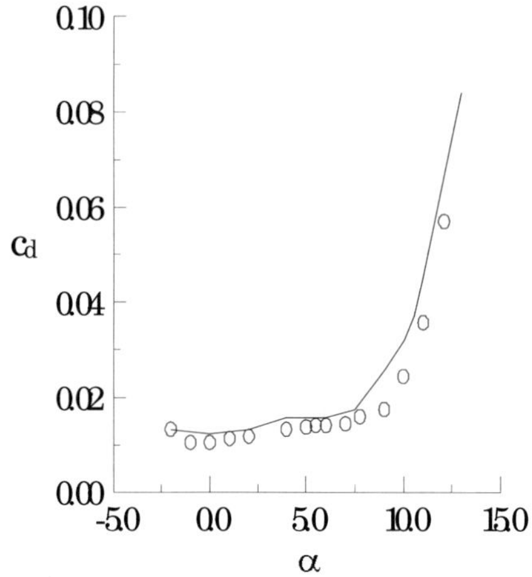

(b)

Fig. 5.20. Comparison of calculated (solid lines) and measured (symbols) (a) lift, and (b) drag coefficients for the Eppler airfoil at $R_c = 2 \times 10^5$.

Table 5.1. Experimental and calculated chordwise laminar separation (LS), and turbulent reattachment (TR), and transition locations on the upper surface of the Eppler airfoil for $R_c = 2 \times 10^5$.

	Calculated			Experiment	
α	LS	$(x/c)_{\rm tr}$	TR	LS	TR
−2	0.56	0.748	0.835	0.53	0.80
0	0.51	0.688	0.785	0.48	0.74
2	0.46	0.624	0.716	0.43	0.67
4	0.415	0.564	0.65	0.40	0.62
5	0.40	0.526	0.60	0.38	0.59
6	0.39	0.467	0.52	0.37	0.55

have been derived using boundary-layer and inviscid-flow theories together with the calculus of variations. It is characterized by an upper surface pressure distribution which consists of a rooftop plus a pressure recovery region based on the theory of Stratford [18] which avoids separation by a constant specified margin along its entire length. Next, an inviscid inverse flow method developed by James [19] is used to calculate the corresponding airfoil shape. Since an arbitrarily prescribed pressure distribution cannot be guaranteed a priori to provide a corresponding close and non-reentrant airfoil, some iteration between the pressure distribution and the inverse program is required. However, convergence to the desired design is quite rapid. A typical airfoil design is shown in Fig. 5.21. Finally, a boundary-layer method developed by Cebeci, similar to the one described in Chapter 3 for standard mode, is used to theoretically determine

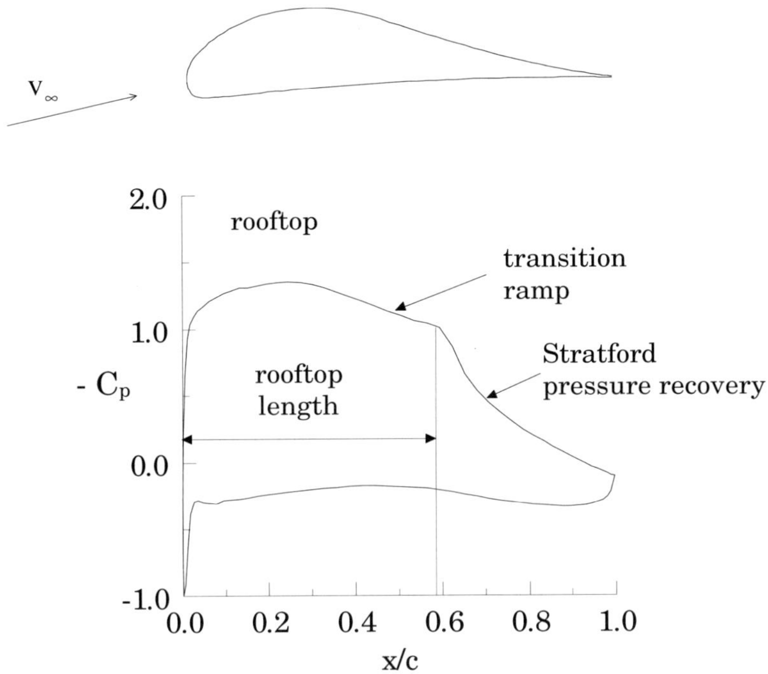

Fig. 5.21. A typical Liebeck airfoil and its pressure distribution.

the off-design aerodynamic characteristics of the airfoil. The basic airfoil design procedure is sketched in Fig. 5.23.

To demonstrate the accuracy of the calculation method for Liebeck airfoils, results are presented for two of his airfoils, designated as LTR201 and LTR217 at chord Reynolds numbers corresponding to 2.5×10^5 and 5×10^5 and for a range of angles of attack up to stall. These airfois represent two distinct philosophies in their design: airfoil LTR201 is conventional design with the Stratford pressure recovery terminating smoothly to the trailing-edge pressure, and airfoil LTR217 is an aft-loaded design where the Stratford recovery is followed by a short but severe pressure rise at the trailing edge. Geometry and design pressure distributions of the airfoils are given in Figs. 5.23a and 5.23b.

Figures 5.24 and 5.25 present the measured lift and drag coefficients of the two airfoils together with the calculated results obtained with the present method with n corresponding to 8 and with the ISES code of [20] with n corresponding to 9. Overall agreement of drag polar and lift prediction is good except for the case of airfoil LTR201 at a Reynolds number of 2.5×10^5 where both theoretical methods overestimate the effects of the laminar separation bubble and lead to less accurate drag coefficients.

These early Leibeck airfoils have performance characteristics which are sensitive to the location of laminar separation due to the nature of its surface

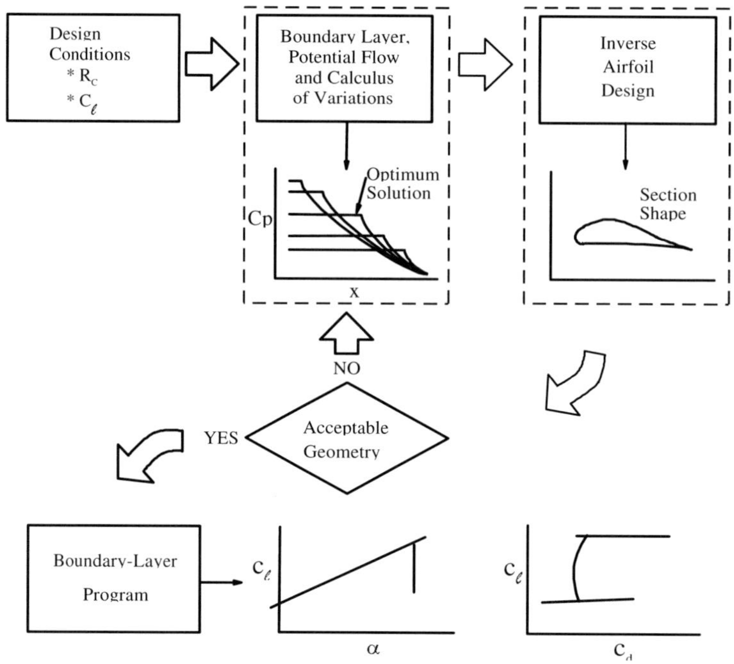

Fig. 5.22. Design procedure used in Liebeck airfoils.

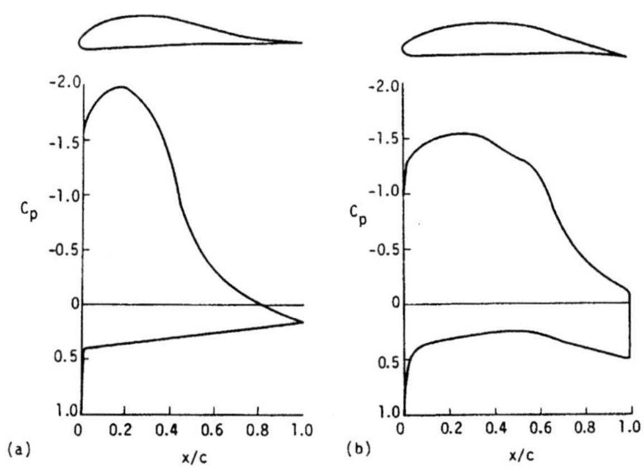

Fig. 5.23. Geometry and design pressure-coefficient distributions of two Liebeck airfoils; (a) LTR201, $(c_l)_{\text{design}} = 1.16$, (b) LTR217, $(c_l)_{\text{design}} = 1.38$.

pressure distribution. Unlike the Eppler airfoil which has a separation bubble on its upper surface only, both Leibeck airfoils have separation bubbles on their upper and lower surfaces at very low angles of attack. This causes difficulties in the calculation procedure and in the case of ISES code, does not allow the cal-

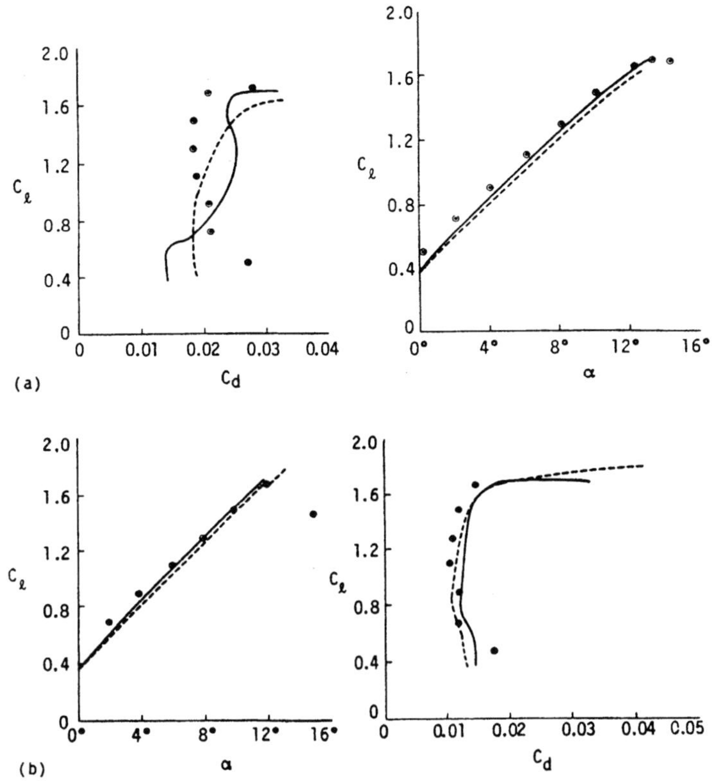

Fig. 5.24. Comparison of calculated (dashed lines – present method, solid lines – ISES code) and measured (symbols) lift and drag coefficients of LTR201 airfoil at (a) $R_c = 2.5 \times 10^5$, (b) $R_c = 5 \times 10^5$.

culations to be performed for angles of attack less than 0 degree for the LTR217 airfoil. It is interesting to note that the calculation times of both methods are essentially the same.

Figures 5.26a and 5.26b allow comparison between measured and calculated distributions of pressure coefficient of the two airfoild at a chord Reynolds number of 2.5×10^5. In both cases, the calculations were made by matching the experimental lift coefficient with the computed value. The results in Fig. 5.26a are for the LTR201 airfoil at a lift coefficient of 1.30 and indicate that, while the predicted lower surface pressure distributions are in good agreement with data, those for the upper surface are not. Both methods do not predict the distribution near the leading edge and do not capture the constant values of pressure coefficient in the separation region. Once the flow reattaches, however, the predictions of both methods improve and show better agreement with data. The results in Fig. 5.26b, which are for LTR217 airfoil at a lift coefficient of 1.23 on the other hand, show good agreement with data.

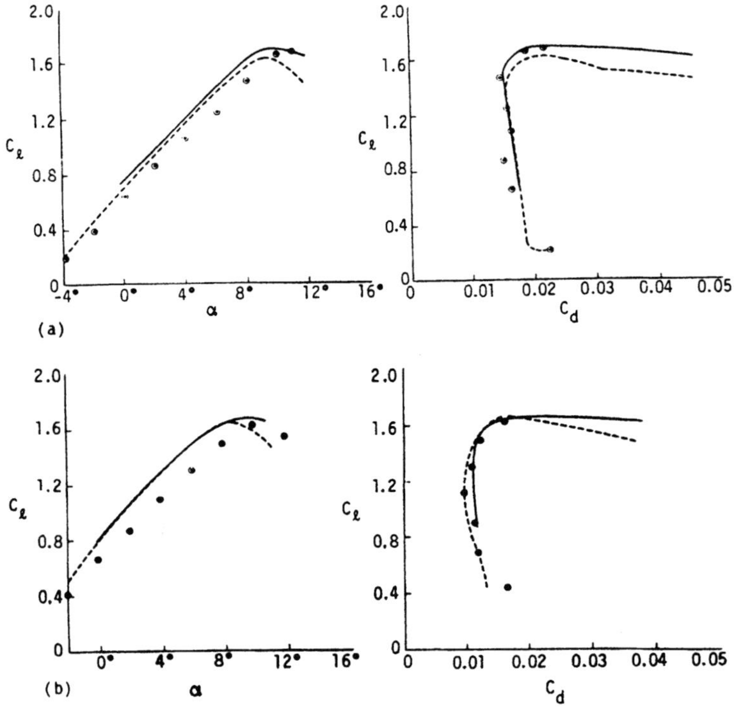

Fig. 5.25. Comparison of calculated (dashed lines – present method, solid lines – ISES code) and measured (symbols) lift and drag coefficients of LTR217 airfoil at (a) $R_c = 2.5 \times 10^5$, (b) $R_c = 5 \times 10^5$.

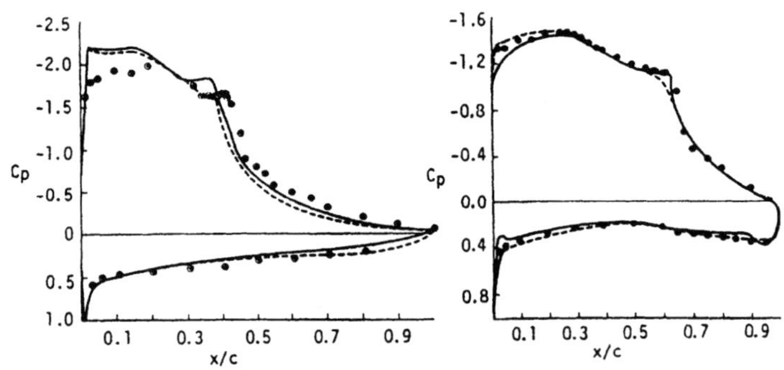

Fig. 5.26. Comparison of calculated (dashed lines – present method, solid lines – ISES code) and measured (symbols) pressure-coefficient distribution of (a) LTR201, $(c_l)_{\text{design}} = 1.16$, (b) LTR217, $(c_l)_{\text{design}} = 1.38$.

5.4 Airfoils at Low Mach Numbers

As will be discussed in Chapter 6, the calculation method of Chapters 2 to 4 can also be used to compute the performance characteristics of airfoils in transonic and even in low supersonic flows. In this case, the panel method must be replaced by an Euler method, and the incompressible boundary-layer method replaced with a compressible boundary-layer method. For lower Mach numbers, where there are either weak or no shocks, it is prudent to use an inviscid method based on the solution of the full potential equation which has greater reduced computer time requirements than Euler solutions and can be used for flows at very small Mach numbers, whereas the Euler solutions begin to exhibit "problems" if the Mach number is very low. At even smaller Mach numbers, it is common to make compressibility corrections to the panel method solutions. These corrections, which relate incompressible and compressible flows about different shapes, are very easy to make and avoid the need to solve either the full potential or Euler equations. Their drawback, however, is their restriction to very low Mach numbers.

These compressibility corrections depend upon the linearized form of the compressible velocity potential equation and are based on the assumption of small perturbations and thin airfoils. A simple correction formula given by Karman-Tsien allows the compressible pressure coefficient and velocity to be calculated from the incompressible pressure coefficient and velocity as given by Eqs. (2.4.2) and (2.4.3).

We now apply the compressibility corrections to the panel method of Chapter 2 to study the effect of compressibility on the pressure distribution and effect on the stall angle. In addition, we use the full potential method of [21] and introduce viscous corrections to it by coupling it to the compressible boundary-layer method discussed in Chapter 6 which is an extension of the incompressible boundary-layer method discussed in Chapter 3. In this way the interactive boundary-layer calculations are performed with the same inverse boundary-layer method using an inviscid method which is either a panel method or a full-potential method, both containing the same viscous corrections.

Figures 5.27 to 5.29 show the results for the NACA 0012 airfoil for $R_c = 4 \times 10^6$, $M_\infty = 0.30$ reported in [22]. Figures 5.27 and 5.28 show a comparison between the calculated and measured pressure coefficient distributions at two angles of attack. As can be seen, there is very little difference between the predictions of the panel method with compressibility corrections and full potential method.

Figure 5.29 shows the variation of lift coefficient with angle of attack with inviscid flow calculated with compressibility corrections in the panel method and with the full potential method. At low and moderate angles of attack, both calculated results agree with each other and with experimental data [23]. At higher angles of attack ($\alpha > 10°$), the lift coefficient is slightly over-predicted with the panel method, Agreement is excellent with the full potential method.

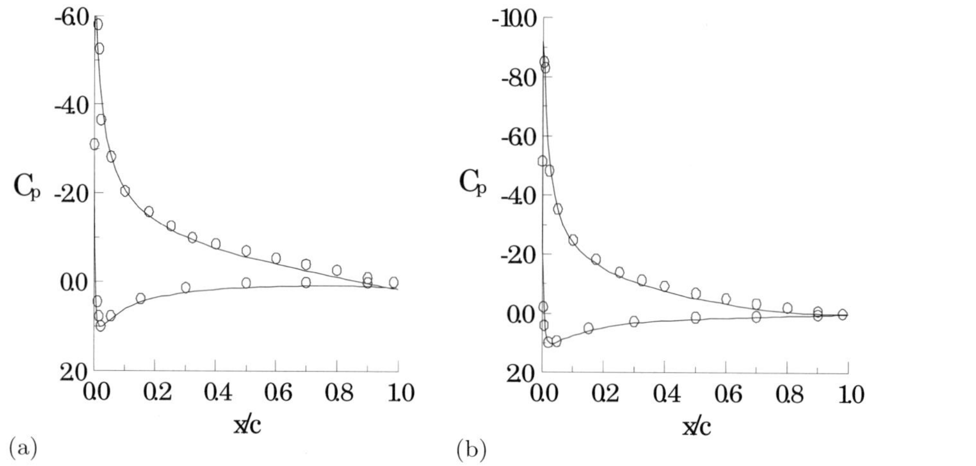

Fig. 5.27. Comparison of measured (symbols) and calculated (continuous line) pressure coefficient distributions with the panel method for the NACA 0012 airfoil at $M_\infty = 0.30$ and $R_c = 4 \times 10^6$ at (a) $\alpha = 10.0°$ and (b) $\alpha = 13.3°$.

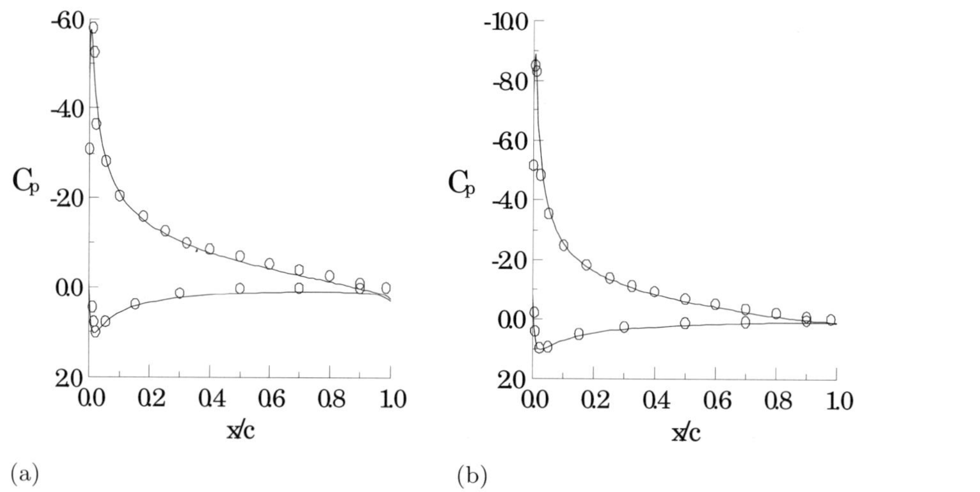

Fig. 5.28. Comparison of measured (symbols) and calculated (continuous line) pressure coefficient distributions with the full potential flow method for the NACA 0012 airfoil at $M_\infty = 0.30$ and $R_c = 4 \times 10^6$ at (a) $\alpha = 10.0°$ and (b) $\alpha = 13.3°$.

5.5 Prediction of Ice Shapes on Airfoils

In the following section, the calculation method of Chapters 2 to 4 is applied to iced airfoils and to airfoils with leading-edge roughness. In the former case, it is also necessary to predict the ice shapes on the airfoil in order to determine

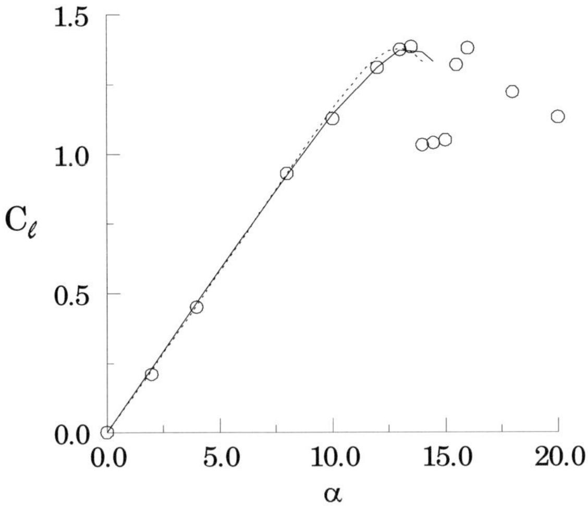

Fig. 5.29. Compressibility effect on the NACA 0012 airfoil. $R_c = 4 \times 10^6$, $M_\infty = 0.30$. Symbols denote experimental data, solid line denotes the predictions with inviscid flow obtained with the full potential flow method and dashed line those with the modified panel method.

their effects on performance. For this reason, in this section we present a brief description of a popular method, LEWICE, for predicting ice shapes on airfoils.

LEWICE is a two-dimensional ice accretion computer program which was initially developed in 1983 at the University of Dayton Research Institute and later modified by Ruff and Berkowitz [24]. This program provides a basis for the determination of the ice buildup on the leading edges of airfoils by specifying the atmospheric parameters of temperature and pressure, the flight velocity and the meteorological parameters of liquid water content (LWC), droplet diameter, and relative humidity. The three major elements of this code are: (1) the flowfield calculation, (2) the particle trajectory calculation, and (3) the energy balance and ice accretion. Each of these elements are discussed in some detail in subsections 5.5.1 to 5.5.3 and in detail in [24]. In subsection 5.5.4, LEWICE is used to calculate a sample of ice shapes on the leading edge of an airfoil in order to demonstrate its accuracy for predicting ice shapes.

5.5.1 Flowfield Calculation

The calculation of the flowfield around an airfoil can be achieved in several ways. One way is to solve the inviscid flow equations by the panel method described in Chapter 2. Another way is to incorporate the viscous effects by an interactive boundary-layer method discussed in Chapters 2 and 3. A third way is to solve the Navier-Stokes equations. All three alternatives require that the surface be free of the discontinuities which are known to exist and which influence the flowfield. The irregular ice shape can lead to multiple stagnation points with subsequent numerical difficulties, including a breakdown of the trajectory calculations that are necessary to determine the spatial distribution of water droplets.

According to experimental data – see, i.e. Bragg and Coirier [25] – ice accretion has an important effect on the pressure distribution near the leading edge of an airfoil. Ice shapes corresponding to glaze ice can lead to rather large regions of flow separation near the leading edge, even at very small angles of attack, and can cause the inviscid velocity distribution to be substantially different from the measured one. The effect of rime ice is less pronounced but becomes important at higher angles of attack.

While the experiments confirm the need to represent viscous effects for accurate calculation of all aerodynamic properties, including the pressure distribution, studies using either the present interactive boundary-layer approach [26] or the Navier-Stokes approach [27] show that none of these methods are able to predict the pressure distribution accurately near the leading edge. The reasons for the failure of either approach are not clear; in the case of the interactive boundary-layer approach, it may be due to the formulation used near the leading edge and how the viscous effects are introduced into the inviscid method. In the case of the Navier-Stokes approach, it may be due to the numerical accuracy and to the turbulence model.

At the present time, LEWICE computes the flowfield by using the panel method discussed in Chapter 2. The numerical difficulties caused by multiple stagnation points are avoided by a smoothing procedure due to Cebeci et al. [28] without loss of important flow characteristics.

5.5.2 Particle Trajectory Calculation

The purpose of the particle trajectory calculation is to determine the regions of droplet impingement and the distribution of the mass of liquid on the surface of the body within this impingement region. The equations of particle motion, to be discussed in subsection 12.4.1, are used to calculate droplet trajectories from far upstream to impingement on the body and, if it occurs, to determine the total and local collection efficiencies. The total collection efficiency is defined as the ratio of the actual mass of impinging water to the maximum value that would occur if the droplets followed straight-line trajectories. If we denote the upper-surface and lower-surface impingement limits by S_u and S_l, the forward projection of the airfoil height by H (see Fig. 5.26) then the local and total collection efficiencies, β and E_m, respectively, are computed from

$$\beta = \frac{dy}{ds}, \quad E_m = \frac{1}{H}[(y_0)_{su} - (y_0)_{sl}] = \frac{1}{H}\int_{s_l}^{s_u} \beta \, ds \tag{5.5.1}$$

where y_0 is the vertical distance between the droplet release points for the upper $(y_0)_{su}$ and lower $(y_0)_{sl}$ surface tangent trajectories.

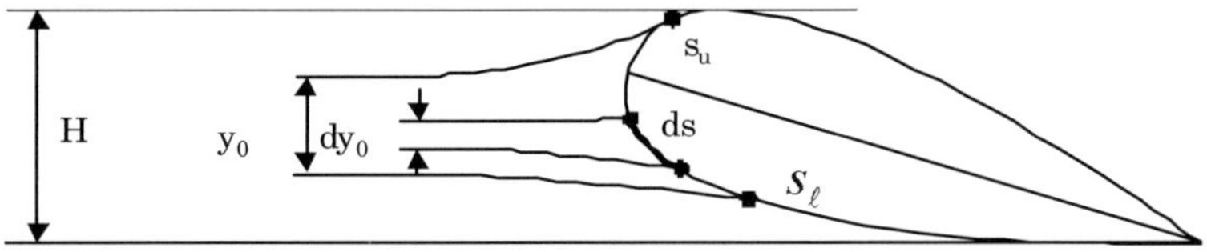

Fig. 5.30. Definition of total and local collection efficiency.

5.5.3 The Energy Balance and Ice Accretion

Without any doubt, the heart of any ice accretion computer program is the physical model used for the ice accretion process. In LEWICE this is characterized by the presence of supercooled droplets entrained in the flow about a body. These droplets follow trajectories that cause them to either be carried past a body or impinge upon the body. Upon impact with a clean surface, the droplets coalesce into larger surface drops under the effects of surface tension and flow along the surface as dictated by airflow along the surface of the body. These surface drops then either freeze on the surface or are shed off the surface because of the aerodynamic forces on the drops. The ice accretions formed by this initial freezing form a rough surface which enhances the convective heat transfer and local collection efficiency of the surface which, therefore, continue the ice accretion process.

We now present the equations that model the thermodynamics of the freezing process on a body undergoing icing and postpone the discussion about the accuracy of the physical model used in LEWICE to Chapter 13 where the extension of LEWICE to three-dimensional flows is described.

We consider a control volume located on the surface of the body and which extends from outside the boundary-layer to the surface as shown in Fig. 5.31a. The lower boundary of the control volume is initially on the surface of the clean geometry and moves outward with the surface as the ice accretes. Computationally a control volume is placed over each panel defining the body geometry as shown in Fig. 5.31b. The conservation of mass requires that the mass flow rate entering the control volume is equal to the mass flow rate leaving the control volume, that is

$$\dot{m}_c + \dot{m}_{r_{\rm in}} = \dot{m}_i + \dot{m}_e + \dot{m}_{r_{\rm out}} \tag{5.5.2}$$

Here $\dot{m}_c$ is the mass flow rate of the impinging water entering the control volume and $\dot{m}_{r_{\rm in}}$ is the mass flow rate of the water along the surface exiting from the adjacent upstream control volume. The leaving mass flow rates consist of the "accumulated ice", $\dot{m}_i$, loss through evaporation, $\dot{m}_e$, and water flowing out of the control volume, $\dot{m}_{r_{\rm out}}$. The three mass flow rates $\dot{m}_i$, $\dot{m}_c$, and $\dot{m}_{r_{\rm in}}$ are related to each other by the concept of freezing factor f, defined as the fraction of the total mass flow rate of water entering the control volume that freezes,

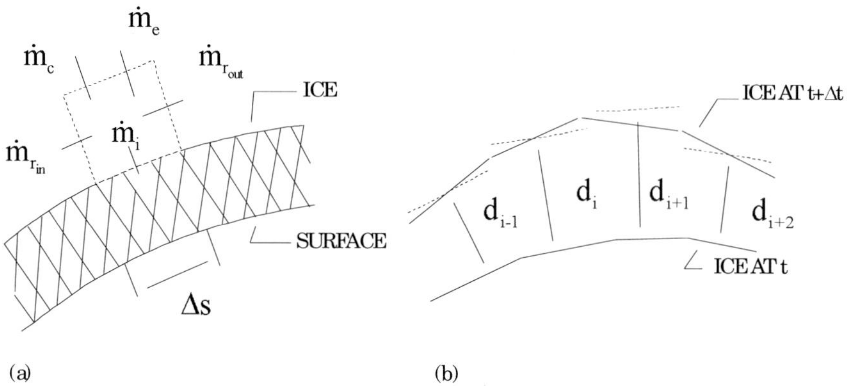

Fig. 5.31. Control volume for the energy balance (a) single control volume on the icing surface, (b) control volumes over each panel which defines the airfoil geometry.

$$f = \frac{\dot{m}_i}{\dot{m}_c + \dot{m}_{r_{\text{in}}}} \tag{5.5.3}$$

Solving for $\dot{m}_i$ and substituting the resulting expression into Eq. (5.5.2), we can obtain $\dot{m}_{r_{\text{out}}}$ from

$$\dot{m}_{r_{\text{out}}} = (\dot{m}_c + \dot{m}_{r_{\text{in}}})(1 - f) - \dot{m}_e \tag{5.5.4}$$

once f, $\dot{m}_c$, $\dot{m}_{r_{\text{in}}}$ and $\dot{m}_e$ are known.

The mass flow rate of the impinging water $\dot{m}_c$ is calculated from

$$\dot{m}_c = \text{LWC} \times V_\infty \times \beta \times \Delta s \tag{5.5.5}$$

where LWC is the liquid water content of the air and β is the local collection efficiency.

The mass flow rate due to evaporation is calculated from

$$\dot{m}_e = q\,\Delta B\,\Delta s \tag{5.5.6}$$

where g is the mass transfer coefficient obtained from

$$g = \frac{h_c}{C_{p_a}}\left(\frac{\text{Pr}}{\text{Sc}}\right)^{2/3} \tag{5.5.7}$$

Here h_c is the film heat transfer coefficient, C_{p_a} the specific heat of air ($\equiv 1002.32$ m^2/s^2/K), Pr the Prandtl number, Sc the Schmidt number with Sc/Pr $= 0.875$ and ΔB is the mass transfer driving potential based on concentration differences given by

$$\Delta B = \frac{\dfrac{P_v(T_s)}{T_s} - R_h\left(\dfrac{P_e}{T_e}\right)\dfrac{P_v(T_\infty)}{P_\infty}}{\dfrac{1}{0.622}\left(\dfrac{P_e}{T_e}\right) - \dfrac{P_v(T_s)}{T_s}} \tag{5.5.8}$$

where $P_v(T_s)$ is the vapor pressure over the surface of water $P_v(T_w)$ or ice $P_v(T_i)$ and R_h is the relative humidity. The variation of $P_v(T)$ with temperature is expressed by $P_v(T) = e^S$. For water S is given by

$$\begin{aligned} S = & -\frac{0.53728198 \times 10^4}{T_w} + 0.2112026 \times 10^2 + 0.23145873 \times 10^{-1} T_w \\ & - 0.64600326 \times 10^{-5} T_w^2 - 0.97969796 \times 10^{-7} T_w^3 + 0.194525335 \ln(T_w) \end{aligned}$$

for $223.15 \leq T_w < 258.149$ and by

$$\begin{aligned} S = & -\frac{0.29912729 \times 10^4}{T_w^2} - \frac{0.6170128 \times 10^4}{T_w} + 0.188764385 \times 10^2 \\ & - 0.28354721 \times 10^{-1} T_w + 0.17838301 \times 10^{-4} T_w^2 - 0.84150417 \times 10^{-9} T_w^3 \\ & + 0.44412543 \times 10^{-12} T_w^4 + 0.2858487 \times 10 \ln(T_w) \end{aligned}$$

for $273.149 \leq T_w \leq 323.15$. In the range $258.143 \leq T_w < 273.149$, the variation of $P_v(T_w)$ with temperature is given by

$$P_v(T_w) = A - \frac{\Delta T[42.164 + \Delta T(2.8482 + \Delta T \times 0.05802)]}{7.50062}$$

where $A = P_v(T_i)$ evaluated at $T_i = T_w$ and

$$\Delta T = T_w - 273.14$$

For ice, $P_v(T_i)$ is equal to

$$P_v(T_i) = 0.0008$$

for $T_i > 273.16$ or $T_i < 213.15$. For $213.15 \leq T_i \leq 273.16$, the variation of $P_v(T_i)$ with temperature is expressed again by $P_v(T_i) = e^S$ with S now given by

$$\begin{aligned} S = & -\frac{0.58653696 \times 10^4}{T_i} + 0.22241033 \times 10^2 + 0.137490 \times 10^{-1} T_i \\ & - 0.34031775 \times 10^{-4} T_i^2 + 0.26967687 \times 10^{-7} T_i^3 + 0.6918651 \ln(T_i) \end{aligned}$$

As we shall see shortly, Eq. (5.5.4) is applied to each surface element on the airfoil where ice accumulates. Calculations start from the stagnation point where $\dot{m}_{r_{\text{in}}}$ because of symmetry and proceed in the downstream direction on both upper and lower surfaces. At the first surface element, for an assumed f, Eq. (5.5.4) contains only one unknown, which is $\dot{m}_{r_{\text{out}}}$ since $\dot{m}_c$ is computed from Eq. (5.5.5) and $\dot{m}_e$ from Eq. (5.5.6). At the next surface element $\dot{m}_{r_{\text{out}}}$ of the first segment becomes $\dot{m}_{r_{\text{in}}}$ for the second segment and with $\dot{m}_c$ and $\dot{m}_e$ computed as before, $\dot{m}_{r_{\text{out}}}$ is computed from Eq. (5.5.4) and this process is repeated for each surface segment.

The energy balance is applied to the control volume by using the first law of thermodynamics which requires that the energy inflow rate is equal to the energy outflow rate. All energy fluxes are written in terms of total enthalpy in order to allow for the compressibility effects, which are really not needed since all calculations are for incompressible flows. None of the expressions for boundary-layer development and formulas for heat transfer coefficient account for compressibility effects. In addition, flowfield calculations for heat-transfer coefficients identify laminar and turbulent flows and compute transition when a Reynolds number based on local roughness, R_k, exceeds 600. Partly, we believe that heat-transfer rates for laminar flow are negligible in comparison to those for turbulent flow and partly because transition will occur instantly on a rough surface, and partly to make the description of the discussion on heat balance shorter, we write the energy equation for turbulent flow. Application of the energy balance to the control volume of Fig. 5.31b gives

$$\dot{m}_c H_{t,T_\infty} + \dot{m}_{r_{\text{in}}} H_{w,s_{i-1}} + \dot{q}_k \Delta s = \dot{m}_e H_{v,s} + \dot{m}_i H_{i,s} + \dot{m}_{r_{\text{out}}} H_{w,s_i} + \dot{q}_c \Delta s \quad (5.5.9)$$

Here H_w represents the droplet enthalpy and is computed from

$$H_w = C_{p_w}(T_\infty - 273.15) + \frac{V_\infty^2}{2} \quad (5.5.10)$$

on the assumption that the droplet is brought to rest. In the above equation, C_{p_w} is the specific heat of water and 273.15 K represents the reference temperature for water. The variation of C_{p_w} with temperature is given by

$$C_{p_w} = 4186.8[0.9648266 - 0.53229597 \times 10^{-2}(T - 273.15)]$$

for $223.15 < T < 273.149$ and by

$$\begin{aligned} C_{p_w} &= 4186.8[1.005783 - 0.337458 \times 10^{-3}(T - 273.15)^2 \\ &\quad + 0.2091696 \times 10^{-8}(T - 273.15)^4] \end{aligned}$$

for $273.149 < T < 323.15$.

The enthalpy of water flow into and out of control volume, that is, $H_{w,s_{i-1}}$ and H_{w,s_i}, respectively, are calculated from

$$H_{w,s_{i-1}} = C_{p_w}(T_{s_{i-1}} - 273.15) \quad (5.5.11a)$$

and

$$H_{w,s} = C_{p_w}(T_{s_i} - 273.15) \quad (5.5.11b)$$

where $T_{s_{i-1}}$ and T_{s_i} denote the surface temperature of the upstream and downstream control volumes, respectively, and C_{p_w} is evaluated at corresponding temperatures.

The third term on the left-hand side of Eq. (5.5.9), $\dot{q}_k \Delta s$, corresponds to the heat conduction from the skin and represents heat transfer to the initial water layer on the airfoil surface before freezing. At subsequent times, after ice formation, the skin and ice reach equilibrium temperature and there is little heat transfer between them. Unless detailed knowledge of temperature distribution in the skin and the supporting structure is available, it is difficult to evaluate and is therefore omitted in the heat-balance equation.

$H_{v,s}$ denotes the enthalpy for the evaporative energy flux and is computed from

$$H_{v,s} = C_{p_w}(T_s - 273.15) + L_v \tag{5.5.12}$$

where L_v is the latent heat of evaporation and is a function of surface temperature and is given by m^2/s^2 units by

$$L_v = \begin{cases} 2{,}500{,}450 & 0 \le f \le 1 \\ 2{,}326[1092.6 - 0.99(T_s - 255.3722)] & f < 0 \\ 2{,}887{,}301.3 & f > 1 \end{cases}$$

If all of the water freezes on impact, water vapor can leave the surface through sublimation; in that case L_v should be replaced by the latent heat of sublimation. It is not, however, used in LEWICE.

$H_{i,s}$ denotes the enthalpy of ice and is computed from

$$H_{i,s} = C_{p_{i,s}}(T_s - 273.15) - L_f$$

where L_f is the latent heat of fusion and is equal to 334,944 m^2/s^2.

The last term on the right-hand side of Eq. (5.5.9) represents the convective heat transport by the flowing air and the kinetic heating of the surface through friction with $\dot{q}_c$ expressed in terms of the film heat transfer coefficient by

$$\dot{q}_c = h_c\left(T_s - T_e - \frac{V_e^2}{2C_{p_a}}\right) \tag{5.5.13}$$

where subscript e refers to the local freestream outside the boundary-layer and h_c represents the local heat-transfer coefficient for turbulent flows and is computed from

$$h_c = \mathrm{St}\,\rho u_e C_p \tag{5.5.14}$$

by assuming that the iced surface is an arbitrary rough surface and its characteristics can be related to an equivalent sand-grain roughness, k_s. The roughness correlation between k_s and iced surfaces has been established empirically from a set of ice accretion calculations which best matched experimental data. In this correlation, k_s/c is a function of the freestream speed V_∞, the static stream temperature T_s, and the liquid water content LWC, of the air.

In Eq. (5.5.14) St represents the Stanton number and is computed with an expression developed for fully rough flow on a flat plate,

$$\mathrm{St} = \frac{c_f/2}{\mathrm{Pr}_t + \sqrt{c_f/2}\,(1/\mathrm{St}_k)} \tag{5.5.15}$$

where c_f is the local skin-friction coefficient, St_k a roughness Stanton number, and Pr_t the turbulent Prandtl number, usually taken equal to 0.9. The local skin-friction coefficient is computed from

$$\frac{c_f}{2} = \left\{ \frac{0.41}{\ln[864(\theta_t/k_s) + 2.568]} \right\}^2 \tag{5.5.16}$$

with the momentum thickness θ_t computed from the momentum integral formula,

$$\theta_t = \left[\frac{0.0156\nu^{0.25}}{u_e^{4.11}} \int_0^s u_e^{3.86} ds \right]^{0.8} \tag{5.5.17}$$

The roughness Stanton number formula developed for a rough surface formed by uniform size closely packed spheres and $\mathrm{Pr}_t = 0.90$ is used in Eq. (5.5.15):

$$\mathrm{St}_k = 1.16 \left(\frac{u_\tau k_s}{\nu} \right)^{-0.2} \tag{5.5.18}$$

Here u_τ is the friction velocity given by

$$u_\tau = \sqrt{\frac{c_f}{2}}\, u_e \tag{5.5.19}$$

In terms of the expressions defined by Eqs. (5.5.10) to (5.5.13), Eq. (5.5.9) can be written as

$$\begin{aligned} &\dot{m}_c \left[C_{p_{w,\infty}} (T_\infty - 273.15) + \frac{V_\infty^2}{2} \right] + \dot{m}_{r_{\text{in}}} C_{p_{w,s_{i-1}}} (T_{s_{i-1}} - 273.15) \\ &= \dot{m}_e [C_{p_{w,s}} (T_s - 273.15) + L_v] + (\dot{m}_c + \dot{m}_{r_{\text{in}}}) f [C_{p_{i,s}} (T_s - 273.15) - L_f] \\ &\quad + \dot{m}_{r_{\text{out}}} C_{p_{w,s}} (T_s - 273.15) + h_c \left[T_s - T_e - \frac{V_e^2}{2C_{p_a}} \right] \Delta s \end{aligned} \tag{5.5.20}$$

where C_{p_i} is the specific heat of ice equal to 1967.796 $\mathrm{m^2/s^2/K}$. With the mass flow rates given by Eq. (5.5.4), the only unknowns in the above equation are surface temperature T_s and freezing factor f. The procedure for computing T_s is to set $T_s = 273.15$ K initially and with $\dot{m}_e$ and $\dot{q}_c$ evaluated at $T_s = 273.15$ K, calculate f from

$$f = \frac{-\dot{m}_c H_{w,T_\infty} - \dot{m}_{r_{\text{in}}} H_{w,s_{i-1}} + \dot{m}_e L_v + \dot{q}_e \Delta s}{(\dot{m}_c + \dot{m}_{r_{\text{in}}}) L_f} \tag{5.5.21}$$

and check for the following conditions:

$f < 0$ implies surface temperature greater than 273.15, set $f = 0$ and calculate T_s.

$0 \le f \le 1$ the initial assumption was correct.

$f > 1$ implies surface temperature less than 273.15, set $f = 1.0$ and calculate T_s.

Note that the calculation of the surface temperature T_s of each segment by Eq. (5.5.20) involves successive approximations, starting with an initial guess, say $T_s = T_g$. An improved solution is obtained from

$$T_s = \frac{\dot{m}_c H_{w,T_\infty} + \dot{m}_{r_{\text{in}}} H_{w,s_{i-1}} - \dot{m}_e L_v - h_c[273.15 - T_e - (V_e^2/2C_{pa})] + f(\dot{m}_c + \dot{m}_{r_{\text{in}}})L_f}{(1-f)(\dot{m}_c + \dot{m}_{r_{\text{in}}})C_{pw} + f(\dot{m}_c + \dot{m}_{r_{\text{in}}})C_{p_i} + h_c \Delta s} + 273.15 \tag{5.5.22}$$

where $\dot{m}_e$, L_v, C_{p_w}, and C_{p_i} are evaluated at T_g. Note that f is either less than zero or equal to one in the above equation. The iteration process is continued until two successive values of T_s differ by less than 0.005.

Once T_s and f are determined, then $\dot{m}_i$ can be calculated from Eq. (5.5.3) on each surface element on the airfoil where ice accumulates. Height of ice growth, Δy_i, on each surface segment during a time interval of Δt is calculated from

$$\Delta y_i = \frac{\dot{m}_i \Delta t \Delta s}{\rho_i} \tag{5.5.23}$$

where ρ_i (kg/m^3) denotes the ice density. For ΔT ($\equiv T_i - 273.15$) greater than -5, ρ_i is set equal to 917. For ΔT less than -5, with d denoting the diameter of water droplets in microns, it is first computed from

$$B = \frac{dV_\infty}{2|-\Delta T|}$$

so that ρ_i can be determined from

$$\rho_i = \begin{cases} 110(B)^{0.76} & 1 < B \leq 17 \\ 100 & B \leq 1 \end{cases}$$

These Δy increments are then added normal to the original surface segments as to create a new airfoil on which the ice accumulation calculations are repeated for the next time interval. Thus, calculations for each time step involves recomputing the inviscid flow, determining the water impingement pattern for the new configuration, and reevaluating the heat balance equation.

5.5.4 Prediction of Ice Shapes

To demonstrate the accuracy of LEWICE discussed in this section, we consider the experimental data obtained by Shin and Bond [29] in the NASA Lewis Icing Research Tunnel (IRT) on a 6-ft span, 21-in. chord, NACA-0012 airfoil with a fiberglass skin. During all icing runs, the model was set at an angle of attack of 4°. The icing conditions correspond to a freestream velocity of $V_\infty = 150$ mph, a liquid water content (LWC) of 1.0 g/m^3, a water droplet size (MVD) of 20 μm,

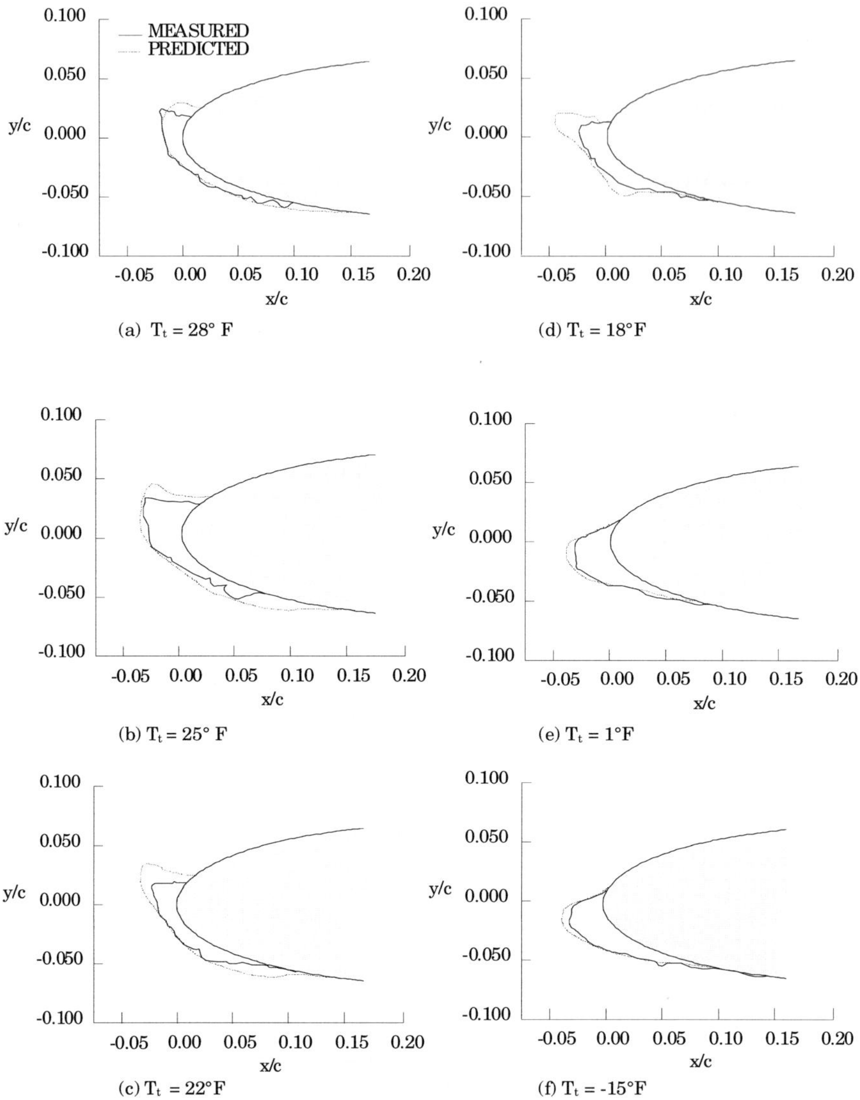

Fig. 5.32. Comparison of calculated and experimental ice shapes on a NACA 0012 airfoil.

and an ice accretion time of 6 minutes. Total temperature was varied from 28 °F to −15 °F to cover glaze, rime and transitional icing regimes.

Figure 5.32 shows a comparison between the calculated and measured ice shapes with time steps of one minute for six total temperatures in which the ice shape changes from white, opaque rime ice to slushy, clear glaze ice as the temperature is increased [30]. At low temperatures, where the ice shapes correspond to rime ice, the agreement between the calculated and measured ice shapes is good, and is consistent with the previous study of Shin et al. [31]. Icing

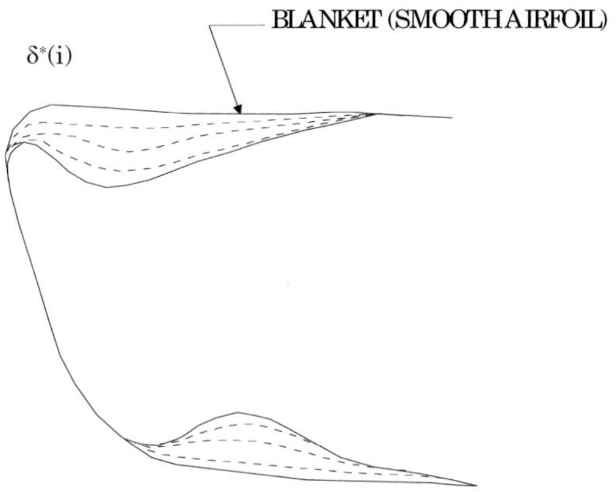

Fig. 5.33. Boundary-layer model used in the viscous-flow calculations.

limits are predicted well for temperatures below 18 °F. At warmer temperatures, the calculated results predict more water run back, which results in more ice accretion beyond the measured icing limits. While the calculated results predict the direction of horn growth of the glaze ice, in general the size of the predicted ice shape is larger than the measured size. Additional comparisons between the calculated and experimental ice shapes are given in [30,31].

5.6 Airfoils with Leading-Edge Roughness

The calculation method of Chapters 2 to 4 can also be used to compute the performance characteristics of airfoils with leading-edge roughness. In this case, changes are required in the turbulence model to account for the roughness effects. In the CS model this is done with the procedure of Cebeci and Chang [32] in which the modified mixing length in Eq. (3.1.29a) is written as

$$l = \kappa(y + \Delta y)\{1 - \exp[-(y + \Delta y)/A]\} \tag{5.6.1}$$

where Δy is a function of an equivalent sand-grain roughness k_s. In terms of dimensionless quantities, with $k_s^+ = k_s u_\tau/\nu$ and $\Delta y^+ = \Delta y\, u_\tau/\nu$,

$$\Delta y^+ = \begin{cases} 0.9[\sqrt{k_s^+} - k_s^+ \exp(-k_s^+/6)] & 5 < k_s^+ \leq 70 \\ 0.7(k_s^+)^{0.58} & 70 \leq k_s^+ \leq 2000 \end{cases} \tag{5.6.2}$$

With the above changes, the calculation method, with an additional modification, can also be used for iced airfoils where the roughness is due to ice. The additional modification involves a procedure to deal with computational difficulties caused by the leading-edge ice shapes since ice on airfoils can introduce substantial geometric changes to their leading edges in a short period and can cause rapid variations in the flow properties. As a result, the inviscid and viscous-flow calculations may have difficulty in producing satisfactory solutions. In the procedure described in [26], the iced airfoil was regarded as a smooth or rough surface obtained by covering the leading-edgeregion with a "blanket" as shown in Fig. 5.33, and boundary-layer calculations were performed with a continuation method in which the initial calculations are performed for the smooth airfoil and subsequent ones for a series of shapes that fall between the smooth and iced airfoils. For each shape, the blowing velocity v_{iw} is computed from

$$v_{iw} = \frac{d}{dx}[u_e(\delta^* - \delta)] \tag{5.6.3}$$

where δ^* corresponds to the displacement thickness obtained from the boundary-layer solutions for the shape whose geometrical difference from the smooth airfoil is $\delta^{(i)}(x)$ and where the δ^* surface is outside the singularity surface. This allows the viscous effects to be incorporated into the inviscid-flow solutions gradually, at each time step, thus reducing the sensitivity of the viscous-flow solutions to the rapid changes in the pressure distribution near the leading edge. For further details, see [26].

In this section, the calculation method of Chapters 2 to 4 is applied to airfoils with leading-edge roughness, which may also be due to ice accretion. In subsection 5.6.1, results are presented for a sample of airfoils with leading-edge roughness. In subsections 5.6.2 and 5.6.3, the calculation method is first applied to iced airfoils corresponding to those in which ice shapes are computed with LEWICE, as discussed in subsection 5.5.4 and then used to demonstrate the effect of ice on airfoil stall at high Reynolds numbers.

5.6.1 Results for Airfoils with Leading-Edge Roughness

The accuracy of the calculation method is demonstrated in Figs. 5.35 to 5.37 for airfoils having 0.011-inch carborundum grains spread evenly over the surface length of 0.08 chord from the leading edge on both surfaces. This roughness was converted into equivalent sand-grain roughness by using the procedure of Kaups and Smith [33] in which the ratio of the equivalent sand-grain roughness to the roughness of the applied elements k_s/k, was assumed to be a function of the concentration and shape of the roughness elements, see Fig. 5.34. In all cases, the shape of the elements was approximated by a sphere, and the concentration which represents the mean value of the area covered by the roughness elements was taken as 0.075 and the equivalent sand-grain roughness height was obtained from Fig. 5.34 to be

$$k_s/c = 0.00094$$

Figure 5.35 shows the results for the NACA 4412 airfoil. It can be seen from the lift coefficient results of Fig. 5.35a that the computations performed

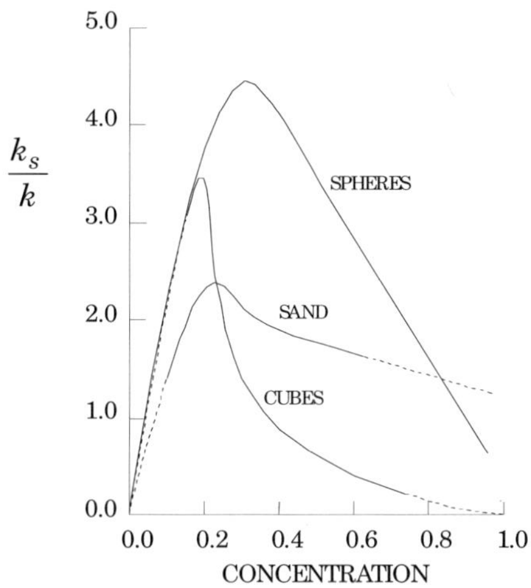

Fig. 5.34. Equivalent sand-grain roughness for uniform three-dimensional roughness as a function of concentration. Dashed lines are extrapolation of experimental data.

up to the stall angle agree well with measurements for the smooth airfoil as well as for the airfoil with leading-edge roughness, although near stall and post stall the lift coefficients are over-predicted. The drag curves of Fig. 5.35b also show good agreement between computations and measurements, although for the clean airfoil the drag is slightly underpredicted at higher angles of attack.

The NACA 23012 airfoil, like the NACA 4412, has camber and the results of Fig. 5.36 again demonstrate good agreement between the computations and measurements, and the drag curve is in even better agreement than that for the NACA 4412. These results, together with the results shown in Fig. 5.37 for the NACA 0012 airfoil quantify the extent to which the calculation method of Chapters 2 to 4 can represent the flows over airfoils with roughened surfaces. It is evident that the calculation method correctly represents this effect and the trends of the lift and drag curves.

5.6.2 Results for Iced Airfoils

To investigate the increase in drag due to ice, several studies have been made with the calculation method of this section [30]. Here a sample of results are presented for the NACA 0012 airfoil on which ice shapes were computed with LEWICE at different icing conditions discussed in subsection 5.5.4. Table 5.2 and Fig. 5.38 show a comparison between the calculated and measured drag coefficients for the ice shapes shown in Fig. 5.32. As can be seen, the values of the total drag coefficient remain nearly constant up to around 12 °F for both speeds before they increase sharply with increasing temperature as the ice shape changes to that of glaze ice.

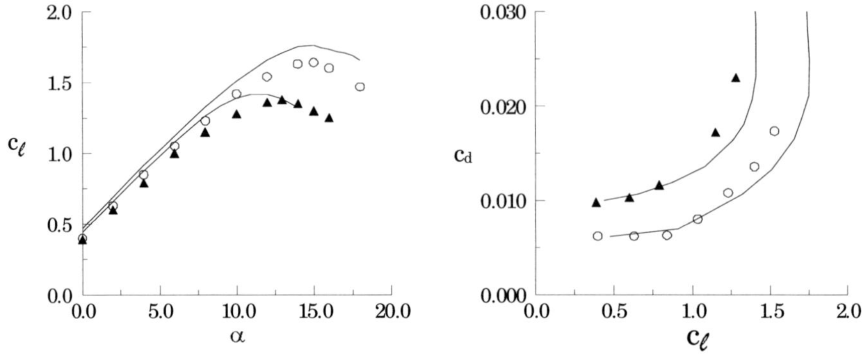

(a) (b)

Fig. 5.35. Comparison of calculated and experimental (symbols) results for the NACA 4412 airfoil at $R_c = 6 \times 10^6$.

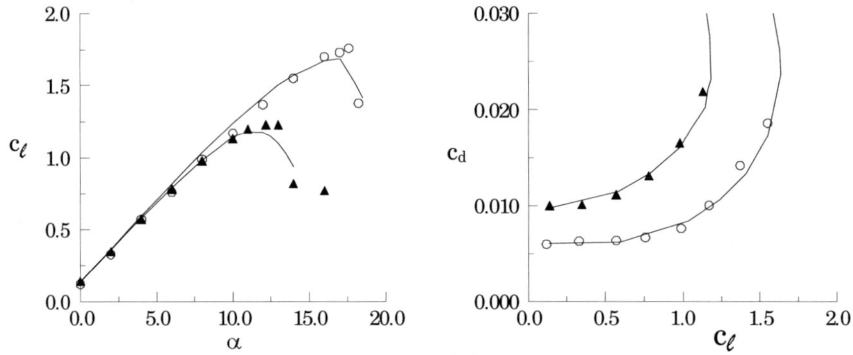

(a) (b)

Fig. 5.36. Comparison of calculated and experimental (symbols) results for the NACA 23012 airfoil at $R_c = 6 \times 10^6$.

Table 5.2. Effect of total temperature on the drag coefficient of the NACA 0012 airfoil at $\alpha = 4°$.

Total temperature °F	Experimental drag coefficient	Calculated drag coefficient
28	0.0578	0.0346
25	0.0540	0.0372
22	0.0315	0.0392
18	0.0271	0.0351
12	0.0229	0.0217
1	0.0229	0.0209
−15	0.0233	0.0202

The adjustment made by extending the range of roughness on the airfoil allows the calculated drag values to agree well with experimental data at low temperatures. The computed drag coefficients begin to rise sharply at around

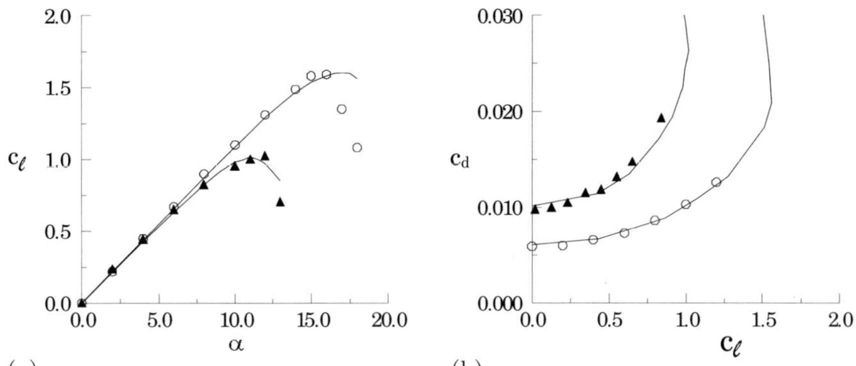

(a) (b)

Fig. 5.37. Comparison of calculated and experimental (symbols) results for the NACA 0012 airfoil at $R_c = 6 \times 10^6$.

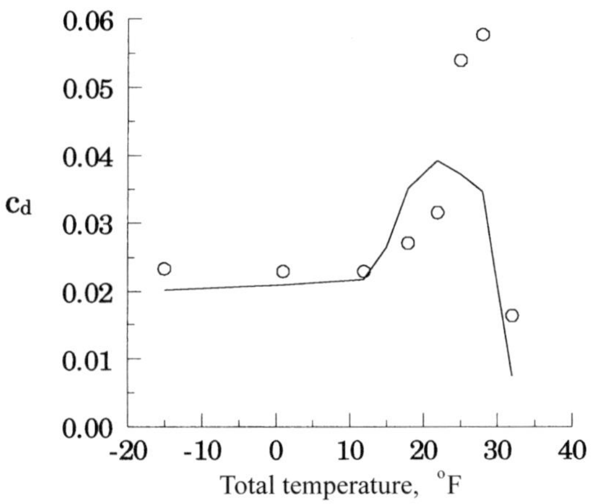

Fig. 5.38. Effect of total air temperature on the drag coefficient of the NACA 0012 airfoil for the experimental conditions in Table 5.2. Solid line denotes calculations and symbols the experimental data.

18 °F and reach a peak at 22 °F. Up to this temperature, the trend of calculated results and their magnitudes are in good agreement with measurements. However, for temperatures higher than 22 °F, while the computed drag coefficients begin to decrease, the measured values continue to rise sharply until 28 °F before they start decreasing rather sharply.

Figure 5.39 shows the variation of the drag coefficient with angle of attack for a fixed ice shape computed at $\alpha = 4°$, $T_t = 22$ °F (Fig. 5.32c). The calculated drag coefficients for angles of attack from 0° to 4° agree well with experimental data, and the calculations exhibit no numerical difficulties. In the angle of attack range of 4° to 8° the calculations exhibited some numerical difficulties. Even

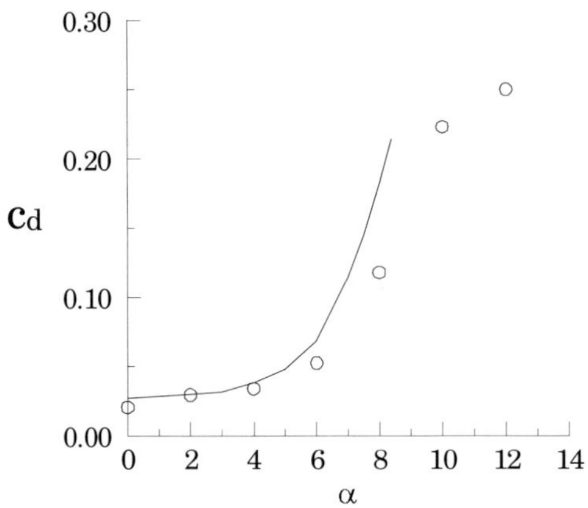

Fig. 5.39. Variation of the drag coefficient of the NACA 0012 airfoil with angle of attack with ice shape corresponding to glaze ice shown in Fig. 5.28c. Solid line denotes calculations and symbols the experimental data.

then, however, the calculated drag coefficients indicate good agreement with measured values, as shown in Fig. 5.39.

5.6.3 Effect of Ice on Airfoil Stall at High Reynolds Numbers

The ability of the calculation method to predict the effect of ice on the stall and post-stall behavior of an airfoil with leading edge rime ice was investigated by first computing the ice shape on a NACA 0012 airfoil for given atmospheric conditions and for a given ice accretion time. The roughness effects in the turbulence model was accounted for by using the modifications to the Cebeci-Smith model, as given by Eqs. (5.6.1) and (5.6.2). The equivalent sand-grain roughness k_s for ice was determined from the expressions used in LEWICE.

Figure 5.40 shows the ice shapes accumulated on the airfoil leading edge after 15, 30 and 60 seconds for atmospheric conditions corresponding to a static temperature, $T_s = 260.44$ K, ambient pressure, $p_\infty = 90,754.8$ Pa, liquid-water content, LWC $= 0.50$ g/m^2, freestream velocity, $V_\infty = 129$ m/s, and droplet size of 20 μm. These results, obtained with LEWICE, show that in practically all cases the extent of the leading edge ice is about 5% of the chord.

Figure 5.41 shows the effect of ice shapes on the lift coefficient at various angles of attack, including stall and post-stall, for a chord Reynolds number of 1.5×10^6. The dashed lines, computed with the calculation method of this section, correspond to the lift coefficient of the iced airfoil and show the strong effect of ice on the lift loss. With very small ice accretion, the stall angle decreases from about 15° to between 7° and 8° with a corresponding lift loss of almost 50%. It is interesting to note that the stall angle is not very sensitive to the size of the rime ice. Even one second ice is sufficient to cause a drastic change in the stall angle.

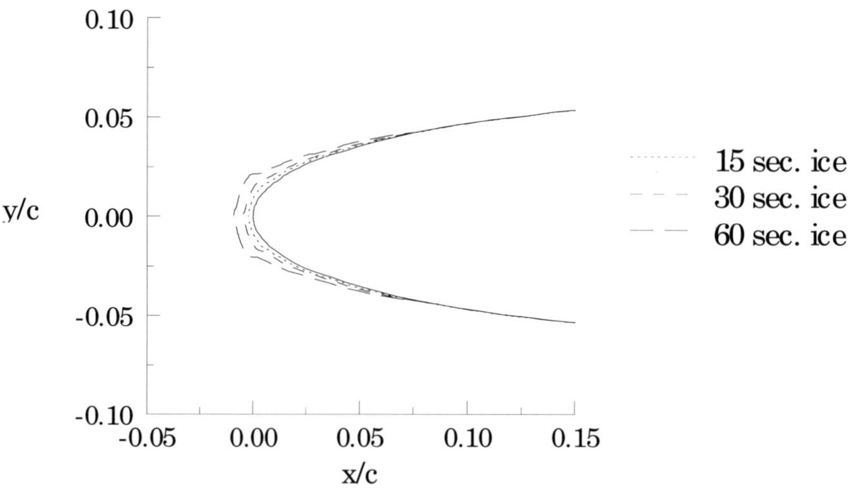

Fig. 5.40. Ice shapes for 15, 30 and 60 seconds on the leading edge of the NACA 0012 airfoil.

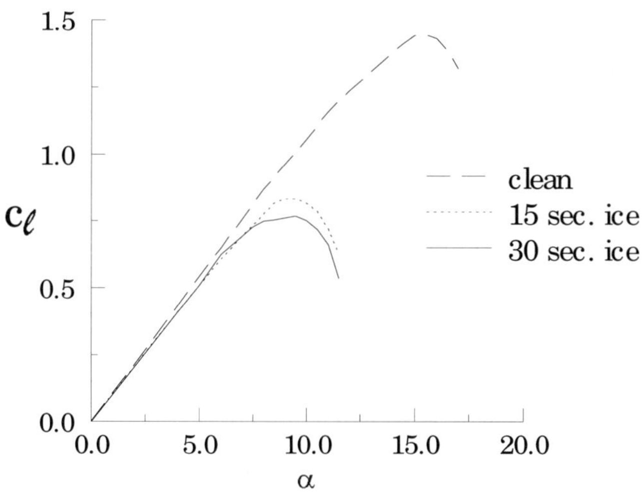

Fig. 5.41. Effect of leading-edge ice on the lift coefficient of the NACA 0012 airfoil. The solid line denotes the lift coefficient of the clean airfoil.

As discussed in Section 5.3, the calculation method of Chapters 2, 3 and 4 also indicates good agreement between calculations and measurements on clean airfoils at low Reynolds numbers. It will be interesting to extend this study to an iced airfoil operating at low Reynolds numbers. For example, typical Reynolds numbers for the blades of a rotorcraft near the hub may be around 2×10^5. For the NACA 0012 airfoil, the stall angle without ice accretion is 11° at this Reynolds number (Fig. 5.17). The question would be, will a small leading edge

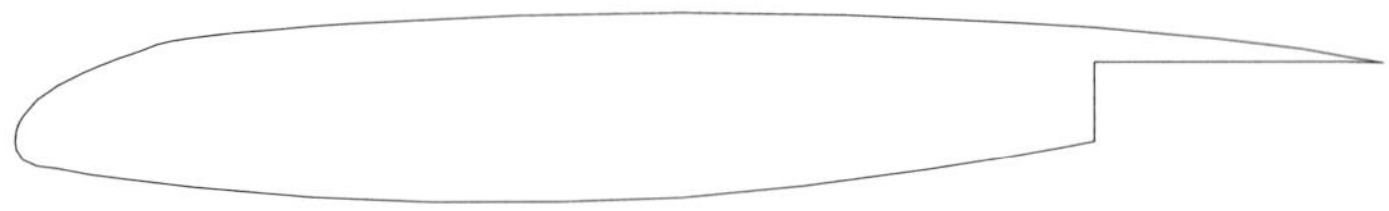

Fig. 5.42. Single airfoil with flap-well cut.

ice reduce the stall angle as drastically as it does in the higher Reynolds number case, and what will that stall angle be?

5.7 Multielement Airfoils

In this section the calculation method of Chapters 2 to 4 is applied to high-lift flows. Again the inviscid flow is computed with a panel method, this time with the multielement panel program discussed in [34–36], and the viscous effects are accounted for by using a procedure similar to that used for single airfoils. The presence of multielement airfoils, a flap well in the main airfoil and possible merging of shear layers from each component require additional changes to the interaction law, turbulence model and the solution procedure used in the calculation method. These modifications are necessary because the interaction law given by Eq. (3.0.4c) and the CS turbulence model in subsection 3.1.4 are appropriate only for airfoils and those parts of airfoils without surface discontinuities, such as flap wells (Fig. 5.42). Where flap wells occur, a different formulation of the inverse procedure and turbulence model are required, as described below. The merging of shear layers add additional complexities to the solution procedure and the turbulence model and are not discussed here.

5.7.1 Interaction Law

The calculation of the flow in the flap-well region is similar to that over a backward-facing step. A large portion of the flow separates immediately after the sudden change of the geometry, and the size of the reversed-flow region depends mainly on the step height, on the gap size, and the overlap. The flow reattaches and gradually recovers downstream in the flap-well region or in the wake. The calculation of flows of this kind is difficult, and potential theory is not adequate because of the singularity that occurs at the geometry discontinuity and the strong viscous effects in the separated flow field. Thus, an initial distribution of displacement thickness is assumed near the flap-well region and the relaxation formula given by Eq. (1.3.1) is used in the inverse method to replace the Hilbert integral formulation of the external boundary condition. The edge boundary condition given by Eq. (3.0.4c) is replaced by the specified displacement thickness with $\delta^*(x)$, computed from Eq. (1.3.1) and with the external velocities $u_{e_v}(x)$ and $u_{e_i}(x)$ computed by the boundary-layer and inviscid methods, respectively; ω is a relaxation parameter commonly taken to be 1.25. At the end of the flap-well region, the solution procedure reverts to the Hilbert-integral approach.

5.7.2 Turbulence Model

In the flap-well region, the CS eddy-viscosity formulation of subsection 3.1.4 used in the applications of the calculation method discussed in the previous sections, is modified by defining the ν_t-formulation as

$$\nu_t = \nu_t^{(I)} + [\nu_t^{(F)} - \nu_t^{(I)}][1 - e^{-(x-x_0)/\lambda L}] \tag{5.7.1}$$

Here $\nu_t^{(I)}$ denotes the eddy viscosity corresponding to the velocity profile above the separated region and $\nu_t^{(F)}$ includes the total region from the wall. The parameter λ is a relaxation parameter (usually around 10), L is a characteristic length and x_0 is the beginning of the flap-well. The expressions for $\nu_t^{(I)}$ and $\nu_t^{(F)}$, with y_0 denoting the location where $u = 0$, are given by:

$$\nu_t^{(I)} = \begin{cases} \{0.4(y - y_0)[1 - e^{-(y-y_0)/A}]\}^2 \left|\dfrac{\partial u}{\partial y}\right| \gamma_{\text{tr}} & y_0 < y < y_c \\ \alpha u_e \left|\displaystyle\int_{y_0}^{\infty} \left(1 - \frac{u}{u_e}\right) dy\right| \gamma\gamma_{\text{tr}} & y > y_c \\ 0 & y < y_0 \end{cases} \tag{5.7.2a}$$

$$\nu_t^{(F)} = \begin{cases} \{0.4y[1 - e^{-y/A}]\}^2 \left|\dfrac{\partial u}{\partial y}\right| \gamma_{\text{tr}} \\ \alpha u_e \left|\displaystyle\int_{0}^{\infty} \left(1 - \frac{u}{u_e}\right) dy\right| \gamma\gamma_{\text{tr}} \end{cases} \quad y > 0 \tag{5.7.2b}$$

5.7.3 Solution Procedure

The larger regions of flow separation associated with the flap well and the near wake require modification in the numerical solution procedure. Since a linearized form of the boundary-layer equations is being solved, it is necessary that the calculations at station x_i have initial profiles which are usually assumed to correspond to those at a previous x-station, x_{i-1}. With increasing flow separation, the effect of the initial profiles on the solutions at x_i increases and can lead to breakdown. A remedy to this problem is to use the continuation method (see subsection 4.2.4) by defining the velocity profile at x_i to be of the form

$$u = u_{\text{ref}} + n(u_a - u_{\text{ref}}) \tag{5.7.3}$$

where u_a denotes the velocity profile at x_{i-1} and u_{ref} to a profile which allows solutions to be obtained at x_i. The iteration process at x_i proceeds with values of n ranging from 0 to unity.

The sequence of the calculations is as follows. The panel method provides an external velocity distribution based on a body shape in which the flap-well region is assumed to be absent. The boundary-layer method provides solutions on the upper surface from the stagnation point through the regions of laminar, transitional and turbulent flow to the trailing edge and into the upper wake. Similarly, it provides results for the lower surface up to the beginning of the flap-well. A displacement thickness distribution, $\delta^*(x)$, is assumed in the flap well and, with the continuation method described above and with the initial velocity profile similar to that of a backward-facing step, calculations proceed to the trailing edge and into the lower wake. In the flap-well region, the blowing velocity v_{iw} is defined by

$$v_{iw} = \frac{d}{dx}(u_e \delta_m^*) \tag{5.7.4}$$

where $\delta_m^* = \delta^* - \delta_t$. Here δ_t corresponds to the body shape assumed to exist over the flap-well. Elsewhere, the blowing velocity is given by Eqs. (5.0.1).

With the blowing velocity distribution known everywhere in the flowfield, a new distribution of external velocity $u_{ei}(x)$ is obtained from the panel method. As before, the boundary-layer solutions on the upper and lower surfaces of the airfoil are obtained with the Hilbert integral in which the edge boundary condition, Eq. (3.2.9), is now written as

$$u_e(x) = u_e^\kappa(x_i) + C_{ii}D_i + \sum_{j=1}^{i-1} C_{ij}(D_j - D_j^\kappa) + \sum_{j=i+1}^{N} C_{ij}(D_j - D_j^\kappa) - C_{ii}D_i^\kappa \tag{5.7.5}$$

with κ indicating the iteration cycle. Equation (3.2.12b) also changes to

$$g_i = u_e^\kappa(x_i) + \sum_{j=1}^{i-1} C_{ij}(D_j - D_j^\kappa) + \sum_{j=i+1}^{N} C_{ij}(D_j - D_j^\kappa) - C_{ii}D_i^\kappa \tag{5.7.6}$$

In the flap-well, with u_{ev} known from the previous flap-well calculation, a new δ^*-distribution is available from Eq. (1.3.1) and is used to obtain solutions up to the trailing edge. This sequence of calculations is repeated for the whole flowfield until convergence is achieved. The continuation method given by Eq. (5.7.3) is involved within this sequence of calculations where necessary.

5.7.4 Results for an Airfoil with Flap-Well

An appraisal of the calculation method, as applied to the flow in and around a flap well, required corresponding measurements and a parallel experimental program which was carried out at the California State University, Long Beach,

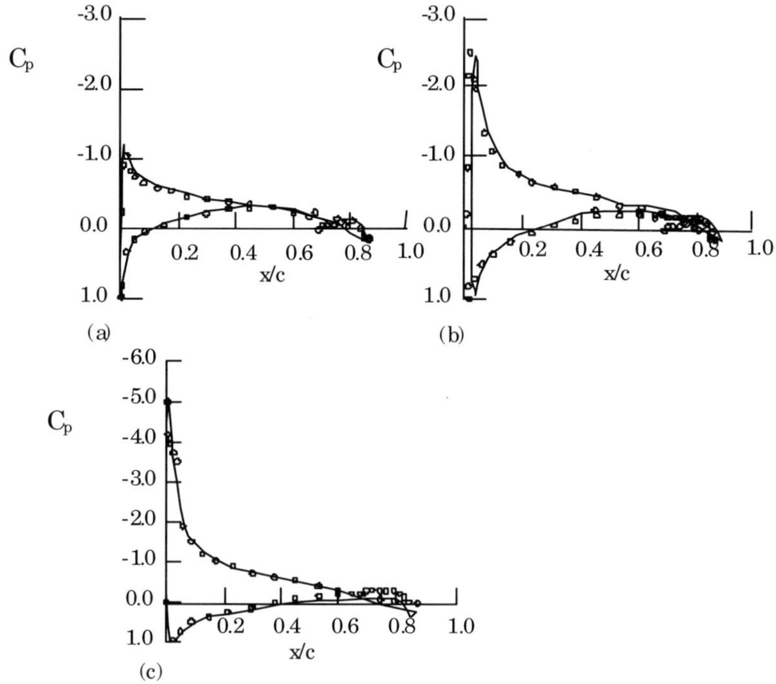

Fig. 5.43. Comparison of calculated pressure distributions (solid lines) with data (symbols) for $R_c = 0.5 \times 10^6$ at (a) $\alpha = 5.0$, (b) $\alpha = 8.0$, (c) $\alpha = 12.0$.

with the single airfoil shown in Fig. 5.42. The experiment described in [37] used this 12-inch chord airfoil, which corresponds to the main element of the three-element configuration tested in the low-turbulence tunnel of the NASA Langley Research Center [38]. The chord Reynolds number was 0.5×10^6, and the surface pressures were measured for angles of attack up to 14 degrees with local velocity information in the flap well at an angle of attack of 5 degrees. Figure 5.43 presents the measured and calculated surface-pressure distributions for angles of attack of 5, 8 and 12 degrees with transition tripped at 0.25c, and the agreement is generally good. The calculated upper-surface pressure peak close to the leading-edge reflects the better spatial resolution of the calculation method. Close to the trailing edge there is a some small disagreement which may stem from the flap-well results. Nevertheless, the pressure-coefficient distributions represent closely the measurements in the flap-well with the near constant values indicating the region of recirculation followed, as can be seen, by a rapid increase in pressure coefficient after reattachment [35].

Figure 5.44 shows the velocity profiles and the distribution of skin-friction coefficients within the flap well. The profiles differ increasingly from the measurements in the near-wall region, although reattachment occurs at about the same location. The reasons for the discrepancy are likely to be associated with

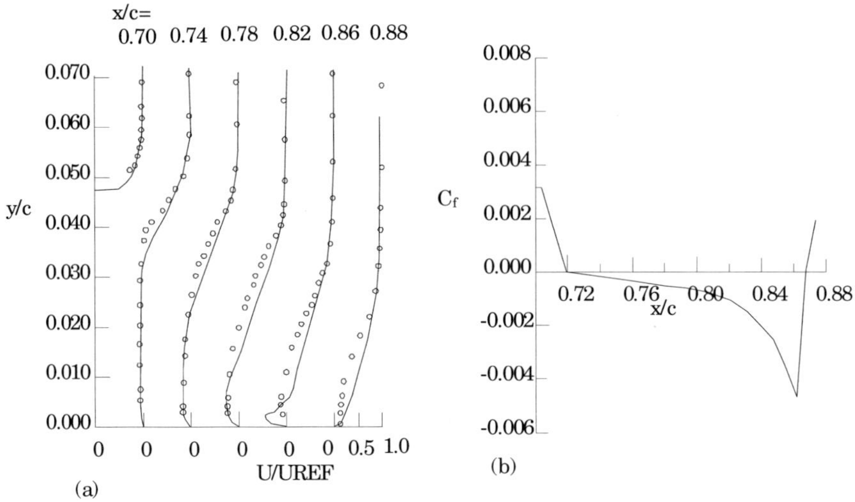

Fig. 5.44. Results for $R_c = 0.5 \times 10^6$ at $\alpha = 5.0$ in the flap-well region. (a) Velocity profiles, (b) local skin-friction coefficient.

the turbulence model that is used here to represent a near-wall flow, which undoubtedly involves low-Reynolds-number characteristics together with a switch from a wall jet in the negative to a boundary-layer in the positive direction. It is unlikely that these local differences will greatly affect the outer region, particularly since the negative velocities and momentum are small. The distribution of skin-friction coefficient, Fig. 5.44b, indicates that flow reattachment occurs at 0.87c.

5.7.5 Results for Two-Element Airfoils

The accuracy of the calculation method for high-lift flows has been investigated in several studies [35–37, 40]. Some of these studies made use of the inviscid method based on the conformal mapping developed by Halsey [40] and some based on the extension of the panel method of Chapter 2 from a single airfoil to multi-element airfoils. In the latter case, the vorticity strength τ_j was taken to be constant on all panels ($\tau_j = \tau$) and its single value was again adjusted to satisfy the condition associated with the specification of circulation. Two-element and three-element airfoil configurations computed with the calculation method employing the multi-element panel method with viscous effects showed that, while the calculated results were in good agreement with experimental data for conventional airfoils, those for airfoils with thin trailing edges and/or supercritical airfoils were not [36] in good agreement. To improve the results for such cases, the multi-element panel method was modified and two options were included for the vorticity distribution in addition to the option which assumes the vorticity to be constant. One option is based on the assumption

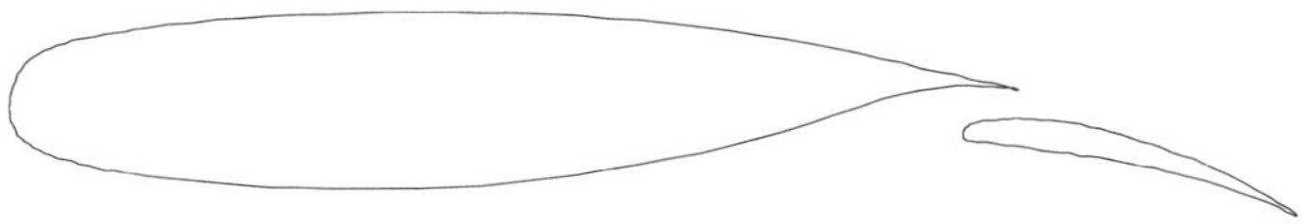

Fig. 5.45. NLR 7301 airfoil with flap.

that vorticity varies quadratically with surface distance and the second option that it varies quadratically in a small region near the trailing edge. In this second option, to be referred to as the partially parabolic vorticity distribution option, the size of the region is specified in the panel code.

Recent studies conducted with the calculation method employing this modified panel method with two options showed a much better agreement with experimental data than those obtained with constant vorticity option and are discussed here for two-element airfoils [36]. Additional results and results for three-element airfoils can be found in [35].

Figure 5.45 presents the airfoil/flap configuration of Van den Berg and Oskam [41] which corresponds to a supercritical main airfoil (NLR 7301) with a flap of 32% of the main chord at a deflection angle of 20 degrees and with a gap of 2.6% chord. Measurements of surface pressure and velocity profiles were obtained at a chord Reynolds number of 2.51×10^6 and for angles of attack of 6 and 13.1 degrees. Lift coefficients were deduced for five angles of attack.

The results shown in Figures 5.46 and 5.47 were obtained with the partially parabolic vorticity distribution option since the trailing edge of the main airfoil is rather thin [36]. Figure 5.46a and 5.46b shows a comparison between measured and computed pressure distributions at $\alpha = 6^\circ$ and 13.1°, respectively. The results for $\alpha = 13.1^\circ$ confirm the absence of separation and the calculated lift coefficients of Fig. 5.47 show very good agreement with measurements at low angles of attack. While some discrepancies exist at higher angles of attack, the stall is predicted accurately. It is believed that these discrepancies are due to the merging of the airfoil shear layer with the boundary-layer on the upper surface of the flap which is not considered in the present calculation method.

Figure 5.48 shows a two-element airfoil with an NACA 4412 profile for the main element and an NACA 4415 flap profile deflected by 10 degrees which Olson and Orloff tested at NASA Ames [42]. The freestream Mach number was 0.06 and the chord Reynolds number 1.3×10^6. The pressure distribution was obtained at $\alpha = 2.2^\circ$.

The trailing edges of both airfoils are not particularly thin. For this reason, the results shown in Figures 5.49 were obtained using the constant vorticity option. Figure 5.49a indicates good agreement between the computed and measured pressure distributions. The transition location on the upper surface of the main element was computed at $x/c = .186$ and on the upper surface of the flap at $x/c = .63$. Table 5.3 shows a comparison of the measured and computed lift

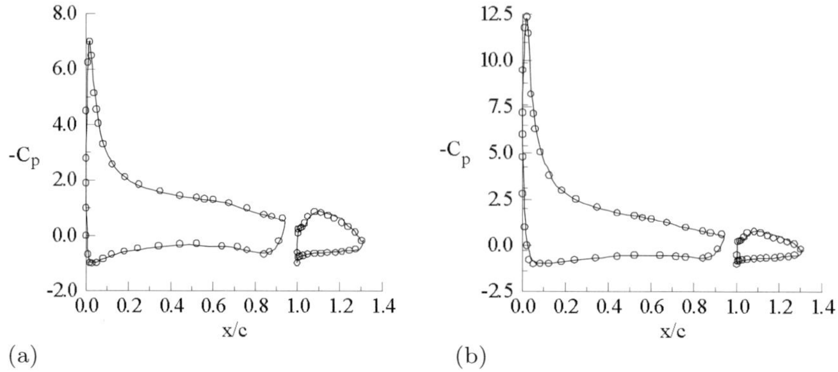

Fig. 5.46. Calculated and measured pressure distributions for the NLR 7301 airfoil with flap (a) $\alpha = 6°$, (b) $\alpha = 13.1°$.

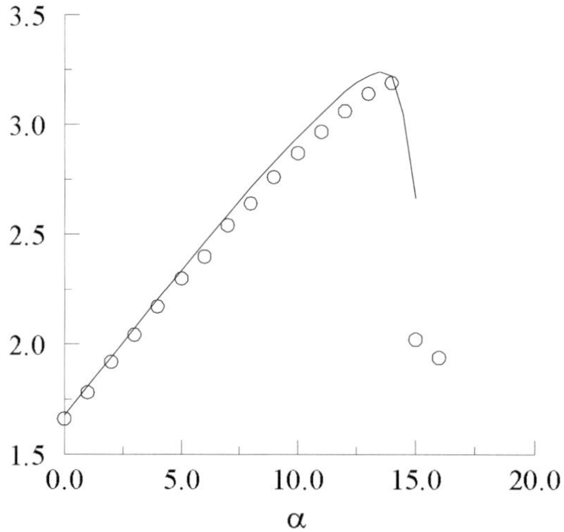

Fig. 5.47. Calculated and measured lift coefficients for the NLR 7301 airfoil with flap.

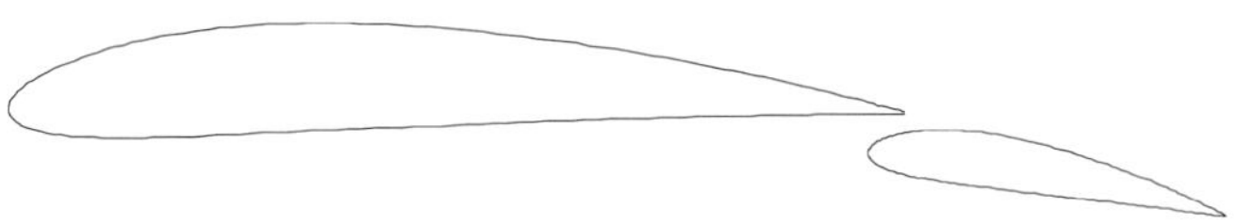

Fig. 5.48. NACA 4412/4415 airfoil/flap configuration.

coefficients for each element. The lift coefficient for the main airfoil is computed accurately. However, there exists some discrepancy on the flap, which suggests that the merging of the shear layers, shown in Fig. 5.49b and not modeled in the present method, has an influence on the flow field computation. Overall, however, the prediction with the present method is satisfactory.

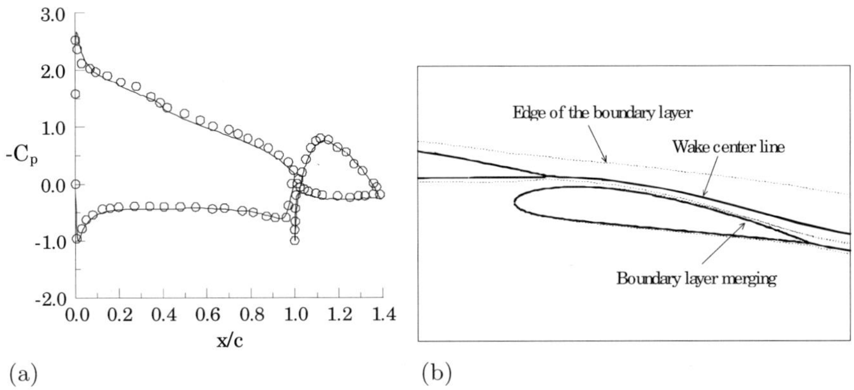

Fig. 5.49. Results for the NACA 4412/4415 airfoil/flap configuration at $\alpha = 2.2°$. (a) calculated and measured pressure distributions, and (b) boundary-layer growth.

Table 5.3. Measured and computed lift coefficients for the NACA 4412/4415 configuration.

	Main element	Flap	Total
Calculated	1.6979	0.2417	1.9396
Measured	1.69	0.22	1.91

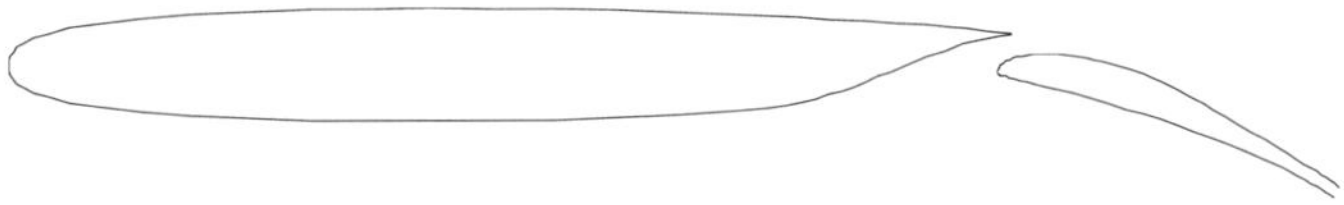

Fig. 5.50. Boeing airfoil/flap configuration.

The calculation method was also applied to a set of supercritical airfoils tested in the 36 × 36 in. wind tunnel of the Boeing Research Laboratories at a Mach number of 0.201 and chord Reynolds number of 2.83×10^6 [3, 43].

Figure 5.50 represents the airfoil/flap configuration, and Fig. 5.51 the pressure distributions for $\alpha = 0.01°$ and $8.93°$, with calculations using the modified multi-element panel method with Prandtl-Glauert compressibility correction which relates velocities for a given airfoil at different Mach numbers without changing its shape. In this case, the incompressible flow ($M_\infty = 0$) is computed and the resulting incompressible pressure coefficient c_{p_i} is scaled by the factor β to simulate the compressible pressure coefficient, that is

$$c_p = \frac{c_{p_i}}{\beta} \tag{5.7.7}$$

where

$$\beta = \sqrt{1 - M_\infty^2} \tag{2.4.4}$$

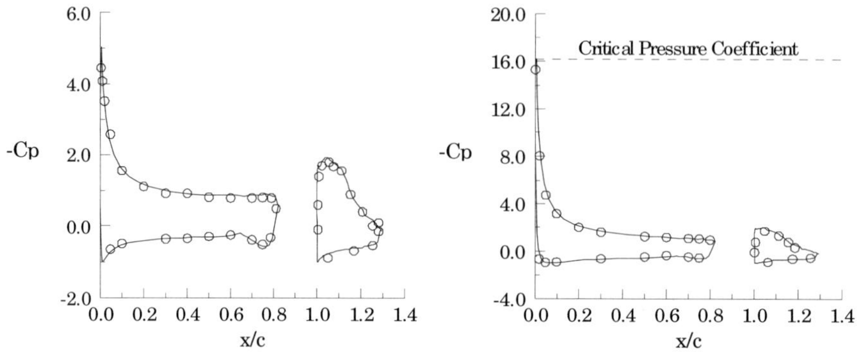

(a) (b)

Fig. 5.51. Calculated and measured pressure distributions for the Boeing airfoil/flap configuration (a) $\alpha = 0.01°$ and (b) $\alpha = 8.93°$.

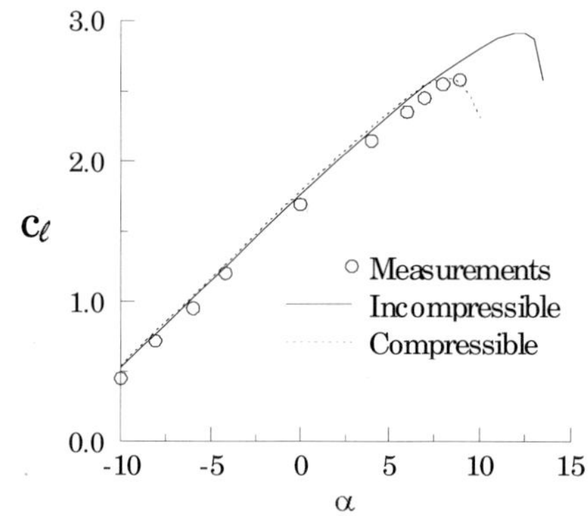

Fig. 5.52. Calculated and measured lift coefficients for the Boeing airfoil/flap configuration.

Figure 5.52 shows a comparison between the calculated and measured total lift coefficients together with the variation of lift coefficient of the main airfoil and flap for incompressible and compressible flows.

As can be seen, the agreement between measured and computed pressure coefficients on the main element is excellent regardless of whether the calculations are performed for an incompressible flow or for a compressible flow with compressibility corrections. However, there exists some discrepancy between measured and calculated values on the flap's upper surface and this becomes more visible when the total calculated lift coefficient is compared with the experimental data in Fig. 5.52. This difference can be explained by the merging of the shear layers observed in the tests and reported in [3].

Figure 5.53 shows a comparison between measured and computed wake center lines and wake thicknesses for $\alpha = 0°$ and $8°$. The predictions of the calcu-

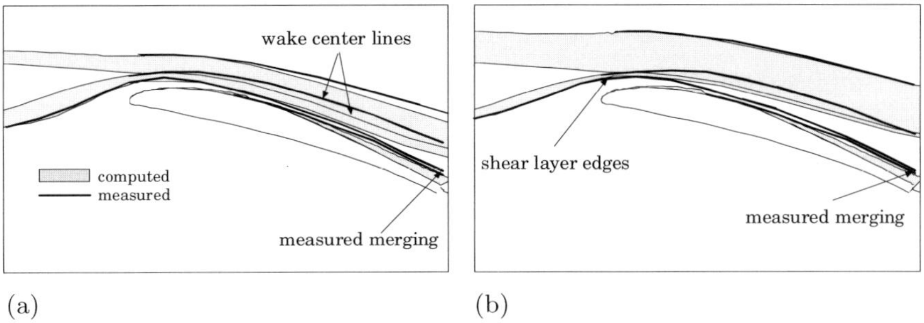

Fig. 5.53. Calculated and measured wake center lines and wake thicknesses for the Boeing airfoil/flap configuration at (a) $\alpha = 0°$, and (b) $\alpha = 8°$.

lation method are satisfactory for the upper wakes and surfaces, but not for the lower wake which is under predicted. This is probably due to the approximation used to calculate the wake "center line", computed inviscidly as the dividing streamline in the calculation method. Also, the assumption of no interaction locally between upper and lower wakes may be insufficient.

We also note from the results in Fig. 5.52 that for this configuration, the predicted incompressible stall is different from the compressible one. The critical pressure coefficient for the freestream Mach number of 0.201 is indicated on Fig. 5.52b. At $\alpha = 8.93°$, the angle to the measured stall, there exists a small region of supersonic flow and a shock wave may occur. Even though this shock cannot be predicted by the panel method, the compressibility corrections to the panel method and boundary-layer method provide a substantial improvement to the stall prediction. However, it must be noted that these compressibility corrections are based on small perturbation theory and therefore are only applicable at small angles of attack. A truly compressible inviscid method should be used in order to predict the stall accurately in such conditions.

Figure 5.54 shows the slat/airfoil configuration. Since the slat is a thin element, the inviscid flow calculations were performed for incompressible flows with the parabolic vorticity option. Figure 5.55a presents a comparison between the measured and computed pressure distributions at $\alpha = 26.88°$. The agreement is excellent on the slat. However, the pressure peak on the main airfoil is slightly underpredicted. Also, a closer examination of the pressure distribution near the trailing edge suggests that the extent of the predicted separated flow is greater than the one observed in the experiments. This may be the reason for the slightly underpredicted maximum lift coefficient as shown in Fig. 5.55b. However, calculated results are still in good agreement with the data.

5.7.6 Summary

The results presented in this section show that the calculation method for high-lift flows with its consideration of the flap-well region, leads to pressure coef-

Fig. 5.54. Boeing slat/airfoil configuration.

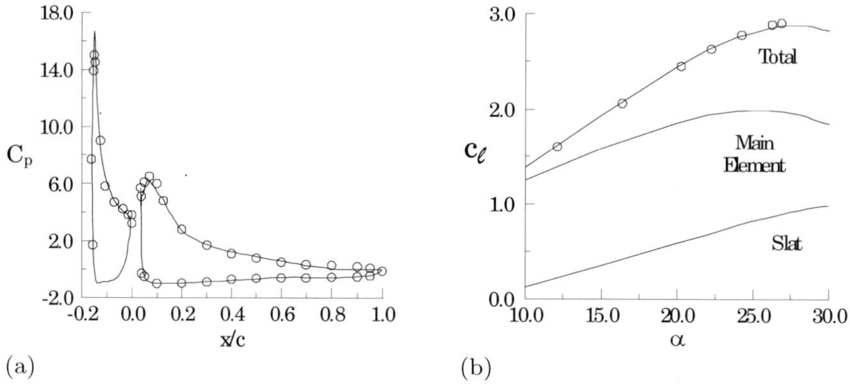

Fig. 5.55. Calculated and measured (a) pressure distributions at $\alpha = 26.88°$ and (b) lift coefficients for the Boeing slat/airfoil configuration.

ficient distributions and values of lift and drag coefficients which are in good agreement with experiment for single and multielement configurations for a range of angles of attack, including stall.

The calculation method neglects the confluent boundary-layers, which become important as the distance between the main airfoil and the flap becomes smaller or where the shear layer thickens with angle of attack. Also, a close examination of the turbulence model in the vicinity of this merging region is desirable. The incorporation of improvements to consider these aspects together with inclusion of compressibility effects is likely to lead to a more generally applicable calculation method. The extension of this method to wings and three-dimensional high-lift configurations is discussed in Sections 13.3 and 13.4.

References

[1] Cebeci, T., Clark, R.W., Chang, K.C., Halsey, N.D. and Lee, K., "Airfoils with Separation and the Resulting Wakes," J. Fluid Mech, Vol. 163, pp. 323–347, 1986.
[2] Abbott, J.H. and von Doenhoff, A.E., Theory of Wing Sections. Dover, 1959.
[3] Omar, E., Zierten, T. Hahn, M., Szpiro, E., and Mahal, A., "Two-Dimensional Wind-Tunnel Tests of a NASA Supercritical Airfoil with Various High-Lift Systems Vol. II – Test Data," NASA CR-2215, 1977.
[4] Wazzan, A.R., "Spatial Stability of Tollmien-Schlichting Waves," Progress in Aerospace Sciences, Vol. 16, No. 2, pp. 99–127, Pergamon Press, N.Y., 1975.
[5] Nayfeh, A.H., Ragab, S.A. and Al-Maaitah, A., "Effect of Roughness on the Stability of Boundary-layers," AIAA Paper No. 86-1044.

[6] Fage, A., "The Smallest Size of Spanwise Surface Corrugation which Affects Boundary-Layer Transition on an Airfoil," R&M No. 2120, Brit. A.R.C., 1943.
[7] Cebeci, T. and Egan, D., "Prediction of Transition Due to Isolated Roughness," AIAA J., Vol. 27, pp. 870–875, 1989.
[8] Cebeci, T., "Essential Ingredients of a Method for Low Reynolds-Number Airfoils," AIAA J., Vol. 27, pp. 1680–1688, 1983.
[9] Gault, D.E., "An Experimental Investigation of Regions of Separated Laminar Flow," NACA TN 3505, 1955.
[10] Cousteix, J. and Pailhas, G., "Etude Exploratorie d'un Processus de Transition Laminair-Turbulent au Voisinage du Decollemente d'une Couche Limite Laminaire," T.P. No. 1979-86. Also LeRecherche Aerospatiale No. 1979-3, pp. 213–218.
[11] Hoheisel, H., Hoeger, M., Meyer, P. and Koerber, G., "A Comparison of Laser-Doppler Anemometry and Probe Measurements Within the Boundary-layer of an Airfoil at Subsonic Flow," in Laser Anemometry in Fluid Mechanics – II, Selected Papers from the Second Intl. Symp. on Applications of Laser Anemometry to Fluid Mechanics, Lisbon, Portugal, July 1984. LADOAN, pp. 143–157.
[12] Lissaman, P.B.S., "Low-Reynolds-Number Airfoils," Annual Review of Fluid Mech., Vol. 15, 1983, pp. 223–239.
[13] Mueller, T.J., eds., Proceedings of the Conference on Low Reynolds Number Airfoil Performance, UNDAS-CP-778123, June 1985.
[14] Mueller, T.J., eds., Proceedings of the Conference on Low Reynolds Number Aerodynamics. University of Notre Dame, June 1989.
[15] Proceedings of the International Conference on Aerodynamics at Low Reynolds Numbers. The Royal Aeronautical Society, October 1986.
[16] McGhee, R.J., Jones, G.S. and Jouty, R., "Performance Characteristics from Wind-Tunnel Tests of a Low Reynolds-Number Airfoil," AIAA paper 88-0607, 1988.
[17] Liebeck, R.H., "A Class of Airfoils Designed for High Lift Without Separation in Incompressible Flow," Douglas Aircraft Report No. MDC-1097/01, Dec. 1971.
[18] Stratford, B.S., "The Prediction of Separation of the Turbulent Boundary-layer," J. Fluid Mech., Vol. 5, 1959.
[19] James, R.M., "A New Look at Two-Dimensional Incompressible Airfoil Theory," Douglas Aircraft Report No. MDC-J0918/01, May 1971.
[20] Liebeck, R.H. and Camacho, P.P., "Airfoil Design at Low Reynolds Number with Constrained Pitching Moment," Proceedings of the Conference on Low Reynolds Number Airfoil Performance, UNDAS-CP-778123, June 1985.
[21] Bharadvaj, B.K., "Computation of Steady and Unsteady Control Surface Loads in Transonic Flow," AIAA J., Vol. 29, No. 11, pp. 1906–1922, 1991.
[22] Cebeci, T., Hefazi, H., Roknaldin, F. and Carr, L.W., "Prediction of Compressibility Effect on Airfoil Stall at Low Mach Numbers," AIAA paper No. 94-0290, Jan. 1994.
[23] Carr, L.W., McCroskey, W.J., McAlister, K.W., Pucci, S.L. and Lambert, O., An Experimental Study of Dynamic Stall on Advanced Airfoil Sections, Vol. 3, Hot-Wire and Hot-Film Measurements. NASA TM 84245.
[24] Ruff, G.A. and Berkowitz, B.M., "User's Manual for the NASA Lewis Ice Accretion Code (LEWICE)," NASA CR-185129, May 1990.
[25] Bragg, M.B. and Coirier, W.J., "Aerodynamic Measurements of an Airfoil with Simulated Glaze Ice," AIAA Paper 86-0484, Jan. 1986.
[26] Cebeci, T., "Calculation of Flow over Iced Airfoils," AIAA J., Vol. 27, p. 853, 1989.
[27] Potapczuk, M.G., "Navier-Stokes Analysis of Airfoils with Leading-Edge Ice Accretions," Ph.D. Dissertation, The University of Akron, May 1989.
[28] Cebeci, T., Chen, H.H. and Alemdaroglu, N., "Fortified LEWICE with Viscous Effects," J. of Aircraft, Vol. 28, No. 9, pp. 565–571, Sept. 1991.
[29] Shin, J. and Bond, T.H., "Experimental and Computational Ice Shapes and Resulting Drag Increase for an NACA Airfoil," Proceedings of the 5th Symposium on Numerical and Physical Aspects of Aerodynamic Flows, California State University, Long Beach, CA, Jan. 1992.

[30] Shin, J., Berkowitz, B., Chen, H.H. and Cebeci, T., "Prediction of Ice Shapes and Their Effect on Airfoil Performance," J. Aircraft, Vol 31, No. 2, pp. 263–270, 1994.
[31] Shin, J. and Bond, T., "Results of an Icing Test on a NACA 0012 Airfoil in the NASA Lewis Icing Research Tunnel," AIAA Paper No. 92-0647, Jan. 1992.
[32] Cebeci, T. and Chang, K.C., "Calculation of Incompressible Rough-Wall Boundary-Layer Flows," AIAA J., Vol. 16, p. 730, 1978.
[33] Smith, A.M.O. and Kaups, K., "Aerodynamics of Surface Roughness and Imperfections." SAE Paper 680198, April 1968.
[34] leading edge XXXXXXXXXX
[35] Cebeci, T., Jau, J. and Vitiello, D., "An Interactive Boundary-Layer Approach to Multielement Airfoils at High Lift," AIAA paper no. 92-0404, 1992
[36] Cebeci, T., "Calculation of Multielement Airfoils and Wings at High Lift," AGARD Conference Proceedings 515 on High-Lift Aerodynamics, Paper No. 24, Banff, Alberta, Canada, Oct. 1992.
[37] Cebeci, T., Besnard, E. and Chen, H.H., "Calculation of Multielement Airfoil Flows, Including Flap Wells," AIAA paper no. 96-0056.
[38] Alemdaroglu, N.A., "Experiemental Investigation of Flow Around a Multielement Airfoil," AGARD Conference Proceedings 515 on High-Lift Aerodynamics, Paper No. 2, Banff, Alberta, Canada, Oct. 1992.
[39] Nakayama, A., Kreplin, H.P., Morgan, H.L., "Experimental Investigation of Flowfield About a Multielement Airfoil," AIAA J., Vol. 28, No. 1, pp. 14–21, Jan. 1990.
[40] Cebeci, T., Besnard, E. and Messié, S., "A Stability Transition Interactive-Boundary-Layer Aproach to Multielement Wings at High Lift," AIAA Paper No. 94-0292, 1994.
[41] Halsey, N.D., "Potential Flow Analysis of Multielement Airfoils Using Conformal Mapping," AIAA J., Vol. 17, pp. 1281–1288, 1979.
[42] Van de Berg, B. and Oskam, B., "Boundary-Layer Measurements on a Two-Dimensional Wing with Flap and a Comparison with Calculations," AGARD CP-271, Paper 18, 1979.
[43] Olson, L.E. and Orloff, K.L., "On the Structure of Turbulent Wakes and Merging Shear Layers of Multielement Airfoils," AIAA Paper No. 81-1238, June 1981.
[44] Omar, E., Zierten, T., and Mahal, A., "Two-Dimensional Wind-Tunnel Tests of a NASA Supercritical Airfoil with Various High-Lift Systems. Vol. I – Data Analysis," NASA CR-2214, 1973.

6 Two-Dimensional Compressible Flows

6.0 Introduction

The calculation method described in Chapters 2 to 4 for airfoils in incompressible flows can also be extended to compressible flows with changes in inviscid, boundary-layer and transition methods. To calculate inviscid flows, the panel method of Chapter 2 must be replaced with a compressible inviscid method which can be either a full-potential or Euler method. For flows at low Mach numbers, the former choice has advantages over the latter since Euler methods have slower convergence rates than full potential methods at low Mach numbers, say around 0.3, which is a typical freestream Mach number for high-lift flows (Section 5.4). At higher Mach numbers, and especially in the presence of shocks, the choice is clear; Euler methods are the appropriate ones.

To calculate viscous flows, the boundary-layer method of Chapter 3 must be replaced with a compressible version of the boundary-layer method. This requires the solution of the continuity and momentum equations with the energy equation providing the temperature field needed to calculate the fluid properties in the continuity and momentum equations, as discussed in this chapter.

To calculate transition with the e^n-method, the linear stability equation of Chapter 4 must be replaced with a compressible version of the linear stability equations. In this case, the linear stability equations for compressible flow do not reduce to a fourth-order single equations but to a system of equations involving continuity, momentum and energy equations, as discussed in Chapter 12.

The solution procedure of the calculation method for compressible flows is identical to that in Fig. 1.1; again the inviscid and viscous flows are computed with the chosen methods and transition is obtained with the e^n-method. The viscous effects are introduced into the inviscid method through the blowing velocity concept prescribed on the airfoil and on its wake dividing streamline. The newly computed inviscid flowfield with viscous effects is then used to solve

the boundary-layer equations with the onset of transition location provided by the e^n-method so that a new blowing velocity distribution can be determined; this procedure is repeated until convergence.

In this chapter we do not describe inviscid methods and their solution procedure with and without viscous effects. This is partly because of the wide choice of methods that can be used for this purpose, and partly because the basic ideas of introducing viscous effects is essentially the same as for panel methods. There are some small differences and they are discussed in Chapter 13.

We also do not describe the transition method for compressible flows. This is partly because the equations and the solution procedure are presented and discussed in Chapter 12 for three-dimensional compressible flows and partly because the basic ideas in computing transition with the e^n-method for compressible flows is identical to that for incompressible flows. Again the linear stability equations are solved and the n-factor is computed from Eq. (4.1.3).

The emphasis in this chapter is on the extension of the boundary-layer method of Chapter 3 to compressible flows. Although the calculation method with inviscid flow either obtained from a full-potential flow method or a Euler method and the viscous flow from the boundary-layer method of this chapter, its application has not been as extensive as for incompressible flows.

6.1 Boundary-Layer Equations

For two-dimensional compressible flows, the boundary-layer equations for laminar and turbulent flows are well known, see for example [1]. With the eddy viscosity and turbulent Prandtl number concepts, they can be written in the form

$$\frac{\partial}{\partial x}(\rho u) + \frac{\partial}{\partial y}(\overline{\rho v}) = 0 \tag{6.1.1}$$

$$\rho u \frac{\partial u}{\partial x} + \overline{\rho v}\frac{\partial u}{\partial y} = \rho_e u_e \frac{du_e}{dx} + \frac{\partial}{\partial y}\left[(\mu + \rho\nu_t)\frac{\partial u}{\partial y}\right] \tag{6.1.2}$$

$$\rho u \frac{\partial H}{\partial x} + \overline{\rho v}\frac{\partial H}{\partial y} = \frac{\partial}{\partial y}\left[\left(k + \rho c_p \frac{\nu_t}{\mathrm{Pr}_t}\right)\frac{\partial T}{\partial y} + u(\mu + \rho\nu_t)\frac{\partial u}{\partial y}\right] \tag{6.1.3}$$

where H denotes total enthalpy,

$$H = c_p T + \frac{u^2}{2} \tag{6.1.4a}$$

Pr_t the turbulent Prandtl number,

$$-\rho\overline{v'T'} = \rho\frac{\nu_t}{\mathrm{Pr}_t}\frac{\partial T}{\partial y} \tag{6.1.4b}$$

T the static temperature; $\overline{\rho v}$ denotes $\rho v + \overline{\rho' v'}$.

The boundary conditions on the airfoil for an adiabatic surface or for a surface with prescribed temperature in the standard mode are

$$y = 0, \quad u = 0, \quad v = v_w(x), \quad \frac{\partial H}{\partial y} = 0 \quad \text{or} \quad H = H_w(x) \tag{6.1.5a}$$

$$y = \delta, \quad u = u_e, \quad H = H_e \tag{6.1.5b}$$

In the inverse mode, similar to Eq. (3.0.4c), the edge boundary condition for u_e is given by

$$y = \delta, \quad u = u_e = u_e^0 + \frac{1}{\rho_e \pi} \int_{x_a}^{x_b} \frac{d}{d\sigma}(\rho_e u_e \delta^*) \frac{d\sigma}{x - \sigma} \tag{6.1.6}$$

where δ^* is the displacement thickness for compressible flows and is given by

$$\delta^* = \int_0^\delta \left(1 - \frac{\rho u}{\rho_e u_e}\right) dy \tag{6.1.7}$$

In the absence of the normal pressure gradient the boundary conditions in the wake, where a dividing line at $y = 0$ is required to separate the upper and lower parts of the inviscid flow, are

$$y = 0, \quad u = v = 0, \quad \partial H/\partial y = 0 \tag{6.1.8a}$$

$$y \to \infty, \quad u \to u_e(x), \quad H \to H_e \tag{6.1.8b}$$

6.1.1 Transformed Equations: Standard Mode

As in Section 3.2, we use the Falkner-Skan transformation to express the boundary-layer equations in the standard mode. The compressible version of this transformation is

$$d\eta = \sqrt{\frac{u_e}{\nu_e x}} \frac{\rho}{\rho_e} dy, \quad \psi(x, y) = \sqrt{\rho_e \mu_e u_e x}\, f(x, \eta) \tag{6.1.9}$$

With the definition of a stream function that statisfies Eq. (6.1.1),

$$\rho u = \frac{\partial \psi}{\partial y}, \quad \overline{\rho v} = -\frac{\partial \psi}{\partial x} + (\rho v)_w \tag{6.1.10}$$

the momentum and energy equations given by Eqs. (6.1.2) and (6.1.3) can now be written in a form where the momentum equation is similar to Eq. (3.2.3) for incompressible flows. In the following notation primes denote differentiation with respect to η and H is replaced by $H = H_e S$.

$$(bf'')' + m_1 f f'' + m_2[c - (f')^2] - m_3 f'' = \xi \left(f' \frac{\partial f'}{\partial \xi} - f'' \frac{\partial f}{\partial \xi}\right) \tag{6.1.11}$$

$$(eS' + df'f'')' + m_1 fS' - m_3 S' = \xi \left(f' \frac{\partial S'}{\partial \xi} - S' \frac{\partial f}{\partial \xi} \right) \tag{6.1.12}$$

where

$$b = C(1 + \nu_t^+), \quad C = \frac{\rho\mu}{\rho_e \mu_e}, \quad c = \frac{\rho_e}{\rho}, \quad \nu_t^+ = \frac{\nu_t}{\nu} \tag{6.1.13a}$$

$$e = \frac{C}{\mathrm{Pr}} \left(1 + \nu_t^+ \frac{\mathrm{Pr}}{\mathrm{Pr}_t} \right), \quad d = \frac{Cu_e^2}{H_e} \left[1 - \frac{1}{\mathrm{Pr}} + \nu_t^+ \left(1 - \frac{1}{\mathrm{Pr}_t} \right) \right], \quad R_x = \frac{u_e x}{\nu_e} \tag{6.1.13b}$$

$$m_1 = \frac{1}{2} \left[1 + m_2 + \frac{\xi}{\rho_e \mu_e} \frac{d}{d\xi} (\rho_e \mu_e) \right], \quad m_2 = \frac{\xi}{u_e} \frac{du_e}{d\xi}$$

$$m_3 = \frac{(\rho v)_w}{\rho_e u_e} \sqrt{R_x}, \quad \xi = \frac{x}{L} \tag{6.1.13c}$$

Including a transpiration mass flow rate velocity at the wall, $(\rho v)_w$, the boundary conditions given by

$$y = 0, \quad \rho v = (\rho v)_w(\xi), \quad u = 0,$$

$$H = H_w(\xi) \quad \text{or} \quad \left(\frac{\partial H}{\partial y} \right)_w = -\frac{C_{p_w}}{k_w} \dot{q}_w \tag{6.1.14a}$$

$$y = \delta, \quad u = u_e(\xi), \quad H = H_e \tag{6.1.14b}$$

can be written in terms of transformed variables as

$$\eta = 0, \quad f = 0, \quad f' = 0 \tag{6.1.15a}$$

$$S = S_w(\xi), \quad \text{or} \quad S'_w = -\frac{\mathrm{Pr}_w}{C_w} \frac{\dot{q}_w}{\rho_e u_e} \frac{\sqrt{R_x}}{H_e} \tag{6.1.15b}$$

$$\eta = \eta_e, \quad f' = 1, \quad S = 1 \tag{6.1.15c}$$

6.1.2 Transformed Equations: Inverse Mode

As in Section 3.2, we modify the Falkner-Skan transformation used for the standard mode for application in the inverse mode; we write the similarity variable η and dimensionless stream function $f(x, \eta)$ given by Eq. (6.1.9) as

$$dY = \sqrt{\frac{u_0}{\nu_0 x} \frac{\rho}{\rho_0}} dy, \quad \psi(x, y) = \sqrt{u_0 \mu_0 \rho_0 x} F(x, Y) \tag{6.1.16}$$

Here, and in the following discussion, u_0, ν_0, ρ_0 and H_0 denote reference values of velocity, kinematic viscosity, density and total enthalpy, respectively. Often these reference conditions are chosen, as in incompressible flows, equal to the freestream values.

Using the transformation given by Eq. (6.1.16) and with a procedure similar to that used to obtain Eqs. (6.1.11) to (6.1.15), the boundary-layer equations given by Eqs. (6.1.1) to (6.1.3) can be written as

$$(bF'')' + \frac{1}{2}FF'' - m_3F'' = -c\xi w\frac{dw}{d\xi} + \xi\left(F'\frac{\partial F'}{\partial \xi} - F''\frac{\partial F}{\partial \xi}\right) \tag{6.1.17}$$

$$(eS' + dF'F'')' + \frac{1}{2}FS' - m_3S' = \xi\left(F'\frac{\partial S}{\partial \xi} - S'\frac{\partial F}{\partial \xi}\right) \tag{6.1.18}$$

Here a prime denotes differentiation with respect to Y and the dimensionless parameters b, C, c, w, F', e and d are defined by

$$b = C(1+\nu_t^+), \quad C = \frac{\rho\mu}{\rho_0\mu_0}, \quad c = \frac{\rho_0}{\rho}, \quad w = \frac{u_e}{u_0}$$
$$F' = \frac{u}{u_0}, \quad S = \frac{H}{H_e} \tag{6.1.19a}$$

$$e = \frac{C}{\Pr}\left(1+\nu_t^+\frac{\Pr}{\Pr_t}\right), \quad d = \frac{Cu_0^2}{H_0}\left[1-\frac{1}{\Pr}+\nu_t^+\left(1-\frac{1}{\Pr_t}\right)\right],$$
$$m_3 = \frac{(\rho v)_w}{\rho_0 u_0}\sqrt{\frac{u_0 x}{\nu_0}} \tag{6.1.19b}$$

The wall boundary conditions with $R_L = \frac{u_0 L}{\nu_0}$, are

$$Y = 0, \quad F = 0, \quad F' = 0$$
$$S = S_w(\xi) \quad \text{or} \quad S'_w = -\frac{L}{k_0}\frac{C_{p_w}}{H_0}\left(\frac{k_0}{k_w}\right)\left(\frac{\rho_0}{\rho_w}\right)\dot{g}_w\sqrt{\xi}\frac{c_w}{\sqrt{R_L}} \tag{6.1.20}$$

The edge boundary conditions for the momentum equation is again obtained from Eq. (3.2.9). In terms of transformed variables defined by Eq. (6.1.16), the parameter D becomes

$$\bar{D} = \frac{D}{\rho_0 u_0 L} = \sqrt{\frac{\xi}{R_L}}\left(w\int_0^{Y_e}\frac{\rho_e}{\rho}dY - F_e\right) \tag{6.1.21}$$

and the relation between the external velocity u_e and displacement thickness δ^* provided by the Hilbert integral [see Eq. (6.1.6)] can then be written in dimensionless form as

$$Y = Y_e, \quad \left(\frac{\rho_e}{\rho_0}\right)^i F'_e(\xi^i) - \lambda\left[F'_e(\xi^i)\int_0^Y\left(\frac{\rho_e}{\rho}\right)dY - F_e(\xi^i)\right] = g_i \tag{6.1.22}$$

where

$$\lambda = C_{ii}\sqrt{\xi^i/R_L} \tag{6.1.23a}$$

$$g_i = \left(\frac{\rho_e}{\rho_0}\right)^i \overline{u_e^0}(\xi^i) + \sum_{j=1}^{i-1}C_{ij}\bar{D}_j + \sum_{j=i+1}^{N}C_{ij}\bar{D}_j \tag{6.1.23b}$$

The edge boundary condition for the dimensionless total enthalpy is the same as in Eq. (6.1.15b), that is, $S = 1$.

6.1.3 Transformed Equations: General Form

As in incompressible flows, the transformed equations for the standard and inverse modes can be written in one general form,

$$(bf'')' + \lambda_1\{m_1 ff'' - m_2[c - (f')^2]\} - m_3 f'' + \lambda_2 \frac{1}{2} ff'' \\ = \xi \left\{ \frac{\partial}{\partial \xi}\left[\frac{(f')^2}{2}\right] - f'' \frac{\partial f}{\partial \xi} - \lambda_2 c \frac{d}{d\xi}\left(\frac{w^2}{2}\right)\right\} \tag{6.1.24}$$

$$(eS' + df'f'')' + \lambda_1 m_1 fS' + \lambda_2 \frac{1}{2} fS' - m_3 S' = \xi \left(f' \frac{\partial S}{\partial \xi} - S' \frac{\partial f}{\partial \xi}\right) \tag{6.1.25}$$

where for $\lambda_1 = 1$ and $\lambda_2 = 0$ we recover Eqs. (6.1.11) and (6.1.12) which are subject to the boundary conditions given by (6.1.14) with dimensionless parameters defined by Eqs. (6.1.13). For $\lambda_1 = 0$ and $\lambda_2 = 1$ we recover Eqs. (6.1.17) and (6.1.18) provided that we note that F now corresponds to f, primes denote differentiation with respect to Y.

6.2 Numerical Method

In solving the compressible boundary-layer equations, two approaches can be pursued. Since the continuity and momentum equations have fluid properties and they depend on the solution of the energy equation, the first choice is to solve the continuity, momentum and energy equations simultaneously, in contrast to the second choice in which the energy equation is solved separately after the continuity and momentum equations are solved. For two-dimensional flows, the first choice requires the solution of five first-order equations and seven for three-dimensional flows. For two-dimensional flows, the second choice requires the solution of three first-order equations for the continuity and momentum equations and two for the energy equation. For three-dimensional flows, the second choice requires five first-order equations for the continuity and two momentum equations and again two for the energy equation.

The first choice provides better convergence rate and is preferable to the second choice. The second choice, on the other hand, is preferable to the first choice due to savings in computer time associated with solving small numbers of first-order equations.

In this chapter we shall adopt the first choice and solve the equations of the previous section simultaneously. The savings in solving three first-order equations over five first-order equations is relatively small. However, in three-dimensional flows, where the first choice requires seven first-order equations, it is preferable to adopt the second choice, as we shall discuss in Chapter 11.

6.2.1 Numerical Formulation

The transformed momentum and energy equations can be expressed as a first-order system by introducing new dependent variables $u(\xi,\eta)$, $v(\xi,\eta)$ and $p(\xi,\eta)$ and by defining

$$f' = u \tag{6.2.1a}$$

$$u' = v \tag{6.2.1b}$$

$$g' = p \tag{6.2.1c}$$

Equations (6.1.24) and (6.1.25) become

$$\begin{aligned}(bv)' + \lambda_1[m_1 fv + m_2(c-u^2)] + \lambda_2 \frac{1}{2} fv - m_3 v \\ = \xi\left[\frac{\partial}{\partial\xi}\left(\frac{u^2}{2}\right) - v\frac{\partial f}{\partial \xi} - \lambda_2 c\frac{d}{d\xi}\left(\frac{w^2}{2}\right)\right]\end{aligned} \tag{6.2.1d}$$

$$(ep + duv)' + \lambda_1 m_1 fp + \lambda_2\frac{1}{2}fp - m_3 p = \xi\left(u\frac{\partial g}{\partial\xi} - p\frac{\partial f}{\partial \xi}\right) \tag{6.2.1e}$$

where g is the total-enthalpy ratio defined by S in Eq. (6.1.25).

The boundary conditions for the standard mode follow from Eqs. (6.1.15)

$$\eta = 0, \quad f = 0, \quad u = 0, \quad g = g_w(\xi) \quad \text{or} \quad p = p_w(\xi) \tag{6.2.2a}$$

$$\eta = \eta_e, \quad u = 1, \quad g = 1 \tag{6.2.2b}$$

Similarly, for the inverse mode, the boundary conditions are

$$\eta = 0, \quad f = 0, \quad u = 0, \quad g = g_w(\xi) \quad \text{or} \quad p = p_w(\xi) \tag{6.2.3a}$$

$$\eta = \eta_e, \quad \lambda f + \left[\frac{\rho_e}{\rho_0} - \lambda\int_0^{\eta_e}\left(\frac{\rho_e}{\rho}\right)d\eta\right] w = g_i, \quad g = 1 \tag{6.2.3b}$$

We now consider the same net rectangle in the ξ,η plane shown in Fig. 3.2 and the net points defined by Eq. (3.3.4). As in subsection 3.3.1 where we used the box method, the finite-difference approximations to the three first-order ordinary differential equations (6.2.1a) to (6.2.1c) are written for the midpoint $(\xi^n, \eta_{j-1/2})$ of the segment P_1P_2 shown in Fig. 3.2 and, as in Eq. (3.3.1c), the finite-difference approximations to the two first-order partial differential equations (6.2.1d) and (6.2.1e) are written for the midpoint $(\xi^{n-1/2}, \eta_{j-1/2})$ of the rectangle $P_1P_2P_3P_4$. This procedure gives

$$\frac{f_j^n - f_{j-1}^n}{h_j} = \frac{u_j^n + u_{j-1}^n}{2} \equiv u_{j-1/2}^n \tag{6.2.4a}$$

$$\frac{u_j^n - u_{j-1}^n}{h_j} = \frac{v_j^n + v_{j-1}^n}{2} \equiv v_{j-1/2}^n \tag{6.2.4b}$$

$$\frac{g_j^n - g_{j-1}^n}{h_j} = \frac{p_j^n + p_{j-1}^n}{2} \equiv p_{j-1/2}^n \tag{6.2.4c}$$

$$\begin{aligned}&\frac{b_j^n v_j^n - b_{j-1}^n v_{j-1}^n}{h_j} + (\alpha_1)(fv)_{j-1/2}^n - (\alpha_2)(u^2)_{j-1/2}^n - m_3^n v^n \\ &+ \alpha^n[v_{j-1/2}^{n-1} f_{j-1/2}^n - f_{j-1/2}^{n-1} v_{j-1/2}^n + \lambda_2 c_{j-1/2}^{n-1/2}(u_J^2)^n] = R_{j-1/2}^{n-1}\end{aligned} \tag{6.2.4d}$$

$$\begin{aligned}&\frac{e_j^n p_j^n - e_{j-1}^n p_{j-1}^n}{h_j} + \frac{d_j^n u_j^n v_j^n - d_{j-1}^n u_{j-1}^n v_{j-1}^n}{h_j} + (\alpha_1)(fp)_{j-1/2}^n - m_3^n p^n \\ &- \alpha^n[(ug)_{j-1/2}^n + u_{j-1/2}^{n-1} g_{j-1/2}^n - g_{j-1/2}^{n-1} u_{j-1/2}^n + f_{j-1/2}^{n-1} p_{j-1/2}^n \\ &- p_{j-1/2}^{n-1} f_{j-1/2}^n] = T_{j-1/2}^{n-1}\end{aligned} \tag{6.2.4e}$$

where α^n is given by Eq. (3.3.6a) and

$$\alpha_1 = \lambda_1 m_1^n + \frac{\lambda_2}{2} + \alpha^n \tag{6.2.5a}$$

$$\alpha_2 = \lambda_1 m_2^n + \alpha^n \tag{6.2.5b}$$

$$R_{j-1/2}^{n-1} = -L_{j-1/2}^{n-1} + \alpha^n[(fv)_{j-1/2}^{n-1} - (u^2)_{j-1/2}^{n-1} + \lambda_2 c_{j-1/2}^{n-1/2}(u_J^2)^{n-1}] - \lambda_1 m_2^n c_{j-1/2}^n \tag{6.2.6a}$$

$$\begin{aligned}L_{j-1/2}^{n-1} = &\frac{b_j^{n-1} v_j^{n-1} - b_{j-1}^{n-1} v_{j-1}^{n-1}}{h_j} + \left(\lambda_1 m_1^{n-1} + \frac{\lambda_2}{2}\right)(fv)_{j-1/2}^{n-1} \\ &+ \lambda_1 m_2^{n-1}[c_{j-1/2}^{n-1} - (u^2)_{j-1/2}^{n-1}] - m_3^{n-1} v_{j-1/2}^{n-1}\end{aligned} \tag{6.2.6b}$$

$$T_{j-1/2}^{n-1} = -M_{j-1/2}^{n-1} + \alpha^n[(fp)_{j-1/2}^{n-1} - (ug)_{j-1/2}^{n-1}] \tag{6.2.6c}$$

$$\begin{aligned}M_{j-1/2}^{n-1} = &\frac{e_j^{n-1} p_j^{n-1} - e_{j-1}^{n-1} p_{j-1}^{n-1}}{h_j} + \frac{d_j^{n-1} u_j^{n-1} v_j^{n-1} - d_{j-1}^{n-1} u_{j-1}^{n-1} v_{j-1}^{n-1}}{h_j} \\ &+ \left(\lambda_1 m_1^{n-1} + \frac{\lambda_2}{2}\right)(fp)_{j-1/2}^{n-1} - m_3^{n-1} p_{j-1/2}^{n-1}\end{aligned} \tag{6.2.6d}$$

To account for the mixed wall boundary conditions for the energy equation, we write the wall boundary conditions for the momentum and energy equations as

$$f_0^n = 0, \quad u_0^n = 0, \quad \gamma_0 g_0^n + \gamma_1 p_0^n = \gamma_2^n \tag{6.2.7a}$$

and the edge boundary conditions as

$$\lambda f_J^n + (1 - \lambda \eta_J) u_J^n = g_i, \quad g_J^n = 1 \tag{6.2.7b}$$

When the wall temperature is specified, we set $\gamma_0 = 1$, $\gamma_1 = 0$ and since the enthalpy ratio at the surface, $g_0(\xi)$ is given, $\gamma_2 = g_0(\xi)$. When the wall heat flux is specified, we set $\gamma_0 = 0$, $\gamma_2 = 1$ and γ_2 equals the dimensionless total-enthalpy gradient at the wall. For the standard mode $\lambda = 0$, $g_i = 1$ and for the inverse mode $\lambda = 1$ and g_i is computed from Eq. (6.1.23b).

6.2.2 Newton's Method

We use the procedure described in subsection 3.3.2 to linearize the incompressible momentum equation. We assume f_j^{n-1}, u_j^{n-1}, v_j^{n-1}, g_j^{n-1}, and p_j^{n-1}, to be known for $0 \le j \le J$; Eqs. (6.2.4) and (6.2.7) then form a system of $5J + 5$ nonlinear equations for the solution of $5J + 5$ unknowns $(f_j^n, u_j^n, v_j^n, g_j^n, p_j^n$, $j = 0, 1, \ldots, J$. With iterates given by Eqs. (3.3.9) and

$$g_j^{(\nu+1)} = g_j^{(\nu)} + \delta g_j^{(\nu)}, \quad p_j^{(\nu+1)} = p_j^{(\nu)} + \delta p_j^{(\nu)}$$

and using a procedure similar to that used to obtain Eqs. (3.3.10), the nonlinear system given by Eqs. (6.2.4) can be expressed in the following linear system (the superscript ν in δ quantities is dropped for simplicity):

$$\delta f_j - \delta f_{j-1} - \frac{h_j}{2}(\delta u_j + \delta u_{j-1}) = (r_1)_j \tag{6.2.8a}$$

$$\delta u_j - \delta u_{j-1} - \frac{h_j}{2}(\delta v_j + \delta v_{j-1}) = (r_4)_{j-1} \tag{6.2.8b}$$

$$\delta g_j - \delta g_{j-1} - \frac{h_j}{2}(\delta p_j + \delta p_{j-1}) = (r_5)_{j-1} \tag{6.2.8c}$$

$$\begin{aligned}(s_1)_j\delta v_j &+ (s_2)_j\delta v_{j-1} + (s_3)_j\delta f_j + (s_4)_j\delta f_{j-1} + (s_5)_j\delta u_j \\ &+ (s_6)_j\delta u_{j-1} + (s_7)_j\delta g_j + (s_8)_j\delta g_{j-1} + (s_9)_j\delta u_J = (r_2)_j\end{aligned} \tag{6.2.8d}$$

$$\begin{aligned}(\beta_1)_j\delta p_j &+ (\beta_2)_j\delta p_{j-1} + (\beta_3)_j\delta f_j + (\beta_4)_j\delta f_{j-1} \\ &+ (\beta_5)_j\delta u_j + (\beta_6)_j\delta u_{j-1} + (\beta_7)_j\delta g_j \\ &+ (\beta_8)_j\delta g_{j-1} + (\beta_9)_j\delta v_j + (\beta_{10})_j\delta v_{j-1} = (r_3)_j\end{aligned} \tag{6.2.8e}$$

where

$$(r_1)_j = f_{j-1} - f_j + h_j u_{j-1/2} \tag{6.2.9a}$$

$$(r_4)_{j-1} = u_{j-1} - u_j + h_j v_{j-1/2} \tag{6.2.9b}$$

$$(r_5)_{j-1} = g_{j-1} - g_j + h_j p_{j-1/2} \tag{6.2.9c}$$

$$\begin{aligned}(r_2)_j = R_{j-1/2}^{n-1} &- \{h_j^{-1}(b_j v_j - b_{j-1}v_{j-1}) + \alpha_1(fv)_{j-1/2} - \alpha_2(u^2)_{j-1/2} \\ &+ \alpha[v_{j-1/2}f_{j-1/2} - f_{j-1/2}v_{j-1/2} + \lambda_2 c_{j-1/2}^{n-1/2}(u_J^2)]\}\end{aligned} \tag{6.2.9d}$$

$$\begin{aligned}(r_3)_j = T_{j-1/2}^{n-1} &- \{h_j^{-1}(e_j p_j - e_{j-1}p_{j-1} + h_j^{-1}(d_j u_j v_j - d_{j-1}u_{j-1}v_{j-1}) \\ &+ \alpha_1(fp)_{j-1/2} - \alpha^n[(ug)_{j-1/2} + u_{j-1/2}^{n-1}g_{j-1/2} - g_{j-1/2}^{n-1}u_{j-1/2} \\ &+ f_{j-1/2}^{n-1}p_{j-1/2} - p_{j-1/2}^{n-1}f_{j-1/2}] - m_3 p_{j-1/2}\}\end{aligned} \tag{6.2.9e}$$

The coefficients of the momentum equation are

$$(s_1)_j = h_j^{-1}b_j + \left(\frac{1}{2}\alpha_1\right)f_j - \frac{\alpha^n}{2}f_{j-1/2}^{n-1} - \frac{1}{2}m_3 \tag{6.2.10a}$$

$$(s_2)_j = -h_j^{-1} b_{j-1} + \left(\frac{1}{2}\alpha_1\right) f_{j-1} - \frac{\alpha^n}{2} f_{j-1/2}^{n-1} - \frac{1}{2} m_3 \tag{6.2.10b}$$

$$(s_3)_j = \left(\frac{1}{2}\alpha_1\right) v_j + \frac{\alpha^n}{2} v_{j-1/2}^{n-1} \tag{6.2.10c}$$

$$(s_4)_j = \left(\frac{1}{2}\alpha_1\right) v_{j-1} + \frac{\alpha^n}{2} v_{j-1/2}^{n-1} \tag{6.2.10d}$$

$$(s_5)_j = -(\alpha_2) u_j \tag{6.2.10e}$$

$$(s_6)_j = -(\alpha_2) u_{j-1} \tag{6.2.10f}$$

$$(s_7)_j = 0 \tag{6.2.10g}$$

$$(s_8)_j = 0 \tag{6.2.10h}$$

$$(s_9)_j = 2\lambda_2 \alpha^n c_{j-1/2}^{n-1/2} \tag{6.2.10i}$$

The coefficients of the energy equation are

$$(\beta_1)_j = h_j^{-1} e_j + \left(\frac{1}{2}\alpha_1\right) f_j - \frac{\alpha^n}{2} f_{j-1/2}^{n-1} \tag{6.2.11a}$$

$$(\beta_2)_j = -h_j^{-1} e_{j-1} + \left(\frac{1}{2}\alpha_1\right) f_{j-1} - \frac{\alpha^n}{2} f_{j-1/2}^{n-1} \tag{6.2.11b}$$

$$(\beta_3)_j = \left(\frac{1}{2}\alpha_1\right) p_j + \frac{\alpha^n}{2} p_{j-1/2}^{n-1} \tag{6.2.11c}$$

$$(\beta_4)_j = \left(\frac{1}{2}\alpha_1\right) p_{j-1} + \frac{\alpha^n}{2} p_{j-1/2}^{n-1} \tag{6.2.11d}$$

$$(\beta_5)_j = h_j^{-1} d_j v_j - \frac{\alpha^n}{2}(g_j - g_{j-1/2}^{n-1}) \tag{6.2.11e}$$

$$(\beta_6)_j = -h_j^{-1} d_{j-1} v_{j-1} - \frac{\alpha^n}{2}(g_{j-1} - g_{j-1/2}^{n-1}) \tag{6.2.11f}$$

$$(\beta_7)_j = -\frac{\alpha^n}{2}(u_j + u_{j-1/2}^{n-1}) \tag{6.2.11g}$$

$$(\beta_8)_j = -\frac{\alpha^n}{2}(u_{j-1} + u_{j-1/2}^{n-1}) \tag{6.2.11h}$$

$$(\beta_9)_j = h_j^{n-1} d_j u_j \tag{6.2.11i}$$

$$(\beta_{10})_j = -h_j^{n-1} d_{j-1} u_{j-1} \tag{6.2.11j}$$

The boundary conditions, Eqs. (6.2.7), become

$$\delta f_0 = 0, \quad \delta u_0 = 0, \quad \gamma_0 \delta g_0 + \gamma_1 \delta p_0 = 0 \tag{6.2.12}$$

for inverse

$$\lambda \delta f_J + (1 - \lambda \eta_J) \delta u_J = (r_4)_J, \quad \delta g_J = (r_5)_J = 0 \tag{6.2.12b}$$

where

$$(r_4)_J = g_i - \lambda f_J^n - (1 - \lambda \eta_J) u_J^n$$

6.2.3 Block-Elimination Method

The linear system given by Eqs. (6.2.8) and (6.2.12) with $\lambda_2 = 0$ has a block tridiagonal structure and can be expressed in matrix-vector form given by Eq. (3.3.16). As before, the coefficient matrix A and the compound vectors $\vec{\delta}$ and $\vec{r}$ are given by Eq. (3.3.17). The five-dimensional vectors, $\vec{\delta}$ and $\vec{r}$, for each value of j, are

$$\vec{\delta}_j = \begin{vmatrix} \delta f_j \\ \delta u_j \\ \delta v_j \\ \delta g_j \\ \delta p_j \end{vmatrix}, \quad \vec{r}_j = \begin{vmatrix} (r_1)_j \\ (r_2)_j \\ (r_3)_j \\ (r_4)_j \\ (r_5)_j \end{vmatrix}, \quad 0 \le j \le J \tag{6.2.13}$$

and the 5×5 matrices A_j, B_j and C_j are given by

$$A_0 = \begin{bmatrix} 1 & 0 & 0 & 0 & 0 \\ 0 & 1 & 0 & 0 & 0 \\ 0 & 0 & 0 & \gamma_0 & \gamma_1 \\ 0 & -1 & -\frac{h_1}{2} & 0 & 0 \\ 0 & 0 & 0 & -1 & -\frac{h_1}{2} \end{bmatrix},$$

$$A_j = \begin{bmatrix} 1 & -\frac{h_j}{2} & 0 & 0 & 0 \\ (s_3)_j & (s_5)_j & (s_1)_j & (s_7)_j & 0 \\ (\beta_3)_j & (\beta_5)_j & (\beta_9)_j & (\beta_7)_j & (\beta_1)_j \\ 0 & -1 & -\frac{h_{j+1}}{2} & 0 & 0 \\ 0 & 0 & 0 & -1 & -\frac{h_{j+1}}{2} \end{bmatrix}, \quad 1 \le j \le J-1 \tag{6.2.14a}$$

$$A_J = \begin{bmatrix} 1 & -\frac{h_J}{2} & 0 & 0 & 0 \\ (s_3)_J & (s_5)_J & (s_1)_J & (s_7)_J & 0 \\ (\beta_3)_J & (\beta_5)_J & (\beta_9)_J & (\beta_7)_J & (\beta_1)_J \\ 0 & 1 & 0 & 0 & 0 \\ 0 & 0 & 0 & 1 & 0 \end{bmatrix},$$

$$B_j = \begin{bmatrix} -1 & -\frac{h_j}{2} & 0 & 0 & 0 \\ (s_4)_j & (s_6)_j & (s_2)_j & (s_8)_j & 0 \\ (\beta_4)_j & (\beta_6)_j & (\beta_{10})_j & (\beta_8)_j & (\beta_2)_j \\ 0 & 0 & 0 & 0 & 0 \\ 0 & 0 & 0 & 0 & 0 \end{bmatrix}, \quad 1 \le j \le J \tag{6.2.14b}$$

$$C_j = \begin{bmatrix} 0 & 0 & 0 & 0 & 0 \\ 0 & 0 & 0 & 0 & 0 \\ 0 & 0 & 0 & 0 & 0 \\ 0 & 1 & -\frac{h_{j+1}}{2} & 0 & 0 \\ 0 & 0 & 0 & 1 & -\frac{h_{j+1}}{2} \end{bmatrix}, \quad 1 \le j \le J \tag{6.2.14c}$$

Because of the boundary conditions, the vectors $\vec{r}_0$ and $\vec{r}_J$ are defined by

$$\vec{r}_0 = \begin{bmatrix} 0 \\ 0 \\ 0 \\ (r_4)_0 \\ (r_5)_0 \end{bmatrix}, \quad \vec{r}_J = \begin{bmatrix} (r_1)_J \\ (r_2)_J \\ (r_3)_J \\ 0 \\ 0 \end{bmatrix} \tag{6.2.15}$$

The solution of Eq. (3.3.16) with the definitions given by Eqs. (6.2.13) to (6.2.15) can again be obtained efficiently and effectively with the block elimination method discussed in subsection 3.3.3.

6.2.4 Bordering Algorithm

As in the solution of the linear system for the incompressible momentum equation, the linear system given by Eqs. (6.2.8) and (6.2.12) with $\lambda_2 \neq 0$ does not have a block tridiagonal structure and cannot be solved with the block-elimination method directly. As in subsection 3.3.4, we use the bordering algorithm. The relations given by Eqs. (3.3.25) to (3.3.28) remain the same. The matrices D and E, with B_j given by Eq. (6.2.14b), are

$$D = \begin{bmatrix} D_0 \\ - \\ D_j \\ - \\ D_{J-1} \end{bmatrix}, \quad E = \begin{bmatrix} 0 \\ - \\ 0 \\ - \\ B_J^T \end{bmatrix} \tag{6.2.16}$$

$$D_0 = \begin{bmatrix} 0 & 0 & 0 & 0 & 0 \\ 0 & 0 & 0 & 0 & 0 \\ 0 & 0 & 0 & 0 & 0 \\ 0 & 0 & 0 & 0 & 0 \\ 0 & 0 & 0 & 0 & 0 \end{bmatrix}, \quad D_j = \begin{bmatrix} 0 & 0 & 0 & 0 & 0 \\ 0 & (s_9)_j & 0 & 0 & 0 \\ 0 & 0 & 0 & 0 & 0 \\ 0 & 0 & 0 & 0 & 0 \\ 0 & 0 & 0 & 0 & 0 \end{bmatrix}, \quad 1 \le j \le J-2$$

$$D_{J-1} = \begin{bmatrix} 0 & 0 & 0 & 0 & 0 \\ 0 & (s_9)_{J-1} & 0 & 0 & 0 \\ 0 & 0 & 0 & 0 & 0 \\ 0 & 1 & -\frac{h_J}{2} & 0 & 0 \\ 0 & 0 & 0 & 1 & -\frac{h_J}{2} \end{bmatrix} \tag{6.2.17}$$

For convenience, we write the D matrix as

$$D = [\vec{D}_1, \vec{D}_2, \vec{D}_3, \vec{D}_4, \vec{D}_5] \tag{6.2.18}$$

where

$$\vec{D}_1 = \begin{bmatrix} 0 \\ 0 \\ 0 \\ 0 \\ 0 \\ \cdots \\ 0 \\ 0 \\ 0 \\ 0 \\ 0 \\ \cdots \\ 0 \\ 0 \\ 0 \\ 0 \\ 0 \end{bmatrix}, \quad \vec{D}_2 = \begin{bmatrix} 0 \\ 0 \\ 0 \\ 0 \\ 0 \\ \cdots \\ 0 \\ (s_9)_1 \\ 0 \\ 0 \\ 0 \\ \cdots \\ 0 \\ (s_9)_j \\ 0 \\ 0 \\ 0 \\ \cdots \\ 0 \\ (s_9)_{J-1} \\ 0 \\ 0 \\ 0 \end{bmatrix}, \quad \vec{D}_3 = \begin{bmatrix} 0 \\ 0 \\ 0 \\ 0 \\ 0 \\ \cdots \\ 0 \\ 0 \\ 0 \\ 0 \\ 0 \\ \cdots \\ 0 \\ 0 \\ 0 \\ -\frac{h_J}{2} \\ 0 \end{bmatrix}, \quad \vec{D}_4 = \begin{bmatrix} 0 \\ 0 \\ 0 \\ 0 \\ 0 \\ \cdots \\ 0 \\ 0 \\ 0 \\ 0 \\ 0 \\ \cdots \\ 0 \\ 0 \\ 0 \\ 1 \\ 0 \end{bmatrix}, \quad \vec{D}_5 = \begin{bmatrix} 0 \\ 0 \\ 0 \\ 0 \\ 0 \\ \cdots \\ 0 \\ 0 \\ 0 \\ 0 \\ 0 \\ \cdots \\ 0 \\ 0 \\ 0 \\ 0 \\ -\frac{h_J}{2} \end{bmatrix} \tag{6.2.19}$$

The bordering algorithm based on Eqs. (3.3.33) to (3.3.37) was discussed in subsection 3.3.4. With $F = [\vec{F}_1, \vec{F}_2, \vec{F}_3, \vec{F}_4, \vec{F}_5]$, Eq. (3.3.34) can be written as five vector equations. Since $\vec{D}_1$ is zero, it is necessary to solve four vector equations.

$$A\vec{F}_2 = \vec{b}_2, \quad A\vec{F}_3 = \vec{b}_3, \quad A\vec{F}_4 = \vec{b}_4, \quad A\vec{F}_5 = \vec{b}_5 \tag{6.2.20}$$

A'_J and $\vec{r}'_J$ in Eq. (3.3.43) are given by

$$A'_J = \begin{bmatrix} 1 & \alpha'_{12} & \alpha'_{13} & \alpha'_{14} & \alpha'_{15} \\ (s_3)_J & \alpha'_{22} & \alpha'_{23} & \alpha'_{24} & \alpha'_{25} \\ (\beta_3)_J & \alpha'_{32} & \alpha'_{33} & \alpha'_{34} & \alpha'_{35} \\ \lambda & 1-\lambda\eta_J & 0 & 0 & 0 \\ 0 & 0 & 0 & 1 & 0 \end{bmatrix}, \quad \vec{r}'_J = \begin{bmatrix} (r_1)'_J \\ (r_2)'_J \\ (r_3)'_J \\ (r_4)'_J \\ (r_5)'_J \end{bmatrix} \tag{6.2.21}$$

where

$$\alpha'_{12} = -\frac{h_J}{2} + (F_{21})_{J-1} + \frac{h_J}{2}(F_{22})_{J-1}$$

$$\alpha'_{13} = (F_{31})_{J-1} + \frac{h_J}{2}(F_{32})_{J-1}$$

$$\alpha'_{14} = (F_{41})_{J-1} + \frac{h_J}{2}(F_{42})_{J-1}$$

$$\alpha'_{15} = (F_{51})_{J-1} + \frac{h_J}{2}(F_{52})_{J-1}$$

$$\alpha'_{22} = (s_J + s_9)_J - (s_4)_J(F_{21})_{J-1} - (s_6)_J(F_{22})_{J-1} - (s_2)_J(F_{23})_{J-1} - (s_8)_J(F_{24})_{J-1}$$

$$\alpha'_{23} = (s_1)_J - (s_4)_J(F_{31})_{J-1} - (s_6)_J(F_{32})_{J-1} - (s_2)_J(F_{33})_{J-1} - (s_8)_J(F_{34})_{J-1}$$

$$\alpha'_{24} = (s_7)_J - (s_4)_J(F_{41})_{J-1} - (s_6)_J(F_{42})_{J-1} - (s_2)_J(F_{43})_{J-1} - (s_8)_J(F_{44})_{J-1}$$

$$\alpha'_{25} = -(s_4)_J(F_{51})_{J-1} - (s_6)_J(F_{52})_{J-1} - (s_2)_J(F_{53})_{J-1} - (s_8)_J(F_{54})_{J-1}$$

$$\alpha'_{32} = (\beta_5)_J - (\beta_4)_J(F_{21})_{J-1} - (\beta_6)_J(F_{22})_{J-1} - (\beta_{10})_J(F_{23})_{J-1} - (\beta_8)_J(F_{24})_{J-1} - (\beta_2)_J(F_{25})_{J-1}$$

$$\alpha'_{33} = (\beta_9)_J - (\beta_4)_J(F_{31})_{J-1} - (\beta_6)_J(F_{32})_{J-1} - (\beta_{10})_J(F_{33})_{J-1} - (\beta_8)_J(F_{34})_{J-1} - (\beta_2)_J(F_{35})_{J-1}$$

$$\alpha'_{34} = (\beta_7)_J - (\beta_4)_J(F_{41})_{J-1} - (\beta_6)_J(F_{42})_{J-1} - (\beta_{10})_J(F_{43})_{J-1} - (\beta_8)_J(F_{44})_{J-1} - (\beta_2)_J(F_{45})_{J-1}$$

$$\alpha'_{35} = (\beta_1)_J - (\beta_4)_J(F_{51})_{J-1} - (\beta_6)_J(F_{52})_{J-1} - (\beta_{10})_J(F_{53})_{J-1} - (\beta_8)_J(F_{54})_{J-1} - (\beta_2)_J(F_{55})_{J-1}$$

$$(\gamma_1)'_J = (\gamma_1)_J + (w)_{J-1} + \frac{h_J}{2}(w_2)_{J-1}$$

$$(\gamma_2)'_J = (\gamma_2)_J - (s_4)_J(w_1)_{J-1} - (s_6)_J(w_2)_{J-1} - (s_2)_J(w_3)_{J-1} - (s_8)_J(w_4)_{J-1}$$

$$(\gamma_3)'_J = (\gamma_3)_J - (\beta_4)_J(w_1)_{J-1} - (\beta_6)_J(w_2)_{J-1} - (\beta_{10})_J(w_3)_{J-1} - (\beta_8)_J(w_4)_{J-1} - (\beta_2)_J(w_5)_{J-1}$$

$$(\gamma_4)'_J = (\gamma_4)_J$$

$$(\gamma_5)'_J = (\gamma_5)_J$$

A Fortran listing similar to that for solving the 3×3 system given in section 3.4 is given in the following subsection.

6.2.5 Subroutine SOLV5

```
      SUBROUTINE SOLV5
      COMMON /BLC001/ NX,NP,NXTE,NXTT,NXTW,NPT,NTR,NXS,IT,ITV,ISWP,ISWPT
      COMMON /BLCG00/ G1,G2
      COMMON /INPT1/ P1(201),P2(201),P3(201),WW(201),ALFA0,ALFA1
      COMMON /BLGRD1/ X(201),ETA(101),A(101),DETA(101)
      COMMON /BLPRC1/ F(101,2),U(101,2),V(101,2),G(101,2),T(101,2),
     &                B(101,2),C(101,2),D(101,2),E(101,2),BC(101),
     &                RMU(101),TEMP(101)
      COMMON /BLC300/ DELF(101),DELU(101),DELV(101),DELG(101),DELT(101)
      COMMON /BLC600/ S1(101),S2(101),S3(101),S4(101),S5(101),S6(101),
     +                S7(101),S8(101),B1(101),B2(101),B3(101),B4(101),
     +                B5(101),B6(101),B7(101),B8(101),B9(101),B10(101),
     +                S9(101),R(5,101)
      DIMENSION     A11(101),A12(101),A13(101),A14(101),A15(101),
     1              A21(101),A22(101),A23(101),A24(101),A25(101),
     2              A31(101),A32(101),A33(101),A34(101),A35(101),
     3              W1(101,5),W2(101,5),W3(101,5),W4(101,5),W5(101,5),
     3              V1(101,4),V2(101,4),V3(101,4),V4(101,4),V5(101,4)
      DD(B11,B12,B13,B21,B22,B23,B31,B32,B33) = B11*(B22*B33-B32*B23)-
     1                    B21*(B12*B33-B32*B13)+B31*(B12*B23-B22*B13)
C - - - - - - - - - - - - - - - - - - - - - - - - - - - - - - - - - - -
      NPM1=NP-1
      NPM2=NP-2
      A11(1)=1.0
      A12(1)=0.0
      A13(1)=0.0
      A14(1)=0.0
      A15(1)=0.0
      A21(1)=0.0
      IF(NX .GT. NXTE ) THEN
      A22(1)=0.0
      A23(1)=1.0
      ELSE
      A22(1)=1.0
      A23(1)=0.0
      END IF
      A24(1)=0.0
      A25(1)=0.0
      A31(1)=0.0
      A32(1)=0.0
      A33(1)=0.0
      A34(1)=ALFA0
      A35(1)=ALFA1
      DO 10 J=1,NPM1
         DO 20 K=2,4
            V1(J,K)=0.
            V2(J,K)=0.
            V3(J,K)=0.
            V4(J,K)=0.
            V5(J,K)=0.
```

```
   20     CONTINUE
          V1(J,1) = R(1,J)
          V2(J,1) = R(2,J)
          V3(J,1) = R(3,J)
          V4(J,1) = R(4,J)
          V5(J,1) = R(5,J)
          V2(J,2) = S9(J)
   10 CONTINUE
      V2(1,2) = 0.
C  FORWARD SWEEP
      DO 90  J=2,NPM1
      AA1=A(J)*A24(J-1)-A25(J-1)
      AA2=A(J)*A34(J-1)-A35(J-1)
      AA3=A(J)*A12(J-1)-A13(J-1)
      AA4=A(J)*A22(J-1)-A23(J-1)
      AA5=A(J)*A32(J-1)-A33(J-1)
      AA6=A(J)*A14(J-1)-A15(J-1)
      AA7=A(J)*S6(J)-S2(J)
      AA8=S8(J)*A(J)
      AA9=A(J)*B6(J)-B10(J)
      AA10=A(J)*B8(J)-B2(J)
C  ELEMENTS OF TRIANGLE MATRIX
      DET=A11(J-1)*(AA4*AA2-AA1*AA5)-A21(J-1)*(AA3*AA2-AA5*AA6)+
     +         A31(J-1)*(AA3*AA1-AA4*AA6)
      G11=(-(AA4*AA2-AA5*AA1)+A(J)**2*(A21(J-1)*AA2-A31(J-1)*AA1))/DET
      G12=((AA3*AA2-AA5*AA6)-A(J)**2*(A11(J-1)*AA2-A31(J-1)*AA6))/DET
      G13=(-(AA3*AA1-AA4*AA6)+A(J)**2*(A11(J-1)*AA1-A21(J-1)*AA6))/DET
      G14=G11*A12(J-1)+G12*A22(J-1)+G13*A32(J-1)+A(J)
      G15=G11*A14(J-1)+G12*A24(J-1)+G13*A34(J-1)
      G21=(S4(J)*(AA2*AA4-AA1*AA5)+A31(J-1)*(AA1*AA7-AA4*AA8)+
     +           A21(J-1)*(AA5*AA8-AA7*AA2))/DET
      G22=(A11(J-1)*(AA2*AA7-AA5*AA8)+A31(J-1)*(AA3*AA8-AA6*AA7)+
     +           S4(J)*(AA5*AA6-AA2*AA3))/DET
      G23=(A11(J-1)*(AA4*AA8-AA1*AA7)+S4(J)*(AA3*AA1-AA4*AA6)+
     +           A21(J-1)*(AA7*AA6-AA3*AA8))/DET
      G24=G21*A12(J-1)+G22*A22(J-1)+G23*A32(J-1)-S6(J)
      G25=G21*A14(J-1)+G22*A24(J-1)+G23*A34(J-1)-S8(J)
      G31=(B4(J)*(AA4*AA2-AA5*AA1)-AA9*(A21(J-1)*AA2-A31(J-1)*AA1)+
     +           AA10*(A21(J-1)*AA5-A31(J-1)*AA4))/DET
      G32=(-B4(J)*(AA3*AA2-AA5*AA6)+AA9*(A11(J-1)*AA2-A31(J-1)*AA6)-
     +           AA10*(A11(J-1)*AA5-A31(J-1)*AA3))/DET
      G33=(B4(J)*(AA3*AA1-AA4*AA6)-AA9*(A11(J-1)*AA1-A21(J-1)*AA6)+
     +           AA10*(A11(J-1)*AA4-A21(J-1)*AA3))/DET
      G34=G31*A12(J-1)+G32*A22(J-1)+G33*A32(J-1)-B6(J)
      G35=G31*A14(J-1)+G32*A24(J-1)+G33*A34(J-1)-B8(J)
C  ELEMENTS OF TRIANGLE MATRIX
      A11(J)=1.0
      A12(J)=-A(J)-G14
      A13(J)=A(J)*G14
      A14(J)=-G15
      A15(J)=A(J)*G15
```

```
      A21(J)=S3(J)
      A22(J)=S5(J)-G24
      A23(J)=S1(J)+A(J)*G24
      A24(J)=-G25+S7(J)
      A25(J)=A(J)*G25
      A31(J)=B3(J)
      A32(J)=B5(J)-G34
      A33(J)=B9(J)+A(J)*G34
      A34(J)=B7(J)-G35
      A35(J)=B1(J)+A(J)*G35
         DO 90 K=1,2
         V1(J,K) = V1(J,K)-G11*V1(J-1,K)-G12*V2(J-1,K)-G13*V3(J-1,K)-
     1             G14*V4(J-1,K)-G15*V5(J-1,K)
         V2(J,K) = V2(J,K)-G21*V1(J-1,K)-G22*V2(J-1,K)-G23*V3(J-1,K)-
     1             G24*V4(J-1,K)-G25*V5(J-1,K)
         V3(J,K) = V3(J,K)-G31*V1(J-1,K)-G32*V2(J-1,K)-G33*V3(J-1,K)-
     1             G34*V4(J-1,K)-G35*V5(J-1,K)
   90 CONTINUE
      V4(NPM1,2)=1.
      V4(NPM1,3)=-A(NP)
      V5(NPM1,4)=1.
C  BACKWARD SWEEP
      J     = NPM1
      DD1=A13(J)-A12(J)*A(J+1)
      DD2=A23(J)-A22(J)*A(J+1)
      DD3=A33(J)-A32(J)*A(J+1)
      EE1=A15(J)-A14(J)*A(J+1)
      EE2=A25(J)-A24(J)*A(J+1)
      EE3=A35(J)-A34(J)*A(J+1)
      DETT=A11(J)*DD2*EE3+A21(J)*DD3*EE1+A31(J)*DD1*EE2
     +         -A31(J)*DD2*EE1-A21(J)*DD1*EE3-A11(J)*DD3*EE2
C  ELEMENTS OF DELTA-VECTOR
      DO 640 K=1,4
      BB1=-V4(J,K)
      BB2=-V5(J,K)
      CC1=V1(J,K)-A12(J)*BB1-A14(J)*BB2
      CC2=V2(J,K)-A22(J)*BB1-A24(J)*BB2
      CC3=V3(J,K)-A32(J)*BB1-A34(J)*BB2
      W1(J,K)=(CC1*DD2*EE3+CC2*DD3*EE1+CC3*DD1*EE2-CC3*DD2*EE1
     +             -CC2*DD1*EE3-CC1*DD3*EE2)/DETT
      W3(J,K)=(A11(J)*CC2*EE3+A21(J)*CC3*EE1+A31(J)*CC1*EE2-
     +             A31(J)*CC2*EE1-A21(J)*CC1*EE3-A11(J)*CC3*EE2)/DETT
      W5(J,K)=(A11(J)*CC3*DD2+A21(J)*CC1*DD3+A31(J)*CC2*DD1-
     +             A31(J)*CC1*DD2-A21(J)*CC3*DD1-A11(J)*CC2*DD3)/DETT
      W2(J,K)=BB1-A(J+1)*W3(J,K)
      W4(J,K)= BB2-A(J+1)*W5(J,K)
  640 CONTINUE
      DO 650 J =NPM2,1,-1
      DD1=A13(J)-A12(J)*A(J+1)
      DD2=A23(J)-A22(J)*A(J+1)
      DD3=A33(J)-A32(J)*A(J+1)
```

```
      EE1=A15(J)-A14(J)*A(J+1)
      EE2=A25(J)-A24(J)*A(J+1)
      EE3=A35(J)-A34(J)*A(J+1)
      DETT=A11(J)*DD2*EE3+A21(J)*DD3*EE1+A31(J)*DD1*EE2
     +          -A31(J)*DD2*EE1-A21(J)*DD1*EE3-A11(J)*DD3*EE2
C  ELEMENTS OF DELTA-VECTOR
      DO 655 K=1,4
      BB1=W2(J+1,K)-A(J+1)*W3(J+1,K)-V4(J,K)
      BB2=W4(J+1,K)-A(J+1)*W5(J+1,K)-V5(J,K)
      CC1=V1(J,K)-A12(J)*BB1-A14(J)*BB2
      CC2=V2(J,K)-A22(J)*BB1-A24(J)*BB2
      CC3=V3(J,K)-A32(J)*BB1-A34(J)*BB2
      W1(J,K)=(CC1*DD2*EE3+CC2*DD3*EE1+CC3*DD1*EE2-CC3*DD2*EE1
     +             -CC2*DD1*EE3-CC1*DD3*EE2)/DETT
      W3(J,K)=(A11(J)*CC2*EE3+A21(J)*CC3*EE1+A31(J)*CC1*EE2-
     +             A31(J)*CC2*EE1-A21(J)*CC1*EE3-A11(J)*CC3*EE2)/DETT
      W5(J,K)=(A11(J)*CC3*DD2+A21(J)*CC1*DD3+A31(J)*CC2*DD1-
     +             A31(J)*CC1*DD2-A21(J)*CC3*DD1-A11(J)*CC2*DD3)/DETT
      W2(J,K)=BB1-A(J+1)*W3(J,K)
      W4(J,K)= BB2-A(J+1)*W5(J,K)
  655 CONTINUE
  650 CONTINUE
      J = NP
      A11(J)  = 1.
      A12(J)  = -A(J)+W1(J-1,2)+A(J)*W2(J-1,2)
      A13(J)  = W1(J-1,3)+A(J)*W2(J-1,3)
      A14(J)  = W1(J-1,4)+A(J)*W2(J-1,4)
      A15(J)  = -A(J)*(W1(J-1,4)+A(J)*W2(J-1,4))
      A21(J)  = S3(J)
      A22(J)  = S5(J)+S9(J)-S4(J)*W1(J-1,2)-S6(J)*W2(J-1,2)-S2(J)*W3(
     1          J-1,2)-S8(J)*W4(J-1,2)
      A23(J)  = S1(J)-S4(J)*W1(J-1,3)-S6(J)*W2(J-1,3)-S2(J)*W3(J-1,3)-
     1          S8(J)*W4(J-1,3)
      A24(J)  = S7(J)-S4(J)*W1(J-1,4)-S6(J)*W2(J-1,4)-S2(J)*W3(J-1,4)-
     1          S8(J)*W4(J-1,4)
      A25(J)  = A(J)*(S4(J)*W1(J-1,4)+S6(J)*W2(J-1,4)+S2(J)*W3(J-1,4)+
     1          S8(J)*W4(J-1,4))
      A31(J)  = B3(J)
      A32(J)  = B5(J)-B4(J)*W1(J-1,2)-B6(J)*W2(J-1,2)-B10(J)*W3(J-1,2)-
     1          B8(J)*W4(J-1,2)-B2(J)*W5(J-1,2)
      A33(J)  = B9(J)-B4(J)*W1(J-1,3)-B6(J)*W2(J-1,3)-B10(J)*W3(J-1,3)-
     1          B8(J)*W4(J-1,3)-B2(J)*W5(J-1,3)
      A34(J)  = B7(J)-B4(J)*W1(J-1,4)-B6(J)*W2(J-1,4)-B10(J)*W3(J-1,4)-
     1          B8(J)*W4(J-1,4)-B2(J)*W5(J-1,4)
      A35(J)  = B1(J)+A(J)*(B4(J)*W1(J-1,4)+B6(J)*W2(J-1,4)+B10(J)*W3(
     1          J-1,4)+B8(J)*W4(J-1,4)+B2(J)*W5(J-1,4))
      R(1,J)  = R(1,J)+W1(J-1,1)+A(J)*W2(J-1,1)
      R(2,J)  = R(2,J)-S4(J)*W1(J-1,1)-S6(J)*W2(J-1,1)-S2(J)*W3(J-1,1)-
     1          S8(J)*W4(J-1,1)
      R(3,J)  = R(3,J)-B4(J)*W1(J-1,1)-B6(J)*W2(J-1,1)-B10(J)*W3(J-1,1)-
     1          B8(J)*W4(J-1,1)-B2(J)*W5(J-1,1)
```

```
      DELG(NP)=0.
      DET = -G1*DD(A12(J),A13(J),A15(J),A22(J),A23(J),A25(J),A32(J),
     1      A33(J),A35(J))+G2*DD(A11(J),A13(J),A15(J),A21(J),A23(J),
     2      A25(J),A31(J),A33(J),A35(J))
      DET1 = -R(4,J)*DD(A12(J),A13(J),A15(J),A22(J),A23(J),A25(J),
     1       A32(J),A33(J),A35(J))+G2*DD(R(1,J),A13(J),A15(J),R(2,J),
     2       A23(J),A25(J),R(3,J),A33(J),A35(J))
      DELF(NP) = DET1/DET
      DELU(NP) = (R(4,NP)-G1*DELF(NP))/G2
      RR1 = R(2,NP)-S3(NP)*DELF(NP)-A22(NP)*DELU(NP)
      RR2 = R(3,NP)-B3(NP)*DELF(NP)-A32(NP)*DELU(NP)
      DETT = A23(NP)*A35(NP)-A33(NP)*A25(NP)
      DELV(NP)=(RR1*A35(NP)-RR2*A25(NP))/DETT
      DELT(NP)=(A23(NP)*RR2-A33(NP)*RR1)/DETT
      DO 115 J=1,NPM1
         DELF(J)=W1(J,1)-DELU(NP)*W1(J,2)-DELV(NP)*W1(J,3)+DELT(NP)*W1(
     1           J,4)*A(NP)
         DELU(J)=W2(J,1)-DELU(NP)*W2(J,2)-DELV(NP)*W2(J,3)+DELT(NP)*W2(
     1           J,4)*A(NP)
         DELV(J)=W3(J,1)-DELU(NP)*W3(J,2)-DELV(NP)*W3(J,3)+DELT(NP)*W3(
     1           J,4)*A(NP)
         DELG(J)=W4(J,1)-DELU(NP)*W4(J,2)-DELV(NP)*W4(J,3)+DELT(NP)*W4(
     1           J,4)*A(NP)
         DELT(J)=W5(J,1)-DELU(NP)*W5(J,2)-DELV(NP)*W5(J,3)+DELT(NP)*W5(
     1           J,4)*A(NP)
  115 CONTINUE
      DO 130 J = 1,NP
         F(J,2) = F(J,2)+DELF(J)
         U(J,2) = U(J,2)+DELU(J)
         V(J,2) = V(J,2)+DELV(J)
         G(J,2) = G(J,2)+DELG(J)
         T(J,2) = T(J,2)+DELT(J)
130   CONTINUE
      F(1,2) = 0.0
      IF(NX .GT. NXTE) THEN
      V(1,2) = 0.0
      ELSE
      U(1,2) = 0.0
      END IF
      RETURN
      END
```

6.3 FORTRAN Program

A Fortran program employing the numerical method described in the previous sections for solving the boundary-layer equations for two-dimensional compressible laminar and turbulent flows (CIBLP) can be obtained from the author. As in the boundary-layer program of Section 3.4, IBLP, this program can also be used in either standard or inverse mode.

References

[1] Cebeci, T. and Bradshaw, P., Physical and Computational Aspects of Convective Heat Transfer, Springer-Verlag, NY, 1988.

7 Two-Dimensional Incompressible Unsteady Flows: Inviscid Method

7.0 Introduction

The panel method discussed in Chapter 2 can be extended to compute the flowfield over an airfoil executing unsteady time-dependent motion in an inviscid incompressible medium. The unsteady motion of the airfoil causes continuous vortex shedding at the trailing wake which distinguishes unsteady flows from steady flows. Therefore, the extension of the panel method of Chapter 2 from steady flows to unsteady flows must be formulated to include the vortex shedding process.

In this chapter we describe a panel method for airfoils which are subject to ramp-type or harmonic oscillation motions. This method was developed first for inviscid flows by Teng [1] and was later modified by Jang [2] and Jang and Cebeci [3] for viscous effects so that, as in the calculation method for steady flows, the modified panel method can be used interactively with the boundary-layer method described in Chapter 8. For further details of the inviscid method, the reader is referred to [1] and for the inclusion of viscous effects to [2,3].

7.1 Unsteady Flow Model

As in the panel method of Chapter 2 for steady flows, the airfoil surface is again represented by singularity distributions of source strength q_j $(j = 1, 2, \ldots, N)$ and vorticity τ leading to $(N+1)$ unknowns as in steady flow. For unsteady flows, however, these unknowns are time dependent. For this reason, we introduce a subscript k as the time-step index so that the inviscid flow solutions can be obtained at successive time-steps t_k $(k = 1, 2, \ldots)$, starting from $t_0 = 0$. At each time-step, the airfoil is again represented by surface singularity distributions of source strength $(q_j)_k$ $(j = 1, 2, \ldots, N)$ and vorticity τ_k.

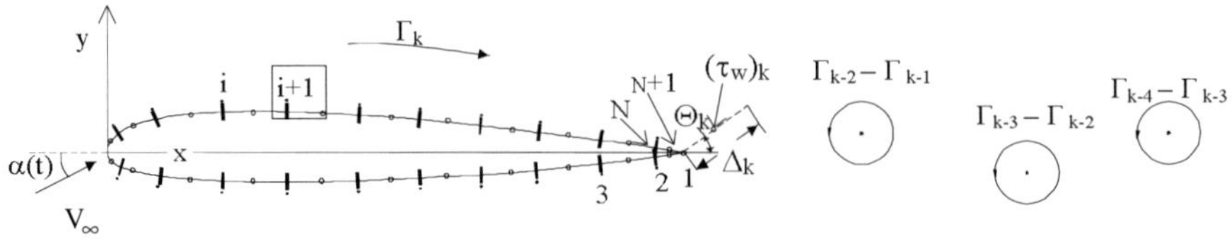

Fig. 7.1. Panel-method representation of unsteady potential flow at time step t_k.

Since the total circulation in the flowfield must be preserved according to the Helmholtz theorem, any changes in the circulation Γ on the airfoil surface must be balanced by an equal and opposite change in the vorticity in the wake which causes vortex shedding at the trailing edge of the airfoil. In the unsteady panel method, the vortex shedding is represented by an additional wake element, often referred to as "the shed vorticity panel" attached to the trailing edge with uniform vorticity distribution $(\tau_w)_k$, see Fig. 7.1.

If the length of the shed vorticity panel is denoted by Δ_k and its inclination angle to the x-axis by Θ_k and the overall circulation on the airfoil surface at time-step t_k by $\Gamma_k(\equiv \tau_k \ell)$ with ℓ denoting the perimeter of the airfoil, then the Helmholtz theorem can be expressed as

$$\Gamma_k + \Delta_k(\tau_w)_k = \Gamma_{k-1}$$

or

$$\Delta_k(\tau_w)_k = \Gamma_{k-1} - \Gamma_k = \ell(\tau_{k-1} - \tau_k) \tag{7.1.1}$$

Here Γ_{k-1} and τ_{k-1} represent the overall circulation on the airfoil surface and the vorticity strength, respectively, already determined at time-step t_{k-1}.

From time-step t_k to t_{k+1}, we assume that the shed vorticity panel is detached from the trailing edge and convects downstream as a concentrated free vortex, with circulation $\Delta_k(\tau_w)_k$. The convection velocity of the free vortex is given by the local flow velocity at the center of the vortex. At the same time, a new shed vorticity panel is formed and the vortex shedding process is repeated. As a result of this continuous vortex shedding, a string of concentrated core vortices is formed in the wake behind the airfoil as shown in Fig. 7.1.

The presence of the shed vorticity panel and the wake core vortices influences the upstream flow. As a result, the wake core vortices are convected under the influence of the freestream and the singularity distributions on the airfoil surface including the shed vorticity panel. This coupling process increases the complexity of the unsteady flow model and the solution procedure recommended is an iterative one.

In addition to the surface singularities $(q_j)_k$ $(j = 1, 2, \ldots, N)$ and τ_k, there are three additional unknowns $(\tau_w)_k$, Δ_k and Θ_k due to the vortex shedding process. The Helmholtz theorem, Eq. (7.1.1) provides one additional relation in conjunction with the flow tangency conditions and the Kutta condition. Two additional relations are required and can be obtained from the assumptions recommended by Basu and Hancock [4]:

1. The shed vorticity panel is oriented along the direction of the local flow velocity at the panel midpoint.
2. The length of the shed vorticity panel is equal to the product of the local flow velocity at the panel midpoint and the size of the time-step.

With these assumptions, the two additional relations can be written as

$$\tan \Theta_k = \frac{(v_w)_k}{(u_w)_k} \tag{7.1.2}$$

and

$$\Delta_k = (t_k - t_{k-1})[(u_w)_k^2 + (v_w)_k^2]^{1/2} \tag{7.1.3}$$

where $(u_w)_k$ and $(v_w)_k$ are the flow velocity components at the panel midpoint in the x- and y-directions, respectively.

The flow tangency conditions remain the same as those for steady flows given by Eq. (2.1.9)

$$[(V^n)_i]_k = 0 \quad i = 1, 2, \ldots, N \tag{7.1.4}$$

However, the Kutta condition must now include the rate of change of velocity potential. According to the unsteady Bernoulli's equation, the Kutta condition can be expressed by [1,2]

$$[(V^t)_1]_k^2 - [(V^t)_N]_k^2 = 2\left[\frac{\partial(\phi_N - \phi_1)}{\partial t}\right]_k = 2\left(\frac{\partial \Gamma}{\partial t}\right)_k \tag{7.1.5}$$

Using a backward finite-difference approximation for the rate of change of total circulation on the airfoil surface, Eq. (7.1.5) can be written as

$$[(V^t)_1]_k^2 - [(V^t)_N]_k^2 = 2\frac{\Gamma_k - \Gamma_{k-1}}{t_k - t_{k-1}} = 2\ell\frac{\tau_k - \tau_{k-1}}{t_k - t_{k-1}} \tag{7.1.6}$$

7.2 Influence Coefficients

The influence coefficients, A_{ij}^n, A_{ij}^τ, B_{ij}^n, B_{ij}^τ for the source and vorticity distributions given by Eqs. (2.1.11)–(2.1.14) for steady flows remain the same for unsteady flows. These coefficients are time-independent since they are functions of geometrical parameters which are fixed for a rigid airfoil. Additional influence coefficients involving the shed vorticity panel and the wake core vortices are introduced for unsteady flows. These coefficients, however, are time-dependent and must be computed at each time-step; they are defined as follows:

$(B_{i,n+1}^n)_k$ = normal velocity component induced at the i-th panel control point by unit strength vorticity distribution on the shed vorticity panel at time-step t_k.

$(B^t_{i,n+1})_k$ = tangential velocity component induced at the i-th panel control point by unit strength vorticity distribution on the shed vorticity panel at time-step t_k.

$(A^x_{n+1,j})_k$ = x-velocity component induced at the shed vorticity panel mid-point by unit strength source distribution on the j-th panel at time-step t_k.

$(A^y_{n+1,j})_k$ = y-velocity component induced at the shed vorticity panel mid-point by unit strength source distribution on the j-th panel at time-step t_k.

$(B^x_{n+1,j})_k$ = x-velocity component induced at the shed vorticity panel mid-point by unit strength vorticity distribution on the j-th panel at time-step t_k.

$(B^y_{n+1,j})_k$ = y-velocity component induced at the shed vorticity panel mid-point by unit strength vorticity distribution on the j-th panel at time-step t_k.

$(A^x_{h,j})_k$ = x-velocity component induced at the center of the h-th core vortex by unit strength source distribution on the j-th panel at time-step t_k.

$(A^y_{h,j})_k$ = y-velocity component induced at the center of the h-th core vortex by unit strength source distribution on the j-th panel at time-step t_k.

$(B^x_{h,j})_k$ = x-velocity component induced at the center of the h-th core vortex by unit strength vorticity distribution on the j-th panel at time-step t_k.

$(B^y_{h,j})_k$ = y-velocity component induced at the center of the h-th core vortex by unit strength vorticity distribution on the j-th panel at time-step t_k.

$(B^x_{h,n+1})_k$ = x-velocity component induced at the center of the h-th core vortex by unit strength vorticity distribution on the shed vorticity panel at time-step t_k.

$(B^y_{h,n+1})_k$ = y-velocity component induced at the center of the h-th core vortex by unit strength vorticity distribution on the shed vorticity panel at time-step t_k.

$(C^n_{i,m})_k$ = normal velocity component induced at the i-th panel control point by unit strength m-th core vortex at time-step t_k.

$(C^t_{i,m})_k$ = tangential velocity component induced at the i-th panel control point by unit strength m-th core vortex at time-step t_k.

$(C^x_{n+1,m})_k$ = x-velocity component induced at the shed vorticity panel mid-point by unit strength m-th core vortex at time-step t_k.

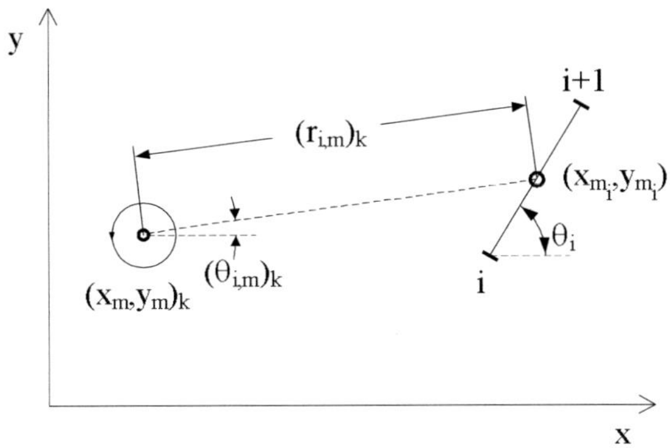

Fig. 7.2. Evaluation of the influence coefficients at (x_{m_i}, y_{m_i}) due to core vortex at $(x_m, y_m)_k$. Note that x_{m_i}, y_{m_i} are the coordinates of the i-th control point as in steady flows (Chapter 2).

$(C^y_{n+1,m})_k$ = y-velocity component induced at the shed vorticity panel mid-point by unit strength m-th core vortex at time-step t_k.

$(C^x_{h,m})_k$ = x-velocity component induced at the center of the h-th core vortex by unit strength m-th core vortex at time-step t_k.

$(C^y_{h,m})_k$ = y-velocity component induced at the center of the h-th core vortex by unit strength m-th core vortex at time-step t_k.

The calculation of the new A and B influence coefficients is similar to the calculation of the old A and B influence coefficients. For example, $(B^n_{i,n+1})_k$ and $(B^i_{i,n+1})_k$ are computed exactly the same way as B^n_{ij} and B^t_{ij} are computed from Eqs. (2.1.11) and (2.1.13) with subscript $n+1$ replacing j. Similarly, $(A^x_{n+1,j})_k$ and $(A^x_{h,j})_k$ are calculated from Eq. (2.1.12) with θ_i set to zero and subscript i appropriately replaced. The relations used to calculate the C coefficients are different from those used for A and B coefficients since they are velocities induced by core vortices.

It follows from Fig. 7.2 that the normal and tangential velocity components induced at the i-th panel control point by unit strength m-th core vortex at time-step t_k are

$$(C^n_{i,m})_k = -\frac{\cos\left[\theta_i - (\theta_{i,m})_k\right]}{2\pi(r_{i,m})_k} \tag{7.2.1}$$

and

$$(C^t_{i,m})_k = -\frac{\sin\left[\theta_i - (\theta_{i,m})_k\right]}{2\pi(r_{i,m})_k} \tag{7.2.2}$$

where

$$(r_{i,m})_k = [(x_{m_i} - x_m)^2 + (y_{m_i} - y_m)^2]^{1/2}$$

$$x_{m_i} = \frac{1}{2}(x_i + x_{i+1})$$

$$y_{m_i} = \frac{1}{2}(y_i + y_{i+1})$$

x_m = x-coordinate of m-th core vortex at time-step t_k. (7.2.3)

y_m = y-coordinate of m-th core vortex at time-step t_k.

$$\theta_i = \tan^{-1}\left(\frac{y_{i+1} - y_i}{x_{i+1} - x_i}\right)$$

$$(\theta_{i,m})_k = \tan^{-1}\left(\frac{y_{m_i} - y_m}{x_{m_i} - x_m}\right)$$

Similarly, $(C^x_{n+1,m})_k$ and $(C^x_{h,m})_k$ are computed by Eq. (7.2.1) while $(C^y_{n+1,m})_k$ and $(C^y_{h,m})_k$ are computed by Eq. (7.2.2) if θ_i is set equal to zero and the subscript i appropriately replaced.

7.3 Solution Procedure

In terms of the influence coefficients, the flow tangency conditions given by Eq. (7.1.4) can be written as

$$\begin{aligned}\sum_{j=1}^{N} A^n_{ij}(q_j)_k + \tau_k \sum_{j=1}^{N} B^n_{ij} + (\vec{V}_{\text{stream}} \cdot \vec{n}_i)_k + (\tau_w)_k (B^n_{i,n+1})_k \\ + \sum_{m=1}^{k-1} (C^n_{i,m})_k (\Gamma_{m-1} - \Gamma_m) = 0 \qquad i = 1, 2, \dots, N\end{aligned} \tag{7.3.1}$$

Here $\vec{V}_{\text{stream}}$ denotes the unsteady stream velocity seen by an observer situated at a coordinate system fixed on the airfoil. For an airfoil executing an unsteady motion, including linear translation and angular rotation about a pivot point,

$$\vec{V}_{\text{stream}} = \vec{V}_\infty + [U\vec{i} + V\vec{j}] + \Omega(\vec{y}_i - \vec{x}_j) \tag{7.3.2}$$

where $-(U\vec{i}+V\vec{j})$ is the translation velocity of the airfoil and Ω is the rotational velocity.

The above equation sums all the induced velocities due to an individual singularity with the stream velocity and sets it to zero to satisfy the flow tangency conditions.

Substituting $(\tau_w)_k$ from Eq. (7.1.1), collecting like terms and rearranging, Eq. (7.3.1) becomes

$$\begin{aligned}\sum_{j=1}^{N} A_{ij}^n (q_j)_k &= \tau_k \left[\frac{\ell}{\Delta_k} (B_{i,n+1}^n)_k - \sum_{j=1}^{N} B_{ij}^n \right] - (\vec{V}_{\text{stream}} \cdot \vec{n}_i)_k \\ &- \left(\frac{\ell}{\Delta_k} \right) \tau_{k-1} (B_{i,n+1}^n)_k - \sum_{m=1}^{k-1} (C_{i,m}^n)_k (\Gamma_{m-1} - \Gamma_m) \\ & \qquad i = 1, 2, \ldots, N \end{aligned} \tag{7.3.3}$$

Eq. (7.3.3) results in a system of algebraic equations which can be written in the form

$$A\vec{x} = \tau_k \cdot \vec{b} + \vec{c} \tag{7.3.4}$$

where A is an $N \times N$ matrix, $\vec{x} = [(q_1)_k, (q_2)_k, \ldots, (q_N)_k]^T$, $\vec{b} = (b_1, b_2, \ldots, b_N)^T$ and $\vec{c} = (c_1, c_2, \ldots, c_N)^T$. The elements of A, the components of $\vec{b}$ and $\vec{c}$ are given by

$$a_{ij} = A_{ij} \tag{7.3.5a}$$

$$b_i = \left[\frac{\ell}{\Delta_k} (B_{i,n+1}^n)_k - \sum_{j=1}^{N} B_{ij}^n \right] \tag{7.3.5b}$$

$$c_i = -(\vec{V}_{\text{stream}} \cdot \vec{n}_i)_k - \left(\frac{\ell}{\Delta_k} \right) \tau_{k-1} (B_{i,n+1}^n)_k - \sum_{m=1}^{k-1} (C_{i,m}^n)_k (\Gamma_{m-1} - \Gamma_m) \tag{7.3.5c}$$

We note that the components of $\vec{b}$ and $\vec{c}$ are known only if the shed vorticity panel at time-step t_k is established, i.e., if Δ_k and Θ_k are known. Therefore, the iterative solution procedure can be formulated as follows:

1. Find the locations of the wake core vortices downstream according to the local flow velocities at their centers with respect to an inertial coordinate system.
2. Compute the coordinates of these core vortices relative to the coordinate system fixed on the airfoil executing an unsteady motion.
3. Start the iteration cycle for the current time-step by assuming values of Δ_k and Θ_k. Except for the first time-step, values at previous time steps can be used as initial guesses.
4. Compute all the influence coefficients needed in Eq. (7.3.3).
5. Solve Eq. (7.3.4) by the Gaussian elimination method (see Section 2.1) to obtain $(q_j)_k$ in terms of τ_k; i.e., find $\vec{x} = \tau_k A^{-1}\vec{b} + A^{-1}\vec{c}$.
6. Solve for τ_k using the Kutta condition of Eq. (7.1.6). Note that Eq. (7.1.6) is a nonlinear algebraic equation with τ_k as the only unknown.
7. Once τ_k is solved, $(q_j)_k$ can be determined. The velocity components $(u_w)_k$ and $(v_w)_k$ at the midpoint of the shed vorticity panel can also be calculated.
8. The values of Δ_k and Θ_k are updated by Eqs. (7.1.2) and (7.1.3).
9. Repeat the iteration cycle from steps 3 to 8 until the values of Δ_k and Θ_k converge.

10. Compute the tangential velocities and the velocity potential at all panel control points to determine the pressure distribution which can be integrated to give force and moments.
11. Compute the local flow velocities at the centers of all the core vortices that are convected downstream. Proceed to the next time-step.

The calculation of tangential velocities at the trailing edge panels, the velocity components $(u_w)_k$ and $(v_w)_k$ at the midpoint of the shed vorticity panel and the local flow velocities at centers of all core vortices required in this iteration process can be obtained by first computing the tangential velocities $[(V^t)_i]_k$ $(i = 1, 2, \ldots, N)$ at the panel control points by:

$$
\begin{aligned}
[(V^t)_i]_k &= \sum_{j=1}^{N} A^t_{ij}(q_j)_k + \tau_k \sum_{j=1}^{N} B^t_{ij} + (\vec{V}_{\text{stream}} \cdot \vec{t}_i)_k \\
&\quad + (\tau_w)_k \cdot (B^t_{i,n+1})_k + \sum_{m=1}^{k-1} (C^t_{i,m})_k (\Gamma_{m-1} - \Gamma_m) \\
&\qquad i = 1, 2, \ldots, N
\end{aligned}
\tag{7.3.6}
$$

The components of the flow velocities at the centers of all core vortices are then calculated with respect to the inertial coordinate system by:

$$
\begin{aligned}
(u_h)_k &= \sum_{j=1}^{N} (A^x_{h,j})_k (q_j)_k \\
&\quad + \tau_k \sum_{j=1}^{N} (B^x_{h,j})_k + (\vec{V}_\infty \cdot \vec{i}\,)_k + (\tau_w)_k (B^x_{h,n+1})_k \\
&\quad + \sum_{m=1,\, m\neq h}^{k-1} (C^x_{h,m})_k (\Gamma_{m-1} - \Gamma_m)
\end{aligned}
\tag{7.3.7}
$$

$$
\begin{aligned}
(v_h)_k &= \sum_{j=1}^{N} (A^y_{h,j})_k (q_j)_k \\
&\quad + \tau_k \sum_{j=1}^{N} (B^y_{h,j})_k + (\vec{V}_\infty \cdot \vec{j}\,)_k + (\tau_w)_k (B^y_{h,n+1})_k \\
&\quad + \sum_{m=1,\, m\neq h}^{k-1} (C^y_{h,m})_k + (\Gamma_{m-1} - \Gamma_m)
\end{aligned}
\tag{7.3.8}
$$

The above equations can also be used to obtain the velocity components $(u_w)_k$ and $(v_w)_k$ by replacing h with $n + 1$ in Eqs. (7.3.7) and (7.3.8).

7.4 Velocity Potential and Pressure Distribution

In a steady flow problem, it is not necessary to calculate the velocity potential ϕ since the pressure distribution can be obtained by evaluating the induced velocity from the concept of influence coefficients. However, this is not the case for an unsteady flow and it is necessary to use the unsteady Bernoulli equation written with respect to the airfoil-fixed coordinates system as [1].

$$C_p = \frac{p - p_\infty}{(1/2)\rho V_\infty^2} = \left(\frac{V_{\text{stream}}}{V_\infty}\right)^2 - \left(\frac{V}{V_\infty}\right)^2 - \frac{2}{V_\infty^2}\frac{\partial \phi}{\partial t} \tag{7.4.1}$$

where $\vec{V}_{\text{stream}}$ is defined according to Eqs. (7.3.2) and $\vec{V}$ is the total velocity with respect to the airfoil-fixed coordinate system computed from

$$V^2 = (V^t)^2 + (V^n)^2 \tag{7.4.2}$$

Eq. (7.4.1) shows that in order to compute the pressure distribution on the airfoil surface, the rate of change of velocity potential must be evaluated. Using a backward finite-difference approximation for $\partial\phi/\partial t$, the pressure coefficients at the i-th panel control point at time-step t_k can be written as

$$[(C_p)_i]_k = \frac{[(V_{\text{stream}})_i]_k^2}{V_\infty^2} - \frac{[(V^t)_i]_k^2}{V_\infty^2} - \frac{2}{V_\infty^2}\frac{(\phi_i)_k - (\phi_i)_{k-1}}{t_k - t_{k-1}} \tag{7.4.3}$$

where $[(V^t)_i]_k$ is obtained from Eq. (7.3.6).

It is necessary to calculate the velocity potential at all the panel control points at each time-step. Instead of solving the Laplace equation, the velocity potential can be evaluated by integrating the velocity field in two stages, first from upstream at infinity to the airfoil leading edge, then along the airfoil surface from the leading edge to each panel control point. Care must be taken here to include only the velocity contribution due to singularities.

The integration of the velocity field can be made by choosing an arbitrary straight line extending upstream to infinity from the leading edge of the airfoil along a direction parallel to $\vec{V}_\infty$ and infinity can be set at ten chord lengths upstream of the leading edge since the velocities at far-field points induced by the singularities are small enough to be neglected. This line is divided into z panels with element lengths near the leading edge comparable to the panel sizes near the leading edge along the airfoil. However, the panel size is progressively increased to take advantage of the inversely decaying induced velocities at larger distances. The tangential components of the induced velocities at the midpoints of these panels are computed using the influence coefficients analogous to those used on the airfoil panels. Using subscript f to denote these panel midpoints, influence coefficients $(A^t_{f,j})_k$, $(B^t_{f,j})_k$, $(B^t_{f,n+1})_k$ and $(C^t_{f,m})_k$ are computed by the same expressions used to calculate the A's, B's and C's coefficients with

$\cos\theta_i$ replaced by $-\cos\alpha$, $\sin\theta_i$ replaced by $-\sin\alpha$ and subscript i replaced by f. With these influence coefficients, the tangential velocities are obtained from

$$\begin{aligned}[(V_\phi^t)_f]_k &= \sum_{j=1}^{N}(A_{f,j}^t)_k(q_j)_k + \tau_k\sum_{j=1}^{N}(B_{f,j}^t)_k + (\tau_w)_k(B_{f,n+1}^t)_k \\ &\quad + \sum_{m=1}^{k-1}(C_{f,m}^t)_k(\Gamma_{m-1}-\Gamma_m) \\ &\qquad\qquad f = 1,2,\ldots,z\end{aligned} \tag{7.4.4}$$

The velocity potential at the airfoil leading edge ($\phi_{\text{l.e.}}$) is the sum of the products of the induced velocity at each panel and the panel length.

$$(\phi_{\text{l.e.}})_k = \sum_{f=1}^{z}[(V_\phi^t)_f]_k[(x_{f+1}-x_f)^2+(y_{f+1}-y_f)^2]^{1/2} \tag{7.4.5}$$

Similarly, for the line integral over the airfoil surface, the tangential component of the induced velocity at the i-th panel control point is computed from

$$\begin{aligned}[(V_\phi^t)_i]_k &= \sum_{j=1}^{N}A_{ij}^t(q_j)_k + \tau_k\sum_{j=1}^{N}B_{ij}^t + (\tau_w)_k(B_{i,n+1}^t)_k \\ &\quad + \sum_{m=1}^{k-1}(C_{i,m}^t)_k(\Gamma_{m-1}-\Gamma_m)\end{aligned} \tag{7.4.6}$$

Performing the line integration by summation, the velocity potential at the i-th boundary point on the airfoil is given by:

$$(\phi_{\text{node}\,i})_k = \begin{cases} (\phi_{\text{l.e.}})_k + \displaystyle\sum_{j=i_{\text{l.e.}}}^{i-1}[(V_\phi^t)_j]_k[(x_{j+1}-x_j)^2+(y_{j+1}-y_j)^2]^{1/2} \\ \qquad\qquad i_{\text{l.e.}} \le i \le N \\ (\phi_{\text{l.e.}})_k + \displaystyle\sum_{j=i}^{i_{\text{l.e.}}-1}[(V_\phi^t)_j]_k[(x_{j+1}-x_j)^2+(y_{j+1}-y_j)^2]^{1/2} \\ \qquad\qquad 1 \le i < i_{\text{l.e.}} \end{cases} \tag{7.4.7}$$

Finally, the velocity potential at the i-th panel control point is

$$(\phi_i)_k = \frac{1}{2}[(\phi_{\text{node}\,i})_k + (\phi_{\text{node}\,i+1})_k] \tag{7.4.8}$$

In summary, the tangential velocity $[(V^t)_i]_k$, the stream velocity $[(V_{\text{stream}})_i]_k$ at the i-th panel control point, the velocity potential $(\phi_i)_k$ are evaluated from Eqs. (7.3.6), (7.3.2) and (7.4.8), respectively. The pressure coefficient C_p is then calculated by Eq. (7.4.3). The aerodynamic coefficients C_ℓ, C_d and C_m are calculated by integrating the pressure distributions as in the steady flow problem.

7.5 Viscous Effects

In unsteady flows, the flowfield around an airfoil at high Reynolds number can again be divided into viscous and inviscid regions as in steady flow. The viscous effects into the panel method can be introduced by replacing the tangency condition given by Eq. (7.1.4) by the blowing condition on the panel control points. At each time-step, t_k, this conditions is written as

$$[(V^n)_i]_k = [(V_b)_i]_k, \quad i = 1, 2, \ldots, N \tag{7.5.1}$$

Also, the Kutta condition is specified at the off-body control points as in steady flows, i.e.

$$(C_p)_{I=1} = (C_p)_{I=N} \tag{7.5.2}$$

which can also be written as

$$\begin{aligned} [(V)_{I=1}]_k^2 - [(V)_{I=N}]_k^2 &= 2\left[\frac{\partial(\phi_{I=N} - \phi_{I=1})}{\partial t}\right]_k \\ &= 2\left(\frac{\partial \Gamma}{\partial t}\right)_k = 2\ell\frac{\tau_k - \tau_{k-1}}{t_k - t_{k-1}} \end{aligned} \tag{7.5.3}$$

The blowing condition of Eq. (7.5.1) can be expressed in terms of the influence coefficients as follows,

$$\begin{aligned} &\sum_{j=1}^{N+nwp} A_{ij}^n \left(q_j\right)_k + \tau_k \sum_{j=1}^{N} B_{ij}^n + (\vec{V}_{\text{stream}} \cdot \vec{n}_i)_k + (\tau_w)_k (B_{i,n+1}^n)_k \\ &+ \sum_{m=1}^{k-1} (C_{i,m}^n)_k (\Gamma_{m-1} - \Gamma_m) = [(V_b)_i]_k \qquad i = 1, 2, \ldots, N \end{aligned} \tag{7.5.4}$$

To satisfy the Kutta condition on the displacement surface at the trailing edge, the total velocity at the 1-st and N-th off-body control points must be computed. If we denote the total velocity on the I-th off-body control point at the time-step t_k as $(V_I)_k$, then we can write

$$(V_I)_k = \{[(V^t)_I]_k^2 + [(V^n)_I]_k^2\}^{1/2} \tag{7.5.5}$$

where

$$\begin{aligned} [(V^t)_I]_k = &\sum_{j=1}^{N+nwp} A_{Ij}^t (q_j)_k \\ &+ \tau_k \sum_{j=1}^{N} B_{Ij}^t + (\vec{V}_{\text{stream}} \cdot \vec{t}_i)_k + (\tau_w)_k (B_{I,n+1}^t)_k \\ &+ \sum_{m=1}^{k-1} (C_{I,m}^t)_k (\Gamma_{m-1} - \Gamma_m) \end{aligned} \tag{7.5.6a}$$

$$
\begin{aligned}
[(V^n)_I]_k = & \sum_{j=1}^{N+nwp} A_{Ij}^n (q_j)_k \\
& + \tau_k \sum_{j=1}^{N} B_{Ij}^n + (\vec{V}_{\text{stream}} \cdot \vec{n}_i)_k + (\tau_w)_k (B_{I,n+1}^n)_k \\
& + \sum_{m=1}^{k-1} (C_{I,m}^n)_k (\Gamma_{m-1} - \Gamma_m)
\end{aligned}
\quad (7.5.6b)
$$

For further details, see, [2,3].

7.6 Fortran Program

The Fortran program for the unsteady panel method with viscous effects is substantially bigger than the Fortran program for the steady panel method described in Section 2.4. For this reason it is not included here. It can be obtained from the author.

References

[1] Teng, N.H., "The development of a computer code (U2DIIF) for the numerical solution of unsteady, inviscid and incompressible flow over an airfoil," M.S. Thesis, Naval Postgraduate School, Monterey, CA 1987.

[2] Jang, H.M., "A viscous-inviscid interactive method for unsteady flows," Ph.D. thesis, University of Michigan, 1990.

[3] Jang, H.M. and Cebeci, T., "An interactive boundary-layer method for unsteady flows," Rept. AE-91-5, Aerospace Engineering Dept., California State University, Long Beach, May 1991.

[4] Badsu, B.C. and Hancock, G.J., "The unsteady motion of a two-dimensional airfoil in incompressible inviscid flow," J. Fluid Mech., Vol. 87, pp. 157–168, 1987.

8 Two-Dimensional Unsteady Incompressible Flows: Boundary-Layer Method

8.0 Introduction

A major problem in the calculation of boundary-layers is to include regions of flow reversal. In a two-dimensional steady flow with a prescribed external velocity distribution, $u_e(x)$ (Chapter 3), the problem is associated with separation; in general the solutions come to an end if and when the skin friction vanishes due to the appearance of a singularity centered at the wall. In a three-dimensional steady flow (Chapter 11) under similar conditions, the problem is associated with cross-flow reversals and with separation. Even though the flow remains attached in the generally accepted terminology and the solutions of the governing equations are not singular at the vanishing of zero cross-flow skin friction, numerical instabilities result from integration opposed to the flow direction and require special numerical procedures be devised to obtain stable and accurate solutions. When the streamwise skin-friction vanishes at the wall, the solutions, as in two-dimensional steady flows, come to an end due to the existence of a singularity. In general this situation occurs after regions of negative crossflow velocity in the flow; an accurate calculation of the flow under these conditions is crucial in defining the flow separation line in three-dimensional flows. As in two-dimensional steady flows, the calculation of separated flows (negative streamwise skin-friction) requires the solution of the governing equations in the inverse mode.

The computational problems associated with unsteady boundary-layers are similar to those for three-dimensional steady boundary-layers. In two-dimensional unsteady flows with no separation, the problem is roughly analogous to a three-dimensional reverse cross-flow problem if we associate time with the direction of a mainstream with a unit velocity component. The unsteady boundary-layer is then the cross-flow velocity. As in three-dimensional steady flows (Section 11.4), the computation of unsteady flows under these conditions requires the development of special procedures to avoid numerical instabilities which result from flow reversal. The flow separation, if there is one, again occurs after some regions of flow reversal develops within the boundary-layer, and unlike steady two-dimensional flows, the breakdown of the solution does not always coincide with the vanishing of the skin-friction nor it is centered at the wall; to avoid the singularity and be able to obtain solutions of separated flows, it is again necessary to solve the unsteady boundary-layer equations in the inverse mode.

In this chapter we extend the boundary-layer method of Chapter 3 from steady airfoil flows to unsteady flows. We again use the eddy viscosity concept to model the Reynolds shear stress term in the momentum equation and, with b given by Eq. (3.1.10), write the unsteady boundary-layer equations for two-dimensional laminar and turbulent incompressible flows and their boundary conditions in the standard mode as

$$\frac{\partial u}{\partial x} + \frac{\partial v}{\partial y} = 0 \tag{8.0.1}$$

$$\frac{\partial u}{\partial t} + u\frac{\partial u}{\partial x} + v\frac{\partial u}{\partial y} = \frac{\partial u_e}{\partial t} + u_e\frac{\partial u_e}{\partial x} + \nu\frac{\partial}{\partial y}\left(b\frac{\partial u}{\partial y}\right) \tag{8.0.2}$$

$$y = 0; \quad u = 0; \quad v = v_w(x) \tag{8.0.3a}$$

$$y = \delta; \quad u = u_e(x,t) \tag{8.0.3b}$$

Equations (8.0.3) are also applicable to the boundary conditions in the inverse mode provided that $u_e(x,t)$ in Eq. (8.0.3b) is written similar to Eq. (3.0.4c), that is,

$$u_e(x,t) = u_e^0(x,t) + \frac{1}{\pi}\int_{x_a}^{x_b} \frac{d}{d\sigma}(u_e\delta^*)\frac{d\sigma}{x-\sigma} \tag{8.0.3c}$$

In addition to the above boundary conditions, Eqs. (8.0.1) and (8.0.2) require initial conditions in the (t,y) and (x,y) planes (Section 8.1) which, depending on the unsteady flow problem under consideration, may have to be calculated by a special procedure such as the one described in subsection 9.1.2.

It is again convenient and efficient to use transformed variables in both standard and inverse modes as discussed in Section 8.2. Sections 8.3 to 8.5 describe the numerical procedures used to solve Eqs. (8.0.1) and (8.0.2) in the contexts of flows without reversal (Section 8.3), flows with reversal (Section 8.4) and flows with separation (Section 8.5).

8.1 Initial Conditions

The determination of initial conditions in the (x, y) and (t, y) planes is important. Sometimes it can be arbitrary and sometimes approximate. For example, if initial conditions in the (x, y) plane are arbitrary, the values of $\partial u/\partial t$ at $t = 0$ are nonzero; this implies an inviscid acceleration and, as a consequence, a slip velocity develops at the wall and is smoothed by an inner boundary-layer initially of thickness $(\nu t)^{1/2}$ in which viscous forces are important. Thus a double structure develops in the boundary-layer and may be treated by the numerical method described in [1]. However, if interest is centered on the solution at large times, this feature may be reduced in importance by requiring that the initial velocity distribution satisfies the steady-state equation with the instantaneous external velocity. In addition, it is necessary to smooth out the external velocity $u_e(x, t)$ so that $\partial u_e/\partial t = 0$ at $t = 0$ and then standard numerical methods may be used and are stable. The use of a smoothing function makes for some loss of accuracy at small values of t, but the error soon decays to zero once the required value of u_e is specified. In this chapter we assume that steady-state conditions prevail at $t = 0$ and generate the initial conditions in the (x, y) plane by the procedure described in Chapter 3.

The calculation of initial conditions in the (t, y) plane at some $x = x_0$ when the conditions at a previous time-line are known, can introduce different problems. In principle, solutions can be determined at the next time-line by an explicit method but, if stability problems are avoided by the use of an implicit method, there arises the problem of generating a starting profile on the desired time-line. This requires the use of a special numerical procedure, such as the one discussed in subsection 9.1.2. An alternative is a quasi-steady approach in the immediate vicinity of the stagnation point and provided there are no flow reversals, this is simpler. For unsteady flows with low frequency, it is expected that the assumption of quasi-steady flow in the region near the stagnation point will not affect the overall solution. Both procedures are described in Chapter 9 where the calculation method is applied to unsteady airfoil flows.

8.2 Transformed Equations

As in two-dimensional steady flows, it is again desirable and convenient to express the unsteady boundary-layer equations in transformed variables.

8.2.1 Transformed Equations: Standard Mode

Near the leading edge region where the initial conditions in the (t, y) plane are generated with a quasi-steady approach, it is convenient to use the Falkner-Skan transformation given by Eq. (3.2.1) and express the continuity and momentum equations and their boundary conditions in a form similar to those given by Eqs. (3.2.3) and (3.2.6),

$$(bf'')' + \frac{m+1}{2} ff'' + m[1-(f')^2] + m_3(1-f') - \frac{1}{2}m_3 f'' \\ = \xi\left(\frac{1}{w}\frac{\partial f'}{\partial \tau} + f'\frac{\partial f'}{\partial \xi} - f''\frac{\partial f}{\partial \xi}\right) \tag{8.2.1}$$

$$\eta = 0, \quad f' = 0, \quad f = 0 \tag{8.2.2a}$$

$$\eta = \eta_e, \quad f' = 1 \tag{8.2.2b}$$

Here the parameters m and m_3 represent the dimensionless pressure gradient terms,

$$m = \frac{\xi}{w}\frac{\partial w}{\partial \xi}, \quad m_3 = \frac{\xi}{w^2}\frac{\partial w}{\partial \tau} \tag{8.2.3}$$

with τ is denoting dimensionless time, w dimensionless velocity and ξ dimensionless streamwise distance,

$$\tau = \frac{tu_0}{L}, \quad w = \frac{u_e}{u_0}, \quad \xi = \frac{x}{L} \tag{8.2.4}$$

We also use this transformation to generate the initial conditions in the (x, y) plane at $t = 0$.

When the initial conditions are generated in the (t, y) plane for $t > 0$ with the procedure described in subsection 9.1.2, it is more convenient to express the continuity and momentum equations in a form different than that given by Eq. (8.2.1). With the definitions given by Eq. (8.2.4) and with the definition of a dimensionless distance Y and stream function $f(\xi, Y)$ by

$$Y = \sqrt{\frac{u_0}{\nu L}}\, y, \quad \psi = (u_0 \nu L)^{1/2} f(\xi, Y) \tag{8.2.5}$$

Eqs. (8.0.1) to (8.0.3) can be written as

$$(bf'')' + \frac{\partial w}{\partial \tau} + w\frac{\partial w}{\partial \xi} = \frac{\partial f'}{\partial \tau} + f'\frac{\partial f'}{\partial \xi} - f''\frac{\partial f}{\partial \xi} \tag{8.2.6}$$

$$Y = 0; \quad f = f' = 0 \tag{8.2.7a}$$

$$Y = Y_e; \quad f' = w \tag{8.2.7b}$$

Here primes denote differentiation with respect to Y which is different than the similarity parameter defined in Eq. (3.2.1). Similarly, the dimensionless stream function $f(\xi, Y)$ in Eq. (8.2.6) is different than that defined in Eq. (8.2.1).

With the definition of

$$\theta = \frac{\partial f}{\partial \xi} \tag{8.2.8}$$

Eq. (8.2.6) can also be written as

$$(bf'')' + \frac{\partial w}{\partial \tau} + w\frac{\partial w}{\partial \xi} + f''\theta = \frac{\partial f'}{\partial \tau} + f'\frac{\partial f'}{\partial \xi} \tag{8.2.9}$$

8.2.2 Transformed Equations: Inverse Mode

When the initial conditions are obtained with a quasi-steady approach, we use the transformation given by Eq. (3.2.2) with u_∞ replaced by u_0 to express the unsteady boundary-layer equations in the inverse mode,

$$(bF'')' + \frac{1}{2}FF'' + \xi w\frac{\partial w}{\partial \xi} + \xi\frac{\partial w}{\partial \tau} = \xi\left(\frac{\partial F'}{\partial \tau} + F'\frac{\partial F'}{\partial \xi} - F''\frac{\partial F}{\partial \xi}\right) \tag{8.2.10}$$

The wall and edge boundary conditions given by Eqs. (3.2.8) and (3.2.11) remain the same.

When the initial conditions are generated in the (t, y) plane for $t > 0$ with the procedure described in subsection 3.4.3, the transformed equations in the inverse mode correspond to Eqs. (8.2.8) and (8.2.9) with wall boundary conditions remaining the same; the edge boundary condition corresponds to Eq. (3.2.9), which in terms of the variables of Eq. (8.2.5), becomes

$$Y = Y_e, \quad f_e'(\xi^i) - \lambda[Y_e f_e'(\xi^i) - f_e(\xi^i)] = g_i \tag{8.2.11}$$

where g_i is given by Eq. (3.2.12b) and

$$\lambda = C_{ii}\sqrt{\frac{\xi^i}{R_L}} \tag{8.2.12}$$

8.2.3 Transformed Equations: General Form

When the initial conditions are obtained with the quasi-steady approach, it is again convenient to write the transformed equations (8.2.1) and (8.2.10) for standard and inverse modes in one general form,

$$\begin{aligned}(bf'')' &+ \frac{(\lambda_1 m + 1)}{2}ff'' + \lambda_1\left[m\{1-(f')^2\} + m_3(1-f') - \frac{m_3}{2}f''\right]\\ &= \xi\left[\frac{\partial}{\partial \xi}\frac{(f')^2}{2} - f''\frac{\partial f}{\partial \xi}\right] + \xi\left(\frac{\lambda_1}{w} + \lambda_2\right)\frac{\partial f'}{\partial \tau} - \lambda_2\xi\left[\frac{\partial}{\partial \xi}\left(\frac{w^2}{2}\right) + \frac{\partial w}{\partial \tau}\right]\end{aligned} \tag{8.2.13}$$

where for $\lambda_1 = 1$ and $\lambda_2 = 0$ we recover Eq. (8.2.1) which is subject to the boundary conditions given by Eq. (8.2.2). For $\lambda_1 = 0$ and $\lambda_2 = 1$, we recover Eq. (8.2.10) subject to the boundary conditions given by Eqs. (8.2.2a) and (8.2.11) provided, of course, we note the distinction between the dimensionless stream functions f and F and dimensionless distances η and Y.

When the initial conditions are obtained with the procedure of Section 9.4, there is no need to place the transformed equations in a general form since Eq. (8.2.6) can be used in both standard and inverse modes.

8.3 Numerical Method: Flows without Reversal

When there is no flow reversal across the boundary-layer, the solutions of the unsteady boundary-layer equations can again be obtained with the box scheme described in Chapter 3. To discuss the numerical procedure, let us assume the quasi-steady approach is being used and discuss the solution of Eq. (8.2.13) in standard and inverse modes.

We first write Eq. (8.2.13) as a first-order system by introducing new variables $u(\xi,\eta)$, $v(\xi,\eta)$ and by defining

$$f' = u \tag{8.3.1a}$$

$$u' = v \tag{8.3.1b}$$

Equation (8.2.13) becomes

$$\begin{aligned}(bv)' + \frac{1}{2}(\lambda_1 m + 1)fv + \lambda_1\left[m(1-u^2) + m_3(1-u) - \frac{m_3}{2}v\right]\\ = \xi\left[u\frac{\partial u}{\partial \xi} - v\frac{\partial f}{\partial \xi}\right] + \xi\left(\frac{\lambda_1}{w} + \lambda_2\right)\frac{\partial u}{\partial \tau} - \lambda_2\xi\left[w\frac{\partial w}{\partial \xi} + \frac{\partial w}{\partial \tau}\right]\end{aligned} \tag{8.3.2}$$

The boundary conditions for the standard and inverse modes are identical to those given by Eqs. (3.3.2) and (3.3.3), respectively.

To write the difference equations for the system given by Eqs. (8.3.1) and (8.3.2), we consider the net cube shown in Fig. 8.1 and denote the net points by

$$\begin{aligned}\xi^0 &= 0, \quad & \xi^n &= \xi^{n-1} + k_n, \quad & n &= 1,2,\ldots,N\\ \eta_0 &= 0, \quad & \eta_j &= \eta_{j-1} + h_j, \quad & j &= 1,2,\ldots,J\\ \tau^0 &= 0, \quad & \tau^i &= \tau^{i-1} + r_i, \quad & i &= 1,2,\ldots,I\end{aligned} \tag{8.3.3}$$

where the definitions of k_n, h_j are identical to those in Section 3.3 and $r_i = \Delta\tau_i$.

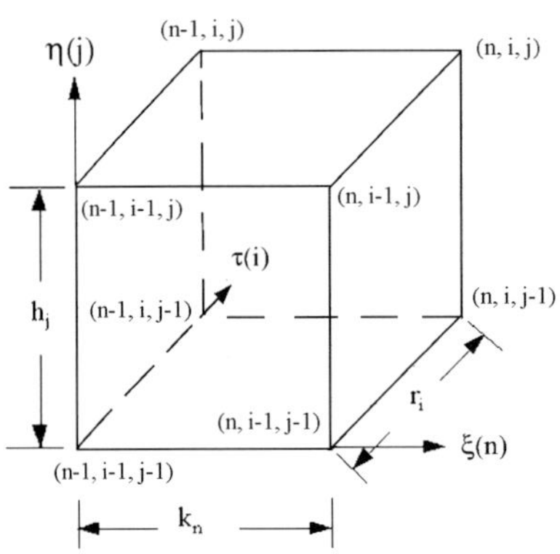

Fig. 8.1. Net cube for the difference equations for two-dimensional unsteady flows.

The difference approximations to Eqs. (8.3.1a,b) are obtained by averaging about the midpoint $(\xi^n, \tau^i, \eta_{j-1/2})$,

$$\frac{f_j^{n,i} - f_{j-1}^{n,i}}{h_j} = \frac{u_j^{n,i} + u_{j-1}^{n,i}}{2} = u_{j-1/2}^{n,i} \tag{8.3.4a}$$

$$\frac{u_j^{n,i} - u_{j-1}^{n,i}}{h_j} = \frac{v_j^{n,i} - v_{j-1}^{n,i}}{2} = v_{j-1/2}^{n,i} \tag{8.3.4b}$$

The difference approximations to Eq. (8.3.2) are obtained by centering all quantities at the center of the cube $(\xi^{n-1/2}, \tau^{i-1/2}, \eta_{j-1/2})$ by taking the values of each, say q_j, at the four corners of the box, that is,

$$\begin{aligned}
\bar{q}_j &= q_j^{n-1/2,i-1/2} \\
&= \frac{1}{4}(q_j^{n,i} + q_j^{n,i-1} + q_j^{n-1,i-1} + q_j^{n-1,i}) \\
&= \frac{1}{4}(q_j^{n,i} + q_j^{234}) \\
\bar{q}_{j-1/2} &= \frac{1}{2}(\bar{q}_j + \bar{q}_{j-1}) = \frac{1}{4}(q_{j-1/2}^{n,i} + q_{j-1/2}^{234})
\end{aligned} \tag{8.3.5}$$

Here by q_j^{234} we mean $q_j^{234} = q_j^{n,i-1} + q_j^{n-1,i-1} + q_j^{n-1,i}$, the sum of the values of q_j at three of the four corners of the face of the box.

In terms of this notation, the finite-difference approximations to Eq. (8.3.2) can be written as:

$$\begin{aligned}
&\frac{(\overline{bv})_j - (\overline{bv})_{j-1}}{h_j} + \frac{1}{2}(\lambda_1 \bar{m} + 1)(\overline{fv})_{j-1/2} \\
&+ \lambda_1 \left[\bar{m}\{1 - (\overline{u^2})_{j-1/2}\} + \bar{m}_3(1 - \bar{u}_{j-1/2}) - \frac{\bar{m}_3}{2}\bar{v}_{j-1/2} \right] \\
&= \xi^{n-1/2} \left[\frac{(\overline{u^2})_{j-1/2}^n - (\overline{u^2})_{j-1/2}^{n-1}}{2k_n} - \bar{v}_{j-1/2}\frac{\bar{f}_{j-1/2}^n - \bar{f}_{j-1/2}^{n-1}}{k_n} \right] \\
&+ \xi^{n-1/2}\left(\frac{\lambda_1}{\bar{w}} + \lambda_2\right) \frac{\bar{u}_{j-1/2}^i - \bar{u}_{j-1/2}^{i-1}}{r_i} \\
&- \lambda_2 \xi^{n-1/2} \left[\frac{(\overline{w^2})^n - (\overline{w^2})^{n-1}}{2k_n} + \frac{\bar{w}^i - \bar{w}^{i-1}}{r_i} \right]
\end{aligned} \tag{8.3.6}$$

where

$$\bar{q}^n_{j-1/2} = q^{n,i-1/2}_{j-1/2} = \frac{1}{4}(q^{n,i}_j + q^{n,i-1}_j + q^{n,i}_{j-1} + q^{n,i-1}_{j-1}) = \frac{1}{2}(q^{n,i}_{j-1/2} + q^{n,i-1}_{j-1/2})$$
$$\bar{q}^i_{j-1/2} = q^{n-1/2,i}_{j-1/2} = \frac{1}{4}(q^{n,i}_j + q^{n-1,i}_j + q^{n,i}_{j-1} + q^{n-1,i}_{j-1}) = \frac{1}{2}(q^{n,i}_{j-1/2} + q^{n-1,i}_{j-1/2})$$
$$\bar{w} = w^{n-1/2,i-1/2} = \frac{1}{4}(w^{n,i} + w^{234}), \quad \bar{w}^i = w^{n-1/2,i} = \frac{1}{2}(w^{n,i} + w^{n-1,i})$$
$$(\overline{w^2})^n = (w^2)^{n,i-1/2} = \frac{1}{2}[(w^2)^{n,i} + (w^2)^{n,i-1}] \tag{8.3.7}$$

Making use of the relations defined by Eqs. (8.3.5) and (8.3.7), after some algebra and rearranging, denoting w by u_J, Eq. (8.3.6) can be written in a form similar to Eq. (3.3.5a) for two-dimensional steady flows,

$$\begin{aligned} &h_j^{-1}[(bv)^{n,i}_j - (bv)^{n,i}_{j-1}] + \alpha_1 (fv)^{n,i}_{j-1/2} - \alpha_2 (u^2)^{n,i}_{j-1/2} \\ &- \alpha_3 u^{n,i}_{j-1/2} - \alpha_4 v^{n,i}_{j-1/2} + \frac{\alpha^n}{2}[v^{n,i}_{j-1/2} f^{n,i}_{j-1/2} + v^{234}_{n,i} f^{n,i}_{j-1/2}] \\ &+ \alpha_5 (u_J^2)^{n,i} + \alpha_6 u_J^{n,i} = -\beta_1 \end{aligned} \tag{8.3.8}$$

where α^n is given by Eq. (3.3.6a) and

$$\alpha_1 = \frac{1+\lambda_1 \bar{m}}{2}, \quad \alpha_2 = \lambda_1 \bar{m} + \alpha^n, \tag{8.3.9a}$$

$$\alpha_3 = \lambda_1 \bar{m}_3 + 2\frac{\xi^{n-1/2}}{r_i}\left(\frac{\lambda_1}{\bar{w}} + \lambda_2\right), \tag{8.3.9b}$$

$$\alpha_4 = \frac{\lambda_1 \bar{m}_3}{2} - \frac{a^n}{2}(f^{n,i-1}_{j-1/2} - 2\bar{f}^{n-1}_{j-1/2}), \tag{8.3.9c}$$

$$\alpha_5 = \lambda_2 \alpha^n, \quad \alpha_6 = 2\lambda_2 \frac{\xi^{n-1/2}}{r_i}, \tag{8.3.9d}$$

$$\alpha^n = \frac{\xi^{n-1/2}}{k_n}$$

$$\begin{aligned} \beta_1 = {} & h_j^{-1}[(bv)^{234}_j - (bv)^{234}_{j-1}] + \alpha_1 (fv)^{234}_{j-1/2} \\ &+ 4\lambda_1 \left[\bar{m}\left\{1 - \tfrac{1}{4}(u^2)^{234}_{j-1/2}\right\} + \bar{m}_3 \left\{1 - \tfrac{1}{4} u^{234}_{j-1/2}\right\} - \tfrac{\bar{m}_3}{8} v^{234}_{j-1/2}\right] \\ &- \tfrac{\alpha^n}{2}[2(u^2)^{n,i-1}_{j-1/2} - 4(\overline{u^2})^{n-1}_{j-1/2} - v^{234}_{j-1/2}(f^{n,i-1}_{j-1/2} - 2\bar{f}^{n-1}_{j-1/2})] \\ &- 2\tfrac{\xi^{n-1/2}}{r_i}\left(\tfrac{\lambda_1}{\bar{w}} + \lambda_2\right)(u^{n-1,i}_{j-1/2} - 2\bar{u}^{i-1}_{j-1/2}) \\ &+ \lambda_2 \alpha^n [(u_J^2)^{n,i-1} - 2(\overline{u_J^2})^{n-1}] + 2\lambda_2 \tfrac{\xi^{n-1/2}}{r_i}(u_J^{n-1,i} - 2\bar{u}_J^{i-1}) \end{aligned} \tag{8.3.9e}$$

Equations (8.3.4) and (8.3.8) are again imposed for $j = 1, 2, \ldots, J$ at given n, i. They are accompanied with boundary conditions for the standard mode given by Eqs. (3.3.7) and for the inverse mode by Eqs. (3.3.8).

The nonlinear system is again linearized by Newton's method and written in a form similar to the linear system for steady flows. The linearized equations for Eqs. (8.3.4) are given by Eqs. (3.3.10a,b). For Eq. (8.3.8), it is given in the same form as Eq. (3.3.10c)

$$(s_1)_j\delta v_j + (s_2)_j\delta v_{j-1} + (s_3)_j\delta f_j + (s_4)_j\delta f_{j-1} + (s_5)_j\delta u_j + (s_6)_j\delta u_{j-1} + (s_7)_j\delta u_J = (r_2)_j \tag{3.3.10c}$$

with different coefficients for $(r_2)_j$ and $(s_1)_j$ to $(s_7)_j$. They are given by

$$\begin{aligned}(r_2)_j = & -\beta_1 - \Big[h_j^{-1}\{(bv)_j^{(\nu)} - (bv)_{j-1}^{(\nu)}\} + \alpha_1(fv)_{j-1/2}^{(\nu)} \\ & - \alpha_2(u^2)_{j-1/2}^{(\nu)} - \alpha_3 u_{j-1/2}^{(\nu)} - \alpha_4 v_{j-1/2}^{(\nu)} \\ & + \frac{\alpha^n}{2}\{v_{j-1/2}^{(\nu)} f_{j-1/2}^{(\nu)} + v_{j-1/2}^{234} f_{j-1/2}^{(\nu)}\} + \alpha_5(u_J^{(\nu)})^2 + \alpha_6 u_J^{(\nu)}\Big]\end{aligned} \tag{8.3.11}$$

$$(s_1)_j = h_j^{-1} b_j^{(\nu)} + \frac{\alpha_1}{2} f_j^{(\nu)} - \frac{\alpha_4}{2} + \frac{\alpha^n}{4} f_{j-1/2}^{(\nu)} \tag{8.3.12a}$$

$$(s_2)_j = -h_j^{-1} b_{j-1}^{(\nu)} + \frac{\alpha_1}{2} f_{j-1}^{(\nu)} - \frac{\alpha_4}{2} + \frac{\alpha^n}{4} f_{j-1/2}^{(\nu)} \tag{8.3.12b}$$

$$(s_3)_j = \frac{\alpha_1}{2} v_j^{(\nu)} + \frac{\alpha^n}{4}(v_{j-1/2}^{(\nu)} + v_{j-1/2}^{234}) \tag{8.3.12c}$$

$$(s_4)_j = \frac{\alpha_1}{2} v_{j-1}^{(\nu)} + \frac{\alpha^n}{4}(v_{j-1/2}^{(\nu)} + v_{j-1/2}^{234}) \tag{8.3.12d}$$

$$(s_5)_j = -\alpha_2 u_j^{(\nu)} - \frac{\alpha_3}{2} \tag{8.3.12e}$$

$$(s_6)_j = -\alpha_2 u_{j-1}^{(\nu)} - \frac{\alpha_3}{2} \tag{8.3.12f}$$

$$(s_7)_j = 2\alpha_5 u_J^{(\nu)} + \alpha_6 \tag{8.3.12g}$$

The boundary conditions for the standard mode and inverse mode are identical to those in Eqs. (3.3.13) to (3.3.15) provided that λ is defined by Eq. (8.2.12) rather than by Eq. (3.2.12a).

The solution of the linear system given by Eqs. (3.3.10) subject to the boundary conditions in Eqs. (3.3.13) to (3.3.14) is identical to the procedure described in subsection 3.3.3 for the standard mode and in subsection 3.3.4 for the inverse mode, provided that initial conditions in the (x, y) plane at $\tau = 0$ and those in the (t, y) plane at $\xi = \xi_0$ are specified.

In the quasi-steady approach, the solutions in the (x, y) plane at $\tau = 0$ are obtained by solving the steady flow equations, Eq. (3.2.3), which is a reduced form of the time-dependent equation (8.2.1) at $\tau = 0$. The initial conditions in the (t, y) plane are also obtained from the solution of Eq. (3.2.3) for a given external velocity distribution for a specified range of ξ values and for all desired time intervals at $\tau > 0$.

8.4 Numerical Method: Flows with Reversal

When there is flow reversal across the shear layer, it is necessary to modify the standard box of the previous section in order to avoid the numerical instabilities resulting from integration opposed to the flow direction. A convenient procedure is to include the zig-zag formulation of Krause et al. [2]. In common with the often-used Crank-Nicolson method, this scheme is easy to employ, particularly since the orientation of the numerical mesh is chosen a priori. This advantage has a corresponding and potentially dangerous drawback in the presence of large reverse flows, since the mesh ratio must be related to the velocity according to the famous Courant, Friedricks, Lewey (CFL) condition [3] if stability is to be achieved. For a fixed grid chosen a priori, this condition may be violated as the flow velocities are determined in ever increasing computational domains. Thus a natural boundary limiting the domain in which stable computations can be made is also determined a priori.

One way to avoid the above limitation is to allow the grid to be determined along with the flow calculations and we shall describe a numerical scheme, [4], which does this. The grid spacings and orientation are adjusted depending upon the magnitude and direction of the local velocity so that the CFL condition is satisfied. The scheme is thus in some sense intelligent in that it maximizes the domain in which the computations can be carried out. For completeness we first describe the zig-zag scheme and then proceed with the intelligent scheme which will be referred to as the characteristic box scheme.

8.4.1 Zig-Zag Box

To solve Eq. (8.2.13) with the zig-zag box scheme, we start with the first-order system given by Eqs. (8.3.1) and (8.3.2). The main difference between the standard box of the previous section and the zig-zag box scheme of this section lies on how the difference equations are written for Eq. (8.3.2); the remaining two equations, Eqs. (8.3.1a,b), and boundary conditions remain unchanged.

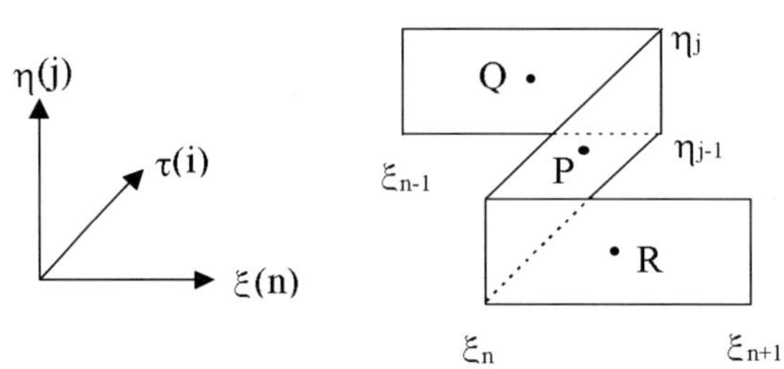

Fig. 8.2. Finite-difference molecule for the zig-zag box.

To write the difference equations for Eq. (8.3.2) centered at P (see Fig. 8.2), we use quantities centered at P, Q and R, where

$$P \equiv (\xi^n, \tau^{i-1/2}, \eta_{j-1/2}), \quad Q \equiv (\xi^{n-1/2}, \tau^i, \eta_{j-1/2}), \\ R \equiv (\xi^{n+1/2}, \tau^{i-1}, \eta_{j-1/2}) \tag{8.4.1}$$

and, for convenience, use the following model equation for Eq. (8.3.2),

$$(bv)' + \frac{\lambda_1 m + 1}{2} fv = \xi \left[u\frac{\partial u}{\partial \xi} - v\frac{\partial f}{\partial \xi} \right] + \xi \left(\frac{\lambda_1}{w} + \lambda_2 \right) \frac{\partial u}{\partial \tau} \tag{8.4.2}$$

and write Eq. (8.4.2) in the form

$$\begin{aligned} &(bv)'(P) + \tfrac{1}{2}[\lambda_1 m(P) + 1] fv(P) \\ &= \xi(P) \left[\theta u(Q) \frac{\partial u}{\partial \xi}(Q) + \phi u(R) \frac{\partial u}{\partial \xi}(R) - \theta v(Q) \frac{\partial f}{\partial \xi}(Q) - \phi v(R) \frac{\partial f}{\partial \xi}(R) \right] \\ &\quad + \xi(P) \left[\frac{\lambda_1}{w(P)} + \lambda_2 \right] \frac{\partial u}{\partial \tau}(P) \end{aligned} \tag{8.4.3}$$

where

$$\theta = \frac{\xi^{n+1} - \xi^n}{\xi^{n+1} - \xi^{n-1}}, \quad \phi = \frac{\xi^n - \xi^{n-1}}{\xi^{n+1} - \xi^{n-1}} \tag{8.4.4}$$

The difference equations for Eq. (8.4.3) are

$$\begin{aligned} &\frac{(bv)_j^{n,i-1/2} - (bv)_{j-1}^{n,i-1/2}}{h_j} + \left(\frac{\lambda_1 m^{n,i-1/2} + 1}{2} \right) (fv)_{j-1/2}^{n,i-1/2} \\ &= \xi^n \left[\theta u_{j-1/2}^{n-1/2,i} \frac{u_{j-1/2}^{n,i} - u_{j-1/2}^{n-1,i}}{k_n} + \phi u_{j-1/2}^{n+1/2,i-1} \frac{u_{j-1/2}^{n+1,i-1} - u_{j-1/2}^{n,i-1}}{k_{n+1}} \right. \\ &\quad \left. - \theta v_{j-1/2}^{n-1/2,i} \frac{f_{j-1/2}^{n,i} - f_{j-1/2}^{n-1,i}}{k_n} - \phi v_{j-1/2}^{n+1/2,i-1} \frac{f_{j-1/2}^{n+1,i-1} - f_{j-1/2}^{n,i-1}}{k_{n+1}} \right] \\ &\quad + \xi^n \left(\frac{\lambda_1}{w^{n,i-1/2}} + \lambda_2 \right) \frac{u_{j-1/2}^{n,i} - u_{j-1/2}^{n,i-1}}{r_i} \end{aligned} \tag{8.4.5}$$

A similar procedure can obviously be applied to Eq. (8.3.2). The resulting algebraic system is again nonlinear and can be linearized with Newton's method. The linear system is identical to the linear system for steady flows discussed in Chapter 3 and to unsteady flows discussed in the previous section, except for the coefficients $(r_2)_j$ and $(s_1)_j$ to $(s_7)_j$ in Eq. (3.3.10c). They are given by

$$\begin{aligned} (r_2)_j = -\beta_1 - \Big\{ &h_j^{-1}[(bv)_j^{(\nu)} - (bv)_{j-1}^{(\nu)}] + \alpha_1 (fv)_{j-1/2}^{(\nu)} \\ &- \alpha_2 (u^2)_{j-1/2}^{(\nu)} - \alpha_3 u_{j-1/2}^{(\nu)} - \alpha_4 v_{j-1/2}^{(\nu)} \\ &+ \theta \frac{\xi^n}{k_n} [v_{j-1/2}^{(\nu)} f_{j-1/2}^{(\nu)} + v_{j-1/2}^{n-1,i} f_{j-1/2}^{(\nu)}] + \alpha_5 (u_J^{(\nu)})^2 + \alpha_6 u_J^{(\nu)} \Big\} \end{aligned} \tag{8.4.6}$$

$$(s_1)_j = h_j^{-1} b_j^{(\nu)} + \frac{\alpha_1}{2} f_j^{(\nu)} - \frac{\alpha_4}{2} + \frac{\theta}{2} \frac{\xi^n}{k_n} f_{j-1/2}^{(\nu)} \tag{8.4.7a}$$

$$(s_2)_j = h_j^{-1} b_{j-1}^{(\nu)} + \frac{\alpha_1}{2} f_{j-1}^{(\nu)} - \frac{\alpha_4}{2} + \frac{\theta}{2} \frac{\xi^n}{k_n} f_{j-1/2}^{(\nu)} \tag{8.4.7b}$$

$$(s_3)_j = \frac{\alpha_1}{2} v_j^{(\nu)} + \frac{\theta}{2} \frac{\xi^n}{k_n} [v_{j-1/2}^{(\nu)} + v_{j-1/2}^{n-1,i}] \tag{8.4.7c}$$

$$(s_4)_j = \frac{\alpha_1}{2} v_j^{(\nu)} + \frac{\theta}{2} \frac{\xi^n}{k_n} [v_{j-1/2}^{(\nu)} + v_{j-1/2}^{n-1,i}] \tag{8.4.7d}$$

$$(s_5)_j = -\alpha_2 u_j^{(\nu)} - \frac{\alpha_3}{2} \tag{8.4.7e}$$

$$(s_6)_j = -\alpha_2 u_{j-1}^{(\nu)} - \frac{\alpha_3}{2} \tag{8.4.7f}$$

$$(s_7)_j = 2\alpha_5 u_J^{(\nu)} + \alpha_6 \tag{8.4.7g}$$

where

$$\begin{aligned}
\beta_1 = {} & h_j^{-1}[(bv)_j^{n,i-1} - (bv)_{j-1}^{n,i-1}] + \alpha_1 (fv)_{j-1/2}^{n,i-1} \\
& + 2\lambda_1 \left\{ m^{n,i-1/2} \left[1 - \frac{1}{2}(u^2)_{j-1/2}^{n,i-1} \right] + m_3^{n,i-1/2} \left[1 - \frac{1}{2} u_{j-1/2}^{n,i-1} \right] - \frac{m_3^{n,i-1}}{4} v_{j-1/2}^{n,i-1} \right\} \\
& + \frac{\xi^n}{k_n} [\theta (u^2)_{j-1/2}^{n-1,i} - 2\phi u_{j-1/2}^{n+1/2,i-1} (u_{j-1/2}^{n+1,i-1} - u_{j-1/2}^{n,i-1}) - \theta v_{j-1/2}^{n-1,i} f_{j-1/2}^{n-1,i} \\
& + 2\phi v_{j-1/2}^{n+1/2,i-1} (f_{j-1/2}^{n+1,i-1} - f_{j-1/2}^{n,i-1})] + 2\frac{\xi^n}{r_i} \left(\frac{\lambda_1}{w^{n,i-1/2}} + \lambda_2 \right) u_{j-1/2}^{n,i-1} \\
& - \lambda_2 \frac{\xi^n}{k_n} [\theta (u_J^2)^{n-1,i} - 2\phi u_J^{n+1/2,i-1} (u_J^{n+1,i-1} - u_J^{n,i-1})] - 2\lambda_2 \frac{\xi^n}{r_i} u_J^{n,i-1}
\end{aligned}$$

$$\alpha_1 = \frac{1 + \lambda_1 m^{n,i-1/2}}{2}, \qquad \alpha_2 = \lambda_1 m^{n,i-1/2} + \theta \frac{\xi^n}{k_n},$$

$$\alpha_3 = \lambda_1 m_3^{n,i-1/2} + 2\frac{\xi^n}{r_i} \left(\frac{\lambda_1}{w^{n,i-1/2}} + \lambda_2 \right), \qquad \alpha_4 = \frac{\lambda_1 m_3^{n,i-1/2}}{2} + \theta \frac{\xi^n}{k_n} f_{j-1/2}^{n-1,i},$$

$$\alpha_5 = \lambda_2 \theta \frac{\xi^n}{k_n}, \qquad \alpha_6 = 2\lambda_2 \frac{\xi^n}{\gamma_i}$$

8.4.2 Characteristic Box

The characteristic box scheme, developed by Cebeci and Stewartson [5], is based on the solution of the governing equations along local streamlines. This method uses Keller's method and allows the grid to be determined along with the flow calculations. As will be discussed in Chapter 11, it has been applied to three-dimensional boundary-layers on wings, bodies of revolution at incidence and ship hulls with good success. It is described here for two-dimensional unsteady flows with flow reversal and separation.

To solve Eqs. (8.0.1) and (8.0.2) subject to their boundary conditions with the characteristic box scheme, we consider Eq. (8.2.10) and with u' defined by Eq. (8.3.1b) and θ' by

$$\theta' = \frac{1}{2}u + \xi\frac{\partial u}{\partial \xi} \tag{8.4.9a}$$

write Eq. (8.2.10) as

$$(bv)' + v\theta + \xi\left(\frac{\partial w}{\partial \tau} + w\frac{\partial w}{\partial \xi}\right) = \xi\left(\frac{\partial u}{\partial \tau} + u\frac{\partial u}{\partial \xi}\right) \tag{8.4.9b}$$

Noting the definition of local streamlines, which in our notation

$$\frac{d\tau}{1} = \frac{d\xi}{u} \tag{8.4.10}$$

and denoting the streamline distance by ψ and the angle that it makes with the τ-axis by ϕ (see Fig. 8.3), we write Eq. (8.4.9b) as

$$(bv)' + \xi\beta + v\theta = \xi\lambda\frac{\partial u}{\partial \psi} \tag{8.4.11}$$

where

$$\lambda = \sqrt{1+u^2} \tag{8.4.12a}$$

$$\phi = \tan^{-1} u \tag{8.4.12b}$$

$$\beta = \frac{\partial w}{\partial \tau} + \frac{1}{2}\frac{\partial w^2}{\partial \xi} \tag{8.4.12c}$$

The solution of the transformed boundary layer equations are obtained by defining two regions depending on the sign of the streamwise velocity u. In the region 1, u is negative, and the solution of the boundary layer equations is obtained

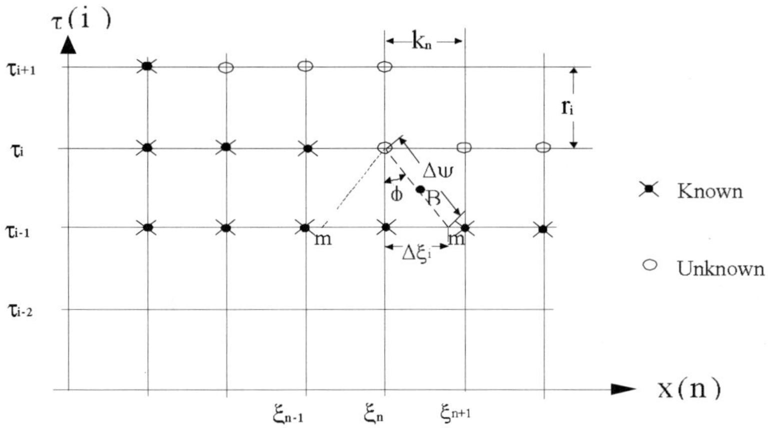

Fig. 8.3. Notation for the characteristic box scheme.

from Eqs. (8.3.1b), (8.4.9a) and (8.4.11) subject to the wall boundary conditions given by Eq. (8.4.23a). The difference approximations to Eq. (8.3.1b) is given by Eq. (8.3.4b) and Eq. (8.4.9a) by

$$\left(\frac{\theta_j^{n-1/2,i-1/2} - \theta_{j-1}^{n-1/2,i-1/2}}{h_j}\right) \equiv \left(\frac{\theta_j - \theta_{j-1}}{h_j}\right)$$
$$= \frac{1}{8}(u_{j-1/2}^{n,i} + u_{j-1/2}^{n-1,i} + u_{j-1/2}^{n,i-1} + u_{j-1/2}^{n-1,i-1}) \tag{8.4.13}$$
$$+ \xi^{n-1/2}\frac{(u_{j-1/2}^{n,i} - u_{j-1/2}^{n-1,i} + u_{j-1/2}^{n,i-1} - u_{j-1/2}^{n-1,i-1})}{2k_n}$$

$$\frac{u_j^{n,i} - u_{j-1}^{n,i}}{h_j} = v_{j-1/2}^{n,i}$$

The difference approximations to Eq. (8.4.11) are obtained along the streamline distance $\Delta\psi_j$ at point B (see Fig. 8.3).

$$\frac{(bv)_j^{n,i} - (bv)_{j-1}^{n,i}}{h_j} + \frac{(bv)_j^{m,i-1} - (bv)_{j-1}^{m,i-1}}{h_j}$$
$$+ (v_{j-1/2}^{n,i} + v_{j-1/2}^{m,i-1})\theta_{j-1/2}^B + 2\xi^B\beta^B \tag{8.4.14}$$
$$= \xi^B(\lambda_{j-1/2}^{n,i} + \lambda_{j-1/2}^{m,i-1})\frac{u_{j-1/2}^{n,i} - u_{j-1/2}^{m,i-1}}{\Delta\psi_{j-1/2}}$$

where

$$\Delta\psi_{j-1/2} = \frac{r_i}{\cos\phi_{j-1/2}} \tag{8.4.15}$$

The relation between $\theta_{j-1/2}^B$ and those values of θ centered at $(n-1/2, i-1/2)$ and $(n-3/2, i-1/2)$ are

$$\theta_{j-1/2}^B = \frac{\theta_{j-1/2}^{n-1/2,i-1/2} - \theta_{j-1/2}^{n-3/2,i-1/2}}{\xi^{n-1/2} - \xi^{n-3/2}}(\xi^B - \xi^{n-3/2}) + \theta_{j-1/2}^{n-3/2,i-1/2} \tag{8.4.16a}$$

or

$$\theta_{j-1/2}^B = \frac{1}{2}(\theta_j + \theta_{j-1})l_1 + l_2 \tag{8.4.16b}$$

where

$$l_1 = \frac{\xi^B - \xi^{n-3/2}}{\xi^{n-1/2} - \xi^{n-3/2}}, \quad l_2 = (1 - l_1)\theta_{j-1/2}^{n-3/2,i-1/2} \tag{8.4.17}$$

The pressure gradient term β is given by

$$\beta^B = \left(\frac{\partial w}{\partial \tau}\right)^B + \frac{1}{2}\left(\frac{\partial w^2}{\partial \xi}\right)^B \tag{8.4.18a}$$

$$\left(\frac{\partial w}{\partial \tau}\right)^B = A_1\frac{w^{n,i}-w^{n,i-1}}{r_i} + A_2\frac{w^{n-1,i}-w^{n-1,i-1}}{r_i} + A_3\frac{w^{n-2,i}-w^{n-2,i-1}}{r_i} \quad (8.4.18b)$$

$$\left(\frac{\partial w^2}{\partial \xi}\right)^B = B_1\frac{(w^2)^{n,i}+(w^2)^{n,i-1}}{2} + B_2\frac{(w^2)^{n-1,i}+(w^2)^{n-1,i-1}}{2} + B_3\frac{(w^2)^{n-2,i}+(w^2)^{n-2,i-1}}{2} \quad (8.4.18c)$$

$$A_1 = \frac{(\xi^B-\xi^{n-1})(\xi^B-\xi^{n-2})}{(\xi^n-\xi^{n-1})(\xi^n-\xi^{n-2})}$$
$$A_2 = \frac{(\xi^B-\xi^{n})(\xi^B-\xi^{n-2})}{(\xi^{n-1}-\xi^{n})(\xi^{n-1}-\xi^{n-2})} \quad (8.4.19a)$$
$$A_3 = \frac{(\xi^B-\xi^{n})(\xi^B-\xi^{n-1})}{(\xi^{n-2}-\xi^{n})(\xi^{n-2}-\xi^{n-1})}$$

$$B_1 = \frac{2\xi^B-\xi^{n-1}-\xi^{n-2}}{(\xi^n-\xi^{n-1})(\xi^n-\xi^{n-2})}$$
$$B_2 = \frac{2\xi^B-\xi^{n}-\xi^{n-2}}{(\xi^{n-1}-\xi^{n})(\xi^{n-1}-\xi^{n-2})} \quad (8.4.19b)$$
$$B_3 = \frac{2\xi^B-\xi^{n}-\xi^{n-1}}{(\xi^{n-2}-\xi^{n})(\xi^{n}-\xi^{n-1})}$$

As before the system given by Eqs. (8.4.3b), (8.4.13), (8.4.14) and the boundary conditions are linearized with Eq. (8.3.4b) given by Eq. (3.3.10b), the linearized form of Eq. (8.4.13) is

$$\frac{\delta\theta_j-\delta\theta_{j-1}}{h_j} + b_1(\delta u_j+\delta u_{j-1}) = (r_3)_{j-1} \quad (8.4.20)$$

where

$$b_1 = -\left(\frac{1}{16}+\frac{\xi^{n-1/2}}{4k_n}\right)$$
$$(r_3)_{j-1} = -\left\{\left(\frac{\theta_j-\theta_{j-1}}{h_j}\right) - \frac{1}{2}u_{j-1/2}^{u-1/2,i-1/2} - \xi^{n-1/2}\frac{(u_{j-1/2}^{n,i}-u_{j-1/2}^{n-1,i}+u_{j-1/2}^{n,i-1}-u_{j-1/2}^{n-1,i-1})}{2k_n}\right\} \quad (8.4.21)$$

After linearization, Eq. (8.4.14) is expressed in the same form as Eq. (3.3.10c) with different coefficients for $(r_2)_j$ and $(s_1)_j$ to $(s_7)_j$ where

$$(s_1)_j = \frac{b_j}{h_j} + \frac{1}{2}\theta_{j-1/2}^B \quad (8.4.22a)$$

$$(s_2)_j = -\frac{b_j}{h_j} + \frac{1}{2}\theta^B_{j-1/2} \tag{8.4.22b}$$

$$(s_3)_j = \frac{C_1}{2}(v^{n,i}_{j-1/2} + v^{m,i-1}_{j-1/2}) \tag{8.4.22c}$$

$$(s_4)_j = (s_3)_j \tag{8.4.22d}$$

$$(s_5)_j = -\frac{1}{2}\xi^B \left(\frac{\lambda^{n,i}_{j-1/2} + \lambda^{m,i-1}_{j-1/2}}{\Delta\psi_{j-1/2}}\right) - \frac{1}{2}\xi^B \frac{u_j}{\lambda_j}\left(\frac{u^{n,i}_{j-1/2} - u^{m,i-1}_{j-1/2}}{\Delta\psi_{j-1/2}}\right) \tag{8.4.22e}$$

$$(s_6)_j = -\frac{1}{2}\xi^B \left(\frac{\lambda^{n,i}_{j-1/2} + \lambda^{m,i-1}_{j-1/2}}{\Delta\psi_{j-1/2}}\right) - \frac{1}{2}\xi^B \frac{u_{j-1}}{\lambda_{j-1}}\left(\frac{u^{n,i}_{j-1/2} - u^{m,i-1}_{j-1/2}}{\Delta\psi_{j-1/2}}\right) \tag{8.4.22f}$$

$$(s_7)_j = 2\xi^B\left(\frac{A_1}{\gamma_i} + \frac{B_1 u_J}{2}\right) \tag{8.4.22g}$$

$$\begin{aligned}(r_2)_j = &-\left\{\left[\frac{(bv)^{n,i}_j - (bv)^{n,i}_{j-1}}{h_j}\right] + \left[\frac{(bv)^{m,i-1}_j - (bv)^{m,i-1}_{j-1}}{h_j}\right]\right.\\ &+ (v^{n,i}_{j-1/2} + v^{m,i-1}_{j-1/2})\theta^B_{j-1/2} + 2\xi^B\beta^B\\ &\left.- \xi^B(\lambda^{n,i}_{j-1/2} + \lambda^{m,i-1}_{j-1/2})\left(\frac{u^{n,i}_{j-1/2} - u^{m,i-1}_{j-1/2}}{\Delta\psi_{j-1/2}}\right)\right\}\end{aligned} \tag{8.4.23}$$

The wall boundary conditions are

$$\begin{aligned}\delta\theta_0 &= 0\\ \delta u_0 &= 0\end{aligned} \tag{8.4.24}$$

In the region 2, u is positive and the solution of the boundary layer equations is obtained by applying the regular box scheme to Eqs. (8.3.16), (8.4.9a,b) subject to the edge boundary conditions given by (8.2.11). After linearization, they are expressed in a form similar to those in region 1. For example, Eq. (8.4.9b) can be written as

$$\begin{aligned}&(bv)' + v\theta + \xi\frac{\partial w}{\partial \tau} + \frac{\xi}{2}\frac{\partial w^2}{\partial \xi} = \xi\frac{\partial u}{\partial \tau} + \frac{\xi}{2}\frac{\partial u^2}{\partial \xi}\\ &\frac{(\overline{bv})_j - (\overline{bv})_{j-1}}{h_j} + \overline{v}_{j-1/2}\theta_{j-1/2} + \xi^{n-1/2}\frac{w^{n-1/2,i} - w^{n-1/2,i-1}}{r_i}\\ &+\frac{\xi^{n-1/2}}{2}\frac{(w^2)^{n,i-1/2} - (w^2)^{n-1,i-1/2}}{k_n}\\ &= \xi^{n-1/2}\frac{u^{n-1/2,i}_{j-1/2} - u^{n-1/2,i-1}_{j-1/2}}{r_i} + \frac{\xi^{n-1/2}}{2}\frac{(u^2)^{n,i-1/2}_{j-1/2} - (w^2)^{n-1,i-1/2}_{j-1/2}}{k_n}\end{aligned} \tag{8.4.25}$$

where

$$(\bar{\ }) = [(\ )^{n,i} + (\ )^{n-1,i} + (\ )^{n,i-1} + (\ )^{n-1,i-1}]/4$$

After linearization, it again can be expressed in the same form as Eq. (3.3.10c) with the $(s_1)_j$ to $(s_7)_j$ and $r_2)_j$ given by with $w = u_J$,

$$(s_1)_j = \frac{b_j}{h_j} + \frac{1}{2}\theta_{j-1/2} \tag{8.4.26a}$$

$$(s_2)_j = -\frac{b_j}{h_j} + \frac{1}{2}\theta_{j-1/2} \tag{8.4.26b}$$

$$(s_3)_j = 2\overline{v}_{j-1/2} \tag{8.4.26c}$$

$$(s_4)_j = (s_3)_j \tag{8.4.26d}$$

$$(s_5)_j = -\frac{\xi^{n-1/2}}{r_i} - \frac{\xi^{n-1/2}}{k_n}u_j \tag{8.4.26e}$$

$$(s_6)_j = -\frac{\xi^{n-1/2}}{r_i} - \frac{\xi^{n-1/2}}{k_n}u_{j-1} \tag{8.4.26f}$$

$$(s_7)_j = 2\left(\frac{\xi^{n-1/2}}{r_i} + \frac{\xi^{n-1/2}}{k_n}u_J\right) \tag{8.4.26g}$$

$$\begin{aligned}(r_2)_j = -4\Bigg[&\frac{\overline{(bv)}_j - \overline{(bv)}_{j-1}}{h_j} + \overline{v}_{j-1/2}\theta_{j-1/2} \\ &+ \xi^{n-1/2}\frac{w^{n-1/2,i} - w^{n-1/2,i-1}}{r_i} \\ &+ \frac{\xi^{n-1/2}}{2}\frac{(w^2)^{n,i-1/2} - (w^2)^{n-1,i-1/2}}{k_n} \\ &- \xi^{n-1/2}\frac{u_{j-1/2}^{n-1/2,i} - u_{j-1/2}^{n-1/2,i-1}}{r_i} \\ &- \frac{\xi^{n-1/2}}{2}\frac{(u^2)_{j-1/2}^{n,i-1/2} - (u^2)_{j-1/2}^{n-1,i-1/2}}{k_n}\Bigg]\end{aligned} \tag{8.4.27}$$

The wall boundary conditions are given by Eq. (8.4.25) and the edge conditions are given by Eq. (3.3.14b) with D and λ now defined by

$$D = u_e\delta^*\sqrt{R_L} \tag{8.4.28a}$$

$$\lambda = C_{ii}\sqrt{\xi} \tag{8.4.28b}$$

The solution of the linear system given by Eqs. (3.3.10b), (8.4.20), (3.3.10c) with coefficients $(s_1)_j$ to $(s_7)_j$ and $(r_2)_j$ given by Eqs. (8.4.22) and (8.4.23) for region 1 and by Eqs. (8.4.27) and (8.4.28) for region 2, subject to the wall edge boundary conditions given by Eqs. (8.4.25) and (3.3.14b) is obtained with the bordering algorithm described in subsection 3.3.4.

References

[1] Cebeci, T., Stewartson, K. and Williams, P.G., "On the Response of a Stagnation Boundary Layer to a Change in the External Velocity," SIAM J. Appl. Math., Vol. 36, p. 190, 1979.
[2] Krause, I., Hirschel, E.H. and Bothmann, Th., Die numerische Integration der Bewegungsgleichungen dreidimensionaler laminarer kompressibler Grenzschichten. Fachtagung Aerodynamik, Verling, 1968, DGLR-Fachzeitschrift, Band 3.
[3] Isaacson, E. and Keller, H.B., Analysis of Numerical Methods, John Wiley, NY, 1966
[4] Cebeci, T., "Unsteady Boundary Layers with an Intelligent Numerical Scheme," J. Fluid Mech., Vol. 163, pp. 129-140, 1986.
[5] Bradshaw, P., Cebeci, T. and Whitelaw, J.H., Engineering Calculation Methods for Turbulent Flows, Academic Press, NY, 1981.

9 Application of the Calculation Method: Airfoils in Incompressible Unsteady Flow

9.0 Introduction

The flowfield of an airfoil in an unsteady flow is considerably more complex than that of an airfoil in a steady flow. Of the many studies in aerodynamic flows, the dynamic stall, which is defined as the aerodynamic response of a fluid to a streamlined body pitching past its static stall angle of attack, is one of the most important practical problems and challenges both experimental and numerical investigations [1].

Early papers on dynamic stall describe experimental work extending back to the 1960's when aerodynamicists attempted to provide a basis to predict helicopter flight performance [2]. These studies began with an emphasis on the retreating blade stall of a helicopter rotor that occurs in high-speed forward flight, and they later shifted to the use of lift and drag augmentation during dynamic stall, which improves aircraft maneuverability [1]. They have also shown that the rapid pitching of an airfoil can delay the occurrence of the stall to angles of attack higher than that of the static stall angle and also produce a maximum lift significantly greater than that obtained in steady flow. For example, the dynamic stall process described by McCroskey et al. [3] and Carr and Chandrasekhara [4] and illustrated in Fig. 9.1, demonstrates the flowfield structure as well as the lift and pitching moment variations when an NACA 0012 airfoil performs one cycle of harmonic oscillation with $\alpha = 15^\circ + 10^\circ \sin\omega t$ at $\omega c/2u_\infty = 0.15$, $R_c = 2.5 \times 10^6$, where c is the chord length, u_∞, the freestream velocity, and R_c the chord Reynolds number. From this figure, it can be seen that the upward pitching of the airfoil suppresses the thick separation region on the upper surface to a thin flow reversal region at an angle of attack beyond its static stall angle, which is one of the important factors relating to the stall-delay phenomenon. This visualization of the effect of pitching rate on flow structure has also been clearly shown by Walker et al. [5]. The other factor

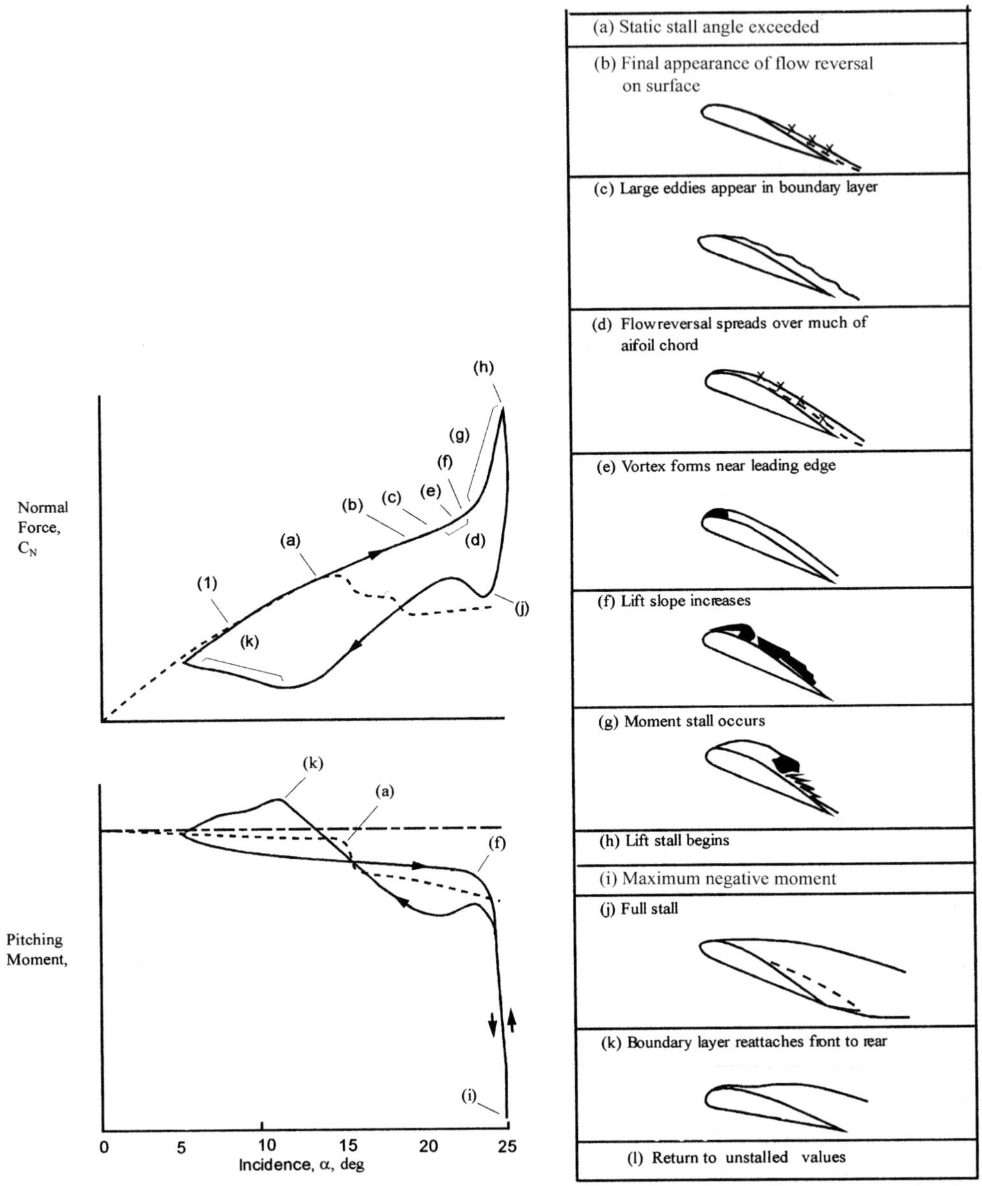

Fig. 9.1. Lift and moment versus angle of attack for an airfoil experiencing dynamic stall [4].

that causes the further delay of stall and significant augmentation of lift and pitching moment is the formation and shedding of a strong vortex, known as a dynamic stall vortex from the leading-edge region. Dynamic stall is not easy to reproduce by numerical methods and, although a conceptual simulation of the dynamic stall process by inviscid flow theory has been provided by Ham

[6], the initiation and termination of the leading-edge vortex sheddings were treated empirically and remained to be predicted by the viscous-flow theory. Due to the complexity of involving the viscosity effect, the early numerical solutions of the full Navier-Stokes equations for a laminar flow did not appear until the late 1970's (see, for example [7]). Since then, due to the improvement of algorithms and the increasing power of computers, the development of the numerical method for solving the full Navier-Stokes equations has progressed towards higher Reynolds numbers with turbulence models, higher Mach numbers and compressibility, lower pitch rates, and more complex airfoils [8–11]. Theoretically the full Navier-Stokes equations completely describe the fluid behavior within the flowfield, but their time-averaged forms require assumptions for the transition location and turbulence models. Numerical solution methods harbor many difficulties, such as the choice of numerical algorithm, type of grid and its density, boundary-layer resolution, and time-step size [1].

The calculation method described in Chapters 7 and 8 for unsteady flows is not a general method for predicting airfoil performance for a wide range of angles of attack and unsteady effects, including dynamic stall; it is limited to relatively small pitching rates and angles of attack less than those that correspond to dynamic stall. Since the stall process depends strongly on flow conditions which exist well before separation, as demonstrated in Chapter 5 for steady flows on airfoils, the simplicity and accuracy of the boundary-layer equations combined with an inviscid method in describing the boundary-layer development before stall, reveal the potential of developing a calculation method as an engineering tool for rapid prediction of the onset of stall. The Navier-Stokes equations will then be valuable for computing the flowfield after stall has begun.

Our knowledge of unsteady flows, particularly those for boundary-layer equations, is less sound than that of steady ones. As was discussed in Chapter 1, in the classical boundary-layer theory, the point of zero wall shear stress coincides with the point of separation, and the equations have singular behavior, but this is not the case in the unsteady theory. As described in [12, 13], for example, flow reversals can occur in unsteady flows without flow separation. It is useful to distinguish one from the other so that flow separation can properly be identified in an unsteady flow when the boundary-layer calculations are performed for a prescribed pressure distribution. It is also useful and necessary to explore the accuracy of the numerical method used to solve the time-dependent boundary-layer equations, especially in regions of flow reversal and separation.

In the following section, we consider a model problem for a thin oscillating airfoil and discuss the evolution of the unsteady boundary layers in the vicinity of its leading edge and the ability of the calculation method to deal with the movement of the stagnation point and with regions of reverse and separated flow. In Section 9.2, we discuss the application of the calculation method to several oscillating airfoils and airfoils subject to ramp-type of motion and compare the calculated results with experimental data.

9.1 Separation and Reattachment Near the Leading Edge of a Thin Oscillating Airfoil

The lift and drag characteristics of airfoils at moderate Reynolds numbers can be affected by separation bubbles which occur close to the leading edge as discussed in Chapter 5 and, at high angles of attack, can increase in size to cause stall. The added complexity of unsteady motion such as that associated with the rotor blades of helicopters implies that the flow characteristics are influenced by amplitude and frequency and that, in particular, the stall characteristics can be considerably modified. The investigations reported in [1–4] examined these effects over limited ranges of the parameters, and that of [4, 14] provides detailed information on the mechanism of dynamic stall of an oscillating airfoil. It appears that stall is associated with flow reversals in the unsteady boundary-layer and that these may translate downstream or upstream depending upon various parameters including the radius of the leading edge of the airfoil. At some stage in the cycle, stall occurs and is preceded by a vortex which forms close to the surface and is probably associated with a breakdown of the unsteady boundary-layer.

The above physical problems involve laminar, transitional and turbulent flow and their representation requires a calculation procedure that can provide accurate solutions to conservation equations in all regions of flow as well as appropriate transition and turbulence models. Here in this section we are concerned with the ability and accuracy of the calculation method of Chapter 8 to represent the regions of reverse flow and its use in examining the nature of solutions for parameters close to those associated with stall. The emphasis is on regions of flow close to the leading edge of a model problem corresponding to a thin oscillating airfoil with calculations performed (a) for a prescribed pressure gradient and (b) with interaction between solutions of the viscous and inviscid flow equations. With the configuration chosen, an analytical solution for the potential flow equations is available and the inviscid flow method of Chapter 7 is not needed.

9.1.1 Model Problem

Local regions of separated flow are common on thin airfoils at angle of attack and can be important particularly because of their association with the phenomenon of stall. For thin airfoils, the laminar boundary layer grows from the stagnation point and is subjected to an adverse pressure gradient near the leading edge, which causes separation and subsequent transition to turbulent flow.

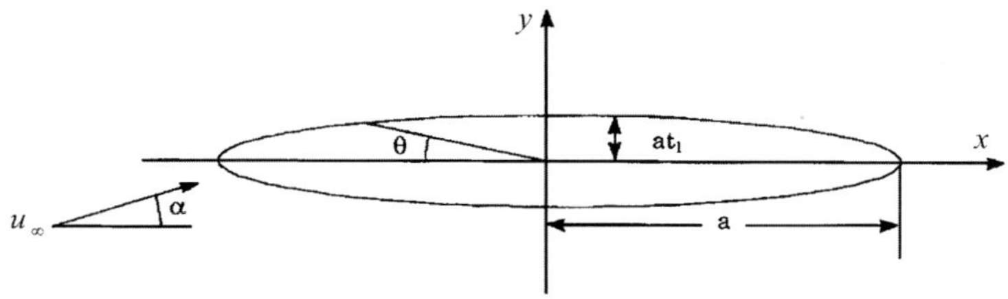

Fig. 9.2. Flow configuration for the model problem.

Reattachment occurs and a turbulent boundary-layer extends to the trailing edge. As the angle of attack is increased, the separation region moves towards the leading edge, and eventually the bubble of the separated flow bursts. Reattachment does not occur and stall sets in.

To study the leading edge separation problem on an airfoil, we consider an ellipse whose center is located at the origin and at an angle of incidence α in a uniform stream of velocity u_∞, and with θ denoting the polar angle as shown in Fig. 9.2. The ellipse is represented by

$$x = -a\cos\theta, \quad y = at_1\sin\theta, \quad -\pi \le \theta \le \pi \tag{9.1.1}$$

and inviscid flow theory provides the external velocity distribution

$$u_e^0 = u_\infty(1+t_1)\frac{\sin(\theta+\alpha)+\sin\alpha}{[t_1^2+(1-t_1^2)\sin^2\theta]^{1/2}} \tag{9.1.2}$$

for a thickness ratio, t_1 ($\equiv b/a$). In the region close to the leading edge, Eq. (9.1.1) becomes

$$x = -a\{1-\tfrac{1}{2}\theta^2[1+O(t_1^2)]\}, \quad y = at_1\theta[1+O(t_1^2)]$$

Neglecting terms where the relative error is $O(t_1^2)$, the ellipse can be approximated by the parabola

$$x = -a(1-\tfrac{1}{2}\theta^2), \quad y = at_1\theta \tag{9.1.3}$$

and, with

$$\xi = \frac{\theta}{t_1} \tag{9.1.4}$$

the relations given by Eq. (9.1.3) can be written as

$$x = -a(1-\tfrac{1}{2}\xi^2 t_1^2), \quad y = at_1^2\xi \tag{9.1.5}$$

Expanding Eq. (9.1.2) and discarding second-order terms leads to

$$\frac{u_e^0}{u_\infty(1+t_1)} = \frac{\xi+\xi_0}{(1+\xi^2)^{1/2}} \tag{9.1.6}$$

where $\xi_0 = 2\alpha/t_1$.

The surface distance s or the dimensionless distance ζ ($\equiv \frac{s}{a}$) required for the boundary-layer equations comes from Eq. (9.1.5) as

$$\zeta = t_1^2 \int_0^\xi (1+\xi^2)^{1/2} d\xi \tag{9.1.7}$$

The above model for the leading edge of a thin airfoil can be extended for an oscillating airfoil by writing Eq. (9.1.6) in the form

$$\overline{u_e^0}(\xi, t) = \frac{u_e^0}{u_\infty(1+t_1)} = \frac{\xi + \xi_0(1 + A\sin\omega^* t)}{(1+\xi^2)^{1/2}} \tag{9.1.8}$$

where A and ξ_0 denote parameters that need to be specified and ω^* is a dimensional frequency. The parameter ξ_0 can also be regarded as a reduced angle of attack. The term

$$\xi_0(1 + A\sin\omega^* t)$$

can be interpreted as an effective angle of attack, $\alpha_{\text{eff}}(t)$. With this definition, Eq. (9.1.8) becomes

$$\overline{u_e^0}(\xi, t) = \frac{\xi + \alpha_{\text{eff}}(t)}{(1+\xi^2)^{1/2}} \tag{9.1.9}$$

9.1.2 Initial Conditions

In order to solve the unsteady boundary-layer equations (8.0.1) and (8.0.2) subject to the boundary conditions given by Eqs. (8.0.3), initial conditions are needed in the (x, y) and (t, y) planes as discussed in Section 8.1. If steady-flow conditions prevail at $t = 0$, the initial conditions in the (x, y) plane can be obtained easily for both surfaces of the airfoil by solving the steady boundary-layer equations, Eqs. (3.0.2) and (3.0.3), for laminar flow. There is no problem with the initial conditions for these equations since the calculations start at the stagnation point where they admit similarity solutions.

The generation of the upstream boundary conditions for Eqs. (8.0.1) and (8.0.2) requires a special numerical procedure. Since the complete velocity-profile distribution on a previous time line is known, solutions can be determined on the next time line by an explicit method. If we wish to avoid stability problems, however, an implicit method is required. Then, however, the generation of a starting profile on the new time line becomes a problem.

In order to explain the problem further, it is instructive to see what happens to the stagnation point as a function of time. For this purpose let us consider Eq. (9.1.8). Since $\overline{u_e^0} = 0$ at the stagnation point, its location, ξ_s, based on the external streamlines is given by

$$\xi_s = -\xi_0(1 + A\sin\omega^* t) \tag{9.1.10}$$

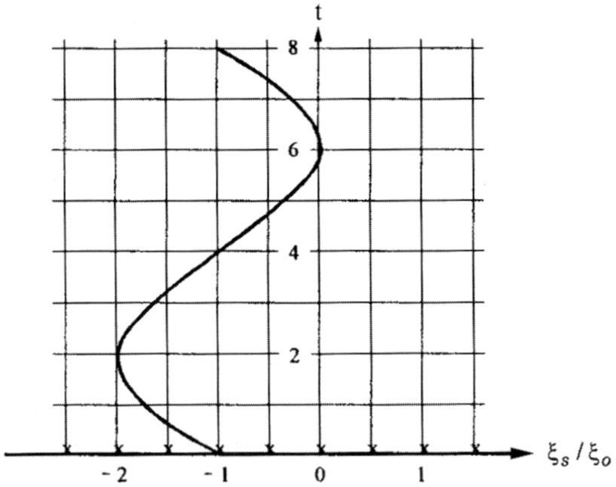

Fig. 9.3. Variation of stagnation point with time for one cycle according to Eq. (9.1.10) with $A = 1$, $\omega^* = \frac{1}{4}\pi$.

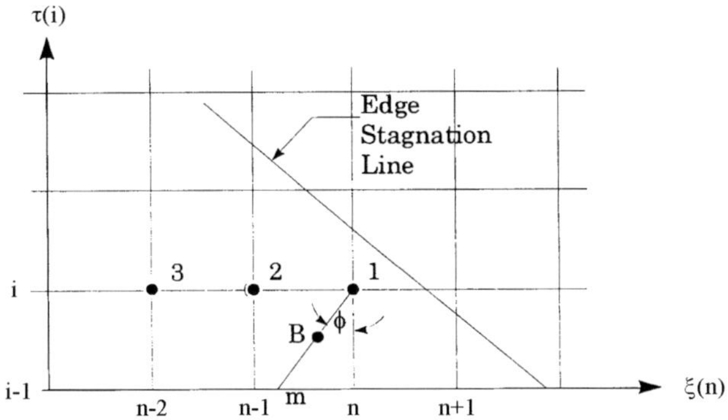

Fig. 9.4. Notation and finite-difference molecule for the characteristic box 2.

Figure 9.2 shows the variation of the stagnation point with time for one cycle according to Eq. (9.1.10), with $A = 1$, $\omega^* = \pi/4$. We see that when $t = 2$, the stagnation point ξ_s is at $-2\xi_0$, and when $t = 6$, it is at 0, etc. If ξ_s were fixed, we could assume that $u = 0$ at $\xi = \xi_s$ for all time and all y, but this is not the case. It is also possible to assume that the stagnation point is coincident with zero u-velocity for a prescribed time, but we should note that the stagnation point defined by Eq. (9.1.10) is based on the vanishing of the external velocity. For a time-dependent flow, this does not imply that the u-velocity must be zero across the layer at a given ξ-location and specified time; indeed flow reversals can occur owing to the movement of the locus of zero u-velocity across the layer and can cause numerical instabilities which require the use of special numerical schemes described below.

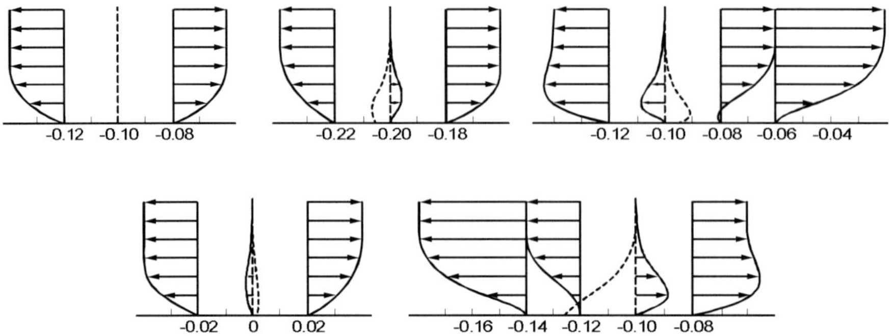

Fig. 9.5. Velocity profiles in the immediate neighborhood of the stagnation line at different times for $\omega = \pi/4$ and $A = 1$. The dashed lines indicate the locus of zero u-velocity across the layer.

To discuss the numerical scheme to obtain solutions on the next time-line in the (t, y) plane at a specified ξ-location, let us consider Fig. 9.4 and denote the edge stagnation line computed according to Eq. (9.1.8). The first velocity profile on either side of the edge stagnation point can be calculated iteratively by using the characteristic box scheme described in subsection 8.4.2. To explain this further, consider the point denoted by 1 on the grid of Fig. 9.4 and assume that the angle ϕ in Eq. (8.4.12b) and $\theta^{n-1/2,i-1}$ are known. The latter assumption decouples the continuity equation (8.4.9b) from Eq. (8.4.11) and reduces the system given by Eqs. (8.4.9a) and (8.4.11) to one given by Eqs. (8.4.9a) and (8.4.11) with u and v being the only unknowns. With the finite-difference approximations to Eqs. (8.4.9a) and (8.4.11) given by Eqs. (8.4.13) and (8.4.21), the matrices resulting from the application of Newton's method are 2×2 matrices. Once an approximate solution at point 1 is obtained, the system given by Eqs. (8.4.9a, b) and (8.4.11) is solved to obtain u_j and v_j at point 2 and θ_j at $i - 1/2, n - 1/2$. This procedure is also repeated at point 3 so that a new estimate for θ_j at B can be obtained from the computed values of θ_j at $(n-3/2, i-1/2)$, and $(n-1/2, i-1/2)$. The 2×2 system is then solved again to obtain an improved solution at point 1. The procedure is repeated until the solutions converge and experience has shown that two to three iterations are sufficient.

Figure 9.5 shows the computed velocity profiles in the immediate neighborhood of the stagnation region at different times when the external stagnation point varies with time according to Fig. 9.3. We note that for steady state the velocity profiles on either side of the stagnation point, defined by the vanishing of external velocity, behave "well" as expected. However, for unsteady flow, the stagnation point varies with time and flow reversals begin to develop in the velocity profiles. The changes in the behavior of the velocity profiles on either side of the vanishing u_e become more drastic as t increases.

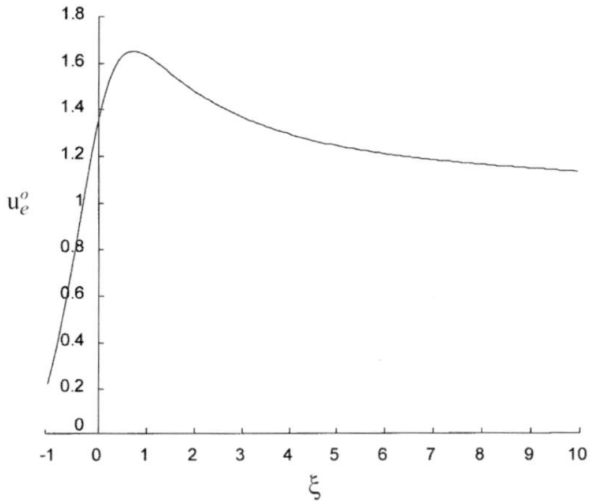

Fig. 9.6. Variation of u_e^0 with ξ.

9.1.3 The Question of Singularity on an Oscillating Airfoil

Before we examine the solutions of the boundary-layer equations for the flow on an oscillating airfoil which can give rise to extensive regions of flow reversal and separation, it is useful to examine the variation of external velocity distribution of the model problem for steady flow and discuss the studies conducted for this problem at several angles of attack [15].

Figure 9.6 shows a typical variation of the external velocity distribution, $\overline{u_e^0}(\xi, t)$ with ξ for a reduced angle of attack of $\xi_0 = 1.3$. As can be seen, the external velocity increases linearly from the stagnation point ($\xi = \xi_s$) to a maximum and then decelerates before it reaches a uniform value. In the region of deceleration, the flow may separate depending on the reduced angle of incidence ξ_0, and the uniform external velocity downstream of the leading edge allows the flow to reattach, provided that ξ_0 is not too large. The studies conducted in [15] showed that the solutions were well behaved and unseparated provided ξ_0 was less than 1.155. At higher angles separation occurred with an associated singularity which was overcome by the use of the inverse boundary-layer procedure described in Chapter 3, and results were obtained for small regions of separated flow. There is, however, a limiting size of separation bubble beyond which solutions could not be obtained and this may be related to the physical phenomenon of non-existence of large laminar separation bubbles as discussed in [16].

To study the question of singularity on an oscillating airfoil, calculations were made for three values of reduced frequency ω in the standard mode with upstream boundary conditions in the (t, y) plane and those in the (x, y) plane at $t = 0$ determined as described in subsection 9.1.2. With $\xi_0 = 1$, $A = -\frac{1}{2}$, dimensionless time τ defined by $\frac{u_\infty(1+t_1)}{at_1^2}t$ and the values of $\omega = 0.001$, 0.01

and 0.10, the maximum value of effective angle of attack, α_{eff} is sufficient to provoke separation with a strong singularity. For example, the maximum value of α_{eff} is 1.5 at $\omega\tau = 270°$ and the flow conditions closely resemble a steady separated flow at the smaller frequencies $\omega = 0.001$ and 0.01. Since the value of α_{eff} corresponding to steady flow separation is 1.155, we would expect the calculations to break down before $\omega\tau = 270°$ owing to the singularity. For the higher frequency case ($\omega = 0.10$), we expect the solutions to break down later than ($\omega\tau = 270°$) with flow reversals occurring in the range $270° < \omega\tau < 360°$.

The calculations were arranged parallel to those previously performed for a circular cylinder [17] and reported in [18]. Thus both the zig-zag and the characteristic-box schemes were used first with time and distance steps that were chosen arbitrarily and subsequently with values in agreement with the stability criterion. The results of Fig. 9.7 for $\omega = 0.10$ and $A = -\frac{1}{2}$ were obtained with the zig-zag box scheme of subsection 8.4.1 for a $\Delta\xi$ ($\equiv \frac{\Delta x}{a t_1^2}$) spacing specified such that $\Delta\xi_n = 0.01$ up to $\xi = 1.7$, $\Delta\xi_n = 0.005$ for $1.7 < \xi < 4$ and $\Delta\xi_n = 0.01$ for $4 < \xi < 8$; the time steps r_i were 10 degrees for $0 < \omega\tau < 260°$, 5 degrees for $260° < \omega\tau < 295°$, and 1.25 degrees for $295° < \omega\tau < 360°$ [17]. The calculations broke down at $\omega\tau = 310°$, indicating flow separation at this location.

Figure 9.7a shows that the variation of the displacement thickness, with R defined by $R = 2au_\infty/\nu$,

$$\tilde{\delta}^* = \frac{\delta^*}{at_1}\left[\frac{R(1+t_1)}{2}\right]^{1/2}$$

is generally smooth except in the neighborhood of $\xi = 2.12$ and for $\omega\tau = 308.75°$. The first sign of irregularity is the steepening of the slope of $\tilde{\delta}^*$ when $\omega\tau = 300°$, and a local maximum of $\tilde{\delta}^*$ occurs at $\xi = 2.12$ when $\omega\tau = 308.75°$. When the same results are plotted for a displacement velocity, $(d/d\xi)(u_e\tilde{\delta}^*)$ (Fig. 9.7b), we observe that the steepening of the displacement velocity near $\xi = 2.12$ is dramatic as the peak moves from $\xi = 2.125$ to 2.08 with $\omega\tau$ changing from 300° to 308.75°. It should be noted that the maximum value of displacement velocity moves towards the separation point with increasing $\omega\tau$ and the same behavior occurs for the circular cylinder as discussed in [16].

As shown in Fig. 9.7c, the wall-shear parameter f''_w shows no signs of irregularity for $\omega\tau \le 308.75°$ but a deep minimum in f''_w occurs near $\xi = 2.15$, i.e. near the peak of $\tilde{\delta}^*$. The calculations that led to the results in Fig. 9.7 were repeated with the characteristic-box scheme using the same coarse variations of r_i and $\Delta\xi_n$ and the results were identical with those obtained with the zig-zag scheme up to $\omega\tau = 280°$. At $\omega\tau = 282.5°$, the solutions of the zig-zag scheme were smooth and free of wiggles but those of the characteristic-box scheme exhibited oscillations in f''_w which led to their breakdown. The solutions with the zig-zag scheme, however, continued without numerical difficulties

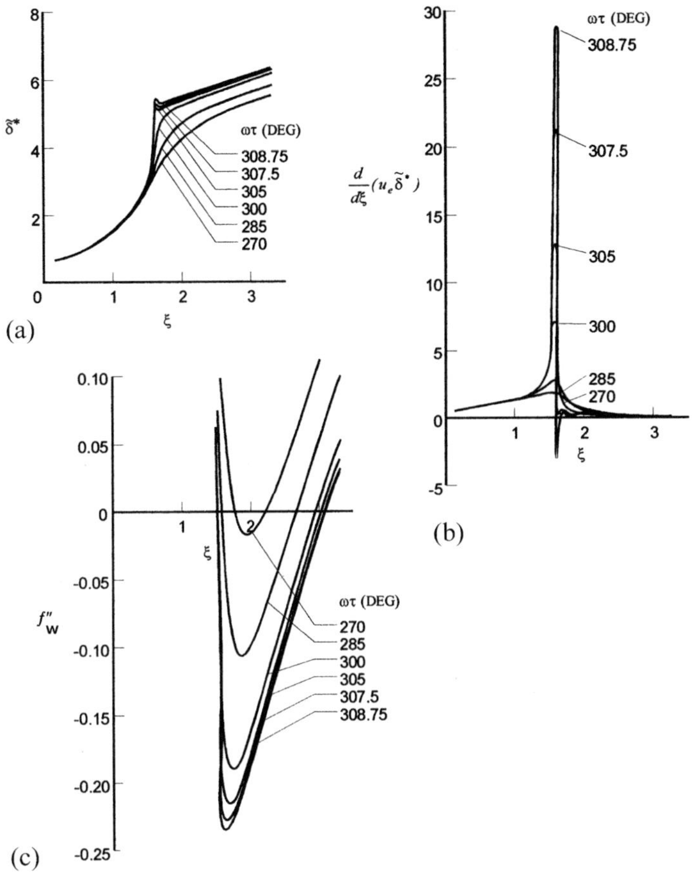

Fig. 9.7. Variation of (a) displacement thickness $\tilde{\delta}^*$, (b) displacement velocity $d/d\xi(u_e\tilde{\delta}^*)$, and (c) wall-shear parameter f''_w with ξ for the oscillating airfoil; $A = \frac{1}{2}$, $\omega = 0.1$.

until $\omega\tau = 310°$, where oscillations appeared and led to the breakdown of the solutions at the next time step.

The characteristic box was used subsequently with values of $\Delta\xi_n$ fixed as before and with values of r_i determined so that $\beta = \frac{\Delta\xi_1}{k_n}$ (see Fig. 8.3), is always less than unity, that is

$$\frac{r_i}{k_n}\tan\phi < 1 \tag{9.1.11}$$

We shall refer to this requirement as the stability criterion of the characteristic box.

The distribution of time steps determined in accord with this stability requirement is shown in Table 9.1. This procedure avoided the breakdown of the solutions and, as can be seen from Figure 9.8, the maximum value of β increases considerably with $\omega\tau$ so that the solutions required correspondingly smaller values of the time step. It is interesting to note that the wall-shear distributions

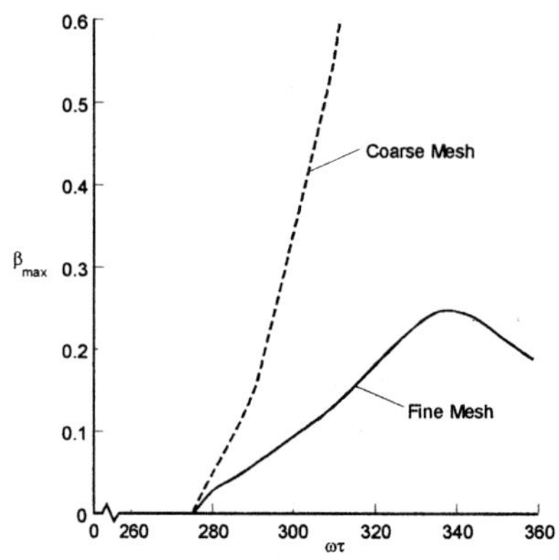

Fig. 9.8. Effect of the coarse and fine meshes on the variation of the stability parameter β with $\omega\tau$.

Table 9.1. The distribution of step sizes in $\omega\tau$ for $\omega = 0.1$ in accordance with the requirements of the stability parameter β.

$\omega\tau$	r_i
0–240°	10°
240–255°	5°
255–261°	3°
261–265°	2°
265–284°	1°
284–305°	0.5°
305–320°	0.25°
320–360°	0.5°

of Figure 9.9 are uninfluenced by the mesh at $\omega\tau = 280°$ and $310°$ but, for $\omega\tau > 310°$, the coarse mesh leads to large values of β and breakdown of the solutions.

Figure 9.10a shows the distributions of displacement thickness for values of $\omega\tau$ from 260° up to 360° and completes the cycle. The results up to 300° were identical with those of Fig. 9.7a with rapid increase of the displacement thickness corresponding to the increasing extent of flow reversal, as shown by the wall-shear distributions of Fig. 9.10b. It can also be seen from this figure that the maximum displacement thickness and minimum wall shear move upstream with increasing $\omega\tau$ for values of $\omega\tau$ up to 324.5°.

Figures 9.11 and 9.12 show the distributions of wall shear and displacement thickness for two smaller frequencies, $\omega = 0.01$ and 0.001. As expected, the critical value of the reduced angle that corresponds to separation is smaller than that for the higher frequency and closer to that of the steady state, $\xi = 1.16$. For $\omega = 0.01$, the breakdown of the solutions occurs at $\omega\tau = 226°$, which corresponds to an effective reduced angle of $\alpha_{\text{eff}} = 1.360$; for $\omega = 0.001$, the corre-

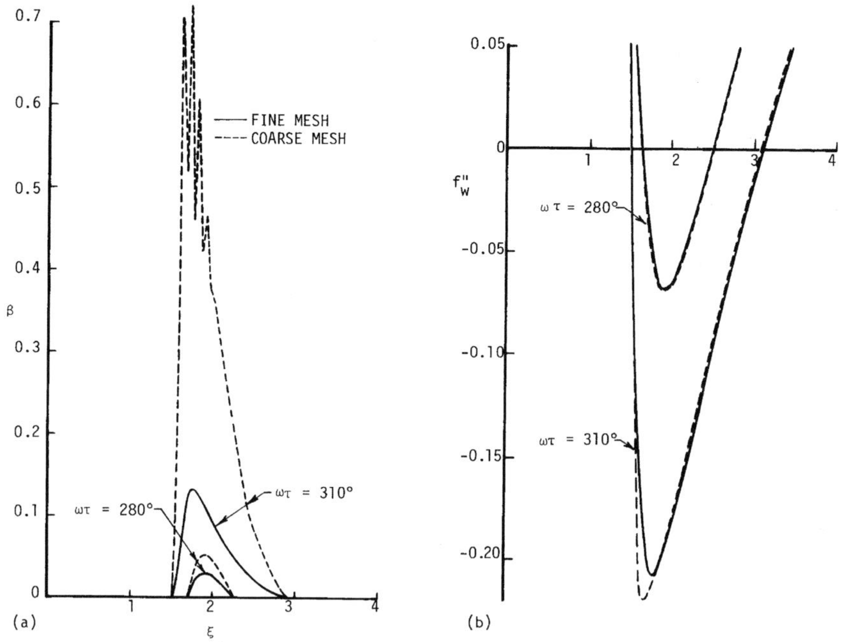

Fig. 9.9. Effect of the coarse and fine meshes on the variation of the (a) stability parameter β and (b) wall shear f_w'' with ξ.

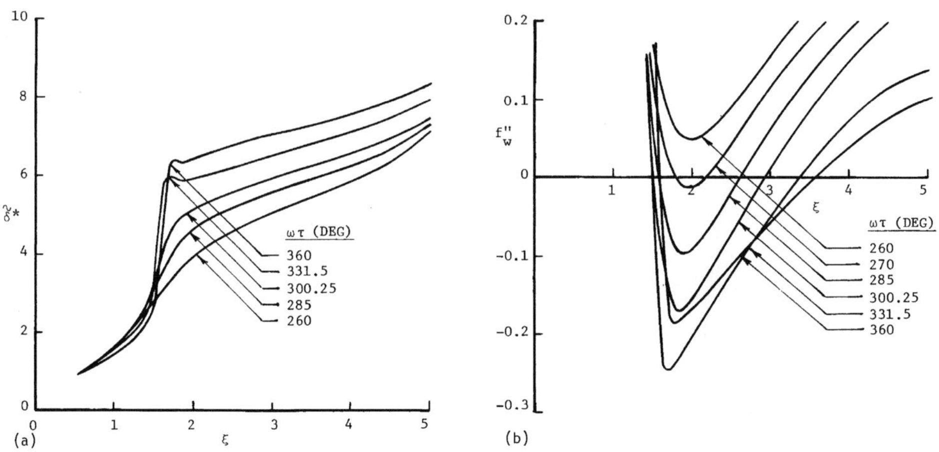

Fig. 9.10. Results obtained with the characteristic-box scheme for $\omega = 0.1$. Variation of (a) displacement thickness $\tilde{\delta}^*$ and (b) wall shear f_w'' with ξ.

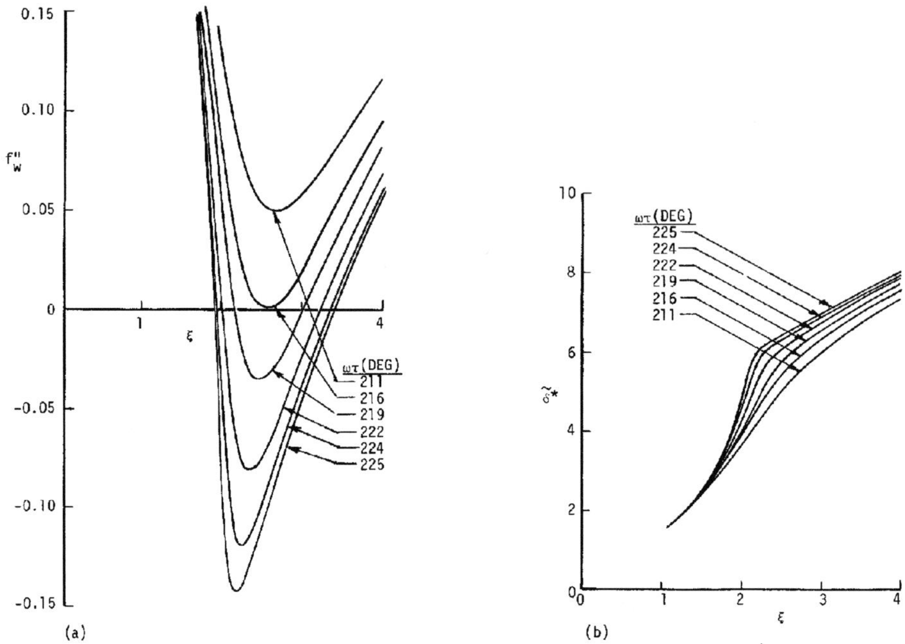

Fig. 9.11. Variation of (a) wall shear f''_w and (b) displacement thickness $\tilde{\delta}^*$ with ξ for $\omega = 0.01$.

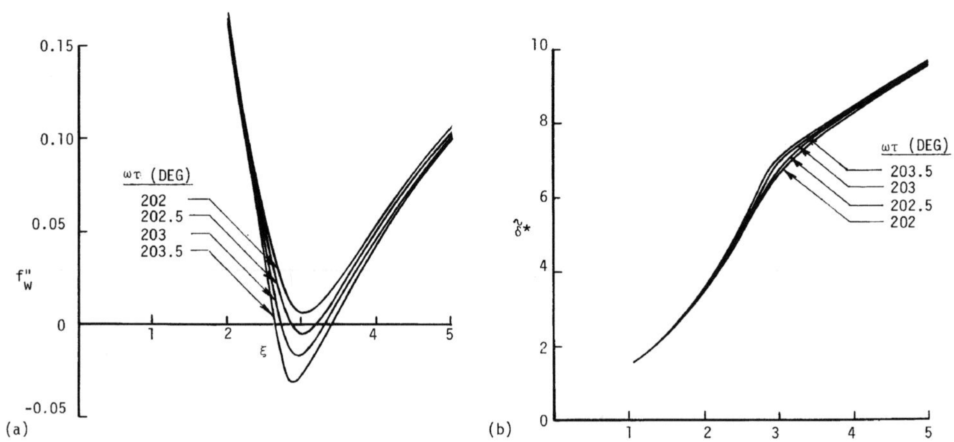

Fig. 9.12. Variation of (a) wall shear f''_w and (b) displacement thickness $\tilde{\delta}^*$ with ξ for $\omega = 0.001$.

sponding values are $\omega\tau = 204°$ and $\alpha_{\text{eff}} = 1.203$. We also note from Figs. 9.11a,b that the flow is a "little" unsteady even at these frequencies, and the solutions do not break down with the appearance of flow reversal, which increases in extent as ω changes from 0.001 and 0.01.

9.1.4 Interaction as an Answer to the Question of Singularity

The interaction procedure discussed in Section 8.5 has been applied to the model problem discussed in the previous subsection with the standard method. Before we present the results for this case, it is useful to point out that again we solve Eq. (8.2.6) subject to the boundary conditions given by Eq. (8.2.7a) and (8.2.11). The dimensionless distance Y and stream function $f(\xi, Y, \tau)$ defined in Eq. (8.2.5) and ξ defined in Eq. (8.2.4) are now defined by

$$\eta = \left[\frac{R(1+t_1)}{2t_1^2}\right]^{1/2} \frac{y}{a}, \quad \xi = \frac{x}{at_1^2}, \quad R = \frac{2au_\infty}{\nu} \tag{9.1.12a}$$

$$\psi = [(1+t_1)au_\infty \nu t_1^2]^{1/2} f(\xi, \eta, \tau) \tag{9.1.12b}$$

with f' and w in Eq. (8.2.6) now denoting

$$f' = \frac{u}{u_\infty(1+t_1)}, \quad w = \frac{u_e}{u_\infty(1+t_1)} \tag{9.1.13}$$

In contrast to the standard problem which makes the implicit assumption of infinite Reynolds number, the interaction requires a finite Reynolds number R and a thickness ratio t_1 in this case. For this reason, the calculations are performed for a specified value ε defined by

$$\varepsilon = \frac{1}{\pi t_1}\left[\frac{2}{R(1+t_1)}\right]^{1/2} \tag{9.1.14}$$

In all cases shown, the calculations made use of time steps determined by the characteristic box scheme in agreement with the stability requirement.

The present calculations were performed in the following way. For all values of time, with $\omega\tau$ ranging from 0° to 360°, the standard method and the leading-edge region procedure described in subsection 9.1.2 were used to generate initial conditions at a short distance from the leading edge, 0.5. With these initial conditions and for each value of $\omega\tau$, the inverse method was used to calculate the unsteady flow from $\xi = 0.5$ to 10, for the specified value of ε. In general three sweeps were required where flow reversal was encountered and a single sweep sufficed where it did not. It is to be expected that the value of ε will influence the number of sweeps and, since it is linked to physical parameters, will affect the singularity and the size of the bubble.

Figures 9.13 and 14 show the results for $\omega = 0.001$ and 0.01 with $\varepsilon = 3\times10^{-3}$. They are nearly the same as those obtained by the standard method and shown

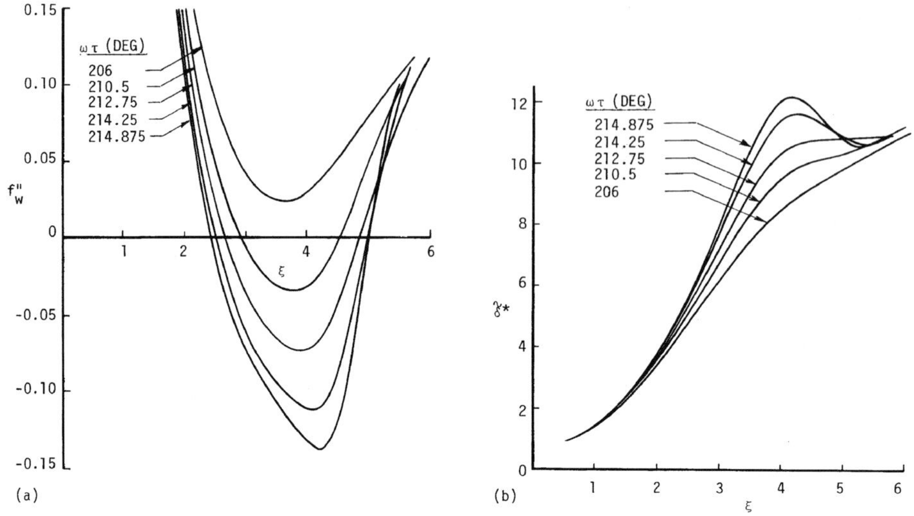

Fig. 9.13. Effect of interaction on the variation of (a) wall shear f''_w and (b) displacement thickness $\tilde{\delta}^*$ for $\omega = 0.001$ and $\varepsilon = 3 \times 10^{-3}$.

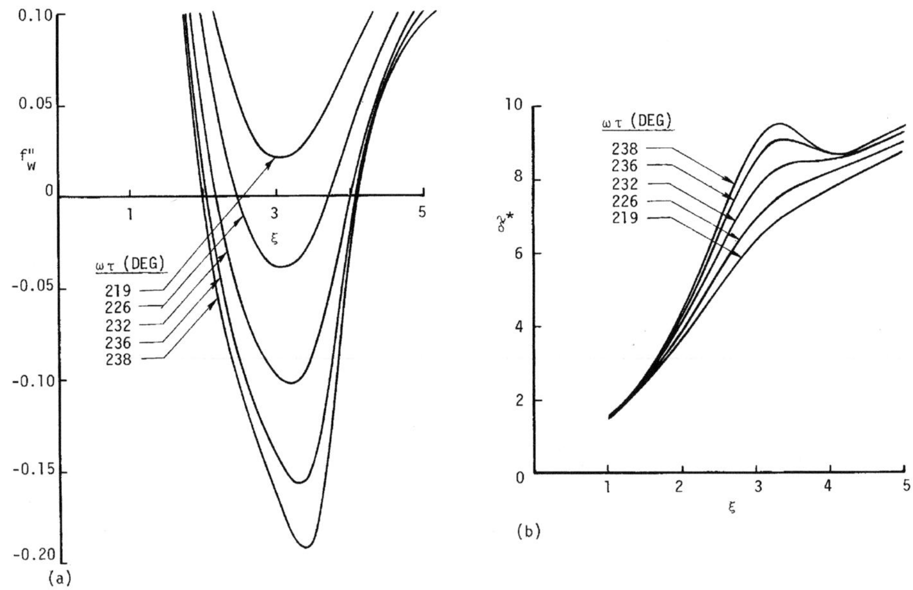

Fig. 9.14. Effect of interaction on the variation of (a) wall shear f''_w and (b) displacement thickness $\tilde{\delta}^*$ with ξ for $\omega = 0.01$ and $\varepsilon = 3 \times 10^{-3}$.

in figures 9.11 and 9.12 prior to flow reversal where the influence of the Reynolds number is small and increase after flow reversal. In the case of $\omega = 0.001$, for example, the standard method predicts flow reversal around $\alpha_{\text{eff}} = 1.19$ (see Fig. 9.12), and with interaction (Fig. 9.13) this effective angle is between 1.219 and 1.254. The maximum negative value of the wall-shear parameter f''_w obtained with the standard method is around -0.03 at $\alpha_{\text{eff}} = 1.199$ and may be compared with the maximum value of f''_w of -0.14 at $\alpha_{\text{eff}} = 1.286$ obtained with interaction. As expected, the interaction allows the calculations to be performed at higher angles of attack than those achieved with the standard method. For $\omega = 0.001$, the maximum α_{eff} for which calculations can be performed with the standard method is 1.199 with breakdown occurring at $\alpha_{\text{eff}} = 1.209$; the corresponding values with interaction are 1.286 and 1.287. Comparison of wall-shear results with both procedures and $\omega = 0.001$ indicates that the extent of the recirculation region $\Delta\xi$ is around 0.5 for the standard case, and around 2.5 for the interactive case. The solutions do not have a singularity in the former case but do contain flow reversals, and this suggests that time-dependent flows can be calculated without using an inverse procedure. As the angle of attack exceeds $\alpha_{\text{eff}} = 1.199$ for $\omega = 0.001$, a singularity develops and requires an inverse procedure as in two-dimensional steady flows. This procedure allows the calculation of larger regions of reverse flow where the flow is now separated.

We see a similar picture with the greater unsteadiness corresponding to $\omega = 0.01$, for which the standard method allows calculations up to an effective angle of attack of 1.354 (Fig. 9.11a), a value considerably higher than 1.199 obtained at $\omega = 0.001$. The first flow reversal occurs shortly after $\alpha_{\text{eff}} = 1.294$ and breakdown occurs at $\alpha_{\text{eff}} = 1.360$ with maximum negative wall shear values of -0.14 at $\alpha_{\text{eff}} = 1.354$ and -0.035 at $\alpha_{\text{eff}} = 1.315$. The extent of the maximum reverse-flow region is now 1.5, considerably larger than $\omega = 0.001$, and indicates that the more unsteady nature of the flow produces a bigger region of reverse flow free from singularities. For this value of ω, the interactive scheme increases the value of α_{eff} for which solutions can be obtained to 1.424 with breakdown occurring shortly after this value, at 1.428 (see Fig. 9.14). The first flow reversal occurs after $\alpha_{\text{eff}} = 1.315$ with maximum negative wall shear equal to -0.19 at $\alpha_{\text{eff}} = 1.424$, and the extent of the recirculation region has now increased by about 30%. Comparison of maximum wall shear values f''_w at a similar value of α_{eff} indicates that those computed with the interactive scheme are lower than those with the standard scheme so that, for example, the interactive scheme gives $(f''_w)_{\max} = -0.04$ at $\alpha_{\text{eff}} = 1.36$ compared with -0.14 at $\alpha_{\text{eff}} = 1.354$ with the standard method (figure 11a).

Figure 9.15 shows that the size of the reverse-flow region increases with Reynolds number but the effective angle of attack for which solutions can be obtained is only slightly reduced, changing from 1.428 for $\varepsilon = 3\times10^{-3}$ to around

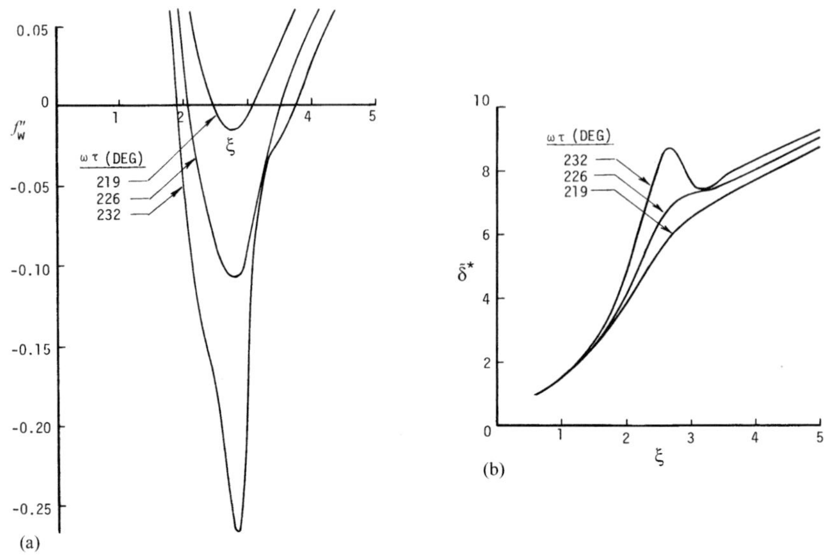

Fig. 9.15. Effect of interaction on the variation of (a) wall shear f''_w and (b) displacement thickness $\tilde{\delta}^*$ with ξ for $\omega = 0.01$ and $\varepsilon = 10^{-3}$.

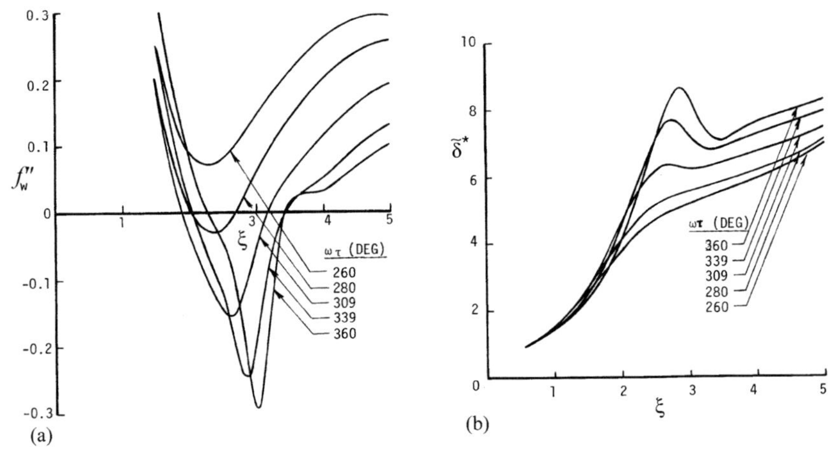

Fig. 9.16. Effect of interaction on the variation of (a) wall shear f''_w and (b) displacement thickness $\tilde{\delta}^*$ with ξ for $\omega = 0.1$ and $\varepsilon = 3 \times 10^{-3}$.

1.415 for $\varepsilon = 10^{-3}$. It is interesting to note that the interactive solutions do not have any flow reversal at $\alpha_{\text{eff}} = 1.315$ with $\varepsilon = 3 \times 10^{-3}$.

Figures 9.16 and 9.17 show the results for $\omega = 0.1$ with values of ε of 3×10^{-3} and 10^{-3} and they are again similar to those obtained by the standard method, as shown in Fig. 9.10, prior to flow reversal where the influence of Reynolds number is small. After flow reversal, the differences between the results obtained

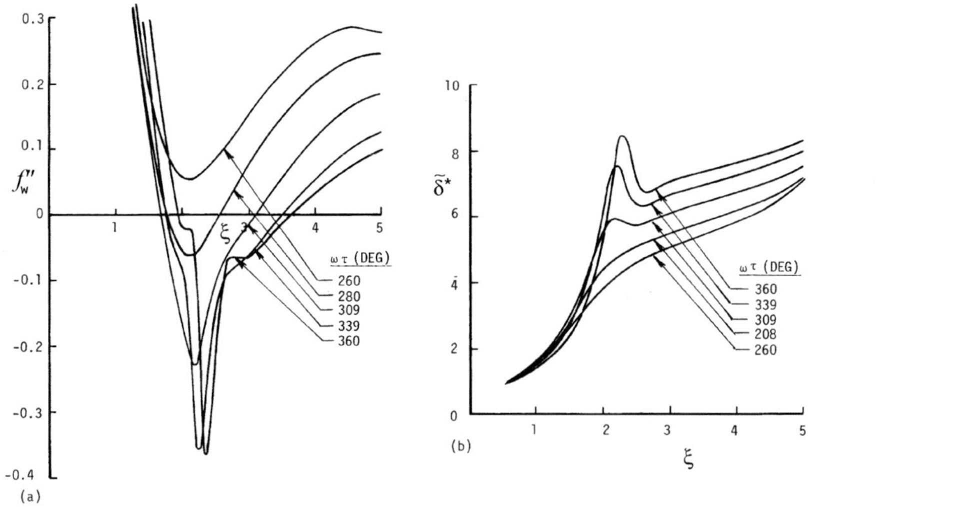

Fig. 9.17. Effect of interaction on the variation of (a) wall shear f''_w and (b) displacement thickness $\tilde{\delta}^*$ with ξ for $\omega = 0.1$ and $\varepsilon = 10^{-3}$.

with the standard and interactive methods increase as the Reynolds number decreases. It is clear that the solutions are free from the numerical "wiggles" encountered when the stability criterion was not met.

Comparison of results obtained at the two Reynolds numbers for $\omega = 0.1$ indicates that the interaction does not reduce the maximum negative value of the wall-shear parameter as it did with lower frequencies. For example, f''_w at $\omega\tau = 360°$ is around -0.19 with the standard scheme, and around -0.30 at $\varepsilon = 3 \times 10^{-3}$, and around -0.35 at $\varepsilon = 10^{-3}$ with the interactive method. The maximum value of negative wall shear calculated with interaction is considerably greater than its corresponding value obtained with the standard method at the end of one complete cycle. Furthermore, the behavior of the wall shear is not monotonic without interaction so that, for example, f''_w reaches a maximum value equal to -0.25 around $\omega\tau = 331°$ and then decreases to -0.195 at $\omega\tau = 360°$. With interaction this is not the case, with the continuously increasing with $\omega\tau$.

The results of figures 9.10, 9.16 and 9.17 are for an unsteady flow and they are free from singularities. For this reason, even though the results in the reverse-flow region are different, owing to the Reynolds number effect, the extent of the reverse-flow region is essentially the same and is consistent with the results obtained at lower frequencies in the absence of flow separation even though the extent of the reverse-flow region is reduced at the lower Reynolds numbers.

Since the calculations began at $\omega\tau = 0°$ with solutions obtained by solving steady-state equations, it was necessary to confirm the extent of their influence. As a consequence, calculations were performed for a second cycle and gave

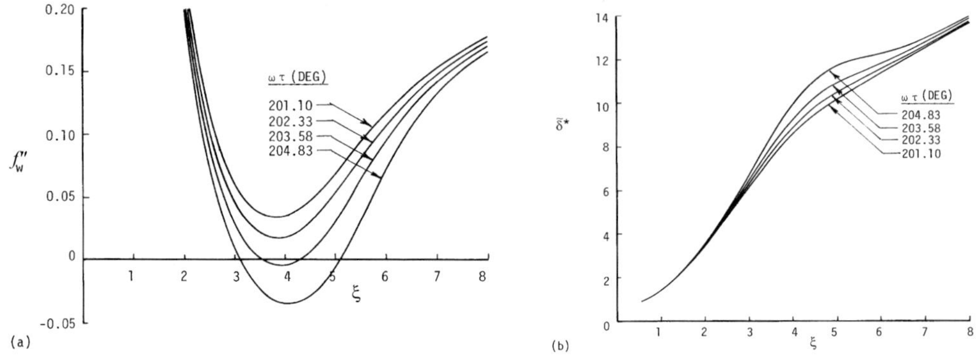

Fig. 9.18. Effect of interaction on the variation of (a) wall shear f''_w and (b) displacement thickness $\tilde{\delta}^*$ for a steady flow at $\varepsilon = 3 \times 10^{-3}$.

results at $\omega\tau = 720°$ that were identical with those at $\omega\tau = 360°$, confirming that the flow is cyclic. Examination of the results showed that the influence of the initial conditions die out rapidly and have no influence on the solutions presented here.

The results obtained with $\omega = 0.001$ can usefully be compared with the steady-state results of [15] shown in Fig. 9.18. We might expect that the small unsteadiness associated with this frequency will lead to results very similar to those of steady state. Inspection of Figs. 9.12 and 9.18 show that, although this is correct in general terms, the answers are quantitatively different. As can be seen, the maximum effective angle at which solutions can be obtained is greater in the unsteady case by some 7%. There are differences in the two calculation procedures but it is unlikely that they are responsible for this difference. On the other hand, it is possible that the difference in the negative wall-shear values may have been influenced by the use of the FLARE approximation in the steady-state solutions. Nevertheless, the unsteady nature of the flow with $\omega = 0.001$ is clear, in spite of this very low reduced frequency.

9.2 Steady and Unsteady Airfoil Flows

We now apply the calculation method of Chapters 7 and 8 to oscillating airfoils and airfoils subject to a ramp-type motion. As in steady flows, we again compute the inviscid velocity distribution with a panel method, this time with the method described in Chapter 7, and solve the boundary layer equations with the method described in Chapter 8 for a specified transition location. The calculations again involve sweeps both on the airfoil and its wake. After each sweep, the displacement thickness is updated so that the upstream and downstream viscous effects can be included in the next sweep. In addition to

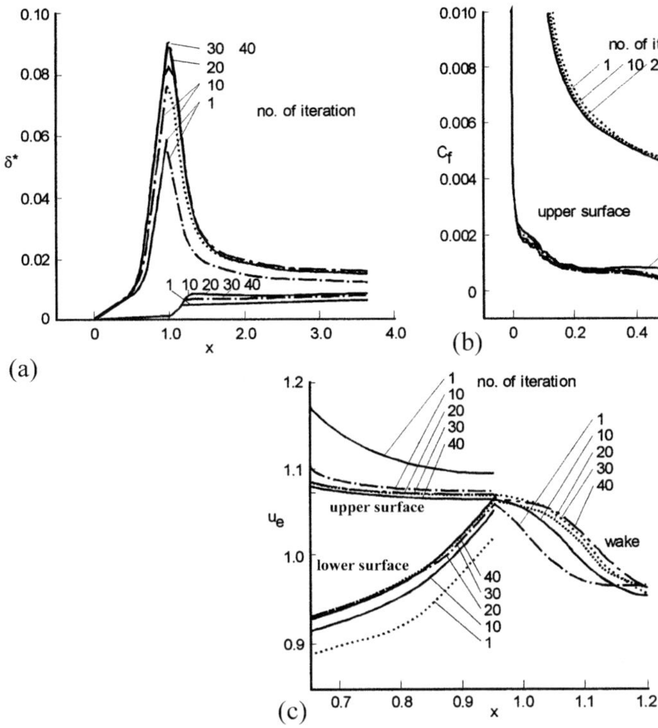

Fig. 9.19. The variation of the (a) displacement thickness, (b) local skin-friction coefficient, and (c) boundary-layer edge velocity (only the trailing edge regions are shown) distributions with iteration number, — steady, $\alpha = 15°$, $R_c = 2 \times 10^6$.

improving the boundary-layer calculations in this way, the viscous effects on the pressure distribution are also accounted for via the blowing velocity in the panel method. Moreover, for the flow over a lifting body, the Kutta condition and wake vorticity effects are related to the pressure distribution on the body surface.

The calculations start with the steady-state conditions and upon the convergence of the solutions, the same iterative procedure shifts to the next time step with the solutions at previous time as initial guesses. A typical evolution of the solutions during the invisicd-viscous iteration procedure is shown in Figs. 9.19 and 9.20. The calculations were performed for the Sikorsky SSC-A09 airfoil section at a chord Reynolds number of 2×10^6 and an angle of attack of 15° which is beyond that of maximum lift [19]. In order to improve the rate of convergence, the calculations began with the displacement thickness and blowing velocity distributions obtained from a converged solution at an angle of attack of 14° so that the solutions indicated by iteration 1 in Fig. 9.19 include viscous effects. Figure 9.19a shows the upper surface, lower surface and wake displacement thickness distributions. The displacement thickness on the lower surface

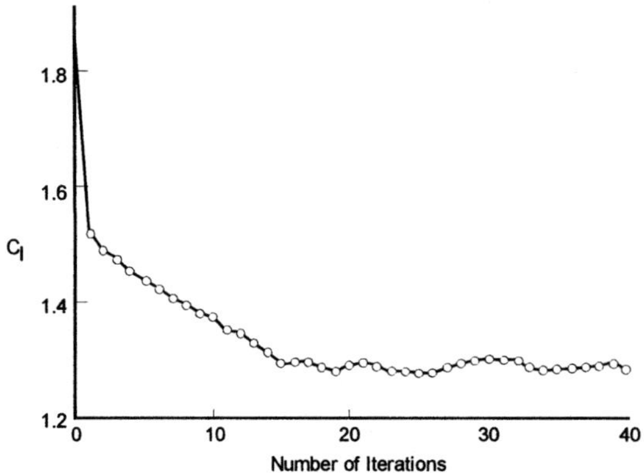

Fig. 9.20. The variation of lift-coefficient with iteration number, — steady flow, $\alpha = 15°$, $R_c = 2 \times 10^6$.

and for the forward 0.4 of chord on the upper surface, do not change with the iterations, whereas the trailing-edge displacement thickness on the upper surface increases by 50 percent during the computation. Figure 9.19b illustrates the change in local skin-friction coefficients with iteration number on the two surfaces; the only important changes are in the vicinity of separation. The variation of the velocity distribution, u_e, with iteration is shown in Fig. 9.19c, and it should be noted that the discontinuities in the u_e and δ^* distributions at the trailing edge are gradually eliminated as the iteration number increases.

The above calculations were reported in [19]. In general, twenty inviscid-viscous iterations were required for convergence of the solutions involving stall, and fewer iterations were sufficient where stall did not occur. The variation of lift coefficient with iteration is shown in Fig. 9.20, and the large change in the first iteration is due to the initial guess.

As discussed in Chapter 5, the interactive boundary-layer calculations can be performed for an airfoil with and without consideration of its wake. Before we present the results for an airfoil operating under unsteady flow conditions, as we did in steady flows, we consider the wake effect on the solutions. To this end, we examine the Sikorsky SSC-A09 airfoil undergoing a harmonic oscillation

$$\alpha = 5° + 7° \sin \omega t$$

at a very low reduced frequency, $k\,(\equiv \frac{\omega c}{2u_\infty}) = 1 \times 10^{-5}$. The lift coefficients of Fig. 9.21a show that for a chord number, R_c, of 2×10^6, the roles of wall boundary layer and viscous wake are opposite to that the former reduces the lift coefficient while the latter increases it, and the magnitudes of the effect increase at high angles of attack. Figure 9.21b shows the displacement thicknesses on the upper and lower surfaces at the trailing edge for one cycle of motion, and Fig. 9.21c shows the displacement thickness distribution on the upper surface

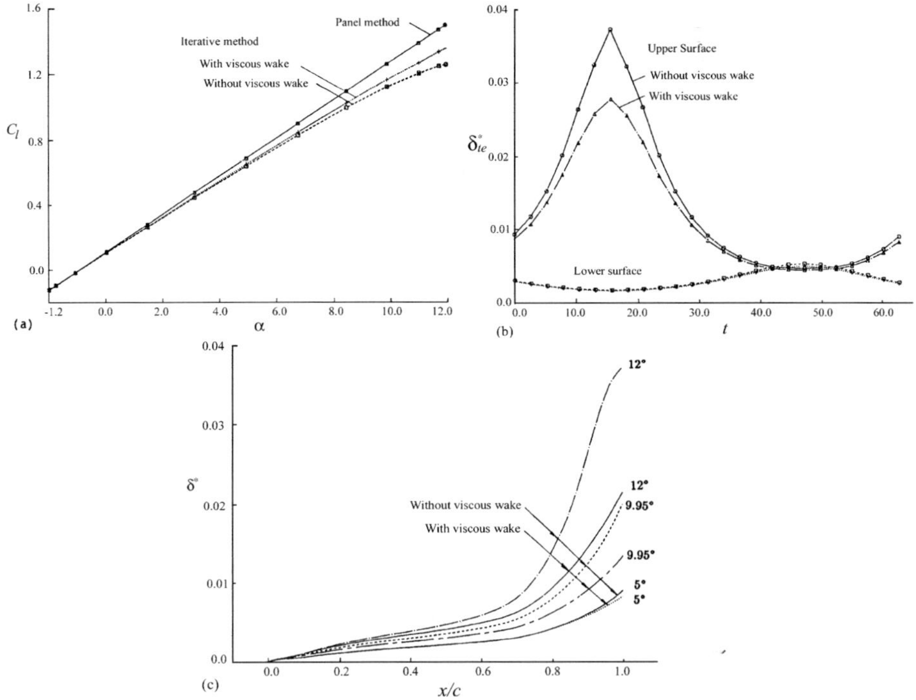

Fig. 9.21. The effect of viscous wake on (a) lift coefficient and (b) displacement-thickness distributions at trailing edge and (c) displacement-thickness distributions on the upper surface.

at various angles of attack. Both figures show that the viscous wake reduces the displacement thickness which is built up by the viscous effects in the wall boundary layer.

The study of the viscous wake effects in the slow motion cases, as described above, provides a quantitative standard for evaluating the viscous wake effects in more severe unsteady flows. When the airfoil performs slow movements, the vorticities shed to the wake are weak and the wake, due to the viscosity, behaves like a distribution of sinks along the dividing line. Increasing the severity of the unsteady motion increases the strength of the vorticities shed to the wake. The wake of a real flow over an airfoil can, therefore, be described as including vorticities and sinks which mix and interact with each other.

To model the unsteady wake by a viscous-inviscid interactive method, some assumptions are necessary. In the study reported in [19], it was assumed that the two elements of the unsteady wake model, vorticities and sinks, are distinct and do not influence each other. This means that the vorticities shed via the potential flow go downstream according to their own local velocities without the influence of sinks, and the sinks computed from the viscous wake distribute on the instantaneous wake dividing line, the location of which is evaluated without the influence of vorticities. This model satisfies the steady-flow condition as the unsteady motion approaches zero and should capture most of the characteristics of the wake as the unsteady motion of the airfoil increases.

9.2.1 Results of Unsteady Flows

The accuracy of an interactive method for simulating the unsteady flow depends on several factors which include the inviscid-flow method, the modification of the inviscid-flow method to accept viscous effects, the boundary-layer scheme as well as the numerical scheme used to solve the equations, and the coupling procedure between the inviscid- and viscous-flow calculations. The results discussed in Chapter 5 show that these factors are properly represented in the steady-flow calculations so that the steady-flow model can now be used as a base for extension to unsteady-flow calculations.

The overall features of a subsonic flow over an airfoil executing an unsteady motion are primarily characterized by the extent of flow reversal (or separation). For testing the performance of the method, calculations are performed in the order of increasing levels of flow complexity, starting first with flows without, and then with reversal and in subsection 9.2.2 are followed by flows with substantial regions of flow separation.

Flows Without Reversal Region

We again consider the same Sikorsky airfoil in which the change in angle of attack takes place according to

$$\alpha(t) = 5^\circ + 5^\circ \sin \omega t$$

at three reduced frequencies, $k = 1 \times 10^{-4}$, 0.1 and 0.5 for a chord Reynolds number of 2×10^6. The interactive flow calculations are performed with the unsteady panel method of Chapter 7 and the inverse boundary-layer method described in Section 8.5.

The results with the interactive method shown in Fig. 9.22 were first obtained with the inverse boundary-layer calculations performed for steady flow calculations. Even though at each time the inviscid flow was calculated with the unsteady panel method, the viscous effects were calculated with the boundary-layer equations in which the time-dependent terms, $\partial u/\partial t$ and $\partial u_e/\partial t$ were neglected. This quasi-steady model is valid when

$$\frac{\partial u}{\partial t} \quad \text{and} \quad \frac{\partial u_e}{\partial t} \ll u_e \frac{\partial u_e}{\partial x}$$

Repeating the calculations in which the viscous effects are computed with the unsteady boundary-layer equations and comparing the results with those obtained with the quasi-steady model showed that the solutions of both interactive methods were identical (Fig. 9.22).

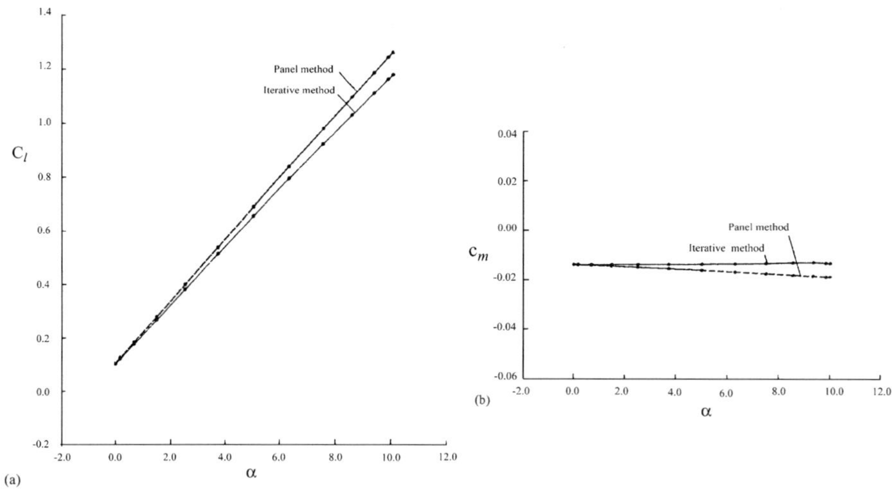

Fig. 9.22. Variations of (a) lift and (b) pitching moment coefficients with angle of attack for the SSC-A09 airfoil. $\alpha(t) = 5^\circ + 5^\circ \sin \omega t$, $k = 10^{-4}$.

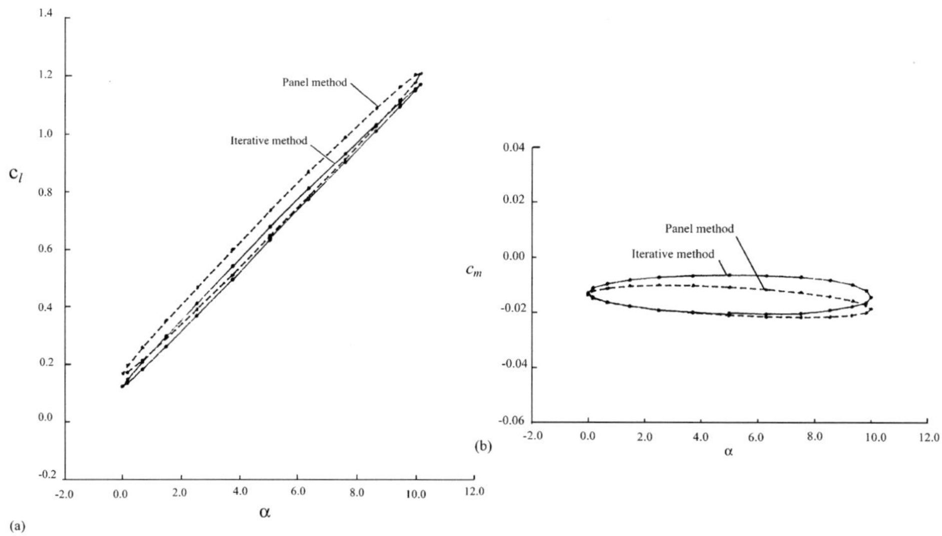

Fig. 9.23. Variations of (a) lift and (b) pitching moment coefficients of SSC-A09 airfoil with angle of attack. $\alpha(t) = 5^\circ + 5^\circ \sin \omega t$, $k = 0.1$.

Figures 9.23 and 9.24 show the lift and pitching moment coefficient curves for the other two frequencies, all computed with viscous effects obtained from the solutions of the unsteady boundary-layer equations. We can see from the results that the hysteresis effects increase with the increase of the reduced frequency. Since the boundary-layer calculations started from steady state and

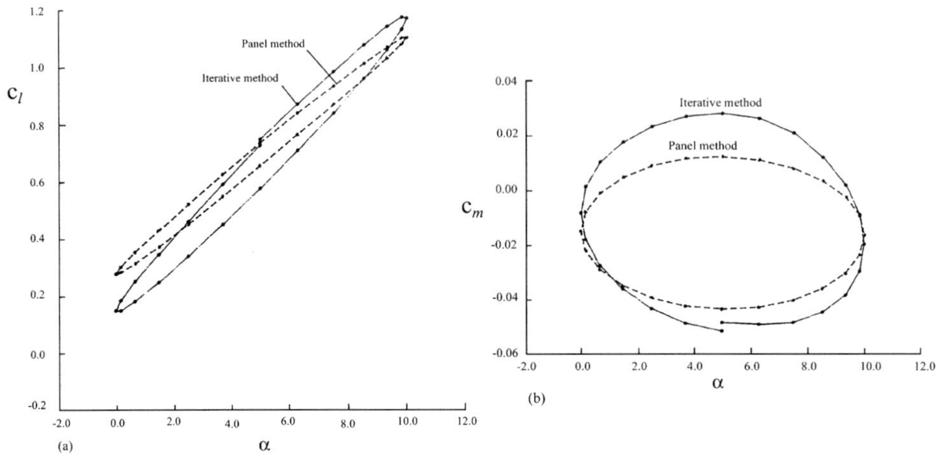

Fig. 9.24. Variations of (a) lift and (b) pitching moment coefficients of SSC-A09 airfoil with angle of attack. Unsteady. $\alpha(t) = 5° + 5° \sin \omega t$, $k = 0.5$.

suddenly joined the inviscid flow calculation, the viscous-inviscid results show a little discontinuity at the beginning and the end of the cycle. One very interesting aspect of the lift coefficient curve is that at $k = 0.1$ the lift coefficient of the upward stroke is lower than that of the downward stroke, but it is just opposite at $k = 0.5$. This phenomenon can be explained by comparing the pressure coefficient distributions in the upward and downward strokes at a certain angle of attack ($\alpha = 7.5°$ purely inviscid flow, for example), as shown in Fig. 9.25 for different frequencies. At $k = 10^{-4}$, no hysteresis effects appear and the pressure-coefficient distributions of upward and downward strokes are identical to the steady one. Increasing k to 0.1, the hysteresis effects appear mainly on the fore part of the airfoil. The pressure peak at $\alpha = 7.5° \uparrow$ is lower than the corresponding steady one, whereas that at $\alpha = 7.5° \downarrow$ is higher, so that the lift coefficient in the downward stroke is higher than that in the upward stroke. Increasing k generates more unsteady motion of the airfoil aft (since it is farther from the pivot point) so that the effects on the pressure distribution there are more enhanced, as shown in Fig. 9.25c, for $k = 0.5$. From the pressure distribution, we can see that in the downward stroke the negative contribution to the lift coefficient in the aft part cannot be balanced by the positive contribution from the fore part so that the c_l in the downward stroke is lower.

When an airfoil performs an unsteady motion, there is energy transfer between the airfoil and its surrounding fluid, and this energy transfer is indicated by the pitching moment. The pitching moment at $k = 0.1$ within one cycle of motion is shown in Fig. 9.23b. The negative pitch moment in the upward stroke indicates that the airfoil motion is against the aerodynamic moment induced by the surrounding fluid, whereas in the downward stroke the aerodynamic moment

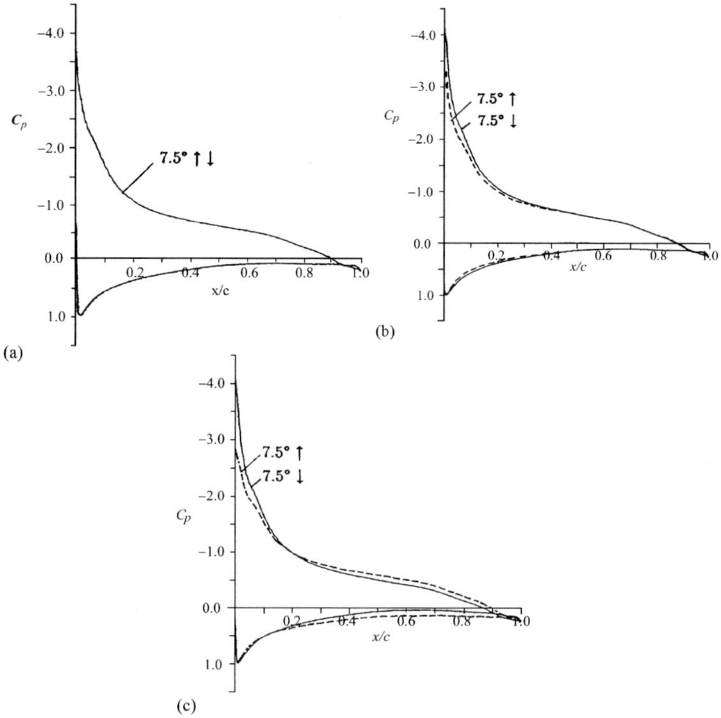

Fig. 9.25. The inviscid-flow pressure-coefficient distributions for the SSC-A09 airfoil at $\alpha = 7.5^\circ$ of the harmonic oscillation, $\alpha = 5^\circ + 5^\circ \sin \omega t$, with $k =$ (a) 10^{-4}, (b) 0.1, and (c) 0.5.

is favorable to the airfoil motion. Therefore, the energy is transferred to and absorbed from the fluid in the upward and the downward strokes, respectively. The area enclosed by the pitch-moment curve indicates the net energy transferred to the fluid in each cycle of harmonic oscillation. The viscosity enhances the hysteresis effects, Fig. 9.26, so that more energy is required to execute a cycle of motion. In the very slow pitching case, $k = 10^{-4}$, both the inviscid and the viscous-inviscid interactive curves show almost no hysteresis effects, Fig. 9.23b, so that the energy transferred to the fluid in one cycle is negligible. At $k = 0.5$, the pitching moment in most of the downward stroke becomes positive, which indicates that the airfoil motion is against the aerodynamic moment not only in the upward stroke but also in most of the downward stroke. In this case, the viscosity also increases the hysteresis effects and the energy transferred to the surrounding fluid in one cycle of motion.

The boundary-layer development can be represented by the displacement thickness distributions; some features of them are displayed in Figs. 9.26 and 9.27. The displacement thickness distributions on the upper and lower sides of the airfoil and wake at angles of attack $\alpha = 5^\circ$, 7.5° and 10° in the upward

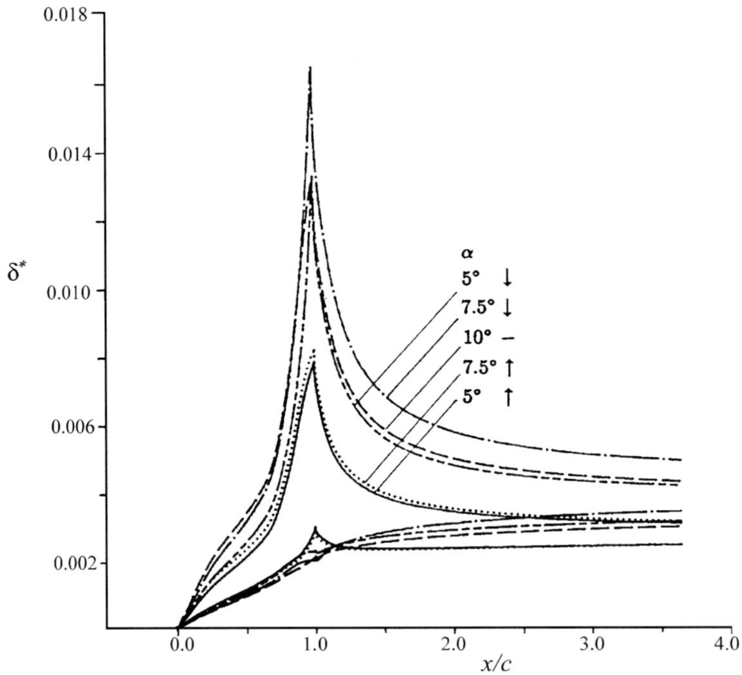

Fig. 9.26. Comparison of the displacement-thickness distributions of the SSC-A09 airfoil in the upward and downward strokes.

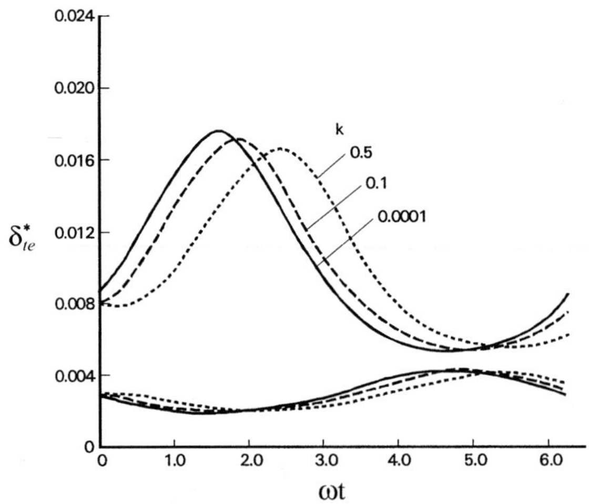

Fig. 9.27. The effect of reduced frequency on the trailing-edge displacement thickness of the SSC-A09 airfoil.

and downward strokes are shown in Fig. 9.26 for $k = 0.5$. At a certain angle of attack, the upper-side displacement thickness in the downward stroke is higher than that in the upward stroke. The influence of the reduced frequency on the displacement thickness distribution can be clearly represented by the trailing-edge value, as shown in Fig. 9.27. The maximum displacement thickness shifts to the downward stroke as the reduced frequency increases, as in steady flows discussed in the previous section.

Flows with Reversal Region

In the upward stroke of the harmonic oscillation, the pitch rate decreases when the airfoil approaches the maximum angle. At high angles of attack, decreasing the pitch rate speeds up the rate of expansion of the trailing-edge reversal or separation region. To suppress this region and delay the occurrence of stall, a high pitch rate must be maintained at high angles of attack. To study the behavior of an airfoil under these conditions we consider the same airfoil executing a ramp-type motion from 5° to 16° with a constant pitch rate of $A(\dot{\alpha}c/u_\infty) = 0.02$.

In Fig. 9.28, the lift-coefficient curves computed by using the unsteady and quasi-steady approaches for $A = 0.02$ are compared with the solution computed by the quasi-steady approach for $A = 10^{-5}$. The quasi-steady solution at $A = 0.02$ has a higher maximum lift at higher angle of attack than that at $A = 10^{-5}$. Due to the retarding effects from the lower angle of attack and smaller streamwise pressure gradient, the unsteady-flow solution shows no stall within the angles of attack computed. Figures 9.28b and 9.28c display the displacement thickness distributions for two angles of attack. At an angle of attack before static stall, $\alpha = 13°$, the difference between the three distributions in Fig. 9.28b are much less than those shown in Fig. 9.28c for the solutions after static stall at $\alpha = 15°$. The variation of displacement thickness with the reduced frequency can be represented by the trailing edge values, Fig. 9.29. From Figs. 9.28b, 9.28c and 9.29, we can conclude that increasing the pitch rate reduces the displacement thickness on the upper side of the flowfield and hence the boundary-layer effects on the outer inviscid flow.

The local skin-friction coefficient distributions on the upper and lower surfaces of the airfoil computed with the quasi-steady approach at $\alpha = 13°$ are shown in Fig. 9.30. The c_f distribution on the lower surface are influenced very little with pitch rate, but those on the upper surface are influenced with the reversal region suppressed by the increasing pitch rate. Figure 9.31 shows the results in which the boundary-layer calculations are performed with unsteady mode for $A = 0.02$. With increase in angle of attack, the friction coefficient on the upper surface decreases, whereas on the lower surface it increases, and a reversal flow region starts growing from the trailing edge at the angle of attack of 12.7°. The wiggle on the c_f distribution near the trailing edge is enhanced as the reversal region increases. The extension of the calculation method requires

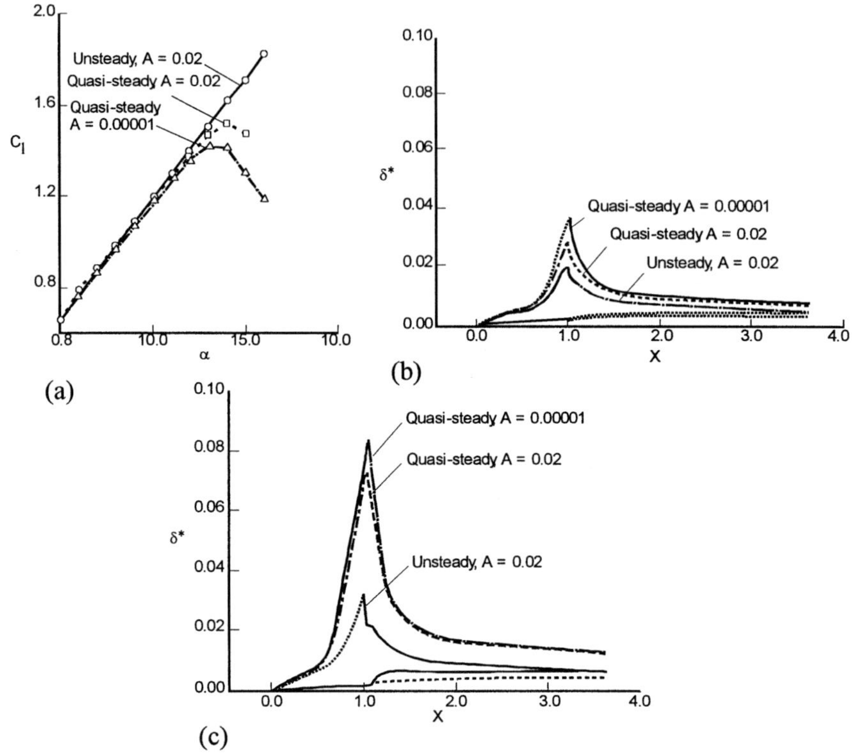

Fig. 9.28. The effects of pitch rate and the boundary-layer approach on the (a) lift coefficient and (b) displacement thickness distributions of the SSC-A09 airfoil subject to ramp-type motion with constant pitch rate $\alpha = 13^\circ$ and (c) $\alpha = 15^\circ$.

a procedure which is able to avoid the wiggles in the solutions as discussed in the next subsection.

9.2.2 Initiation of Dynamic Stall on a Pitching Airfoil

The calculation method of Chapters 7 and 8 has also been applied to study the initiation of dynamic stall on the Sikorsky airfoil subject to a ramp-time motion with a pitch rate of 0.02. The experimental data due to Lorber and Carta [14] indicates that vortex initiates around 18 to 19 degrees of angle of attack. The calculations confirm this and indicate how the trailing-edge separation causes the initiation of the vortex. To elaborate further on this point, let us consider the distribution of the local skin-friction values on the upper surface of this airfoil at several angles of attack, Fig. 9.32. It is clear from this figure that for $\alpha < 12^\circ$, there is no flow separation on the airfoil which has a steady stall angle of around 14°. The unsteadiness causes the stall angle to increase to around 30°, according to experiments. The flow behavior on the airfoil begins to change quickly, however, once the trailing-edge separation takes place for $\alpha > 12^\circ$. At

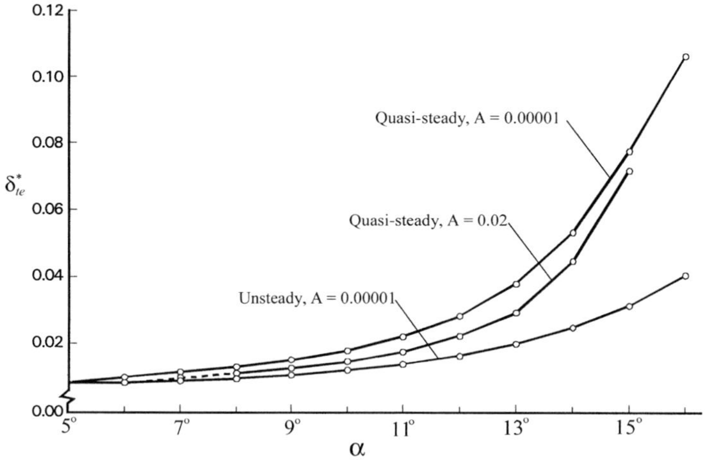

Fig. 9.29. The effects of pitch rate and the boundary-layer approach on the trailing edge displacement thickness distribution of the SSC-A09 airfoil subject to a ramp-type motion with constant pitch rate.

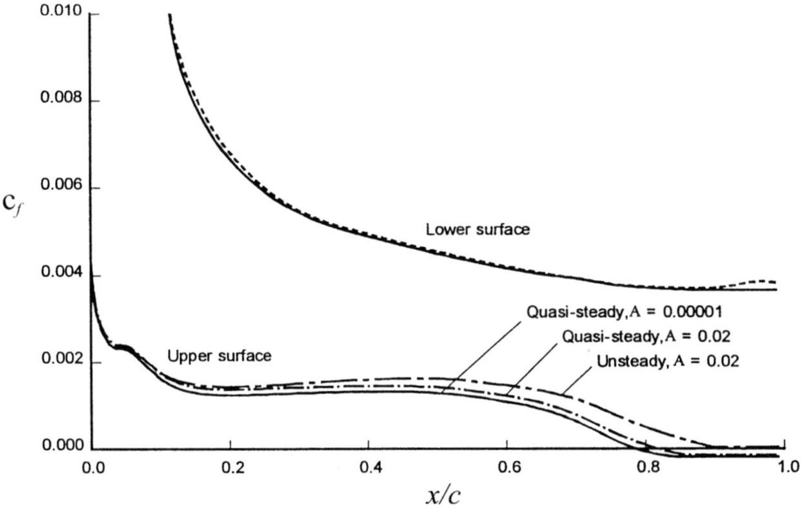

Fig. 9.30. The effects of pitch rate and the boundary-layer approach on the local skin-friction coefficient distribution of the SSC-A09 airfoil subject to a ramp-type with constant pitch rate.

$\alpha = 17.09°$, there is no leading-edge flow separation but only a trailing-edge separation which occurs around 12%. At the next angle of attack, $\alpha = 17.59°$, leading-edge separation takes place close to 12% chord with a bubble reattaching around 33% and is followed by a trailing-edge separation at 70%. The explosive nature of the leading-edge separation bubble, which is all turbulent, becomes

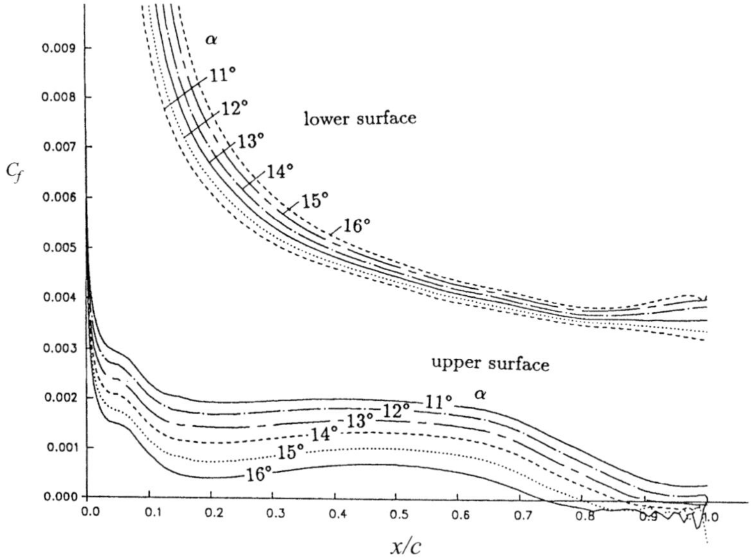

Fig. 9.31. The friction coefficient distribution of the SSC-A09 airfoil subject to a ramp-type motion with constant pitch rate, $A = 0.02$.

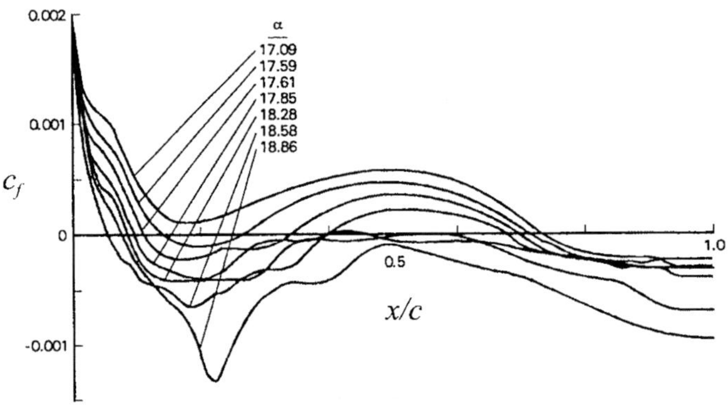

Fig. 9.32. Initiation of leading-edge vortex with trailing-edge separation on the Sikorsky airfoil subject to a ramp-type motion with a pitch rate of 0.02, $R_c = 2 \times 10^6$, α varies from 0 to 30°.

more obvious at the next $\alpha = 17.85°$, where leading-edge separation takes place around 10% but the reattachment of the bubble moves to 50% chord, a bubble of 40% in extent, followed by trailing-edge separation moving to 60% chord. Very shortly thereafter, less than one degree increase in angle of attack, the leading-edge separation bubble disappears with complete flow separation taking place at around 5% chord.

9.2.3 Summary

The following principal conclusions may be drawn from the studies reported in this section.

(a) In the harmonic oscillation cases, hysteresis effects are evident as the reduced frequency increases. The hysteresis phenomena appear in the solutions of both viscous and inviscid flows and are qualitatively represented by the calculation method.
(b) In the case of ramp-type motion with a high constant pitch rate, the solutions correctly show that the thick trailing-edge separation region in steady flow is suppressed to a thin reversal region and, at the same time, stall is delayed.
(c) Due to the neglect of unsteady terms, $\partial u/\partial t$ and $\partial u_e/\partial t$, the quasi-steady approach does not properly simulate unsteady flows at high pitch rates.
(d) The prediction of the onset of dynamic stall with the calculation method agrees with the data of Lorber and Carta for the Sikorsky airfoil and shows that for this airfoil the initiation of dynamic stall is caused by the trailing-edge separation.

References

[1] Patterson, M.T. and Lorber, P.F., "Computational and experimental studies of compressible dynamic stall," Fourth Symposium on Numerical and Physical Aspects of Aerodynamic Flows, Long Beach, CA, 16–19 Jan., 1989.
[2] Carr, L.W., "Progress in analysis and prediction of dynamic stall," J. of Aircraft, Vol. 25, pp. 6–17, 1988.
[3] McCroskey, W.J., McAlister, K.W., Carr, L.W. and Pucci, S.L., "An experimental study of dynamic stall on advanced airfoil sections," NASA TM 84245, 1982.
[4] Carr, L.W. and Chandrasekhara, M.S., "Design and development of a compressible dynamic stall facility," AIAA Paper 89-0647, 1989.
[5] Walker, J.M., Helin, H.E. and Strickland, J.H., "An experimental investigation of an airfoil undergoing large-amplitude pitching motions," AIAA J., Vol. 12, pp. 1141–1142, 1985.
[6] Ham, N.D., "Aerodynamic loading on a two-dimensional airfoil during dynamic stall," AIAA J., Vol. 6, pp. 1927–1934, 1968.
[7] Mehta, U.B., "Dynamic stall of an oscillating airfoil," AGARD Conf. Proc., No. 227, Paper No. 23, 1977.
[8] Wu, J.C., "Zonal solution of unsteady viscous flow problems," AIAA Paper 84-1637, 1984.
[9] Visbal, M.R., "Evaluation of an implicit Navier-Stokes solver for some unsteady separated flows," AIAA Paper 86-1053, 1986.
[10] Visbal, M.R., "Effects of compressibility on dynamic stall of a pitching airfoil," AIAA Paper 88-0132, 1988.
[11] Ekaterinaris, J.A., "Compressible studies on dynamic stall," AIAA Paper 89-0024, 1989.
[12] Sears, W.R. and Telionis, D.P., "Boundary-layer separation in unsteady flow," SIAM J. Appl. Math., Vol. 18, pp. 215–235, 1975.
[13] Cebeci, T., "The laminar boundary layer on a circular cylinder started impulsively from rest," J. Comp. Phys., Vol. 31, pp. 153–172, 1979.

[14] Lorber, P.F. and Carta, F.O., "Unsteady stall penetration experiments at high Reynolds number," United Technologies Research Center Rept. R87-956939-3, 1987.
[15] Cebeci, T., Stewartson, K. and Williams, P.G., "Separation and reattachment near the leading edge of a thin airfoil at incidence," AGARD CP 291, Paper 20, 1980.
[16] Cebeci, T., "Numerical instabilities in the calculation of laminar separation bubbles and their implications," AIAA J., Vol. 27, p. 656, 1988.
[17] Cebeci, T., "Unsteady boundary layers with an intelligent numerical scheme," J. Fluid Mech., Vol. 163, pp. 129–140, 1986.
[18] Cebeci, T., Khattab, A.A. and Schimke, S.M., "Separation and reattachment near the leading edge of a thin oscillating airfoil, " J. Fluid Mech., Vol. 188, pp. 253–274, 1986.
[19] Jang, H.M., "A viscous-inviscid interactive method for unsteady flows," Ph.D. Dissertation, Univ. of Michigan, 1990.
[20] Cebeci, T., Khattab, A.A. and Schimke, S.M., "Can the singularity be removed in time-dependent flows," In Workshop of Unsteady Separated Flow (M.S. Francis and M.W. Luttges, eds.) Colorado Springs, CO, 1984.

10 Three-Dimensional Incompressible Flows: Inviscid Method

10.0 Introduction

In this chapter and the following three chapters, we describe the extension of the calculation method of the previous chapters to three-dimensional subsonic and transonic flows, respectively, with a procedure similar to that used for two-dimensional steady flows. In each case, the computational strategy is essentially the same; the flowfield is computed by an inviscid method and the viscous flowfield by a boundary-layer method formulated in the inverse mode. For transonic flows, an Euler method is used to calculate the inviscid flow and the boundary-layer equations are solved for compressible flows as described in Chapter 11. For subsonic flows, a panel method is used to calculate the inviscid flow and the boundary-layer equations are solved for incompressible flows. An interface program, illustrated in Fig. 10.1, is placed between the inviscid method and the three-dimensional boundary-layer method to process the geometry and inviscid velocity data for input to the boundary-layer program. The onset of the transition location is obtained with an extension of the e^n-method discussed in detail in Chapter 4 for two-dimensional flows and in Chapter 12 for three-dimensional flows.

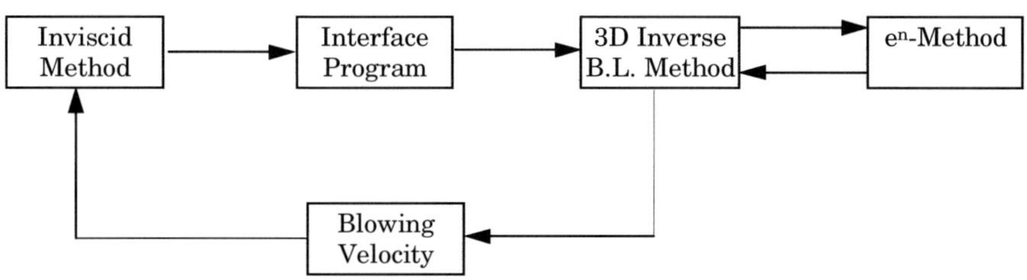

Fig. 10.1. The calculation method for three-dimensional flows.

While the calculation method is applicable to both incompressible and compressible flows, a detailed description of the inviscid method is given only for a panel method. Inviscid methods based on full potential or Euler equations are beyond the scope of this book, and the reader is referred to a rather rich literature describing these methods, see for example [1–6].

In this chapter we discuss the extension of the panel method of Chapters 2 and 7, developed by Hess [7], to three-dimensional flows. As discussed in [7], the representation of a three-dimensional lifting flow as an inviscid flow presents certain logical difficulties. However, all investigators in the field of computational fluid dynamics have agreed on a model of such flows, which will be accepted here without further discussion. Figure 10.2 shows a relatively simple configuration, which contains all the essential features of the problem. The configuration consists of a lifting portion (e.g., a wing) having a sharp trailing edge from which issues a zero-thickness trailing vortex wake and along which a Kutta condition is applied. A nonlifting portion (e.g., a fuselage) lacks a trailing edge and as a result does not need a Kutta condition.

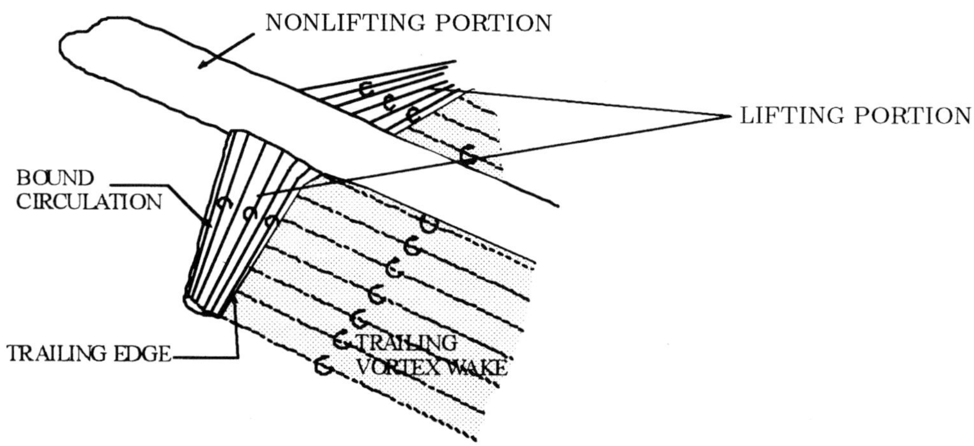

Fig. 10.2. Inviscid flow model of a lifting three-dimensional configuration.

The inviscid flow problem for steady three-dimensional lift requires the same conditions as in two-dimensional steady flows (Chapter 2). They are: (a) the velocity potential satisfies Laplace's equation outside the body and wake, (b) the disturbance velocity due to the body vanishes at infinity, (c) the normal component of velocity is zero on the body, (d) the Kutta condition is satisfied along the trailing edge, and (e) the wake is a stream surface of the flow with equal pressure on both sides. The last condition indicates that the wake position is initially unknown and must be computed as part of the solution. In the vast

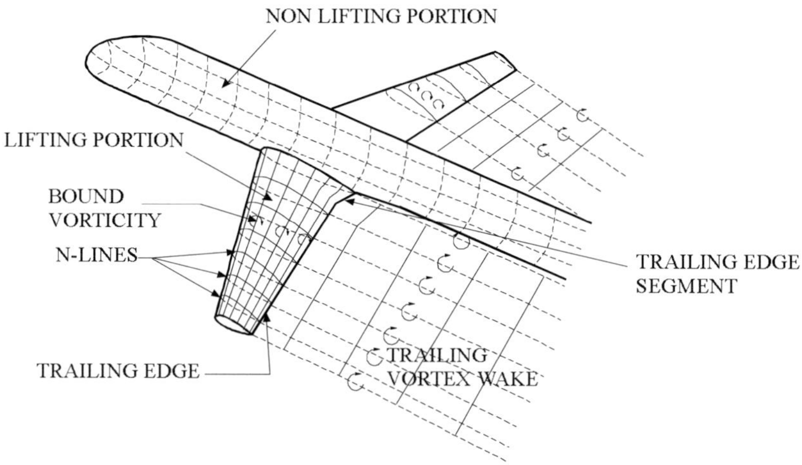

Fig. 10.3. Discretization of a lifting three-dimensional configuration.

majority of cases the flow quantities of interest are not sensitive to details of the wake shape and an assumed shape is used (as in the present discussion).

The inviscid flow problem outlined above may be solved numerically by means of a panel method with the body and wake represented by plane quadrilateral panels, as shown in Figure 10.3. On the lifting portion, the panels are organized into lifting strips wrapped around the wing section and containing a trailing edge segment. A lifting strip lies between consecutive N-lines, which are essentially surface coordinates along which the points defining the body are input (see subsection 10.1.1). On the panels are distributions of singularity, whose flow-fields automatically satisfy condition (a) and (b) above, and whose strengths are adjusted to satisfy conditions (c) and (d). The means of accomplishing this are highly non-unique, and many investigators have put forward their favorites. This selection will not be discussed here. Instead one particular method, due to Hess [7], will be presented, which has proven very satisfactory in applications. In this, the so-called first order source method, each panel – both lifting and nonlifting – has a constant value of source density. The vorticity distribution (which in the calculations is replaced by an equivalent dipole distribution) is taken as zero on the nonlifting portion of the configuration and on the lifting portion consists of a set of distributions, each of which has a constant value on one lifting strip and is zero on the other lifting strips. Thus, there is an unknown vorticity strength for trailing-edge segment, and in the final solution these represent the spanwise load distribution on the wing.

10.1 General Features of the Method of Solution

10.1.1 Order of Input Points

The body, about which flow is to be computed, is defined by a set of points on its surface. These defining points are input along curves lying in the body surface, which have some of the features of surface coordinates. These curves are designated "N-lines" (Fig. 10.3). First all points along a certain N-line are input from beginning to end, and then the same is done for the adjacent N-line. To obtain outward normal vectors, if successive points on a N-line are seen by an observer in the fluid to go upwards, then the next N-line must lie to the right. This matter is discussed at length in [8]. Two adjacent N-lines bound a "strip" of panels of approximately constant width (Fig. 10.3).

Panels are taken as plane trapezoids. Each of the two parallel sides is formed from two conservative points on the same N-line. Thus the parallel sides are approximately along the N-lines. Of course, in the general case the four input points that are associated to form a panel do not even lie in the same plane, much less form a trapezoid. They must be "adjusted" to do this. Adjacent panels have two input points in common, but the adjustment that these points are subject to is usually different for the two panels. Thus, in general, after adjustment the sides of adjacent panels are not coincident, and there are gaps between the panels. Errors from this source should be unimportant. For some bodies the gaps between panels vanish. For lifting bodies the important case for which this occurs is an untwisted wing, possibly swept and tapered, having the same airfoil section at all spanwise locations. The centroids of the panels are used as control points where boundary conditions are applied and where velocities are calculated. Thus, for each lifting strip the locus of control points is approximately midway between the two N-lines used to generate the strip. N-lines are extended into the wake, and wake panels are formed just like surface panels.

10.1.2 Use of a Dipole Distribution to Represent Vorticity

As mentioned at the end of Section 10.0, on lifting sections there is a constant vorticity distribution on each strip, which in the final solution varies from strip to strip, i.e., a step function in the spanwise direction. This requires that trailing vortex filaments lie along the N-lines and continue onto the N-lines of the wake. Formulas that express the velocity induced by such a vorticity distribution are required. Derivation of such expressions is complicated by the fact that in general the surface vorticity strength is a vector that varies in both magnitude and direction. Furthermore, care must be taken to insure that the vorticity distribution gives rise to a potential flow, i.e., that the individual infinitesimal vortex lines either form closed curves or go to infinity. Use of a surface dipole distribution circumvents these complications, because the dipole strength is a scalar and any arbitrary dipole distribution gives rise to a potential flow. A general result giving the relationship between dipole sheet and a vortex sheet is given in [9]. It may be summarized as follows: A variable-strength dipole sheet is equivalent to the sum of: (1) a variable-strength vortex sheet on the same surface as the dipole sheet whose vorticity has a direction at right angles to the gradient of the dipole strength and a magnitude equal to the magnitude of this gradient, and (2) a concentrated vortex filament around the edge of the sheet whose strength is everywhere equal to the local edge value of dipole strength. This relation, which is a straightforward generalization of the well-known two-dimensional result, does not appear explicitly in the literature. The proof of this relation in [9] is apparently the first. In the present method all formulas are derived in terms of dipole distributions and the above relationship is used to interpret this situation in terms of the more physically significant vorticity. In particular "chordwise" dipole variation is equivalent to "spanwise" vorticity and "spanwise" dipole variation to "chordwise" vorticity. Also, if a dipole sheet terminates with a nonzero strength, it results in a concentrated vortex filament.

10.1.3 Lift Carry-Over: The Extra Strip

A special situation occurs when a lifting portion of the body intersects a non-lifting portion at a finite angle (often nearly normal). The important case of this is the wing-fuselage intersection, as illustrated in Figure 10.4. As is well-known, the local "section lift coefficient" on the wing does not fall to zero at the fuselage intersection. Thus the dipole strength on the N-line lying along the intersection is not zero. If the lifting section is simply terminated, the result is a concentrated vortex filament right on the surface. Since no such vortex filament

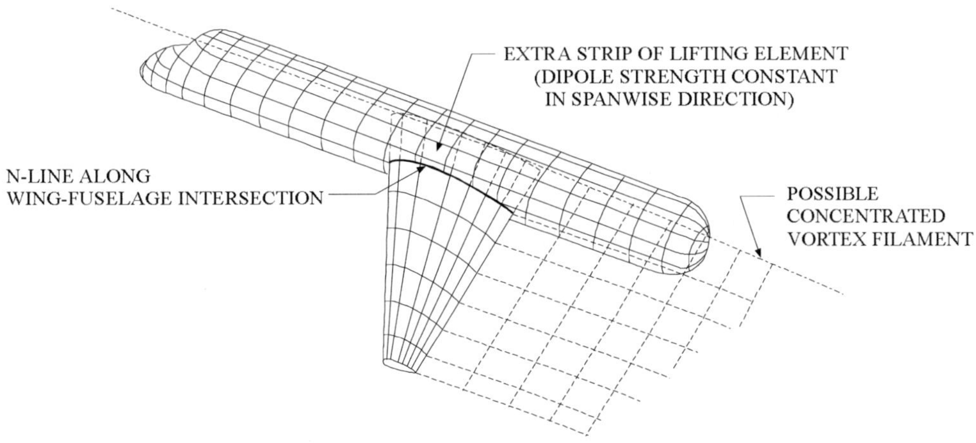

Fig. 10.4. Lift carry-over: the extra strip.

is in fact present, the fuselage panels would have to produce the equivalent of an equal and opposite vortex filament. This would require very dense fuselage panels and even then would be very difficult. To ease the burden placed on the numerics, an additional or "extra" lifting strip is added to the lifting section (Fig. 10.4). It is either the first or the last strip of the lifting section. The extra strip lies inside the nonlifting body and is a complete lifting strip including wake. No source densities or normal- velocity boundary conditions are applied to the panels of the extra strip. The dipole strength is taken constant in the "spanwise" direction across the extra strip. The value of the dipole strength on the extra strip is such that the dipole strength is continuous across the N-line lying along the wing-fuselage intersection. The interior edge of the extra strip has nonzero dipole strength and may lead to a concentrated vortex in the streamwise direction. For example, as shown in Fig. 10.4, the vortex may lie along the fuselage centerline and its downstream extension. If the lifting configuration has a right- and-left of symmetry, e.g., a fuselage with both wings, and if the flow is also symmetric, e.g., zero yaw, the extra strips for the right and left sides have the same strengths on their interior edges. Thus, in this case there is no discontinuity of dipole strength and no concentrated vortex. If, however, the lift is not symmetric, there will be a concentrated vortex. This is unavoidable because it is physically real. An example is the hub vortex of a propeller. This also occurs at a tip tank, which is essentially a small fuselage with only one wing.

10.1.4 The Matrix of Induced Velocities

Every panel has on it a constant source density. Lifting panels also have a dipole distribution. Formulas have been derived that enable velocities induced by the panels at points in space to be calculated (Section 10.3). For each panel the velocities induced by its constant source density at all control points are computed and saved. If there are symmetry planes, velocities induced by the reflections of a panel are added to the velocities due to the panel itself. This is the vector matrix of induced velocities. For each panel of a lifting section velocities induced by its dipole distribution at all control points are computed. These, however, are not saved individually. Instead, dipole velocities for all panels of a lifting strip, including wake panels, are added to obtain velocities due to the entire strip. Velocities due to an extra strip are added to those of the adjacent strip outside the nonlifting body (Figure 10.4). Thus, if there are N source panels, at whose control points the normal-velocity boundary condition is to be applied, and L lifting strips, there is a $N \times N$ matrix of induced source velocities and an $N \times L$ matrix of induced dipole velocities.

10.2 Formation of the Panels from Input Points

Let the coordinates of the input points that are used to form a panel be denoted x_k^i, y_k^i, z_k^i, $k = 1, 2, 3, 4$. These coordinates are with respect to the reference coordinate system, the system in which the physical lifting configuration is defined. It simplifies the equations to use vector notation, so define

$$\vec{x}_k^i = x_k^i \vec{i} + y_k^i \vec{j} + z_k^i \vec{k} \tag{10.2.1}$$

where $\vec{i}, \vec{j}, \vec{k}$ are unit vectors along the axes of the reference coordinate system. Recall that a panel is formed from points on two consecutive N-lines. The input points $k = 1$ and 2 are on one N-line, the "first" N-line, and the input pints $k = 3$ and 4 are on the "second" N-line. In what follows, subscripts F and S are used to denote quantities associated with the first and second N-lines, respectively. The numbering $k = 1, 2, 3, 4$ is cyclic around the panel. The adjustment of the input points, which is shown in Figure 10.5, is as follows.

First form the N-line vectors.

$$\vec{P}_F = \vec{x}_2^i - \vec{x}_1^i, \quad \vec{P}_S = \vec{x}_3^i - \vec{x}_4^i \tag{10.2.2}$$

The two parallel sides of the trapezoid are taken as parallel to the weighted average of these two vectors. In the coordinate system of the panel this is also the direction of the x-axis. The unit vector parallel to the two parallel sides of the trapezoid is denoted $\vec{i}_E$ to show it is also the unit vector along the x or ξ axis of the panel coordinate system. It is computed from

$$\vec{i}_E = \frac{\vec{P}_F + \vec{P}_S}{|\vec{P}_F + \vec{P}_S|} \tag{10.2.3}$$

where $|\vec{v}|$ means absolute magnitude of the vector $\vec{v}$, i.e., the square root of the sum of the squares of its components. The parallel sides have the direction of $\vec{i}_E$. The calculation insures that each parallel side has the same midpoint and

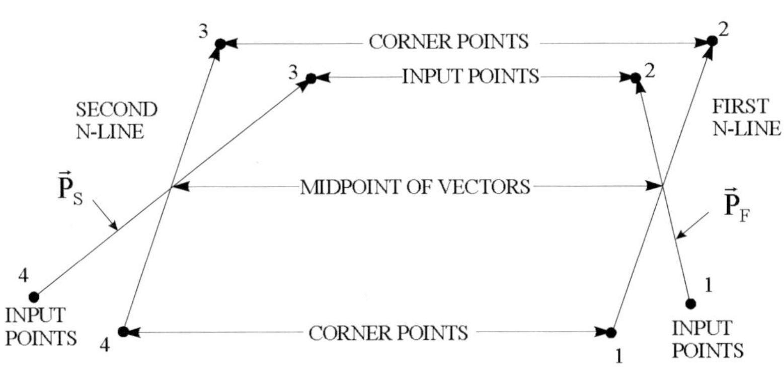

Fig. 10.5. Adjustment of the input points to form a plane trapezoidal panel.

the same length as the segment of N-line from which it is formed. In fact, once the panels are formed the original N-line segments are replaced by these parallel sides. The side lengths are

$$d_F = |\vec{P}_F|, \quad d_S = |\vec{P}_S| \tag{10.2.4}$$

The midpoints in vector form are

$$\vec{x}_F = \frac{1}{2}(\vec{x}_1^i + \vec{x}_2^i), \quad \vec{x}_S = \frac{1}{2}(\vec{x}_3^i + \vec{x}_4^i) \tag{10.2.5}$$

The endpoints of the two parallel sides, which are thus the corner points of the trapezoidal panel are, in vector form,

$$\begin{aligned} \vec{x}_1 &= \vec{x}_F - \frac{1}{2}d_F\vec{i}_E, \quad \vec{x}_2 = \vec{x}_F + \frac{1}{2}d_F\vec{i}_E \\ \vec{x}_3 &= \vec{x}_S + \frac{1}{2}d_S\vec{i}_E, \quad \vec{x}_4 = \vec{x}_S - \frac{1}{2}d_S\vec{i}_E \end{aligned} \tag{10.2.6}$$

The normal vector to the plane of the panel is

$$\vec{N} = (\vec{x}_4 - \vec{x}_2) \times (\vec{x}_3 - \vec{x}_1) \tag{10.2.7}$$

The unit normal vector is

$$\vec{n} = \frac{\vec{N}}{|\vec{N}|} \tag{10.2.8}$$

This is also the unit vector along the z-axis of the panel coordinate system. The unit vector along the y or η axis of the panel coordinate system is

$$\vec{j}_E = \vec{n} \times \vec{i}_E \tag{10.2.9}$$

In component form the three unit vectors are

$$\begin{aligned} \vec{i}_E &= a_{11}\vec{i} + a_{12}\vec{j} + a_{13}\vec{k} \\ \vec{j}_E &= a_{21}\vec{i} + a_{22}\vec{j} + a_{23}\vec{k} \\ \vec{n} = \vec{k}_E &= a_{31}\vec{i} + a_{32}\vec{j} + a_{33}\vec{k} \end{aligned} \tag{10.2.10}$$

The 3×3 array of a's is the transformation matrix that is used to transform coordinates of points and components of vectors between the reference and panel coordinate systems in the manner described in [8].

Temporarily the origin of the panel coordinate system is taken as the point whose coordinates are the averages of those of the input points. (The same averages are obtained using the corner points.) In vector notation, the average point is

$$\vec{x}_{av} = \frac{1}{2}(\vec{x}_F + \vec{x}_S) \tag{10.2.11}$$

With this origin, the panel coordinates of the corner points are

$$\begin{aligned}\xi_k^* &= a_{11}(x_k - x_{av}) + a_{12}(y_k - y_{av}) + a_{13}(z_k - z_{av}) \\ \eta_k^* &= a_{21}(x_k - x_{av}) + a_{22}(y_k - y_{av}) + a_{23}(z_k - z_{av})\end{aligned} \quad k = 1, 2, 3, 4 \qquad (10.2.12)$$

where in accordance with the vector notation, x_k, y_k, z_k are the coordinates of $\vec{x}_k$ from (10.2.6). It will turn out that

$$\eta_1^* = \eta_2^* \quad \text{and} \quad \eta_3^* = \eta_4^* = -\eta_1^* \qquad (10.2.13)$$

The width of the panel is

$$w = \eta_1^* - \eta_3^* = 2\eta_1^* \qquad (10.2.14)$$

The slopes of the nonvertical sides of the panel (Fig. 10.6) are

$$m_{32} = \frac{\xi_2^* - \xi_3^*}{w}, \qquad m_{41} = \frac{\xi_1^* - \xi_4^*}{w} \qquad (10.2.15)$$

with respect to the η-axis. The coordinates of the centroid are

$$\begin{aligned}\eta_0 &= \frac{w^2}{6} \frac{m_{32} - m_{41}}{\xi_3^* + \xi_2^* - \xi_1^* - \xi_4^*} \\ \xi_0 &= \frac{m_{32} + m_{41}}{2} \eta_0\end{aligned} \qquad (10.2.16)$$

The reference coordinates of the centroid are

$$\begin{aligned}x_0 &= x_{av} + a_{11}\xi_0 + a_{21}\eta_0 \\ y_0 &= y_{av} + a_{12}\xi_0 + a_{22}\eta_0 \\ z_0 &= z_{av} + a_{13}\xi_0 + a_{23}\eta_0\end{aligned} \qquad (10.2.17)$$

The centroid is now taken as the origin of the panel coordinate system and replaces the average point in all subsequent calculations. With respect to the centroid as origin, the panel coordinates of the corner points are

$$\begin{aligned}\xi_k &= \xi_k^* - \xi_0 \\ \eta_k &= \eta_k^* - \eta_0\end{aligned} \qquad (10.2.18)$$

where

$$\eta_2 = \eta_1 \quad \text{and} \quad \eta_4 = \eta_3 \qquad (10.2.19)$$

These are the corner point coordinates used in all subsequent calculations.

Several other geometric quantities are needed in future calculations. These are now computed. The intercepts where the sides intersect the x or ξ axis (Fig. 10.6) are

$$b_{32} = \frac{\xi_3\eta_2 - \xi_2\eta_3}{w}, \qquad b_{41} = \frac{\xi_4\eta_1 - \xi_1\eta_4}{w} \qquad (10.2.20)$$

The maximum diagonal of the panel is

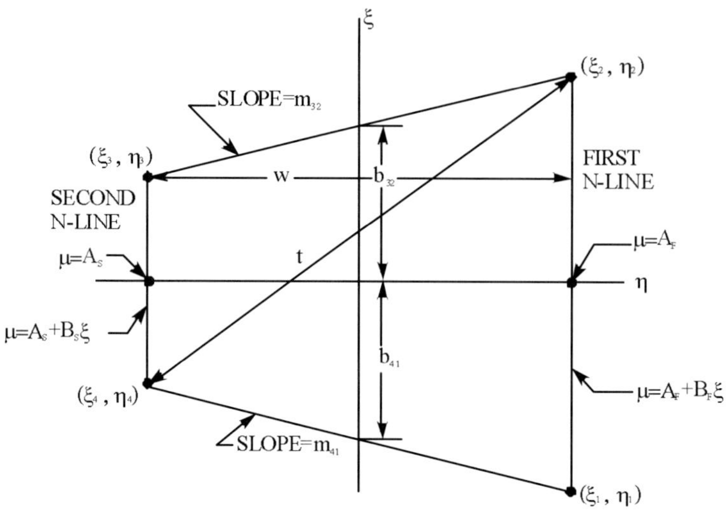

Fig. 10.6. A plane trapezoidal panel.

$$t = \text{Max} \begin{cases} \sqrt{(\xi_2 - \xi_4)^2 + (\eta_2 - \eta_4)^2} \\ \sqrt{(\xi_3 - \xi_1)^2 + (\eta_3 - \eta_1)^2} \end{cases} \tag{10.2.21}$$

The lengths of the sides are

$$\begin{aligned} d_{12} &= d_F, \quad d_{34} = d_S \\ d_{32} &= w\sqrt{1 + m_{32}^2}, \quad d_{41} = w\sqrt{1 + m_{41}^2} \end{aligned} \tag{10.2.22}$$

Also needed are the total arc lengths along the N-lines from the trailing edge up to the panel in question. These are

$$L_F = \sum d_F, \quad L_S = \sum d_S \tag{10.2.23}$$

where sums are over previous panels of the lifting strip. Finally, the area of the panel is

$$A = \frac{w}{2}(\xi_3 - \xi_4)(\xi_2 - \xi_1) \tag{10.2.24}$$

10.3 Induced-Velocity Formulas

10.3.1 Form of the Surface Dipole Distribution

To determine the form of the surface dipole distribution over a panel of a lifting strip, first consider its variation over the N-lines bounding the strip. The dipole strength varies linearly along each N-line, say

$$\mu = B(s + c) \tag{10.3.1}$$

where s is arc length along the N-line measured from the trailing edge. The arc length s is zero for the first point of the N-line at the trailing edge. The final on-body point of the N-line, which is also the trailing edge, has $s = L(\text{tot})$, where, referring to Eq. (10.2.23)

$$L(\text{tot}) = \sum_{\text{strip}} d \tag{10.3.2}$$

The factor B that determines the growth of μ along the N-line is essentially the "spanwise" vorticity strength at that N-line, which is the bound vorticity that gives the lift (see subsection 10.1.2). The constant c in Eq. (10.3.1) is arbitrary. An effective choice is $c = -\frac{1}{2}L(\text{tot})$ so that (10.3.1) becomes

$$\mu = B\left(s - \frac{1}{2}L(\text{tot})\right) \tag{10.3.3}$$

With this choice

$$\begin{aligned} \mu(\text{initial trailing edge}) &= -\frac{1}{2}BL(\text{tot}) \\ \mu(\text{final trailing edge}) &= -\frac{1}{2}BL(\text{tot}) \end{aligned} \tag{10.3.4}$$

Equation (10.3.3) gives the variation of μ around the wing section. In the wake downstream of the trailing edge μ equals the difference of its two trailing edge values, i.e.,

$$\mu(\text{wake}) = BL(\text{tot}) \tag{10.3.5}$$

10.3.2 Variation over a Trapezoidal Panel

Consider now a typical trapezoidal panel of a lifting strip as shown in Fig. 10.6. As defined in Section 10.2 the lifting strip is bounded by two N-lines, which are designated the "first" N-line and the "second" N-line and which are represented by subscripts F and S respectively. The variation of dipole strengths along these N-lines are

$$\begin{aligned} \mu &= B\left(s - \frac{1}{2}L_F(\text{tot})\right) \quad \text{on the first } N\text{-line} \\ \mu &= B\left(s - \frac{1}{2}L_S(\text{tot})\right) \quad \text{on the second } N\text{-line} \end{aligned} \tag{10.3.6}$$

On the panel itself the parallel side at $\eta = \eta_1$ is a segment of the first N-line, and the parallel side at $\eta = \eta_3$ is a segment of the second N-line (Fig. 10.6). The linear factor B is associated with the lifting strip. The adjacent lifting strip has an N-line in common with this one, but the value of B is different, and thus the dipole values on adjacent strips are discontinuous. The resulting spanwise variation of B is a step-function. This implies the presence of trailing vortex filaments along the N-lines, which the use of a dipole distribution automatically insures.

In terms of panel coordinates the variation of dipole strength along the parallel sides of the panel are

$$\begin{aligned} \mu &= B(\xi + h_F) \quad \text{on} \quad \eta = \eta_1 \\ \mu &= B(\xi + h_S) \quad \text{on} \quad \eta = \eta_3 \end{aligned} \tag{10.3.7}$$

where from Eqs. (10.3.6) and (10.2.23), and Fig. 10.6

$$\begin{aligned} h_F &= L_F - \xi_1 - \frac{1}{2} L_F(\text{tot}) \\ h_S &= L_S - \xi_4 - \frac{1}{2} L_S(\text{tot}) \end{aligned} \tag{10.3.8}$$

Now a dipole distribution μ on the panel is assumed in the form of a general two-variable first degree polynomial. When conditions (10.3.7) are applied, it turns out that μ must be

$$\mu = \frac{B}{w}[(h_F - h_S)\eta - (\eta_3 - \eta_1)\xi - \eta_3 h_F + \eta_1 h_S] \tag{10.3.9}$$

where $w = \eta_1 - \eta_3$ is the width of the panel.

In the wake μ is constant along the N-lines. From Equation (10.3.5)

$$\begin{aligned} \mu &= BL_F(\text{tot}) \quad \text{on} \quad \eta = \eta_1 \\ \mu &= BL_S(\text{tot}) \quad \text{on} \quad \eta = \eta_3 \end{aligned} \tag{10.3.10}$$

Thus the dipole distribution over a wake panel (or indeed over the entire wake) is

$$\mu = \frac{B}{w}\{[L_F(\text{tot}) - L_S(\text{tot})]\eta + L_S(\text{tot}_1)\eta_1 - L_F(\text{tot}_1)\eta_3\} \tag{10.3.11}$$

10.3.3 Use of a Far-Field Approximation

The basic formulas of the present method are those giving the velocity induced by a panel at points in space. These are applied to obtain the effects of the panels at each other's control points. For a panel of a lifting section on the body surface there are two kinds of induced velocities, that due to the constant source density on the panel and that due to the dipole distribution of the form (10.3.9). For two different ranges of distance between the panel centroid and the point where velocities are evaluated, different sets of formulas are used. The two ranges are denoted: (1) the far-field or point singularity regime and (2) the near field or exact regime. The near-field formulas are obtained by an exact integration over the panels. Such formulas are necessary to obtain the desired accuracy at points near the panel, but are quite time consuming. When the distance between the panel centroid and at the point where velocities are being evaluated exceeds four times the maximum diagonal of the panel, velocities are calculated directly in the reference coordinate system using the point-singularity formulas of the far-field. In the near field the field point where velocities are to be evaluated is transformed into the panel coordinate system using the transformation matrix (10.2.10). Then velocities are computed in panel coordinates, and finally the computed velocities are transformed back to reference coordinates using the transformation matrix. This procedure is well known and will not be discussed further here. A complete description is contained in [8].

10.3.4 Notation for the Induced Velocities

Now notation will be introduced for the velocity calculation. It is assumed that the velocities that are being computed are due to the jth panel and are being evaluated at the control point (centroid) of the ith panel. Clearly, any point could be substituted for the ith control point. The velocity due to the constant unit strength source density is denoted $\vec{V}_{ij}$. In panel coordinates it has components $V_x(\text{source})$, $V_y(\text{source})$, and $V_z(\text{source})$, i.e.,

$$\vec{V}_{ij} = V_x(\text{source})\vec{\imath}_E + V_y(\text{source})\vec{\jmath}_E + V_z(\text{source})\vec{k}_E \tag{10.3.12}$$

For a panel of a nonlifting section, this source velocity is the only induced velocity. For a panel of a lifting section the calculation of source velocities is the same as for a nonlifting panel.

To develop formulas for the velocity induced by the dipole distribution on a panel, some additional notation is required. Furthermore, it simplifies the development to consider the velocity potential initially rather than the velocity components. The potential due to the dipole distribution on the panel at points of space is obtained by integrating over the panel. Namely,

$$\phi = \iint_E \phi(\text{dipole})\mu(\xi,\eta)\, d\xi d\eta \tag{10.3.13}$$

where $\mu(\xi,\eta)$ is given by (10.3.9), the integration is over the area of the panel and where $\phi(\text{dipole})$ is the potential of a unit point dipole with axis normal to the panel.

Now define the auxiliary potentials

$$\phi_{pq} = \iint_E \phi(\text{dipole})\xi^p\eta^q\, d\xi d\eta \tag{10.3.14}$$

where p and q are integers. Now from (10.3.9), the potential of the panel is

$$\phi = \frac{B}{w}[(h_F - h_S)\phi_{01} - (\eta_3 - \eta_1)\phi_{10} - (\eta_3 h_F - \eta_1 h_S)\phi_{00}] \tag{10.3.15}$$

The velocity due to the dipole distribution is

$$\vec{V}_{ij}(\text{dipole}) = -\nabla\phi \tag{10.3.16}$$

where ∇ denotes the gradient operator. In panel coordinates this is

$$\nabla\phi = \frac{\partial\phi}{\partial x}\vec{i}_E + \frac{\partial\phi}{\partial y}\vec{j}_E + \frac{\partial\phi}{\partial z}\vec{k}_E \tag{10.3.17}$$

To evaluate the dipole velocities formulas are needed for the derivatives of $\phi_{00}, \phi_{10}, \phi_{01}$. These formulas are presented in the following sections, along with those for the source velocities.

10.3.5 Far-Field Formulas for the Velocity Induced by a Lifting Panel

First calculate the distance r_0 between the centroid of the panel and the field point where velocities are to be calculated. If the reference coordinates of the centroid are x_0, y_0, z_0 and the reference coordinates of the field point are x', y', z', then

$$r_0 = \sqrt{(x' - x_0)^2 + (y' - y_0)^2 + (z' - z_0)^2} \tag{10.3.18}$$

Now test r_0/t, where t is the maximum diagonal of the panel Eq. (10.2.21). If

$$r_0/t > \rho_1 \tag{10.3.19}$$

where ρ_1 is a certain criterion, then the far-field formulas are used. Currently, ρ_1 is set equal to 4.0. The far-field formulas calculate velocities directly in reference coordinates. First define the vector

$$\vec{r}_0 = (x' - x_0)\vec{i} + (y' - y_0)\vec{j} + (z' - z_0)\vec{k} \tag{10.3.20}$$

The source velocity is

$$\vec{V}_{ij} = \frac{A}{r_0^2}\frac{\vec{r}_0}{r_0} \tag{10.3.21}$$

where the panel area A is given by Eq. (10.2.24). The dipole velocity is

$$\vec{V}_{ij}(\text{dipole}) = \frac{B}{w}\frac{A}{r_0^3}(\eta_3 h_F - \eta_1 h_S)\vec{D} \tag{10.3.22}$$

where the vector $\vec{D}$ is

$$\vec{D} = 3\left(\frac{\vec{n}\cdot\vec{r}_0}{r_0}\right)\frac{\vec{r}_0}{r_0} - \vec{n} \tag{10.3.23}$$

where $\vec{n}$ is the normal vector of the jth panel. A comparison of Eq. (10.3.22) with (10.3.15) and its underlying Eq. (10.3.9) show that the terms in ϕ_{10} and ϕ_{01} have dropped out. Integrals of the first powers of ξ and n are zero because the centroid is used as origin of coordinates.

10.3.6 Near-Field Formulas for the Velocity Induced by a Lifting Panel

If $r_0/t < \rho_1$, the near-field or exact formulas are used to compute induced velocities. The calculation starts with the panel coordinates x, y, z of the field point and the geometric quantities associated with the panel that are discussed in Section 10.2. In particular, the corner point coordinates ξ_k, η_k, $k = 1, 2, 3, 4$ are needed, together with the width w from (10.2.14), the slopes m_{32} and m_{41} from (10.2.15), the intercepts b_{32} and b_{41} from (10.2.20), the maximum diagonal t from (10.2.21), and the side lengths $d_{12}, d_{32}, d_{34}, d_{41}$, from (10.2.22). These quantities are illustrated in Fig. 10.6.

Preliminary quantities to be calculated are:

$$
\begin{aligned}
r_k &= \sqrt{(x-\xi_k)^2 + (y-\eta_k)^2 + z^2}, \quad k = 1,2,3,4 \\
\alpha_k &= \frac{x-\xi_k}{r_k}, \quad \beta_k = \frac{y-\eta_k}{r_k}, \quad y_k = \frac{z}{r_k}, \quad k = 1,2,3,4
\end{aligned}
\tag{10.3.22}
$$

$$
\begin{aligned}
p_k^{(32)} &= m_{32}[z^2 + (y-\eta_k)^2] - (x-\xi_k)(y-\eta_k), \quad k = 3 \text{ or } 2 \\
p_k^{(41)} &= m_{41}[z^2 + (y-\eta_k)^2] - (x-\xi_k)(y-\eta_k), \quad k = 4 \text{ or } 1
\end{aligned}
\tag{10.3.23}
$$

The basic functions are

$$
L^{(mn)} = \log \frac{r_m + r_n - d_{mn}}{r_m + r_n + d_{mn}}, \quad \begin{array}{l} m, n \text{ consecutive, i.e.,} \\ mn = 12, 23, 34, \text{ or } 41 \end{array}
\tag{10.3.24}
$$

and

$$
\begin{aligned}
T_k^{(32)} &= \tan^{-1}\left[\frac{p_k^{(32)}}{z r_k}\right], \quad k = 3 \text{ or } 2 \\
T_k^{(41)} &= \tan^{-1}\left[\frac{p_k^{(41)}}{z r_k}\right], \quad k = 4 \text{ or } 1
\end{aligned}
\tag{10.3.25}
$$

Also needed are derivatives of the T's and L's. The derivatives of $T_k^{(32)}$ are

$$
\left.
\begin{aligned}
\frac{\partial T_k^{(32)}}{\partial x} &= -\frac{z(r_k^2\beta_k + p_k^{(32)}\alpha_k)}{D_k^{(32)}} \\
\frac{\partial T_k^{(32)}}{\partial y} &= \frac{z[(2m_{32}\beta_k - \alpha_k)r_k^2 - p_k^{(32)}\beta_k]}{D_k^{(32)}} \\
\frac{\partial T_k^{(32)}}{\partial z} &= \frac{2m_{32}z^2 r_k - p_k^{(32)}(r_k + z\gamma_k)}{D_k^{(32)}} \\
D_k^{(32)} &= z^2 r_k^2 + [p_k^{(32)}]^2
\end{aligned}
\right\}
\tag{10.3.26}
$$

There is an analogous set of formulas for the derivatives of $T_k^{(41)}$.

The derivatives of $L^{(mn)}$ are

$$\frac{\partial L^{(mn)}}{\partial x} = D_{mn}(\alpha_m + \alpha_n), \qquad \frac{\partial L^{(mn)}}{\partial y} = D_{mn}(\beta_m + \beta_n), \qquad mn = 12, 23, 34, 41$$
$$\frac{\partial L^{(mn)}}{\partial z} = D_{mn}(\gamma_m + \gamma_n), \quad D_{mn} = \frac{2d_{mn}}{(r_m + r_n)^2 - d_{mn}^2} \tag{10.3.27}$$

Now in terms of the above functions the source velocities and dipole potential derivatives needed for (10.3.15) can be written.

The source velocities are

$$\begin{aligned}
V_x(\text{source}) &= -\frac{1}{\sqrt{1+m_{32}^2}} L^{(32)} + \frac{1}{\sqrt{1+m_{41}^2}} L^{(41)} \\
V_y(\text{source}) &= -L^{(12)} + L^{(34)} + \frac{m_{32}}{\sqrt{1+m_{32}^2}} L^{(32)} - \frac{m_{41}}{\sqrt{1+m_{41}^2}} L^{(41)} \\
V_z(\text{source}) &= -T_2^{(32)} + T_3^{(32)} + T_1^{(41)} - T_4^{(41)}
\end{aligned} \tag{10.3.28}$$

To evaluate the dipole potential derivatives, the derivatives of V_x and V_y are needed (since $V_z = \phi_{00}$, its derivatives are exactly a potential derivative). The derivatives of V_x and V_y are

$$\begin{aligned}
\frac{\partial V_x(\text{source})}{\partial x} &= -\frac{1}{\sqrt{1+m_{32}^2}} \frac{\partial L^{(32)}}{\partial x} + \frac{1}{\sqrt{1+m_{41}^2}} \frac{\partial L^{(41)}}{\partial x} \\
\frac{\partial V_x(\text{source})}{\partial y} &= -\frac{1}{\sqrt{1+m_{32}^2}} \frac{\partial L^{(32)}}{\partial y} + \frac{1}{\sqrt{1+m_{41}^2}} \frac{\partial L^{(41)}}{\partial y} \\
\frac{\partial V_x(\text{source})}{\partial z} &= -\frac{1}{\sqrt{1+m_{32}^2}} \frac{\partial L^{(32)}}{\partial z} + \frac{1}{\sqrt{1+m_{41}^2}} \frac{\partial L^{(41)}}{\partial z}
\end{aligned} \tag{10.3.29}$$

$$\begin{aligned}
\frac{\partial V_y(\text{source})}{\partial x} &= -\frac{\partial L^{(12)}}{\partial x} + \frac{\partial L^{(34)}}{\partial x} + \frac{m_{32}}{\sqrt{1+m_{32}^2}} \frac{\partial L^{(32)}}{\partial x} - \frac{m_{41}}{\sqrt{1+m_{41}^2}} \frac{\partial L^{(41)}}{\partial x} \\
\frac{\partial V_y(\text{source})}{\partial y} &= -\frac{\partial L^{(12)}}{\partial y} + \frac{\partial L^{(34)}}{\partial y} + \frac{m_{32}}{\sqrt{1+m_{32}^2}} \frac{\partial L^{(32)}}{\partial y} - \frac{m_{41}}{\sqrt{1+m_{41}^2}} \frac{\partial L^{(41)}}{\partial y} \\
\frac{\partial V_y(\text{source})}{\partial z} &= -\frac{\partial L^{(12)}}{\partial z} + \frac{\partial L^{(34)}}{\partial z} + \frac{m_{32}}{\sqrt{1+m_{32}^2}} \frac{\partial L^{(32)}}{\partial z} - \frac{m_{41}}{\sqrt{1+m_{41}^2}} \frac{\partial L^{(41)}}{\partial z}
\end{aligned}$$

Now the potential derivatives are as follows.

$$\begin{aligned}
\frac{\partial \phi_{00}}{\partial x} &= -\frac{\partial T_2^{(32)}}{\partial x} + \frac{\partial T_3^{(32)}}{\partial x} + \frac{\partial T_1^{(41)}}{\partial x} - \frac{\partial T_4^{(41)}}{\partial x} \\
\frac{\partial \phi_{00}}{\partial y} &= -\frac{\partial T_2^{(32)}}{\partial y} + \frac{\partial T_3^{(32)}}{\partial y} + \frac{\partial T_1^{(41)}}{\partial y} - \frac{\partial T_4^{(41)}}{\partial y} \\
\frac{\partial \phi_{00}}{\partial z} &= -\frac{\partial T_2^{(32)}}{\partial z} + \frac{\partial T_3^{(32)}}{\partial z} + \frac{\partial T_1^{(41)}}{\partial z} - \frac{\partial T_4^{(41)}}{\partial z}
\end{aligned} \tag{10.3.30}$$

$$\begin{aligned}
\frac{\partial \phi_{01}}{\partial x} &= -z\frac{\partial V_y(\text{source})}{\partial x} + y\frac{\partial \phi_{00}}{\partial x} \\
\frac{\partial \phi_{01}}{\partial y} &= -z\frac{\partial V_y(\text{source})}{\partial y} + y\frac{\partial \phi_{00}}{\partial y} + V_z(\text{source}) \\
\frac{\partial \phi_{01}}{\partial z} &= -z\frac{\partial V_y(\text{source})}{\partial z} + y\frac{\partial \phi_{00}}{\partial z} - V_y(\text{source})
\end{aligned} \tag{10.3.31}$$

$$\begin{aligned}
\frac{\partial \phi_{10}}{\partial x} &= -z\frac{\partial V_x(\text{source})}{\partial x} + x\frac{\partial \phi_{00}}{\partial x} + V_z(\text{source}) \\
\frac{\partial \phi_{10}}{\partial y} &= -z\frac{\partial V_x(\text{source})}{\partial y} + x\frac{\partial \phi_{00}}{\partial y} \\
\frac{\partial \phi_{10}}{\partial z} &= -z\frac{\partial V_x(\text{source})}{\partial z} + x\frac{\partial \phi_{00}}{\partial z} - V_x(\text{source})
\end{aligned} \tag{10.3.32}$$

10.3.7 The Velocity Induced by a Wake Panel

In the wake the dipole strength is constant along N-lines, and the dipole distribution over a wake panel is given by Eq. (10.3.11). Accordingly the velocity potential due to the dipole distribution on a wake panel (recall that on the wake the source density is zero) is

$$\phi = \frac{B}{w}\{[L_F(\text{tot}) - L_s(\text{tot})]\phi_{01} + [L_s(\text{tot})\eta_1 - L_F(\text{tot})\eta_3]\phi_{00}\} \tag{10.3.33}$$

and the velocity is

$$V_{ij}(\text{dipole}) = -\nabla\phi \tag{10.3.34}$$

In the near field these are evaluated by the formulas of subsection 10.3.6. In the far-field following subsection 10.3.5

$$V_{ij}(\text{dipole}) = \frac{B}{w}\frac{A}{r_0^3}[L_F(\text{tot})\eta_3 - L_S(\text{tot})\eta_1]\vec{D} \tag{10.3.35}$$

10.3.8 Option for Semi-Infinite Last Wake Panel

In most cased of interest the trailing vortex wake extends to infinity. To facilitate accounting for this condition, provision has been made for considering the last panel of the wake to be semi-infinite. A finite panel of the sort shown in Fig. 10.6 is formed at the end of the wake, including all the geometric quantities of Section 10.2. The induced velocity calculation for this panel is performed using the origin of coordinates appropriate to the finite panel, but the formulas used to calculate induced velocity are appropriate for the semi-infinite panel. Naturally, all points in space are in the "near field" with respect to a semi-infinite panel, so it is the formulas of subsection 10.3.6 that apply. These formulas are modified by setting

$$\begin{gathered} m_{32} = 0 \\ \xi_2 \to \infty \quad \xi_3 \to \infty \end{gathered} \tag{10.3.36}$$

This yields immediately

$$\begin{gathered} \alpha_1, \beta_1, \gamma_1, \alpha_4, \beta_4, \gamma_4 \text{ unchanged} \\ \alpha_3 = \alpha_2 = -1, \quad \beta_3 = \beta_2 = \gamma_3 = \gamma_2 = 0 \end{gathered} \tag{10.3.37}$$

The log functions (10.3.24) and their derivatives (10.3.28) are replaced by

$$\begin{aligned} L^{(41)} &= \text{unchanged, all derivatives unchanged} \\ L^{(32)} &= 0, \text{ all derivatives equal to zero} \end{aligned} \tag{10.3.38}$$

$$-L^{(12)} + L^{(34)} = \log \frac{r_4 - (x - \xi_4)}{r_1 - (x - \xi_1)} \tag{10.3.39}$$

$$\begin{aligned} \frac{\partial L^{(34)}}{\partial x} &= \frac{\alpha_4 - 1}{r_4 - (x - \xi_4)} & \frac{\partial L^{(12)}}{\partial x} &= \frac{\alpha_1 - 1}{r_1 - (x - \xi_1)} \\ \frac{\partial L^{(34)}}{\partial y} &= \frac{\beta_4}{r_4 - (x - \xi_4)} & \frac{\partial L^{(12)}}{\partial y} &= \frac{\beta_1}{r_1 - (x - \xi_1)} \\ \frac{\partial L^{(34)}}{\partial z} &= \frac{\gamma_4}{r_4 - (x - \xi_4)} & \frac{\partial L^{(12)}}{\partial z} &= \frac{\gamma_1}{r_1 - (x - \xi_1)} \end{aligned} \tag{10.3.40}$$

The inverse tangent functions (10.3.25) and their derivatives (10.3.26) are replaced by

$$\begin{aligned} T_k^{(32)} &= \tan^{-1}\left(\frac{y - \eta_k}{z}\right), \quad k = 3 \text{ or } 2 \\ T_k^{(41)} &= \text{unchanged}, \quad k = 4 \text{ or } 1 \end{aligned} \tag{10.3.41}$$

$$\frac{\partial T_k^{(32)}}{\partial x} = 0$$
$$\frac{\partial T_k^{(32)}}{\partial y} = \frac{z}{z^2 + (y - \eta_k)^2}, \quad \eta_2 = \eta_1, \quad k = 3 \text{ or } 2 \tag{10.3.42}$$
$$\frac{\partial T_k^{(32)}}{\partial z} = \frac{-(y - \eta_k)}{z^2 + (y - \eta_k)^2}$$

All of the quantities of subsection 10.3.6 are now recalculated using these modified values. The induced velocities from the last wake panel are added to the other dipole velocities of the lifting strip in the ordinary way.

10.4 Assembly of Final Velocities and Solution of the Equations for the Source and Dipole Strengths

10.4.1 Final Velocities

As described in Section 10.3 the induced velocity calculation produces a set source of velocities V_{ij} and a set of dipole velocities V_{ij}(dipole). These are expressed in the reference coordinate system. The source velocities are saved individually to yield an $N \times N$ matrix, where N is the number of on-body control points. The dipole velocities, however, are not save individually but are summed over each lifting strip. Specifically,

$$\vec{V}_j^{(k)} = \sum_j^{\text{strip}\,k} \vec{V}_{ij}(\text{dipole}), \quad k = 1, 2, \ldots, L \tag{10.4.1}$$

where L is the number of lifting strips and B_k is the value of B for the kth strip. The summation of (10.4.1) is over a complete lifting strip including wake panels, and "extra strip" velocities are added where appropriate. Thus Eq. (10.4.1) gives and $N \times L$ matrix of dipole velocities.

The total velocity at the ith control points is

$$\vec{V}_i = \sum_{j=1}^{N} \vec{V}_{ij}\sigma_j + \sum_{k=1}^{L} \vec{V}_i^{(k)} B_k + \vec{V}_\infty \tag{10.4.2}$$

where σ_j is the constant value of the source density on the jth panel. At this stage σ_j and B_k are not known, but when they have been determined (10.4.2) gives the velocity at the ith control point and

$$Cp_i = 1 - V_i^2 \tag{10.4.3}$$

gives the pressure coefficient for a freestream of unit magnitude.

10.4.2 The Kutta Condition

The Kutta condition is applied as a condition of pressure equality on the upper and lower surfaces of the trailing edge. As a numerical approximation the Kutta condition is applied by equating pressures at the control points of the two panels adjacent to the trailing edge on the upper and lower surfaces of the lifting strips. Let the subscript $m = 1, 2, \ldots, L$ denote the mth lifting strip and superscripts U and L denote upper and lower trailing edge panels. Then, in view of Eq. (10.4.3) the condition of pressure equality can be written

$$\vec{V}_m^{(U)} \cdot \vec{V}_m^{(U)} = \vec{V}_m^{(L)} \cdot \vec{V}_m^{(L)} \tag{10.4.4}$$

or

$$\vec{V}_m^{(U)} \cdot [\vec{V}_m^{(U)} + \vec{V}_m^{(L)}] = \vec{V}_m^{(L)} \cdot \left[\vec{V}_m^{(U)} + \vec{V}_m^{(L)}\right] \tag{10.4.5}$$

Finally,

$$(\vec{V}_m^{(U)} - \vec{V}_m^{(L)}) \cdot \vec{V}_{\text{avg}} = 0 \tag{10.4.6}$$

where $\vec{V}_{\text{avg}}$ is the average of the two velocities. Since V_i from (10.4.2) is linear in σ_j and B_k, Eq. (10.4.6) represents L quadratic equations in these quantities.

10.4.3 Iterative Matrix Solution

The velocity at the ith control point is given in Eq. (10.4.2). Taking the scalar products of these velocity vectors with the unit normal vector $\vec{n}_i$ of each control point gives an $N \times N$ matrix of source influence coefficients and an $N \times L$ matrix of dipole influence coefficients defining the normal-velocity influence of each elementary singularity distribution at every control point. If we define the solution vector X whose entries are the source strengths σ_j, $(j = 1, \ldots, N)$ followed by the dipole strengths B_k, $(k = 1, \ldots, L)$, the condition of zero normal velocity on the body can be written in the form

$$AX = R \tag{10.4.7}$$

where A is the $N \times (N + L)$ matrix of source and dipole normal velocity influence coefficients, and the right side is the negative of the normal component of the freestream velocity. It is important to note that A is purely geometric and does not depend on the onset flow, which enters only the right side.

Equation (10.4.7) defines a system of N linear equations in the $(N + L)$ unknown singularity strengths. A further set of L equations therefore are required to complete the formulation of the problem. These equations are provided by the Kutta condition, Eq. (10.4.6). The scheme adopted here for the solution of these equations is a block Gauss-Seidel iterative procedure. This section will outline the details of this iterative scheme. The acceleration scheme which has been adopted in order to improve the speed and the stability of the convergence procedure will be relegated to reference [10].

As pointed out above, the Kutta condition to be applied is nonlinear, and so it must be linearized in some manner consistent with the iterative solution procedure which is to be applied. If we introduce the subscript K to denote quantities evaluated after the Kth iteration, Eq. (10.4.6) can be written as

$$(\vec{V}_m^{(U)} - \vec{V}_m^{(L)})_K \cdot (\vec{V}_{\text{avg}})_{K-1} = 0 \tag{10.4.8}$$

so that the average velocity from the previous iteration is used. For the first iteration, $\vec{V}_{\text{avg}}$ is replaced by a vector along the freestream vector. Thus (10.4.8) is a set of linear equations whose coefficients change as the iterations proceed.

For lifting flow problems there is a strong coupling between the source and dipole strengths for a given lifting strip. Therefore, in order to maintain the diagonal dominance of the matrix, it is necessary to adopt a block Gauss-Seidel scheme. The particular approach used here takes the source strengths and the associated dipole strength for each lifting strip as separate blocks in the solution vector. In this way the normal velocity conditions for a given lifting strip are satisfied simultaneously along with the Kutta condition before proceeding with the solution for the next block. For nonlifting sections of a configuration, the choice of the block structure is less crucial. As the block size for such panels is increased, the computational effort is increased but the rate of convergence is also increased. The use of a block size of 50 has been found to give a good compromise.

This iterative scheme is equivalent to solving a series of quasi-two-dimensional problem corresponding to each block in the matrix. The onset flow for each of these calculations includes the current effects of all the other panels on the body. Therefore, as the solution converges, a fully consistent three-dimensional solution is obtained.

The iterative solution procedure can be broken down into two steps, the first of which involves the calculation of the right-hand side based on the previous solution,

$$RHS_{K-1} = R - A_u X_{K-1} \tag{10.4.9}$$

The second step is the calculation of the new approximation

$$A_\ell X_K = RHS_{K-1} \tag{10.4.10}$$

In these equations X_K is the Kth approximation to the solution, and the matrix A_ℓ represents the diagonal and lower triangular part of A, while A_u represents the upper triangular part. Each of these steps is performed successively for each block of unknown source strengths, each of which involves the direct solution of a small set of simultaneous equations. In addition, for lifting strips, the dipole strength is computed by satisfying the Kutta condition for that particular strip.

The above describes the implementation of a block Gauss-Seidel iteration scheme in a three-dimensional panel method calculation. For many external flow problems, such as the flow past wing-body configurations, this procedure converges very rapidly. However, for more complicated configurations, such as three-dimensional high-lift systems or wing/pylon/nacelle configurations, the convergence of this scheme becomes worse, and in some cases it can fail to converge entirely. A new convergence acceleration scheme has therefore been developed, which converges in all cases. It is described in [10].

References

[1] Murman, E.M., Cole, J.D., "Calculation of Plane Steady Transonic Flows," AIAA J., Vol. 9, 1971, pp. 114–121.

[2] Caughey, D.A., "The Computation of Transonic Potential Flows," Annual Review of Fluid Mechanics, Vol. 14, 1982.

[3] Johnson, F.T., et. al, "Application of the Tranair Rectangular Grid Approach to the Aerodynamic Analysis of Complex Configurations," AGARD-CP-464, 1989.

[4] Rizzi, A., "Damped Euler-Equation Method to Compute Transonic Flow Around Wing-Body Combinations," J. AIAA, Vol. 20, Oct. 1982, pp. 1321–1328.

[5] Jameson, A. and Yoon, S., "Multigrid Solution of the Euler Equations Using Implicit Schemes," J. AIAA, Vol. 24, Nov. 1986, pp. 1737–1743.

[6] Roe, P.L., "Characteristic-Based Schemes for the Euler Equations," Annual Review of Fluid Mechanics, Vol. 18, 1986, pp. 337–365.

[7] Hess, J.L., "The Problem Of Three-Dimensional Potential Flow and Its Solution by Means of a Surface Singularity Distribution," Computer Methods in Applied Mechanics and Engineering, Vol. 4, pp. 283–319, 1974.

[8] Hess, J.L. and Smith, A.M.O., "Calculation of Non-Lifting Potential Flow About Arbitrary Three-Dimensional Bodies," McDonnell Douglas Report No. ES 40622, 15 March 1962.

[9] Hess, J.L., "Calculation of Potential Flow About Arbitrary Three-Dimensional Lifting Bodies" McDonnell Douglas Report No. MDC J5679-01, October 1972.

[10] Hess, J.L., Friedman, D.M. and Clark, R.W., "Calculation of Compressible Flow About Three-Dimensional Inlets with Auxilary Inlets, Slat and Vanes by Means of a Panel Method," NASA CR-174975, June 1985.

11 Three-Dimensional Flows: Boundary-Layer Method

11.0 Introduction

In this chapter we extend the two-dimensional steady boundary-layer methods of Chapters 3 and 6 to three-dimensional steady flows. Section 11.1 presents the boundary-layer equations for a nonorthogonal coordinate system. The geometric parameters such as the metric coefficients and geodesic curvatures that appear in these equations are calculated with the interface program described in Section 11.5. Initial conditions are discussed in Section 11.2 and are followed by a description of the extension of the Cebeci-Smith turbulence model to three-dimensional flows. Section 11.4 presents the interaction law for three-dimensional flows. Transformed equations are given in Section 11.6 for both standard and inverse modes. The numerical solution of the three-dimensional boundary-layer equations are addressed first in the standard mode without flow reversal in subsection 11.7.1, then with flow reversal in subsection 11.7.2, and in the inverse mode in Section 11.8. Modelling of viscous effects in three-dimensional inviscid flows is addressed in Section 11.9.

11.1 Boundary-Layer Equations

In the solution of the boundary-layer equations for three-dimensional flows, the choice of an appropriate coordinate system is essential and a nonorthogonal system provides the necessary flexibility. For a wing defined in a Cartesian coordinate system $\bar{x}, \bar{y}, \bar{z}$ by

$$F(\bar{x}, \bar{y}, \bar{z}) = 0 \tag{11.1.1}$$

where $\bar{x}$ represents the direction of the airplane longitudinal axis, $\bar{y}$ the spanwise direction, and $\bar{z}$ normal to the $\bar{x}, \bar{z}$ plane, a nonorthogonal coordinate system

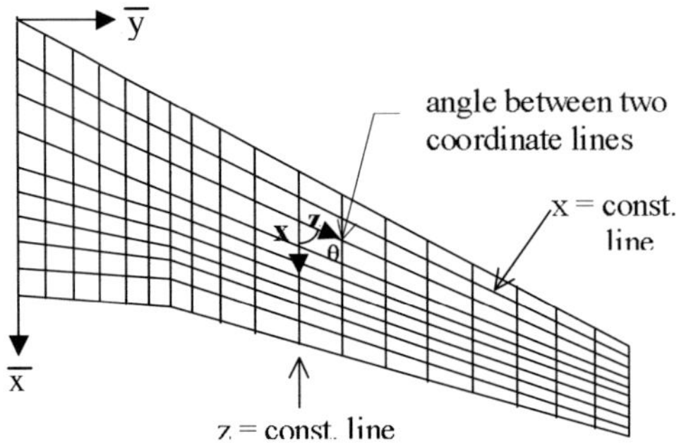

Fig. 11.1. The nonorthogonal coordinate system.

can be chosen by taking x to denote the coordinates along the lines formed by the intersection of the wing surface and planes representing constant percent semispan, z the coordinate along the constant percent chordlines which generate the wing surface, and y the coordinate normal to the wing surface, see Fig. 11.1.

The three-dimensional compressible laminar and turbulent boundary-layer equations for this system are given in [1] and may be expressed in the form

$$\frac{\partial}{\partial x}(\rho u h_2 \sin\theta) + \frac{\partial}{\partial z}(\rho w h_1 \sin\theta) + \frac{\partial}{\partial y}(\overline{\rho v} h_1 h_2 \sin\theta) = 0 \tag{11.1.2}$$

$$\begin{aligned} &\rho\frac{u}{h_1}\frac{\partial u}{\partial x} + \rho\frac{w}{h_2}\frac{\partial u}{\partial z} + \overline{\rho v}\frac{\partial u}{\partial y} - \rho\cot\theta K_1 u^2 + \rho\csc\theta K_2 w^2 + \rho K_{12} uw \\ &\quad = -\frac{\csc^2\theta}{h_1}\frac{\partial p}{\partial x} + \frac{\cos\theta\csc\theta}{h_2}\frac{\partial p}{\partial z} + \frac{\partial}{\partial y}\left(\mu\frac{\partial u}{\partial y} - \rho\overline{u'v'}\right) \end{aligned} \tag{11.1.3}$$

$$\begin{aligned} &\rho\frac{u}{h_1}\frac{\partial w}{\partial x} + \rho\frac{w}{h_2}\frac{\partial w}{\partial z} + \overline{\rho v}\frac{\partial w}{\partial y} - \rho\cot\theta K_2 w^2 + \rho\csc\theta K_1 u^2 + \rho K_{21} uw \\ &\quad = \frac{\cot\theta\csc\theta}{h_1}\frac{\partial p}{\partial x} - \frac{\csc^2\theta}{h_2}\frac{\partial p}{\partial z} + \frac{\partial}{\partial y}\left(\mu\frac{\partial w}{\partial y} - \rho\overline{w'v'}\right) \end{aligned} \tag{11.1.4}$$

$$\rho\frac{u}{h_1}\frac{\partial H}{\partial z} + \rho\frac{w}{h_2}\frac{\partial H}{\partial z} + \overline{\rho v}\frac{\partial H}{\partial y} = \frac{\partial}{\partial y}\left[\frac{\mu}{\Pr}\frac{\partial H}{\partial y} + \mu\left(1 - \frac{1}{\Pr}\right)\frac{\partial}{\partial y}\left(\frac{u_t^2}{2}\right) - \rho\overline{v'H'}\right] \tag{11.1.5}$$

Here $\overline{\rho v} = \rho v + \overline{\rho' v'}$ and $h_1(x,z)$ and $h_2(x,z)$ are the metric coefficients and θ represents the angle between the coordinate lines; $\theta = \pi/2$ for an orthogonal system. The parameters K_1 and K_2 are known as the geodesic curvatures of the curves z = constant and x = constant, respectively, and are given by

$$K_1 = \frac{1}{h_1 h_2 \sin\theta}\left[\frac{\partial}{\partial x}(h_2\cos\theta) - \frac{\partial h_1}{\partial z}\right], \tag{11.1.6a}$$

$$K_2 = \frac{1}{h_1 h_2 \sin\theta}\left[\frac{\partial}{\partial z}(h_1\cos\theta) - \frac{\partial h_2}{\partial x}\right], \tag{11.1.6b}$$

The parameters K_{12} and K_{21} are defined by

$$K_{12} = \frac{1}{h_1 h_2 \sin^2\theta}\left[\left(1+\cos^2\theta\right)\frac{\partial h_1}{\partial z} - 2\cos\theta\frac{\partial h_2}{\partial x}\right] \tag{11.1.6c}$$

$$K_{21} = \frac{1}{h_1 h_2 \sin^2\theta}\left[\left(1+\cos^2\theta\right)\frac{\partial h_2}{\partial x} - 2\cos\theta\frac{\partial h_1}{\partial z}\right] \tag{11.1.6d}$$

The magnitude of the velocity vector u_t in the boundary-layer is given by

$$u_t = (u^2 + w^2 + 2uw\cos\theta)^{1/2} \tag{11.1.7}$$

At the edge of the boundary-layer, Eqs. (11.1.3) and (11.1.4) reduce to

$$\begin{aligned}\rho_e\left(\frac{u_e}{h_1}\frac{\partial u_e}{\partial x} + \frac{w_e}{h_2}\frac{\partial u_e}{\partial z} - \cot\theta K_1 u_e^2 + \csc\theta K_2 w_e^2 + K_{12}u_e w_e\right)\\ = -\frac{\csc^2\theta}{h_1}\frac{\partial p}{\partial x} + \frac{\cot\theta\csc\theta}{h_2}\frac{\partial p}{\partial z}\end{aligned} \tag{11.1.8a}$$

$$\begin{aligned}\rho_e\left(\frac{u_e}{h_1}\frac{\partial w_e}{\partial x} + \frac{w_e}{h_2}\frac{\partial w_e}{\partial z} - \cot\theta K_2 w_e^2 + \csc\theta K_1 u_e^2 + K_{21}u_e w_e\right)\\ = \frac{\cot\theta\csc\theta}{h_1}\frac{\partial p}{\partial x} - \frac{\csc^2\theta}{h_2}\frac{\partial p}{\partial z}\end{aligned} \tag{11.1.8b}$$

The boundary conditions appropriate to Eqs. (11.1.2) to (11.1.5) are

Standard Mode

$$\begin{aligned} &y = 0: \quad u, w = 0, \quad v = v_w(x,z)\\ &H = H_w(x,z) \text{ or } \left(\frac{\partial H}{\partial y}\right)_w = -\frac{C_{p_w}}{k_w}\dot{q}_w\end{aligned} \tag{11.1.9a}$$

$$y = \delta, \quad u = u_e(x,z), \quad w = w_e(x,z), \quad H = H_e \tag{11.1.9b}$$

Inverse Mode

The above boundary conditions are also applicable to those in the inverse mode provided that $u_e(x,z)$ and $w_e(x,z)$ in Eq. (11.1.9b) are written in the form

$$u_e(x,z) = u_e^{\text{o}}(x,z) + \delta u_e(x,z) \tag{11.1.10a}$$

$$w_e(x,z) = w_e^{\text{o}}(x,z) + \delta w_e(x,z) \tag{11.1.10b}$$

with superscript o denoting the inviscid velocity components and the perturbation velocities $\delta u_e(x,z)$ and $\delta w_e(x,z)$ are computed utilizing the interaction law described in Section 11.4.

11.2 Initial Conditions

In general, the solution of the system of equations given in the previous section requires initial conditions in the (x, y)-plane at some $z = z_0$ and initial conditions in the (z, y)-plane at $x = x_0$. It is common to choose them along the wing-fuselage intersection (subsection 11.2.1) and along the wing leading edge aligned with the z-coordinate line (subsection 11.2.2) and often called the attachment line.

11.2.1 Quasi-Three-Dimensional Boundary-Layer Equations

The behavior of the flow along the wing-fuselage intersection is rather complicated. Unless special care is taken, it is likely to separate ahead of the junction generating an eddy and possibly eddies at the wing-root with a strong horseshoe vortex covering the upper and lower surfaces of the wing. Apart from the consequent loss of lift and drag increase due to the eddy, the boundary-layer on the wing receives an injection of fluid from the fuselage via the eddy boundary with unknown consequences for the flow properties on the wing.

Until a more general and soundly based procedure for calculating the wing-root flow has been developed, it is necessary to determine corresponding initial conditions with some approximations. A convenient procedure is to solve the boundary-layer equations that simulate quasi-three-dimensional flow conditions; that is, all derivatives with respect to z are neglected. This approximation leads to the following equations:

$$\frac{\partial}{\partial x}(\rho u h_2 \sin\theta) + \frac{\partial}{\partial y}(\overline{\rho v} h_1 h_2 \sin\theta) = 0 \tag{11.2.1}$$

$$\begin{aligned} \rho\frac{u}{h_1}\frac{\partial u}{\partial x} + \overline{\rho v}\frac{\partial u}{\partial y} - \rho\cot K_1 u^2 + \rho\csc\theta K_2 w^2 + \rho K_{12} u w \\ = \rho_e\left(\frac{u_e}{h_1}\frac{\partial u_e}{\partial x} - \cot\theta K_1 u_e^2 + \csc\theta K_2 w_e^2 + K_{12} u_e w_e\right) \\ + \frac{\partial}{\partial y}\left(\mu\frac{\partial u}{\partial y} - \rho\overline{u'v'}\right) \end{aligned} \tag{11.2.2}$$

$$\begin{aligned} \rho\frac{u}{h_1}\frac{\partial w}{\partial x} + \overline{\rho v}\frac{\partial w}{\partial y} - \rho\cot K_2 w^2 + \rho\csc\theta K_1 u^2 + \rho K_{21} u w \\ = \rho_e\left(\frac{u_e}{h_1}\frac{\partial w_e}{\partial x} - \cot\theta K_2 w_e^2 + \csc\theta K_1 u_e^2 + K_{21} u_e w_e\right) \\ + \frac{\partial}{\partial y}\left(\mu\frac{\partial w}{\partial y} - \rho\overline{w'v'}\right) \end{aligned} \tag{11.2.3}$$

$$\rho\frac{u}{h_1}\frac{\partial H}{\partial x} + \overline{\rho v}\frac{\partial H}{\partial y} = \frac{\partial}{\partial y}\left[\frac{\mu}{\Pr}\frac{\partial H}{\partial y} + \mu\left(1 - \frac{1}{\Pr}\right)\frac{\partial}{\partial y}\left(\frac{u_t^2}{2}\right) - \rho\overline{v'H'}\right] \tag{11.2.4}$$

The boundary conditions are the same as those given by Eqs. (11.1.9) and (11.1.11) except that u_e and w_e are independent of z.

In practice, corner fillets may reduce the problem of separation and, in this case, the above approximations based on boundary-layer theory may represent the details of the flow a little better than the case in which flow has extensive and/or large separation.

11.2.2 Attachment Line Equations

On the attachment line, u and $\partial p/\partial x$ are equal to zero and this makes the x-momentum equation (11.1.3) singular. However, differentiating this equation with respect to x and taking advantage of symmetry conditions

$$\frac{\partial w}{\partial x} = \frac{\partial v}{\partial x} = \frac{\partial^2 u}{\partial x^2} = 0$$

and using Eqs. (11.1.8), the attachment line equations can be written as

$$\rho h_2 \sin\theta u_x + \frac{\partial}{\partial z}(\rho w h_1 \sin\theta) + \frac{\partial}{\partial y}(\overline{\rho v} h_1 h_2 \sin\theta) = 0 \tag{11.2.5}$$

$$\begin{aligned} &\rho\frac{u_x^2}{h_1} + \rho\frac{w}{h_2}\frac{\partial u_x}{\partial z} + \overline{\rho v}\frac{\partial u_x}{\partial y} + \rho K_{12} w u_x \\ &\quad = \rho_e\left(\frac{u_{ex}^2}{h_1} + \frac{w_e}{h_2}\frac{\partial u_{xe}}{\partial z} + K_{12} w_e u_{xe}\right) + \frac{\partial}{\partial y}\left[\mu\frac{\partial u_x}{\partial y} - \rho(\overline{u'v'})_x\right] \end{aligned} \tag{11.2.6}$$

$$\begin{aligned} &\rho\frac{w}{h_2}\frac{\partial w}{\partial z} + \overline{\rho v}\frac{\partial w}{\partial y} - \rho\cot\theta K_2 w^2 \\ &\quad = \rho_e\left(\frac{w_e}{h_2}\frac{\partial w_e}{\partial z} - \cot\theta K_2 w_e^2\right) + \frac{\partial}{\partial y}\left(\mu\frac{\partial w}{\partial y} - \rho\overline{w'v'}\right) \end{aligned} \tag{11.2.7}$$

$$\rho\frac{w}{h_2}\frac{\partial H}{\partial z} + \overline{\rho v}\frac{\partial H}{\partial y} = \frac{\partial}{\partial y}\left[\frac{\mu}{\Pr}\frac{\partial H}{\partial y} + \mu\left(1 - \frac{1}{\Pr}\right)\frac{\partial}{\partial y}\left(\frac{u_t^2}{2}\right) - \rho\overline{v'H'}\right] \tag{11.2.8}$$

Here $u_x = \partial u/\partial x$, $u_{xe} = \partial u_e/\partial x$, $u_t = w$ and the equations are subject to the following boundary conditions

$$y = 0: \quad u_x = w = 0, \quad v = v_w, \quad H = H_w(x,z) \text{ or } \left(\frac{\partial H}{\partial y}\right)_w = -\frac{C_{p_w}}{k_w}\dot{q}_w \tag{11.2.9a}$$

$$y = \delta, \quad u_x = u_{xe}(x,z), \quad w = w_e(x,z), \quad H = H_e \tag{11.2.9b}$$

11.3 Turbulence Model

As in two-dimensional flow, the solution of the three-dimensional boundary-layer equations discussed in the previous sections also require turbulence models for the fluctuating quantities $-\rho\overline{u'v'}$, $-\rho\overline{w'v'}$ and $-\rho\overline{v'H'}$. Again, models based on zero, one or two equations as well as those based on Reynolds stress models may be used to satisfy the closure requirements. Here, as before, we use the zero equation model and extend the Cebeci-Smith model described in subsection 3.3.4 from two- to three-dimensional flows.

With the concepts of eddy viscosity ν_t and turbulent Prandtl number Pr_t, we define

$$-\overline{u'v'} = \nu_t \frac{\partial u}{\partial y}, \quad -\overline{w'v'} = \nu_t \frac{\partial w}{\partial y}, \quad -\overline{v'H'} = \frac{\nu_t}{\mathrm{Pr}_t}\frac{\partial H}{\partial y} \tag{11.3.1}$$

As discussed in [2], a generalization of the CS eddy-viscosity formulation of subsection 3.3.4 for three-dimensional boundary-layers is

$$(\nu_t)_i = l^2 \left[\left(\frac{\partial u}{\partial y}\right)^2 + \left(\frac{\partial w}{\partial y}\right)^2 + 2\cos\theta \frac{\partial u}{\partial y}\frac{\partial w}{\partial y}\right]^{1/2} \gamma_{tr} \quad 0 \le y \le y_c \tag{11.3.2a}$$

$$(\nu_t)_0 = \alpha \left| \int_0^\delta (u_{te} - u_t)\, dy \right| \gamma_{tr}\gamma \qquad y_c \le y \le \delta \tag{11.3.2b}$$

with $\alpha = 0.0168$, for small adverse pressure gradient flows. Its variation with strong pressure gradient flows can again be expressed by a generalization of Eq. (3.1.33) with u_τ defined for three-dimensional flows and with R_t now representing the ratio of the total wall shear to total Reynolds shear stress.

In Eq. (11.3.2a), the mixing length l is given by Eq. (3.1.29a) with A and N defined by Eq. (3.1.29b) except that now

$$u_\tau = \left(\frac{\tau_t}{\rho}\right)^{1/2}_{\max}, \quad \left(\frac{\tau_t}{\rho}\right)_{\max} = \nu \left[\left(\frac{\partial u}{\partial y}\right)^2 + \left(\frac{\partial w}{\partial y}\right)^2 + 2\cos\theta \left(\frac{\partial u}{\partial y}\right)\frac{\partial w}{\partial y}\right]^{1/2}_{\max} \tag{11.3.3}$$

In Eq. (11.3.2b), u_{te} is the total edge velocity defined by

$$u_{te} = (u_e^2 + w_e^2 + 2u_e w_e \cos\theta)^{1/2} \tag{11.3.4}$$

and u_t is the total velocity in the boundary-layer given by Eq. (11.1.7). The parameter θ again denotes the angle between the coordinate lines x and z and is equal to $\pi/2$ for an orthogonal system.

11.4 Interaction Law

The Hilbert integral formula given by Eq. (1.3.2b) can be extended to three-dimensional flows in order to account for the viscous effects. Roget [2] calculates the effects of a perturbation at the wall on inviscid flow solution using Fourier transforms. Equivalently, the presence of viscous effects in the potential flow can also be modelled by a change in boundary condition at the surface ($y = 0$) by imposing a blowing velocity $v_w(x, z)$ defined in a Cartesian coordinate system by

$$v_w(x, z) = \frac{\partial(u_e \delta_x^*)}{\partial x} + \frac{\partial(w_e \delta_z^*)}{\partial z} \tag{11.4.1}$$

where

$$\delta_x^* \equiv \int_0^\delta \left(1 - \frac{\rho u}{\rho_e u_e}\right) dy \tag{11.4.2a}$$

$$\delta_z^* \equiv \int_0^\delta \left(1 - \frac{\rho w}{\rho_e w_e}\right) dy \tag{11.4.2b}$$

If the body curvature is small so that a flat surface can be locally assumed, this boundary condition can be replaced by a source distribution of strength $q_w = 2v_w$ per unit area in the $y = 0$ plane, so that the potential is

$$F(x, z, 0) = \frac{1}{4\pi} \int_{x_a}^{x_b} \int_{z_a}^{z_b} \frac{q_w(\sigma, \lambda) d\sigma d\lambda}{\sqrt{(x - \sigma)^2 + (z - \lambda)^2}} \tag{11.4.3}$$

and the corresponding induced velocity is

$$\mathbf{V} = \nabla F \tag{11.4.4}$$

After integration by parts, assuming that q_w reaches zero on the boundaries of the domain, the perturbation velocity components for an orthogonal coordinate system can be written as

$$\delta u_e(x, z) = \frac{1}{2\pi} \int_{x_a}^{x_b} \int_{z_a}^{z_b} \frac{\partial(v_w(\sigma, \lambda))}{\partial \sigma} \frac{d\sigma d\lambda}{\sqrt{(x - \sigma)^2 + (z - \lambda)^2}} \tag{11.4.5a}$$

$$\delta w_e(x, z) = \frac{1}{2\pi} \int_{x_a}^{x_b} \int_{z_a}^{z_b} \frac{\partial(v_w(\sigma, \lambda))}{\partial \lambda} \frac{d\sigma d\lambda}{\sqrt{(x - \sigma)^2 + (z - \lambda)^2}} \tag{11.4.5b}$$

11.4.1 Quasi-Three-Dimensional Hilbert Integral

Under the quasi-three-dimensional approach, since all derivatives with respect to z can be neglected, the integrals for the perturbation velocities become

$$\delta u_e(x,z) = \frac{1}{2\pi}\int_{x_a}^{x_b}\int_{-\infty}^{+\infty} \frac{\partial(v_w(\sigma))}{\partial\sigma}\frac{d\sigma d\lambda}{\sqrt{(x-\sigma)^2+(z-\lambda)^2}} \tag{11.4.6a}$$

$$\delta w_e(x,z) = 0 \tag{11.4.6b}$$

so that the boundary condition for w_e remains unchanged compared with the direct or standard mode. From the potential equation, Eq. (11.4.3), and Eq. (11.4.4), the perturbation velocity δu_e can also be written as

$$\delta u_e(x,z) = -\frac{1}{2\pi}\int_{x_a}^{x_b} v_w(\sigma)\left[\lim_{a\to\infty}\frac{\partial}{\partial x}\left(\int_{-a}^{a}\frac{d\lambda}{\sqrt{(x-\sigma)^2+(z-\lambda)^2}}\right)\right]d\sigma \tag{11.4.7}$$

Since

$$\int_{-a}^{a}\frac{d\lambda}{\sqrt{(x-\sigma)^2+(z-\lambda)^2}} = \left[\ln\left|\gamma+\sqrt{\gamma^2+(x-\sigma)^2}\right|\right]_{-a_1}^{a_2} \tag{11.4.8}$$

where $a_1 = a - z$ and $a_2 = a - z$. After differentiating with respect to x and developing to first order, i.e.

$$\sqrt{a_i^2+(x-\sigma)^2} = a_i\left(1+\frac{(x-\sigma)^2}{2a_i^2}+O\left(\frac{(x-\sigma)^4}{a_i^4}\right)\right), \quad a_i\to\infty \tag{11.4.9}$$

we get

$$\lim_{a\to\infty}\frac{\partial}{\partial x}\left(\int_{-a}^{a}\frac{d\lambda}{\sqrt{(x-\sigma)^2+(z-\lambda)^2}}\right) = \frac{-2}{x-\sigma} \tag{11.4.10}$$

so that the perturbation velocity in the streamwise direction, δu_e, reduces to its value for the two-dimensional case, Eq. (1.3.2).

11.4.2 Interaction Law for Three-Dimensional Flows

We shall adopt Eq. (1.3.2) also for three-dimensional flows. Since in three-dimensional flows, unlike the displacement thickness of two-dimensional flows corresponds to a displacement surface, it is clear that the interaction law requires consideration of chordwise and spanwise directions at the same time and with the knowledge that the flow variables are influenced by upstream and downstream conditions in both directions. These features impose special problems since the boundary-layer equations are parabolic with a marching solution procedure, and the interaction procedure must be devised to meet these requirements. A plausible approach used here has the following ingredients. First, an initial displacement surface is generated by solving the quasi-three-dimensional boundary-layer equations, Eqs. (11.2.1) to (11.2.4), with the external velocity distribution obtained from three-dimensional inviscid flow equations. The second step involves interaction between the inviscid flow equations and the quasi-three-dimensional flow equations. As in two-dimensional flows, the solutions of the boundary-layer equations are used to compute distributions of blowing velocity on the surface, which are subsequently used to update the inviscid flow solutions. In step three, the fully three-dimensional boundary-layer equations are solved with the external velocity components resulting from step two. The chordwise edge velocity component is expressed, as in two-dimensional flows, as a sum of inviscid external velocity $u_e^0(x)$ and a perturbation velocity $\delta u_e(x)$ due to viscous effects. The spanwise edge velocity component is assumed to correspond to its inviscid value. The viscous flow solutions are obtained by marching in the spanwise direction at each advancing chordwise location. This represents the first phase in an interactive loop that involves the fully three-dimensional boundary-layer equations. In subsequent phases, as before, the blowing velocity distribution is used to obtain improved inviscid flow solutions so that the fully three-dimensional boundary-layer equations can be solved with this interactive procedure.

The neglect of viscous effects in the spanwise component w_e is assumed to be second order. This assumption is contrary to the irrotationality condition, but trial calculations involving variations of both velocity conditions suggest that errors were smaller than those associated with convergence.

11.5 Interface Program

The gemetric parameters such as the metric coefficients and geodesic curvature of the coordinate lines appearing in the three-dimensional boundary-layer equations must be known prior to the boundary-layer calculations for given freestream conditions and inviscid velocity distribution. The interface program described in this section computes the geometric parameters from the definition of wing geometry together with the nonorthogonal coordinate used in the boundary-layer method. The interface program also interpolates the inviscid velocity distribution for the boundary-layer grid which may be different from that calculated by the inviscid method.

11.5.1 Choice of the Surface Coordinate System

When a body-oriented coordinate system is used for three-dimensional boundary-layer calculations, it is necessary to compute the geometric parameters of the coordinate system (x_1, x_2) which correspond to the independent variables x and z defined in the boundary-layer equations given by Eqs. (11.1.2) to (11.1.5). To discuss this further, we consider a swept tapered wing which is defined in the Cartesian coordinate system $\bar{x}, \bar{y}, \bar{z}$ and has a sweep angle λ as shown in Fig. 11.2. Here $\bar{x}$ is in the direction of the airplane longitudinal axis, $\bar{y}$ is in the spanwise direction and $\bar{z}$ is normal to the $(\bar{x}, \bar{y})$ plane. We assume that the wing definition is given by a number of airfoil sections defined in the (x, z) coordinate system in planes $\bar{y} =$ constant. For simplicity we assume that the root and tip airfoil sections are defined in planes $\bar{y} = 0$ and $\bar{y} = b$ and have chord lengths of c_r and c_b, respectively.

The relationship between the two coordinate systems $(\bar{x}, \bar{z})$ and (x, z) for a point on the defining airfoil section (see Fig. 11.3) is given by the following expressions:

$$\bar{x} = \bar{x}_l + x \cos\alpha + z \sin\alpha \tag{11.5.1a}$$

$$\bar{z} = \bar{z}_l - x \sin\alpha + z \sin\alpha \tag{11.5.1b}$$

where α is the angle of attack of the defining airfoil section and, with subscripts l and t denoting leading and trailing edges, respectively, is given by

$$\alpha = \tan^{-1}\left(\frac{\bar{z}_l - \bar{z}_t}{\bar{x}_t - \bar{x}_l}\right) \tag{11.5.2}$$

Noting that the relation between the local airfoil chord c and α, for example,

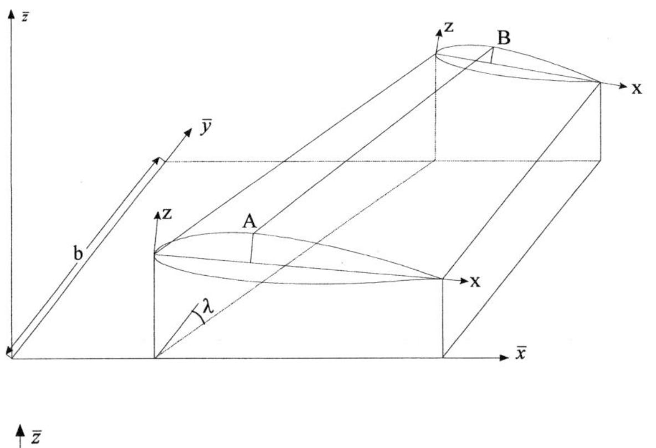

Fig. 11.2. Notation for the wing definition.

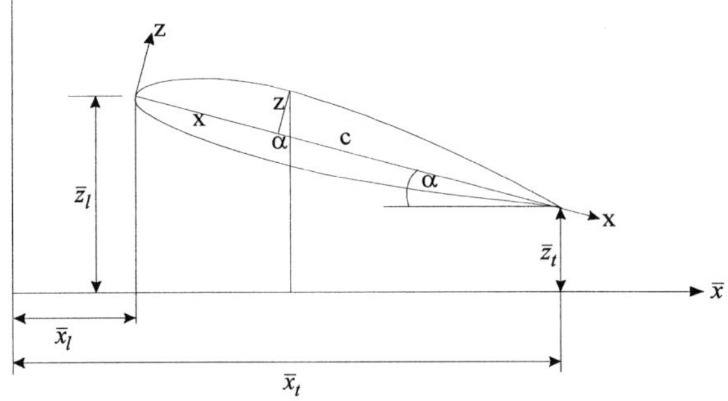

Fig. 11.3. Notation for the $(\bar{x}, \bar{z})$ and (x, z) coordinate systems.

$$c = \frac{\bar{x}_t - \bar{x}_l}{\cos\alpha} \tag{11.5.3}$$

Eqs. (11.5.1) can be written as

$$\bar{x} = \bar{x}_l + \left(\frac{x}{c}\right)(\bar{x}_t - \bar{x}_l) + \left(\frac{z}{c}\right)(\bar{z}_l - \bar{z}_t) \tag{11.5.4a}$$

$$\bar{z} = \bar{z}_l + \left(\frac{x}{c}\right)(\bar{z}_l - \bar{z}_t) + \left(\frac{z}{c}\right)(\bar{x}_t - \bar{x}_l) \tag{11.5.4b}$$

Since for an airfoil section

$$z = z(x)$$

it is clear from Eq. (11.5.4a) that $\bar{x}$ and $\bar{z}$ are functions of x and that they provide a relationship between $\bar{x}$ and x and $\bar{z}$ and z for a defining airfoil section.

From here on, the $\bar{x}$ and $\bar{z}$ coordinates of the defining airfoil sections will be designated by subscripts r and b. The sections in between the defining root and tip sections can be obtained for each dimensionless spanwise location $\bar{y}/b$ by projecting line AB in Fig. 11.2 into the $(\bar{x}, \bar{y})$ and $(\bar{y}, \bar{z})$ planes and by using linear interpolation (see Fig. 11.4).

$$\bar{x} = \bar{x}_r + (\bar{x}_b - \bar{x}_r)\left(\frac{\bar{y}}{b}\right) \tag{11.5.5a}$$

$$\bar{z} = \bar{z}_r + (\bar{z}_b - \bar{z}_r)\left(\frac{\bar{y}}{b}\right) \tag{11.5.5b}$$

Note that in the above equations $\bar{x}$ and $\bar{z}$ denote the coordinates of the airfoils in between the root and tip sections and they are distinct from $\bar{x}$ and $\bar{z}$ in Eqs. (11.5.4) which represent the coordinates of the root and tip sections. Equations (11.5.5) are the parametric equations of a wing surface which can, in principle, be expressed in the form given by Eq. (11.1.1), that is,

$$F(\bar{x}, \bar{y}, \bar{z}) = 0 \tag{11.1.1}$$

Since $\bar{x}$ and $\bar{z}$ are functions of x and $\bar{y}$, the surface of the wing can also be defined by coordinates (x/c) and $\bar{y}/b$, and the above relation can be written in the form

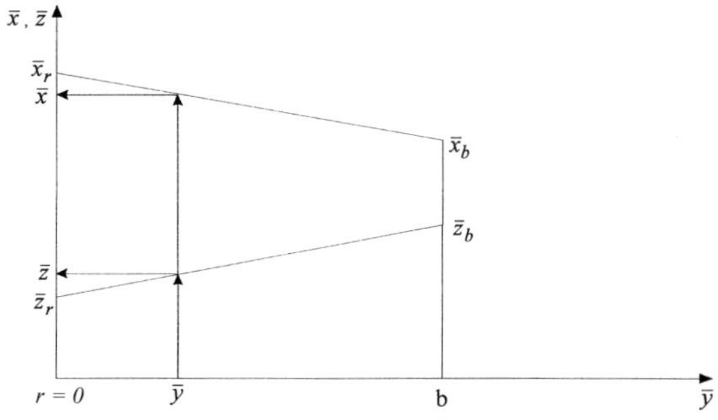

Fig. 11.4. Spanwise interpolation for section coordinates $\bar{x}$ and $\bar{z}$ along constant percent chord lines (x = const) at fixed $\bar{y}$.

$$f\left(\frac{x}{c}, \frac{\bar{y}}{b}\right) = 0 \tag{11.5.6}$$

thus defining a surface coordinate system formed by constant percent chord lines and constant percent spanlines.

The coordinates lines formed by constant (x/c) values are satisfactory provided the upper and lower surfaces of the wing are treated separately. When the stagnation point of the wing is located on the lower surface, the boundary-layer calculations contain a portion of the lower surface, and the meaning of coordinate (x/c) becomes ambiguous since z/c is a dual function of x/c and has a singularity at $x/c = 0$. To avoid this ambiguity and also to stretch the coordinate near the leading edge where flow quantities vary rapidly, we can, for example, relate x/c to a new coordinate x_1 by

$$\frac{x}{c} = 1 - \cos x_1 \tag{11.5.7}$$

Here $x_1 = 0$ corresponds to the chordwise coordinate in Fig. 11.2 where $x_1 = 0$ denotes the leading edge and $x_1 = \pm\frac{\pi}{2}$ to the trailing edge with positive values of x_1 representing the upper surface and negative values the lower surface.

In general, the surface coordinates x_1 and x_2 $(\equiv \bar{y}/b)$ which correspond to the independent variables x and z used in Eqs. (11.1.2) to (11.1.5), constitute a nonorthogonal coordinate system embedded in the wing surface.

11.5.2 Geometric Parameters of the Coordinate System

To discuss the calculation of the geometric parameters, we write the metric coefficient along a curve in space, with P denoting an independent parameter, as

$$h^2 = \left(\frac{\partial \bar{x}}{\partial P}\right)^2 + \left(\frac{\partial \bar{y}}{\partial P}\right)^2 + \left(\frac{\partial \bar{z}}{\partial P}\right)^2 \tag{11.5.8}$$

Taking $P = x_1$ along the curves $x_2 =$ const, we obtain

$$h_1^2 = \left(\frac{\partial \bar{x}}{\partial x_1}\right)^2 + \left(\frac{\partial \bar{z}}{\partial x_2}\right)^2 \tag{11.5.9a}$$

Similarly, taking $P = x_2$ along the curves $x_1 =$ const, we get

$$h_2^2 = b^2 + \left(\frac{\partial \bar{x}}{\partial x_2}\right)^2 + \left(\frac{\partial \bar{z}}{\partial x_2}\right)^2 \tag{11.5.9b}$$

where the derivatives in Eqs. (11.5.9) are obtained numerically.

Since the unit tangent vector $\vec{t}$ along a curve is given by

$$\vec{t} = \frac{d\vec{r}}{ds} = \frac{d\vec{r}}{dP}\frac{1}{(ds/dP)} = \frac{1}{h}\frac{d\vec{r}}{dP} \tag{11.5.10}$$

the unit tangent vector $\vec{t_1}$ along the curve $x_2 =$ const is

$$\vec{t_1} = \frac{1}{h_1}\left[\left(\frac{\partial \bar{x}}{\partial x_1}\right) i + \left(\frac{\partial \bar{z}}{\partial x_1}\right) k\right] \tag{11.5.11a}$$

where i, j, k denote the unit vectors in the coordinate directions $\bar{x}, \bar{y}, \bar{z}$, respectively. Similarly, the unit tangent vector $\vec{t_2}$ along the curve $x_1 = \text{const}$ is

$$\vec{t_2} = \frac{1}{h_2}\left[\left(\frac{\partial \bar{x}}{\partial x_2}\right) i + bj + \left(\frac{\partial \bar{z}}{\partial x_2}\right) k\right] \tag{11.5.11b}$$

The angle between the coordinate lines is then

$$\cos\theta = \vec{t_1} \cdot \vec{t_2} = \frac{1}{h_1 h_2}\left(\frac{\partial \bar{x}}{\partial x_1}\frac{\partial \bar{x}}{\partial x_2} + \frac{\partial \bar{z}}{\partial x_1}\frac{\partial \bar{z}}{\partial x_2}\right) \tag{11.5.12}$$

Once the metric coefficients h_1 and h_2 and their derivatives together with coordinate angle θ are determined, the curvature parameters are calculated from their definitions in Eqs. (11.1.6).

Once the surface coordinate system is selected, its geometric parameters can be calculated by making use of Eqs. (11.5.9), (11.5.12) and (11.1.6), and the relation between the Cartesian coordinates $(\bar{x}, \bar{y}, \bar{z})$ and the surface coordinates (x_1, x_2) can be written in the form:

$$\bar{x} = \bar{x}(x_1, x_2) \qquad \bar{z} = \bar{z}(x_1, x_2) \tag{11.5.13a}$$

$$\bar{y} = bx_2 \tag{11.5.13b}$$

We note that the geodesic curvature of the curve, $x_1 =$ costant, namely, K_2 given by Eq. (11.1.6b) is zero because of the straight-line development of the wing and that all orders of partial derivatives of $\bar{y}$ with respect to x_1 and x_2, except $\partial\bar{y}/\partial x_2 = b$, are zero and $\partial^2\bar{x}/\partial x_2^2$ and $\partial^2\bar{z}/\partial x_2^2$ are also zero because $\bar{x}$ and $\bar{z}$ are linear in x_2 (see Eqs. (11.5.5). Then h_1, h_2 and θ are calculated from Eqs. (11.5.9) and (11.5.12). With $K_2 = 0$, Eqs. (11.1.6a), (11.1.6c) and (11.1.6d) become, with $h_{12} = h_1 h_2 \cos\theta$,

$$K_1 = \frac{1}{h_1 h_2 \sin\theta}\left[\frac{1}{h_1}\frac{\partial h_{12}}{\partial x_1} - \frac{h_{12}}{h_1^2}\frac{\partial h_1}{\partial x_1} - \frac{\partial h_1}{\partial x_2}\right] \tag{11.5.14a}$$

$$K_{12} = \frac{1}{h_1 h_2 \sin^2\theta}\left[(1+\cos^2\theta)\frac{\partial h_1}{\partial x_2} - 2\cos\theta\frac{\partial h_2}{\partial x_1}\right] \tag{11.5.14b}$$

$$K_{21} = \frac{1}{h_1 h_2 \sin^2\theta}\left[(1+\cos^2\theta)\frac{\partial h_2}{\partial x_1} - 2\cos\theta\frac{\partial h_1}{\partial x_2}\right] \tag{11.5.14c}$$

The terms appearing in Eqs. (11.5.14) are obtained from Eqs. (11.5.9) and (11.5.12).

$$\frac{\partial h_1}{\partial x_1} = \frac{1}{h_1}\left(\frac{\partial \bar{x}}{\partial x_1}\frac{\partial^2 \bar{x}}{\partial x_1^2} + \frac{\partial \bar{z}}{\partial x_1}\frac{\partial^2 \bar{z}}{\partial x_1^2}\right) \tag{11.5.15a}$$

$$\frac{\partial h_2}{\partial x_1} = \frac{1}{h_2}\left(\frac{\partial \bar{x}}{\partial x_2}\frac{\partial^2 \bar{x}}{\partial x_1 \partial x_2} + \frac{\partial \bar{z}}{\partial x_2}\frac{\partial^2 \bar{z}}{\partial x_1 \partial x_2}\right) \tag{11.5.15b}$$

$$\frac{\partial h_{12}}{\partial x_1} = \frac{\partial \bar{x}}{\partial x_2}\frac{\partial^2 \bar{x}}{\partial x_1^2} + \frac{\partial \bar{z}}{\partial x_2}\frac{\partial^2 \bar{z}}{\partial x_1^2} + \frac{\partial \bar{x}}{\partial x_1}\frac{\partial^2 \bar{x}}{\partial x_1 \partial x_2} + \frac{\partial \bar{z}}{\partial x_1}\frac{\partial^2 \bar{z}}{\partial x_1 \partial x_2} \tag{11.5.15c}$$

$$\frac{\partial h_1}{\partial x_2} = \frac{1}{h_2}\left(\frac{\partial \bar{x}}{\partial x_1}\frac{\partial^2 \bar{x}}{\partial x_1 \partial x_2} + \frac{\partial \bar{z}}{\partial x_1}\frac{\partial^2 \bar{z}}{\partial x_1 \partial x_2}\right) \tag{11.5.15d}$$

To obtain the partial derivatives in Eqs. (11.5.9), (11.5.14) and (11.5.15), we differentiate Eqs. (11.5.5), which are functions of x/c and $\bar{y}/b$ and are related to x_1 and x_2 through Eqs. (11.5.7) and (11.5.13b). Also from Eqs. (11.5.1) we can obtain relations between $\bar{x}$, $\bar{z}$ and x for the defining root and tip sections, for example, for the root section, z

$$\bar{x}_r = \bar{x}_{tr} + x\cos\alpha_r + z\sin\alpha_r \tag{11.5.17a}$$

$$\bar{z}_r = \bar{z}_{lr} - x\sin\alpha_r + z\cos\alpha_r \tag{11.5.17b}$$

Since x is related to x_1 through Eq. (11.5.7), and z is a function of x, these equations can be used to find the partial derivatives of $\bar{x}_r$ and $\bar{z}_r$ with respect to x_1.

11.5.3 Calculation of Inviscid Velocity Components for Boundary-Layer Grid

Since the inviscid flow calculations use a Cartesian coordinate system and the inviscid velocity vector on the wing surface is available in terms of its Cartesian components $V_{\bar{x}}, V_{\bar{y}}, V_{\bar{z}}$,

$$\vec{V} = iV_{\bar{x}} + jV_{\bar{y}} + kV_{\bar{z}} \tag{11.5.18}$$

it is necessary to express $\vec{V}$ in the nonorthogonal boundary-layer coordinate system (x, z). This can be done by taking a dot product of unit vectors $\vec{t}_1$ and $\vec{t}_2$ with Eq. (11.5.18) to obtain the projections of the resultant velocity on the x and z axis, respectively.

$$V_x = \vec{V}\cdot\vec{t}_1, \qquad V_z = \vec{V}\cdot\vec{t}_2 \tag{11.5.19}$$

The nonorthogonal velocity components for the boundary-layer coordinate system are then obtained from (see Fig. 11.5)

$$u_e = \frac{V_x - V_z\cos\theta}{\sin^2\theta}, \qquad w_e = \frac{V_z - V_x\cos\theta}{\sin^2\theta} \tag{11.5.20}$$

where θ denotes the angle between the coordinate lines. The resultant inviscid velocity u_{te} is given by:

$$u_{te}^2 = u_e^2 + w_e^2 + 2u_e w_e\cos\theta \tag{11.5.21}$$

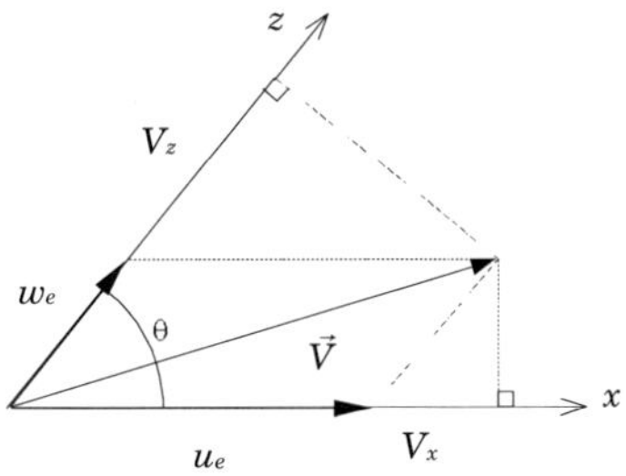

Fig. 11.5. Resolution of the inviscid velocity components for a nonorthogonal coordinate system.

In general, the velocity distribution grid of an inviscid method does not coincide with the grid used in boundary-layer calculations. For this reason, it is necessary to interpolate the inviscid velocity distribution once the velocity components u_e and w_e are computed for the inviscid grid. For simplicity, single variable interpolation (first in the x-direction then in the z-direction) is used.

11.6 Transformed Equations

As in two-dimensional flows, it is again desirable to express the three-dimensional boundary-layer equations in two sets of coordinate transformations, one for the standard mode and the other for the inverse mode.

11.6.1 Transformed Equations: Standard Mode

A convenient transformation for the standard mode is a modified version of the Falkner-Skan transformation given by Eq. (3.2.1). With $ds = h_1 dx$, transformations for the independent variables are

$$x = x, \qquad z = z, \qquad d\eta = \left(\frac{u_e}{\rho_e \mu_e s}\right)^{1/2} \rho dy, \qquad s = \int_0^x h_1 dx \tag{11.6.1}$$

For the dependent variables u, v, w, a two-component vector potential is introduced such that

$$\rho u h_2 \sin\theta = \frac{\partial \psi}{\partial y}, \qquad \rho w h_1 \sin\theta = \frac{\partial \phi}{\partial y}$$

$$\overline{\rho v} h_1 h_2 \sin\theta = -\left(\frac{\partial \psi}{\partial x} + \frac{\partial \phi}{\partial z}\right) + \rho_w v_w h_1 h_2 \sin\theta \tag{11.6.2}$$

together with dimensionless parameters f and g defined by

$$\psi = (\rho_e \mu_e u_e s)^{1/2} h_2 \sin\theta f(x, z, \eta)$$

$$\phi = (\rho_e \mu_e u_e s)^{1/2} u_0 / u_e h_1 \sin\theta g(x, z, \eta) \tag{11.6.3}$$

Thus, with the transformations and definitions given by Eqs. (11.6.1) to (11.6.3), together with the eddy viscosity and turbulent Prandtl number definitions given by Eq. (11.3.1), the three-dimensional boundary-layer equations and their boundary conditions, Eqs. (11.1.2) to (11.1.5) and (11.1.9), for compressible laminar and turbulent flows can be written as

$$\begin{aligned}(bf'')' + f''\theta + m_2(f')^2 + m_5 f'g' + m_8(g')^2 + m_{11}c + m_{13}f'' \\ = m_{10} f' \frac{\partial f'}{\partial x} + m_7 g' \frac{\partial f'}{\partial z}\end{aligned} \tag{11.6.4}$$

$$\begin{aligned}(bg'')' + g''\theta + m_4 f'g' + m_3(g')^2 + m_9(f')^2 + m_{12}c + m_{13}g'' \\ = m_{10} f' \frac{\partial g'}{\partial x} + m_7 g' \frac{\partial g'}{\partial z}\end{aligned} \tag{11.6.5}$$

$$\theta' = m_1 f' + m_6 g' + m_{10} \frac{\partial f'}{\partial x} + m_7 \frac{\partial g'}{\partial z} \tag{11.6.6}$$

$$(e_1 E')' + E'\theta + e_2' + m_{13}E' = m_{10} f' \frac{\partial E}{\partial x} + m_7 g' \frac{\partial E}{\partial z} \tag{11.6.7}$$

$$\eta = 0, \quad f = g = f' = g' = 0, \quad E_w \text{ or } E_w' \text{ given} \tag{11.6.8a}$$

$$\eta = \eta_e, \quad f' = 1, \quad g' = \overline{w_e}, \quad E = 1 \tag{11.6.8b}$$

Here primes denote differentiation with respect to η and

$$f' = u/u_e, \quad g' = w/u_0, \quad E = H/H_e, \quad \bar{u}_e = \frac{u_e}{u_0}, \quad \bar{w}_e = \frac{w_e}{u_0} \tag{11.6.9a}$$

$$b = C\left(1 + \nu_t^+\right), \quad C = \frac{\rho\mu}{\rho_e \mu_e}, \quad c = \frac{\rho_e}{\rho}, \quad \nu_t^+ = \nu_t/\nu \tag{11.6.9b}$$

$$\begin{aligned}e_1 &= \frac{C}{\Pr}\left(1 + \nu_t^+ \frac{\Pr}{\Pr_t}\right), \\ e_2 &= \frac{C u_0^2}{H_e}\left(1 - \frac{1}{\Pr}\right)\left[\bar{u}_e^2 f' f'' + g'g'' + \bar{u}_e \cos\theta (f'g'' + g'f'')\right]\end{aligned} \tag{11.6.9c}$$

The coefficients m_1 to m_{10} are given by

$$m_1 = \frac{s}{h_1}\left(\sqrt{\rho_e\mu_e u_e s}\, h_2 \sin\theta\right)^{-1} \frac{\partial}{\partial x}\left(\sqrt{\rho_e\mu_e u_e s}\, h_2 \sin\theta\right),$$
$$m_2 = \frac{-s}{h_1\bar{u}_e}\frac{\partial \bar{u}_e}{\partial x} + sK_1\cot\theta, \quad m_3 = \frac{s}{\bar{u}_e}\cot\theta K_2,$$
$$m_4 = -sK_{21}, \quad m_5 = -\frac{s}{h_2}\frac{1}{\bar{u}_e^2}\frac{\partial \bar{u}_e}{\partial z} - K_{12}s\left(\bar{u}_e\right)^{-1}$$
$$m_6 = \frac{s}{h_2}\left(\sqrt{\rho_e\mu_e u_e s}\, h_1 \sin\theta\right)^{-1} \frac{\partial}{\partial z}\left(\sqrt{\rho_e\mu_e u_e s}\, h_1 \sin\theta \bar{u}_e^{-1}\right),$$
$$m_7 = \frac{s}{h_2}\bar{u}_e^{-1}, \quad m_8 = -sK_2\csc\theta\left(\bar{u}_e\right)^{-2}, \tag{11.6.10}$$
$$m_9 = -sK_1\csc\theta\bar{u}_e, \quad m_{10} = \frac{s}{h_1},$$
$$m_{11} = -[m_2 + m_5\bar{w}_e + m_8\bar{w}_e^2],$$
$$m_{12} = -\left[m_3\bar{w}_e^2 + m_4\bar{w}_e + m_9 - m_{10}\frac{\partial \bar{w}_e}{\partial x} - m_7\bar{w}_e\frac{\partial \bar{w}_e}{\partial z}\right]$$
$$m_{13} = -\frac{\rho_w v_w}{\rho_e u_e}\sqrt{\frac{u_e s}{\nu_e}}$$

Attachment Line Equations

To transform the attachment-line equations, Eqs. (11.2.5) to (11.2.8), we define the transformed coordinates by

$$x = x, \quad z = z, \quad d\eta = \left(\frac{u_{xe}}{\rho_e\mu_e h_1}\right)^{1/2}\rho dy \tag{11.6.11}$$

the two-component vector potential by

$$\rho u_x h_2 \sin\theta = \frac{\partial\psi}{\partial y}, \quad \rho h_1 \sin\theta = \frac{\partial\phi}{\partial y},$$
$$\overline{\rho v} h_1 h_2 \sin\theta = -\left(\psi + \frac{\partial\phi}{\partial z}\right) + \rho_w v_w h_1 h_2 \sin\theta \tag{11.6.12a}$$

and the dimensionless ψ and ϕ by

$$\psi = (\rho_e\mu_e u_{xe} h_1)^{1/2} h_2 \sin\theta f(z,\eta)$$
$$\phi = (\rho_e\mu_e u_{xe} h_1)^{1/2} h_1 \sin\theta \frac{u_0}{u_{xe}} g(z,\eta) \tag{11.6.12b}$$

With these new variables, the attachment-line equations for a laminar flow can be written as

$$b(f'')' + \theta f'' - (f')^2 + m_5 f'g' + c\,(1 - \bar{w}_e m_5) + m_{13}f'' = m_7\left(g'\frac{\partial f'}{\partial z}\right) \tag{11.6.13}$$

$$(bg'')' + \theta g'' + m_3[(g')^2 - (\bar{w}_e)^2 c] + m_{13} g'' = m_7 \left(g' \frac{\partial g'}{\partial z} - c\bar{w}_e \frac{\partial \bar{w}_e}{\partial z} \right) \quad (11.6.14)$$

$$\theta' = f' + m_6 g' + m_7 \frac{\partial g'}{\partial z} \quad (11.6.15)$$

$$(e_1 E')' + \theta E' + e_2' + m_{13} E' = m_7 \left(g' \frac{\partial E}{\partial z} \right) \quad (11.6.16)$$

Here m_3, m_5, m_6 and m_7 are:

$$\begin{aligned} m_3 &= h_1 K_2 \cot\theta \frac{u_0}{u_{xe}} \\ m_5 &= -h_1 \frac{u_0}{u_{xe}} K_{12} - \frac{h_1}{h_2} \frac{u_0}{u_{xe}^2} \frac{\partial u_{xe}}{\partial z} \\ m_6 &= \frac{(\rho_e \mu_e u_{xe} h_1)^{-1/2}}{h_2 \sin\theta} \frac{\partial}{\partial z} \left(\sqrt{\rho_e \mu_e u_{xe}}\, h_1 \sin\theta \frac{u_0}{u_{xe}} \right) \\ m_7 &= \frac{h_1}{h_2} \frac{u_0}{u_{xe}}, \quad m_{13} = -\frac{\rho_w v_w}{\rho_e u_{xe}} \sqrt{\frac{u_{xe} h_1}{\nu_e}} \end{aligned} \quad (11.6.17)$$

The coefficients e_1 and e_2 are:

$$e_1 = \frac{C}{\Pr} \left(1 + \nu_t^+ \frac{\Pr}{\Pr_t} \right), \quad e_2 = \frac{C u_0^2}{H_e} \left(1 - \frac{1}{\Pr} \right) g' g'' \quad (11.6.18)$$

The boundary conditions are:

$$\eta = 0, \quad f = g = f' = g' = 0, \quad E_w \text{ or } E_w' \text{ given} \quad (11.6.19a)$$

$$\eta = \eta_e, \quad f' = 1, \quad g' = \bar{w}_e, \quad E = 1 \quad (11.6.19b)$$

Quasi-Three-Dimensional Boundary-Layer Equations

The quasi-three-dimensional boundary-layer equations are obtained from Eqs. (11.6.4) to (11.6.7) by neglecting the partial derivatives of the dependent variables with respect to z. They are given by

$$(bf'')' + f''\theta + m_2(f')^2 + m_5 f' g' + m_8 (g')^2 + m_{11} c + m_{13} f'' = m_{10} f' \frac{\partial f'}{\partial x} \quad (11.6.20)$$

$$(bg'')' + g''\theta + m_4 f' g' + m_3 (g')^2 + m_9 (f')^2 + m_{12} c + m_{13} g'' = m_{10} f' \frac{\partial g'}{\partial x} \quad (11.6.21)$$

$$\theta' = m_1 f' + m_6 g' + m_{10} \frac{\partial f'}{\partial x} \quad (11.6.22)$$

$$(e_1 E')' + E'\theta + e_2' + m_{13} E' = m_{10} f' \frac{\partial E}{\partial x} \quad (11.6.23)$$

where m_1 to m_4, m_7 to m_{13} remain the same; m_5 and m_6 become

$$m_5 = -K_{12} s \bar{u}_e^{-1}, \quad m_6 = 0 \quad (11.6.24)$$

11.6.2 Transformed Equations: Inverse Mode

Similar transformations are used to express the equations in the inverse mode except that those in Eqs. (11.6.1) and (11.6.3) are written in the forms:

$$d\eta = \left(\frac{u_0}{\rho_0\mu_0 s}\right)^{1/2}\rho dy$$

$$\psi = (\rho_0\mu_0 u_0 s)^{1/2} h_2 \sin\theta F(x,z,\eta)$$
$$\phi = (\rho_0\mu_0 u_0 s)^{1/2} h_1 \sin\theta G(x,z,\eta) \tag{11.6.25}$$

to avoid the use of u_e, which is treated as an unknown in the calculations. The parameters expressed with subscript o indicate constant reference quantities which are usually taken as corresponding to freestream conditions. The resulting momentum and energy equations and their boundary conditions are written in a form identical to those given by Eqs. (11.6.4) to (11.6.7) with the pressure gradient parameters m_1 and m_{13} defined by

$$m_1 = \frac{s^{1/2}}{h_1 h_2 \sin\theta}\frac{\partial}{\partial x}(s^{1/2}h_2\sin\theta), \quad m_2 = -sK_1\cot\theta,$$

$$m_3 = sK_2\cot\theta, \quad m_4 = -sK_{21}, \quad m_5 = -sK_{12},$$

$$m_6 = \frac{s^{1/2}}{h_1 h_2 \sin\theta}\frac{\partial}{\partial z}(s^{1/2}h_1\sin\theta),$$

$$m_7 = \frac{s}{h_2}, \quad m_8 = -sK_2\csc\theta, \quad m_9 = -sK_1\csc\theta, \quad m_{10} = \frac{s}{h_1},$$

$$-m_{11} = m_2\bar{u}_e^2 + m_5\bar{u}_e\bar{w}_e + m_8\bar{w}_e^2 - m_{10}\bar{u}_e\frac{\partial\bar{u}_e}{\partial x} - m_7\bar{w}_e\frac{\partial\bar{u}_e}{\partial z},$$

$$-m_{12} = m_3\bar{w}_e^2 + m_4\bar{u}_e\bar{w}_e + m_9\bar{u}_e^2 - m_{10}\bar{u}_e\frac{\partial\bar{w}_e}{\partial x} - m_7\bar{w}_e\frac{\partial\bar{w}_e}{\partial z}$$

$$m_{13} = -\frac{\rho_w v_w}{\rho_0 u_0}\sqrt{\frac{u_0 s}{\nu_0}}$$

The parameter e_1, remains the same but e_2 is defined by

$$e_2 = \frac{Cu_0^2}{H_e}\left(1 - \frac{1}{\Pr}\right)[F'F'' + G'G'' + \cos\theta(F'G'' + G'F'')]$$

The boundary conditions are

$$\eta = 0: \quad F = G = f' = G' = E' = 0, \tag{11.6.26a}$$

$$\eta = \eta_e: \quad F' = \bar{u}_e, \quad G' = \bar{w}_e, \quad E = 1 \tag{11.6.26b}$$

Quasi-Three-Dimensional Boundary-Layer Equations

The quasi-three-dimensional boundary-layer equations in the inverse mode, as before, are identical to those given by Eqs. (11.6.20) to (11.6.23). The m's and e's remain the same except

$$m_5 = -K_{12}s(\bar{u}_e)^{-1}, \quad m_6 = 0 \tag{11.6.27b}$$

and m_1, m_2 do not have the derivatives of $\bar{u}_e$ und $\bar{w}_e$ with respect to z.

11.7 Solution of the Three-Dimensional Boundary-Layer Equations: Standard Mode

The solution of the steady three-dimensional boundary-layer equations in the standard mode as in the solution of the unsteady two-dimensional boundary-layer equations, is routine when velocity components u and w are positive and can again be obtained with Keller's box method, discussed in subsection 11.7.1. When the spanwise velocity component contains regions of flow reversal, however, the solution of the steady three-dimensional boundary-layer equations is not so straightforward as unsteady two-dimensional boundary-layer equations with flow reversal and requires special procedures to avoid the numerical instabilities that can result from flow reversals in w. A common choice is to use the characteristic box scheme described in subsection 8.4.1 for two-dimensional unsteady flows; it is discussed here in subsection 11.7.2 for three-dimensional steady flows.

For flows with separation (negative u), the procedure for solving the three-dimensional boundary-layer equations is more involved than the procedure for solving the three-dimensional boundary-layer equations with negative spanwise velocity component. The solution procedure not only requires an inverse formulation such as the one described in Chapter 3 for two-dimensional steady flows, but it also requires that the domains of dependence and regions of influence are satisfied. To explain this further, it is helpful to review briefly the general principles that accurate solutions must obey, as discussed in [3]. These principles follow from the realization that the momentum equations are diffusive normal to the body and wave-like parallel to the body, the direction of propagation being along the local stream direction. Since this direction varies across the boundary-layer, it is possible to identify zones of influence and dependence for any point and to obtain solutions to the three-dimensional boundary-layer equations such that their solutions obey these zones. To explain this further, it is useful to consider the grid of Fig. 11.6a, in which the solutions are known at the points indicated by x and are required at P. The wall and external streamlines have been drawn with the assumption that both the u and w velocity components are positive. The domain of dependence of point P is denoted by the shaded

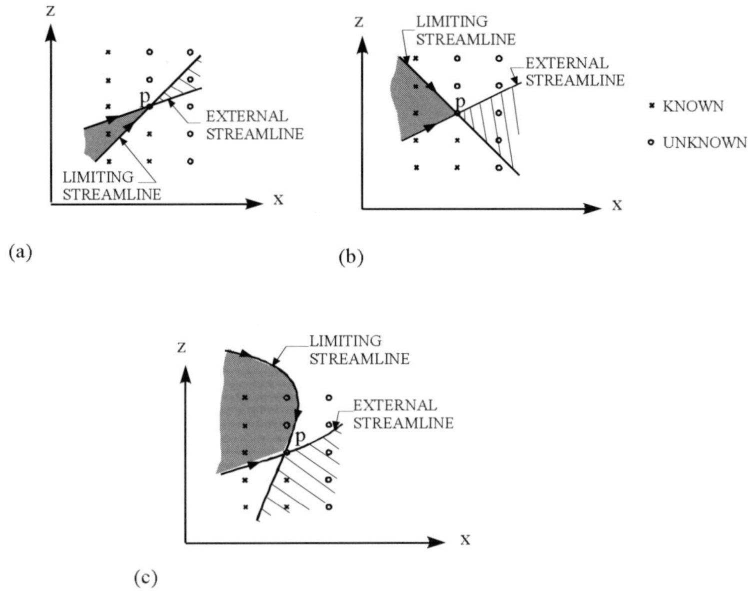

Fig. 11.6. Domains of dependence (shaded area) and regions of influence (hatched area) of point P when (a) both u and w in the boundary-layer are positive, (b) u is positive across the layer and only w has some negative values, and (c) both u and w have some negative values.

area, and the region of influence of point P corresponds to the hatched area in which the solution is altered when a change occurs in the solution at the point P. The information required to obtain a solution at point P comes from the domain of dependence.

When the u velocity is positive across the layer but the w velocity component is negative near the wall, the wall streamlines cause the domains of dependence and regions of influence to change, as shown in Fig. 11.6b. The wall streamlines, which were beneath the external streamlines move above the external streamlines. The angle between these two streamlines widens further when both velocity components of the wall streamlines become negative (Fig. 11.6c) and information comes from a region, which is not accessible from P so that the extent of possible calculations is restricted.

In this section, we consider the solution of the three-dimensional boundary-layer equations for cases that correspond to sketches in Figs 11.6a and 11.6b. In regions of positive u and w velocity profiles, we use the regular box scheme as described in subsection 11.7.1; in regions of negative w but positive u, we use the characteristic box scheme as described in subsection 11.7.2 For separated flows (negative u), as described in Section 11.8, we use the zig-zag box scheme since the solution procedure for separated flows with the characteristic box scheme is still in the development stage.

11.7.1 Three-Dimensional Steady Flows Without Reversal

To solve the momentum and continuity equations and their boundary conditions given by Eqs. (11.6.4) to (11.6.6) and (11.6.8) with the regular box scheme described in Section 8.4 for two-dimensional unsteady flows, again new dependent variables $u(x,z,\eta)$, $w(x,z,\eta)$, $v(x,z,\eta)$ and $t(x,z,\eta)$ are introduced so that the continuity and two momentum equations and their boundary conditions with $f' = u$ and $g' = w$ can be written as

$$u' = v \tag{11.7.1a}$$

$$w' = t \tag{11.7.1b}$$

$$\theta' = m_1 u + m_6 w + m_{10}\frac{\partial u}{\partial x} + m_7 \frac{\partial w}{\partial z} \tag{11.7.1c}$$

$$(bv)' + v\theta + m_2 u^2 + m_5 uw + m_8 w^2 + m_{11} c + m_{13} v = m_{10} u \frac{\partial u}{\partial x} + m_7 w \frac{\partial u}{\partial z} \tag{11.7.1d}$$

$$(bt)' + t\theta + m_4 uw + m_3 w^2 + m_9 u^2 + m_{12} c + m_{13} t = m_{10} u \frac{\partial w}{\partial x} + m_7 w \frac{\partial w}{\partial z} \tag{11.7.1e}$$

$$\eta = 0, \quad u = w = \theta = 0; \quad \eta = \eta_e, \quad u = 1, \quad w = \bar{w}_e \tag{11.7.2}$$

We now consider the net cube shown in Fig. 11.7 and denote the net points by

$$\begin{array}{lll} x^0 = 0 & x^n = x^{n-1} + k_n & n = 1, 2, \ldots, N \\ z^0 = 0 & z^i = z^{i-1} + r_i & i = 1, 2, \ldots, I \\ \eta^0 = 0 & \eta_j = \eta_{j-1} + h_j & j = 1, 2, \ldots, J \end{array} \tag{11.7.3}$$

We approximate the quantities (u, w, θ, v, t) at points (x^n, z^i, η_j) of the net by $(u_j^{n,i}, w_j^{n,i}, \theta_j^{n,i}, v_j^{n,i}, t_j^{n,i}) \equiv \vec{\Lambda}_j^{n,i}$. The difference equations that are to approximate Eqs. (11.7.1a) and (11.7.1b) are obtained by averaging about the midpoint $(x^n, z^i, \eta_{j-1/2})$

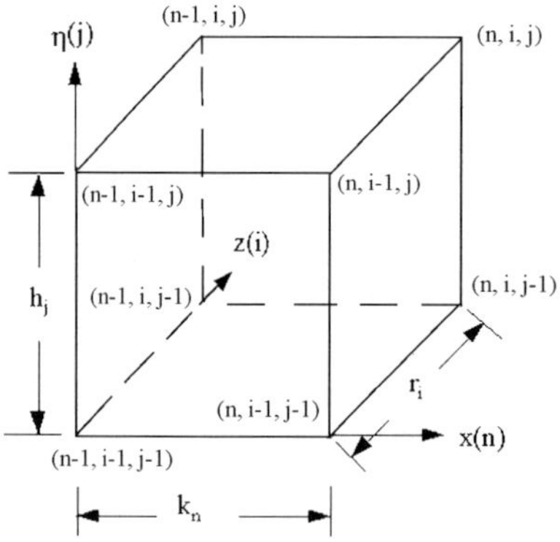

Fig. 11.7. Net cube for the difference equations for three-dimensional flows.

$$\frac{(u_j - u_{j-1})}{h_j} = v_{j-1/2} \tag{11.7.4a}$$

$$\frac{(w_j - w_{j-1})}{h_j} = t_{j-1/2} \tag{11.7.4b}$$

where we have neglected the superscripts n, i on u, v, w and t and defined the averaged $v_{j-1/2}$ by

$$v_{j-1/2} = 1/2(v_j^{n,i} + v_{j-1}^{n,i})$$

The difference approximations to Eq. (11.7.1c) are obtained by using the procedure used for Eq. (8.4.9b), that is, all quantities except θ are centered at the center of the cube $(x^{n-1/2}, z^{i-1/2}, \eta_{j-1/2})$ by taking the values of each parameter at the four corners of the box. The parameter θ is centered as in Eq. (8.4.14).

In terms of this notation, and the notation defined in Eqs. (8.3.5) and (8.3.7), the finite-difference approximations to Eq. (11.7.1c) can be written in the following form:

$$\begin{aligned}&\frac{\theta_j - \theta_{j-1}}{h_j} - \bar{m}_1(\bar{u})_{j-1/2} - \bar{m}_6(\bar{w})_{j-1/2} \\ &= \frac{\bar{m}_{10}}{k_n}(\bar{u}_{j-1/2}^n - \bar{u}_{j-1/2}^{n-1}) + \frac{\bar{m}_7}{r_i}(\bar{w}_{j-1/2}^i - \bar{w}_{j-1/2}^{i-1})\end{aligned} \tag{11.7.5}$$

Using the definitions in Eqs. (8.3.5) and (8.3.7), Eq. (11.7.5) can also be written as

$$\theta_j - \theta_{j-1} - h_j\left[\frac{\bar{m}_1}{4} + \frac{\bar{m}_{10}}{2k_n}\right]u_{j-1/2} - h_j\left[\frac{\bar{m}_6}{4} + \frac{\bar{m}_7}{2r_i}\right]w_{j-1/2} = h_j\lambda_1 \tag{11.7.6}$$

where

$$\begin{aligned}\lambda_1 = {} & \frac{\bar{m}_1}{4}u_{j-1/2}^{234} + \frac{\bar{m}_6}{4}w_{j-1/2}^{234} + \frac{\bar{m}_{10}}{2k_n}(u_{j-1/2}^{n,i-1} - 2\bar{u}_{j-1/2}^{n-1}) \\ & + \frac{\bar{m}_7}{2r_i}(w_{j-1/2}^{n-1,i} - 2\bar{w}_{j-1/2}^{i-1})\end{aligned} \tag{11.7.7}$$

The difference equations that are to approximate Eqs. (11.7.1d,e) are written similar to Eq. (11.7.1c); again all quantities except θ are written at the center of the cube $(x^{n-1/2}, z^{i-1/2}, \eta_{j-1/2})$ by taking the values of each parameter at the four corners of the box. The resulting expressions are rather lengthy and to illustrate the procedure, we consider a model equation for Eqs. (11.7.1d) and (11.7.1e),

$$(bv)' + v\theta + m_8 w^2 = m_{10}u\frac{\partial u}{\partial x} + m_7 w\frac{\partial u}{\partial z} \tag{11.7.8}$$

The finite-difference approximations are

$$h_j^{-1}[(\overline{bv})_j - (\overline{bv})_{j-1}] + \bar{v}_{j-1/2}\theta_{j-1/2} + \bar{m}_8(\overline{w^2})_{j-1/2}$$
$$= \frac{\bar{m}_{10}}{k_n}\bar{u}_{j-1/2}(\bar{u}^n_{j-1/2} - \bar{u}^{n-1}_{j-1/2}) + \frac{\bar{m}_7}{r_i}\bar{w}_{j-1/2}(\bar{u}^i_{j-1/2} - \bar{u}^{i-1}_{i-1/2}) \tag{11.7.9}$$

where, for example, see Eq. (8.3.5),

$$(\overline{w^2})_{j-1/2} = \frac{1}{4}\left[(w^2)^{n,i}_{j-1/2} + (w^2)^{234}_{j-1/2}\right]$$

With the definitions in Eqs. (8.3.5) and (8.3.7), Eq. (11.7.9) can be written as

$$\frac{[(bv)_j - (bv)_{j-1}]}{h_j} + \theta_{j-1/2}(v_{j-1/2} + v^{234}_{j-1/2})$$
$$+ \bar{m}_8(w^2)_{j-1/2} - \frac{\bar{m}_{10}}{2k_n}[(u_{j-1/2})^2 + u_{j-1/2}e_1] \tag{11.7.10}$$
$$- \frac{\bar{m}_7}{2r_i}[w_{j-1/2}u_{j-1/2} + w_{j-1/2}e_3 + u_{j-1/2}w^{234}_{j-1/2}] = \lambda_2$$

where

$$\lambda_2 = -\frac{[(bv)^{234}_j - (bv)^{234}_{j-1}]}{h_j} - \bar{m}_8(w^2)^{234}_{j-1/2}$$
$$+ \frac{\bar{m}_{10}}{2k_n}[u^{234}_{j-1/2}(u^{n,i-1}_{j-1/2} - 2\bar{u}^{n-1}_{j-1/2})]$$
$$+ \frac{\bar{m}_7}{2r_i}[w^{234}_{j-1/2}(u^{n-1,i}_{j-1/2} - 2\bar{u}^{i-1}_{j-1/2})]$$

$$e_1 = u^{234}_{j-1/2} + u^{n,i-1}_{j-1/2} - 2\bar{u}^{n-1}_{j-1/2}, \quad e_3 = u^{n-1,i}_{j-1/2} - 2\bar{u}^{i-1}_{j-1/2}$$

The boundary conditions given by Eq. (11.7.2) become

$$u_0 = w_0 = \theta_0 = 0, \quad u_J = 1, \quad w_J = \bar{w}_e \tag{11.7.11}$$

Before the difference equations (11.7.4a,b), (11.7.6) and the difference equations for Eqs. (11.7.1d,e) can be solved subject to Eq. (11.7.11), the net functions $\vec{\Lambda}^{n-1,i-1}_j$, $\vec{\Lambda}^{n,i-1}_j$ and $\vec{\Lambda}^{n-1,i}_j$ must be known for all values of j. Those corresponding to $\vec{\Lambda}^{n-1,i-1}_j$ and $\vec{\Lambda}^{n-1,i}_j$ are obtained from the initial conditions specified in the spanwise direction and those to $\vec{\Lambda}^{n,i-1}_j$ from the initial conditions specified in the chordwise direction. If we assume that they are known, then the resulting difference equations for Eqs. (11.7.1) subject to Eq. (11.7.2) are a system of $5J+5$ equations for the solution of $5J+5$ unknowns $\vec{\Lambda}^{n,i}_j$, $j = 0, 1, \ldots, J$. To solve this nonlinear system we again use Newton's method. For simplicity of notation the net functions $\vec{\Lambda}^{n,i}_j$ are written as $\vec{\Lambda}_j$ ($\equiv u_j, w_j, \theta_j, v_j, t_j$). Then the system resulting from Eqs. (11.7.4a,b), (11.7.6) and those resulting from Eqs. (11.7.1d,e) can be written as

$$\delta u_j - \delta u_{j-1} - \frac{h_j}{2}(\delta v_j + \delta v_{j-1}) = (r_4)_{j-1} \tag{11.7.12a}$$

$$\delta w_j - \delta w_{j-1} - \frac{h_j}{2}(\delta t_j + \delta t_{j-1}) = (r_5)_{j-1} \tag{11.7.12b}$$

$$(c_1)_j(\delta\theta_j - \delta\theta_{j-1}) + (c_2)_j(\delta u_j + \delta u_{j-1}) + (c_3)_j(\delta w_j + \delta w_{j-1}) = (r_1)_j \tag{11.7.12c}$$

$$\begin{aligned}(s_1)_j\delta v_j + (s_2)\delta v_{j-1} + (s_3)_j\delta u_j + (s_4)_j\delta u_{j-1} + (s_5)_j\delta w_j\\ + (s_6)_j\delta w_{j-1} + (s_7)_j\delta\theta_j + (s_8)_j\delta\theta_{j-1} = (r_2)_j\end{aligned} \tag{11.7.12d}$$

$$\begin{aligned}(\beta_1)_j\delta t_j + (\beta_2)\delta t_{j-1} + (\beta_3)_j\delta u_j + (\beta_4)_j\delta u_{j-1} + (\beta_5)_j\delta w_j\\ + (\beta_6)_j\delta w_{j-1} + (\beta_7)_j\delta\theta_j + (\beta_8)_j\delta\theta_{j-1} = (r_3)_j\end{aligned} \tag{11.7.12e}$$

where

$$(c_1)_j = h_j^{-1}, \quad (c_2)_j = -\frac{1}{2}\left(\frac{\bar{m}_1}{4} + \frac{\bar{m}_{10}}{2k_n}\right), \quad (c_3)_j = -\frac{1}{2}\left(\frac{\bar{m}_6}{4} + \frac{\bar{m}_7}{2r_i}\right) \tag{11.7.13a}$$

$$(r_1)_j = \lambda_1 - h_j^{-1}(\theta_j - \theta_{j-1}) + \left[\frac{\bar{m}_1}{4} + \frac{\bar{m}_{10}}{2k_n}\right]u_{j-1/2} + \left[\frac{\bar{m}_6}{4} + \frac{\bar{m}_7}{2r_i}\right]w_{j-1/2} \tag{11.7.13b}$$

$$(r_4)_{j-1} = h_j v_{j-1/2} - (u_j - u_{j-1}) \tag{11.7.13c}$$

$$(r_5)_{j-1} = h_j t_{j-1/2} - (w_j - w_{j-1}) \tag{11.7.13d}$$

The coefficients $(r_2)_j$, $(r_3)_j$ and $(s_k)_j$ and $(\beta_k)_j$ $(k = 1, \ldots, 8)$ follow from the complete finite-difference equations for Eqs. (11.7.1d,e). For Eq. (11.7.1d), they are

$$\begin{aligned}(r_2)_j = -4\Big\{& h_j^{-1}[(\overline{bv})_j - (\overline{bv})_{j-1}] + \bar{v}_{j-1/2}\theta_{j-1/2} + \bar{m}_2(\overline{u^2})_{j-1/2}\\ &+ \bar{m}_5(\overline{uw})_{j-1/2} + \bar{m}_8(\overline{w^2})_{j-1/2} + \bar{m}_{11}\bar{c}_{j-1/2} + \bar{m}_{13}\bar{v}_{j-1/2}\\ &- \frac{\bar{m}_{10}}{k_n}\bar{u}_{j-1/2}(\bar{u}^n_{j-1/2} - \bar{u}^{n-1}_{j-1/2}) - \frac{\bar{m}_7}{r_i}\bar{w}_{j-1/2}(\bar{u}^i_{j-1/2} - \bar{u}^{i-1}_{j-1/2})\Big\}\end{aligned} \tag{11.7.14}$$

$$(s_1)_j = h_j^{-1}b_j + \frac{1}{2}\theta_{j-1/2} + \frac{\bar{m}_{13}}{2} \tag{11.7.15a}$$

$$(s_2)_j = -h_j^{-1}b_{j-1} + \frac{1}{2}\theta_{j-1/2} + \frac{\bar{m}_{13}}{2} \tag{11.7.15b}$$

$$(s_3)_j = \bar{m}_2 u_j - \frac{\bar{m}_{10}}{k_n}\left\{\frac{1}{2}(\bar{u}^n_{j-1/2} - \bar{u}^{n-1}_{j-1/2}) + \bar{u}_{j-1/2}\right\} + \frac{\bar{m}_5}{2}w_j - \frac{\bar{m}_7}{r_i}\bar{w}_{j-1/2} \tag{11.7.15c}$$

$$(s_4)_j = \bar{m}_2 u_{j-1} - \frac{\bar{m}_{10}}{k_n}\left\{\frac{1}{2}(\bar{u}^n_{j-1/2} - \bar{u}^{n-1}_{j-1/2}) + \bar{u}_{j-1/2}\right\} + \frac{\bar{m}_5}{2}w_{j-1} - \frac{\bar{m}_7}{r_i}\bar{w}_{j-1/2} \tag{11.7.15d}$$

$$(s_5)_j = \frac{\bar{m}_5}{2}u_j + \bar{m}_8 w_j - \frac{\bar{m}_7}{2r_i}(\bar{u}^i_{j-1/2} - \bar{u}^{i-1}_{j-1/2}) \tag{11.7.15e}$$

$$(s_6)_j = \frac{\bar{m}_5}{2}u_{j-1} + \bar{m}_8 w_{j-1} - \frac{\bar{m}_7}{2r_i}(\bar{u}^i_{j-1/2} - \bar{u}^{i-1}_{j-1/2}) \tag{11.7.15f}$$

$$(s_7)_j = 2\bar{v}_{j-1/2} \tag{11.7.15g}$$

$$(s_8)_j = 2\bar{v}_{j-1/2} \tag{11.7.15h}$$

For Eq. (11.7.1e), $(r_3)_j$ and $(\beta_1)_j$ to $(\beta_8)_j$ are

$$\begin{aligned}(r_3)_j = -4\Big\{ & h_j^{-1}[(\overline{bt_j})_j - (\overline{bt_j})_{j-1}] + \bar{t}_{j-1/2}\theta_{j-1/2} + \bar{m}_4(\overline{uw})_{j-1/2} \\ & + \bar{m}_3(\overline{w^2})_{j-1/2} + \bar{m}_9(\bar{u}^2)_{j-1/2} + \bar{m}_{12}\bar{c}_{j-1/2} \\ & + \bar{m}_{13}\bar{t}_{j-1/2} - \frac{\bar{m}_{10}}{k_n}\bar{u}_{j-1/2}(\bar{w}^n_{j-1/2} - \bar{w}^{n-1}_{j-1/2}) \\ & - \frac{\bar{m}_7}{r_i}\bar{w}_{j-1/2}(\bar{w}^i_{j-1/2} - \bar{w}^{i-1}_{j-1/2})\Big\}\end{aligned} \tag{11.7.16}$$

$$(\beta_1)_j = h_j^{-1}b_j + \frac{1}{2}\theta_{j-1/2} + \frac{\bar{m}_{13}}{2} \tag{11.7.17a}$$

$$(\beta_1)_j = h_j^{-1}b_{j-1} + \frac{1}{2}\theta_{j-1/2} + \frac{\bar{m}_{13}}{2} \tag{11.7.17b}$$

$$(\beta_3)_j = \bar{m}_9 u_j + \frac{\bar{m}_4}{2}w_j - \frac{\bar{m}_{10}}{2k_n}(\bar{w}^n_{j-1/2} - \bar{w}^{n-1}_{j-1/2}) \tag{11.7.17c}$$

$$(\beta_4)_j = \bar{m}_9 u_{j-1} + \frac{\bar{m}_4}{2}w_{j-1} - \frac{\bar{m}_{10}}{2k_n}(\bar{w}^n_{j-1/2} - \bar{w}^{n-1}_{j-1/2}) \tag{11.7.17d}$$

$$(\beta_5)_j = \frac{\bar{m}_4}{2}u_j + \bar{m}_3 w_j - \frac{\bar{m}_7}{r_i}\left(\frac{1}{2}\bar{w}^i_{j-1/2} - \bar{w}^{i-1}_{j-1/2} - \bar{w}_{j-1/2}\right) - \frac{\bar{m}_{10}}{k_n}\bar{u}_{j-1/2} \tag{11.7.17e}$$

$$(\beta_6)_j = \frac{\bar{m}_4}{2}u_{j-1} + \bar{m}_3 w_{j-1} - \frac{\bar{m}_7}{r_i}\left(\frac{1}{2}\bar{w}^i_{j-1/2} - \bar{w}^{i-1}_{j-1/2} - \bar{w}_{j-1/2}\right) - \frac{\bar{m}_{10}}{k_n}\bar{u}_{j-1/2} \tag{11.7.17f}$$

$$(\beta_7)_j = 2\bar{t}_{j-1/2} \tag{11.7.17g}$$

$$(\beta_8)_j = 2\bar{t}_{j-1/2} \tag{11.7.17h}$$

The boundary conditions, Eq. (11.7.11), become

$$\delta u_0 = 0 = (r_1)_0 \tag{11.7.18a}$$

$$\delta w_0 = 0 = (r_2)_0 \tag{11.7.18b}$$

$$\delta \theta_0 = 0 = (r_3)_0 \tag{11.7.18c}$$

$$\delta w_J = 0 = (r_4)_J \tag{11.7.18d}$$

$$\delta u_J = 0 = (r_5)_J \tag{11.7.18e}$$

In writing the above linearized system of equations, we have again used a certain order to obtain its solution with the block-elimination method discussed in subsection 3.3.3 for two-dimensional flows. According to this algorithm, the linear system is written in the form given by Eqs. (3.3.16). The vectors $\vec{\delta}_j$ and $\vec{r}_j$ are now

$$\vec{\delta}_j = \begin{bmatrix} \delta u_j \\ \delta v_j \\ \delta w_j \\ \delta t_j \\ \delta \theta_j \end{bmatrix}, \quad \vec{r}_j = \begin{bmatrix} (r_1)_j \\ (r_2)_j \\ (r_3)_j \\ (r_4)_j \\ (r_5)_j \end{bmatrix} \tag{11.7.19}$$

The structure of the coefficient matrix A remains unchanged. The matrices A_j, B_j and C_j are now 5×5 matrices and are given by

$$A_0 = \begin{bmatrix} 1 & 0 & 0 & 0 & 0 \\ 0 & 0 & 1 & 0 & 0 \\ 0 & 0 & 0 & 0 & 1 \\ -1 & -\frac{h_1}{2} & 0 & 0 & 0 \\ 0 & 0 & -1 & -\frac{h_1}{2} & 0 \end{bmatrix}$$

$$A_j = \begin{bmatrix} (c_2)_j & 0 & (c_3)_j & 0 & (c_1)_j \\ (s_3)_j & (s_1)_j & (s_5)_j & 0 & (s_7)_j \\ (\beta_3)_j & 0 & (\beta_5)_j & (\beta_1)_j & (\beta_7)_j \\ -1 & -\frac{h_{j+1}}{2} & 0 & 0 & 0 \\ 0 & 0 & -1 & -\frac{h_{j+1}}{2} & 0 \end{bmatrix} \quad 1 \le j \le J-1$$

$$A_J = \begin{bmatrix} (c_2)_J & 0 & (c_3)_J & 0 & (c_1)_J \\ (s_3)_J & (s_1)_J & (s_5)_J & 0 & (s_7)_J \\ (\beta_3)_J & 0 & (\beta_5)_J & (\beta_1)_J & (\beta_7)_J \\ 0 & 0 & 1 & 0 & 0 \\ 1 & 0 & 0 & 0 & 0 \end{bmatrix}$$

$$B_j = \begin{bmatrix} (c_2)_j & 0 & (c_3)_j & 0 & -(c_1)_j \\ (s_4)_j & (s_2)_j & (s_6)_j & 0 & (s_8)_j \\ (\beta_4)_j & 0 & (\beta_6)_j & (\beta_2)_j & (\beta_8)_j \\ 0 & 0 & 0 & 0 & 0 \\ 0 & 0 & 0 & 0 & 0 \end{bmatrix} \quad 1 \le j \le J$$

$$C_j = \begin{bmatrix} 0 & 0 & 0 & 0 & 0 \\ 0 & 0 & 0 & 0 & 0 \\ 0 & 0 & 0 & 0 & 0 \\ 1 & -\frac{h_{j+1}}{2} & 0 & 0 & 0 \\ 0 & 0 & 1 & -\frac{h_{j+1}}{2} & 0 \end{bmatrix} \quad 0 \le j \le J-1 \tag{11.7.20}$$

11.7.2 Three-Dimensional Steady Flows with Reversal

When there is flow reversal in the spanwise velocity profile, it is necessary to modify the standard box scheme in order to avoid the numerical instabilities resulting from integration opposed to the flow direction. An accurate procedure is to use the characteristic box scheme discussed in subsection 8.4.2 for two-dimensional unsteady flows. In this scheme, the finite-difference approximations for Eqs. (11.7.1d) and (11.7.1e) are written along the streamline direction ψ and are expressed in the following forms

$$(bv)' + v\theta + m_2u^2 + m_5uw + m_8w^2 + m_{11}c + m_{13}v = \lambda_3\frac{\partial u}{\partial\psi} \tag{11.7.21}$$

$$(bt)' + t\theta + m_4uw + m_3w^2 + m_9u^2 + m_{12}c + m_{13}t = \lambda_3\frac{\partial w}{\partial\psi} \tag{11.7.22}$$

where

$$\lambda_3 = \left[(m_{10}u)^2 + (m_7w)^2\right]^{1/2} \tag{11.7.23a}$$

The angle between the streamline direction and the x-axis, ϕ, is given by (see Fig. 11.8)

$$\phi = \tan^{-1}\left(\frac{m_7w}{m_{10}u}\right)^{1/2} \tag{11.7.23b}$$

Note that depending on the sign of w, which changes across the shear layer, the direction of the streamlines ψ may change as discussed in Fig. 11.6.

To illustrate the procedure with the characteristic box scheme for Eqs. (11.7.21) and (11.7.22), again consider a model equation for them in the form

$$(bv)' + v\theta + m_8w^2 = \lambda_3\frac{\partial u}{\partial\psi} \tag{11.7.24}$$

For the notation in Fig. 11.8, the finite-difference approximations for Eq. (11.7.24) at point P are

$$\begin{aligned}
&\frac{[(bv)_j^{n,i} - (bv)_{j-1}^{n,i}]}{2h_j} + \frac{[(bv)_j^{n-1,l} - (bv)_{j-1}^{n-1,l}]}{2h_j} + \frac{1}{2}(v_{j-1/2}^{n,i} + v_{j-1/2}^{n-1,l})\theta_{j-1/2}^B \\
&\quad + \frac{1}{2}(m_8^{n,i} + m_8^{n-1,l})\frac{1}{2}[(w^2)_{j-1/2}^{n,i} + (w^2)_{j-1/2}^{n-1,l}] \\
&= \frac{1}{2}\left\{[(\lambda_3)_{j-1/2}^{n,i} + (\lambda_3)_{j-1/2}^{n-1,l}]\frac{u_{j-1/2}^{n,i} - u_{j-1/2}^{n-1,l}}{\Delta\psi_{j-1/2}}\right\}
\end{aligned} \tag{11.7.25}$$

The relation between θ_j^B and those centered at $(n-1/2, i-1/2)$ and $(n-1/2, i-3/2)$ is identical to that given by Eq. (8.4.19) except that ξ's are replaced by z's, that is,

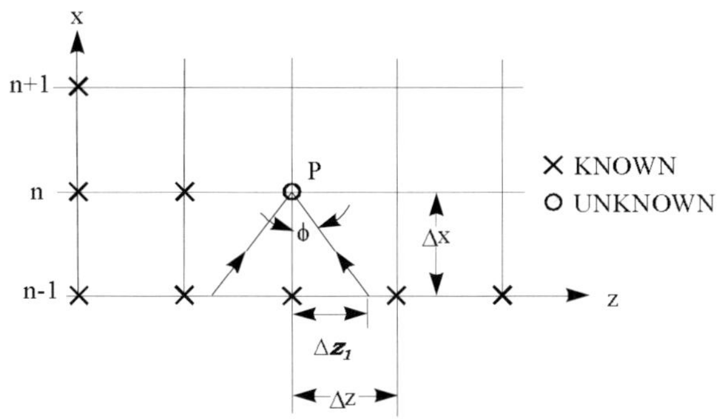

Fig. 11.8. Notation for the characteristic scheme.

$$\theta^B_{j-1/2} = \frac{\theta^{n-1/2,i-1/2}_{j-1/2} - \theta^{n-1/2,i-3/2}_{j-1/2}}{z^{i-1/2} - z^{i-3/2}}(z^B - z^{i-3/2}) + \theta^{n-1/2,i-3/2}_{j-1/2} \tag{11.7.26}$$

The difference approximations for Eq. (11.7.1c) are obtained by using the characteristic box scheme described in subsection 8.4.2. For the finite-difference molecule shown in Fig. 11.8, the finite-difference approximations for Eq. (11.7.1c) are the same as those given by Eq. (11.7.5).

The linearized equations for Eqs. (11.7.21) and (11.7.22) are given by Eqs. (11.7.12c) and (11.7.12d). The coefficients $(r_2)_j$, $(r_3)_j$ and $(s_k)_j$ and $(\beta_k)_j$ ($k = 1$ to 8) are

$$\begin{aligned}
(r_2)_j = & -\Big\{ h_j^{-1}[(bv)^{n,i}_j - (bv)^{n,i}_{j-1}] + h_j^{-1}[(bv)^{n-1,l}_j - (bv)^{n-1,l}_{j-1}] \\
& + (v^{n,i}_{j-1/2} + v^{n-1,l}_{j-1/2})\theta^B_{j-1/2} \\
& + \frac{1}{2}(m_2^{n,i} + m_2^{n-1,l})[(u^2)^{n,i}_{j-1/2} + (u^2)^{n-1,l}_{j-1/2}] \\
& + \frac{1}{2}(m_5^{n,i} + m_5^{n-1,l})[(uw)^{n,i}_{j-1/2} + (uw)^{n-1,l}_{j-1/2}] \\
& + \frac{1}{2}(m_8^{n,i} + m_8^{n-1,l})[(w^2)^{n,i}_{j-1/2} + (w^2)^{n-1,l}_{j-1/2}] \\
& + \frac{1}{2}(m_{11}^{n,i} + m_{11}^{n-1,l})(c^{n,i}_{j-1/2} + c^{n-1,l}_{j-1/2}) \\
& + \frac{1}{2}(m_{13}^{n,i} + m_{13}^{n-1,l})(v^{n,i}_{j-1/2} + v^{n-1,l}_{j-1/2}) \\
& - [(\lambda_3)^{n,i}_{j-1/2} + (\lambda_3)^{n-1,l}_{j-1/2}] \frac{u^{n,i}_{j-1/2} - u^{n-1,l}_{j-1/2}}{\Delta\psi_{j-1/2}} \Big\}
\end{aligned}$$

$$(s_1)_j = h_j^{-1} b_j + \frac{1}{2}\theta^B_{j-1/2} + \frac{1}{4}(m_{13}^{n,i} + m_{13}^{n-1,l})$$

$$(s_2)_j = -h_j^{-1} b_{j-1} + \frac{1}{2}\theta^B_{j-1/2} + \frac{1}{4}(m_{13}^{n,i} + m_{13}^{n-1,l})$$

$$\begin{aligned}(s_3)_j =& \frac{1}{2}(m_2^{n,i} + m_2^{n-1,l})u_j + \frac{1}{4}(m_5^{n,i} + m_5^{n-1,l})w_j \\ & - \frac{1}{2}\frac{[(\lambda_3)^{n,i}_{j-1/2} + (\lambda_3)^{n-1,l}_{j-1/2}]}{\Delta\psi_{j-1/2}} - \frac{1}{2}\frac{(m_{10}^{n,i})^2 u_j}{(\lambda_3)^{n,i}_j}\frac{(u^{n,i}_{j-1/2} - u^{n-1,l}_{j-1/2})}{\Delta\psi_{j-1/2}}\end{aligned}$$

$$\begin{aligned}(s_4)_j =& \frac{1}{2}(m_2^{n,i} + m_2^{n-1,l})u_{j-1} + \frac{1}{4}(m_5^{n,i} + m_5^{n-1,l})w_{j-1} \\ & - \frac{1}{2}\frac{[(\lambda_3)^{n,i}_{j-1/2} + (\lambda_3)^{n-1,l}_{j-1/2}]}{\Delta\psi_{j-1/2}} - \frac{1}{2}\frac{(m_{10}^{n,i})^2 u_{j-1}}{(\lambda_3)^{n,i}_{j-1}}\frac{(u^{n,i}_{j-1/2} - u^{n-1,l}_{j-1/2})}{\Delta\psi_{j-1/2}}\end{aligned}$$

$$\begin{aligned}(s_5)_j =& \frac{1}{4}(m_5^{n,i} + m_5^{n-1,l})u_j + \frac{1}{2}(m_8^{n,i} + m_8^{n-1,l})w_j \\ & - \frac{1}{2}\frac{(m_7^{n,i})^2 w_j}{(\lambda_3)^{n,i}_j}\frac{(u^{n,i}_{j-1/2} - u^{n-1,l}_{j-1/2})}{\Delta\psi_{j-1/2}}\end{aligned}$$

$$\begin{aligned}(s_6)_j =& \frac{1}{4}(m_5^{n,i} + m_5^{n-1,l})u_{j-1} + \frac{1}{2}(m_8^{n,i} + m_8^{n-1,l})w_{j-1} \\ & - \frac{1}{2}\frac{(m_7^{n,i})^2 w_{j-1}}{(\lambda_3)^{n,i}_{j-1}}\frac{(u^{n,i}_{j-1/2} - u^{n-1,l}_{j-1/2})}{\Delta\psi_{j-1/2}}\end{aligned}$$

$$(s_7)_j = \frac{1}{2}\frac{z^B - z^{i-3/2}}{z^{i-1/2} - z^{i-3/2}}(v^{n,i}_{j-1/2} + v^{n-1,l}_{j-1/2})$$

$$(s_8)_j = (s_7)_j$$

$$\begin{aligned}(r_3)_j =& -\Big\{h_j^{-1}[(bt)^{n,i}_j - (bt)^{n,i}_{j-1}] + h_j^{-1}[(bt)^{n-1,l}_j - (bt)^{n-1,l}_{j-1}] \\ & + (t^{n,i}_{j-1/2} + t^{n-1,l}_{j-1/2})\theta^B_{j-1/2} \\ & + \frac{1}{2}(m_4^{n,i} + m_4^{n-1,l})[(uw)^{n,i}_{j-1/2} + (uw)^{n-1,l}_{j-1/2}] \\ & + \frac{1}{2}(m_3^{n,i} + m_3^{n-1,l})[(w^2)^{n,i}_{j-1/2} + (w^2)^{n-1,l}_{j-1/2}] \\ & + \frac{1}{2}(m_9^{n,i} + m_9^{n-1,l})[(u^2)^{n,i}_{j-1/2} + (u^2)^{n-1,l}_{j-1/2}] \\ & + \frac{1}{2}(m_{12}^{n,i} + m_{12}^{n-1,l})(c^{n,i}_{j-1/2} + c^{n-1,l}_{j-1/2}) \\ & + \frac{1}{2}(m_{13}^{n,i} + m_{13}^{n-1,l})(t^{n,i}_{j-1/2} + t^{n-1,l}_{j-1/2}) \\ & - [(\lambda_3)^{n,i}_{j-1/2} + (\lambda_3)^{n-1,l}_{j-1/2}]\frac{w^{n,i}_{j-1/2} - w^{n-1,l}_{j-1/2}}{\Delta\psi_{j-1/2}}\Big\}\end{aligned}$$

$$(\beta_1)_j = (s_1)_j \qquad (\beta_2)_j = (s_2)_j$$

$$(\beta_3)_j = \frac{1}{2}(m_9^{n,i} + m_9^{n-1,l})u_j + \frac{1}{4}(m_4^{n,i} + m_4^{n-1,l})w_j - \frac{1}{2}\frac{(m_{10}^{n,i})^2 u_j}{(\lambda_3)_j^{n,i}}\frac{(w_{j-1/2}^{n,i} - w_{j-1/2}^{n-1,l})}{\Delta\psi_{j-1/2}}$$

$$(\beta_4)_j = \frac{1}{2}(m_9^{n,i} + m_9^{n-1,l})u_{j-1} + \frac{1}{4}(m_4^{n,i} + m_4^{n-1,l})w_{j-1} - \frac{1}{2}\frac{(m_{10}^{n,i})^2 u_{j-1}}{(\lambda_3)_{j-1}^{n,i}}\frac{(w_{j-1/2}^{n,i} - w_{j-1/2}^{n-1,l})}{\Delta\psi_{j-1/2}}$$

$$(\beta_5)_j = \frac{1}{4}(m_4^{n,i} + m_4^{n-1,l})u_j + \frac{1}{2}(m_3^{n,i} + m_3^{n-1,l})w_j - \frac{1}{2}\frac{\left[(\lambda_3)_{j-1/2}^{n,i} + (\lambda_3)_{j-1/2}^{n-1,l}\right]}{\Delta\psi_{j-1/2}} - \frac{1}{2}\frac{(m_7^{n,i})^2 w_j}{(\lambda_3)_j^{n,i}}\frac{(w_{j-1/2}^{n,i} - w_{j-1/2}^{n-1,l})}{\Delta\psi_{j-1/2}}$$

$$(\beta_6)_j = \frac{1}{4}(m_4^{n,i} + m_4^{n-1,l})u_{j-1} + \frac{1}{2}(m_3^{n,i} + m_3^{n-1,l})w_{j-1} - \frac{1}{2}\frac{\left[(\lambda_3)_{j-1/2}^{n,i} + (\lambda_3)_{j-1/2}^{n-1,l}\right]}{\Delta\psi_{j-1/2}} - \frac{1}{2}\frac{(m_7^{n,i})^2 w_{j-1}}{(\lambda_3)_{j-1}^{n,i}}\frac{(w_{j-1/2}^{n,i} - w_{j-1/2}^{n-1,l})}{\Delta\psi_{j-1/2}}$$

$$(\beta_7)_j = \frac{1}{2}\frac{z^B - z^{i-3/2}}{2z^{i-1/2} - z^{i-3/2}}(t_{j-1/2}^{n,i} + t_{j-1/2}^{n-1,l})$$

$$(\beta_8)_j = (\beta_7)_j$$

The boundary conditions are the same as those given by Eqs. (11.7.18a) to (11.7.18c). The solution of the linear system is again obtained with the block elimination method used to solve the linear system discussed in the previous subsection.

11.8 Solution of the Three-Dimensional Boundary-Layer Equations: Inverse Mode

A proper way to solve the three-dimensional boundary-layer equations in the inverse mode is to use the characteristic box scheme. However, its application to separated flows is still in the development stage. For this reason, presently the calculations for three-dimensional flows are being performed either with the zig-zag box scheme described in subsection 8.4.1 for two-dimensional unsteady flows or with the approach used for two-dimensional steady flows by solving the quasi-three-dimensional boundary-layer equations. Both approaches are described in the following two subsections.

11.8.1 Numerical Formulation of the Three-Dimensional Boundary-Layer Equations with the Zig-Zag Scheme

To discuss the solution of the three-dimensional boundary-layer equations with this scheme, we begin with the system of equations given by Eqs. (11.6.26) to (11.6.28). With u', w' and θ' defined by Eqs. (11.7.1a) to (11.7.1c), we write Eqs. (11.6.26) and (11.6.27) in the following form

$$(bv)' + v\theta + m_2[u^2 - c(\bar{u}_e)^2] + m_5[uw - c\bar{u}_e\bar{w}_e] + m_8[w^2 - c(\bar{w}_e)^2] = m_{10}\left(u\frac{\partial u}{\partial x} - c\bar{u}_e\frac{\partial \bar{u}_e}{\partial x}\right) + m_7\left(w\frac{\partial u}{\partial z} - c\bar{w}_e\frac{\partial \bar{u}_e}{\partial z}\right) \tag{11.8.1}$$

$$(bt)' + t\theta + m_3[w^2 - c(\bar{w}_e)^2] + m_4[uw - c\bar{u}_e\bar{w}_e] + m_9[u^2 - c(\bar{u}_e)^2] = m_{10}\left(u\frac{\partial w}{\partial x} - c\bar{u}_e\frac{\partial \bar{w}_e}{\partial x}\right) + m_7\left(w\frac{\partial w}{\partial z} - c\bar{w}_e\frac{\partial \bar{w}_e}{\partial z}\right) \tag{11.8.2}$$

We next consider the finite-difference molecule of Fig. 8.2 with τ replaced by x and ξ by z. With θ and ϕ in Eq. (8.4.4) replaced by

$$c_5 = \frac{z^{i+1} - z^i}{z^{i+1} - z^{i-1}}, \quad c_6 = \frac{z^i - z^{i-1}}{z^{i+1} - z^{i-1}} \tag{11.8.3}$$

and the locations of P, Q and R given by

$$P = (x^{n-1/2}, z^i, \eta_{j-1/2}), \quad Q = (x^n, z^{i-1/2}, \eta_{j-1/2}), \quad R = (x^{n-1}, z^{i+1/2}, \eta_{j-1/2}) \tag{11.8.4}$$

the finite difference approximations for Eq. (11.8.1) are

$$\frac{(bv)_j^{n-1/2,i} - (bv)_{j-1}^{n-1/2,i}}{h_j} + v_{j-1/2}^{n-1/2,i}(c_7\theta_{j-1/2}^{n-1/2,i-3/2} + c_8\theta_{j-1/2}^{n-1/2,i-1/2})$$
$$+ m_2^{n-1/2,i}[(u^2)_{j-1/2}^{n-1/2,i} - c_{j-1/2}^{n-1/2,i}(u_J^2)^{n-1/2,i}]$$
$$+ m_5^{n-1/2,i}[(uw)_{j-1/2}^{n-1/2,i} - c_{j-1/2}^{n-1/2,i}u_J^{n-1/2,i}(\bar{w}_J)^{n-1/2,i}]$$
$$+ m_8^{n-1/2,i}[(w^2)_{j-1/2}^{n-1/2,i} - c_{j-1/2}^{n-1/2,i}(w_J^2)^{n-1/2,i}]$$

$$
\begin{aligned}
&= m_{10}^{n-1/2,i}\left[\frac{(u^2)_{j-1/2}^{n,i}-(u^2)_{j-1/2}^{n-1,i}}{2k_n} - c_{j-1/2}^{n-1/2,i}\left(\frac{(u_J^2)^{n,i}-(u^2)_J^{n-1,i}}{2k_n}\right)\right] \\
&\quad + m_7^{n-1/2,i}\left[c_5 w_{j-1/2}^{n,i-1/2}\left(\frac{u_{j-1/2}^{n,i}-u_{j-1/2}^{n,i-1}}{r_i}\right)\right. \\
&\quad - c_5 c_{j-1/2}^{n-1,i+1/2}\bar{w}_J^{n,i-1/2}\left(\frac{u_J^{n,i}-u_J^{n,i-1}}{r_i}\right) \\
&\quad + c_6 c_{j-1/2}^{n-1,i-1/2} w_{j-1/2}^{n-1,i+1/2}\left(\frac{u_{j-1/2}^{n-1,i+1}-u_{j-1/2}^{n-1,i}}{r_{i+1}}\right) \\
&\quad \left. - c_6 c_{j-1/2}^{n-1,i+1/2}(\bar{w}_J)^{n-1,i+1/2}\left(\frac{u_J^{n-1,i+1}-u_J^{n-1,i}}{r_{i+1}}\right)\right]
\end{aligned} \tag{11.8.5}
$$

where

$$
c_7 = \frac{z_{i-1}-z_i}{z_i - z_{i-2}}, \qquad c_8 = \frac{2z_i - z_{i-1} - z_{i-2}}{z_i - z_{i-2}}
$$

A similar procedure can be used to write the difference approximations to Eq. (11.8.2). The resulting algebraic system is again nonlinear and must be linearized with Newton's method. The linear system again is written in the same form as Eqs. (11.7.12a) to (11.7.12e) with Eqs, (11.7.12d) and (11.7.12e) given by

$$
\begin{aligned}
&(s_1)_j\delta v_j + (s_2)_j\delta v_{j-1} + (s_3)_j\delta u_j + (s_4)_j\delta u_{j-1} + (s_5)_j\delta w_j \\
&\quad + (s_6)_j\delta w_{j-1} + (s_7)_j\delta\theta_j + (s_8)_j\delta\theta_{j-1} + (s_9)_j\delta u_J = (r_2)_j
\end{aligned} \tag{11.8.6a}
$$

$$
\begin{aligned}
&(\beta_1)_j\delta t_j + (\beta_2)_j\delta t_{j-1} + (\beta_3)_j\delta u_j + (\beta_4)_j\delta u_{j-1} + (\beta_5)_j\delta w_j \\
&\quad + (\beta_6)_j\delta w_{j-1} + (\beta_7)_j\delta\theta_j + (\beta_8)_j\delta\theta_{j-1} + (\beta_9)_j\delta u_J = (r_3)_j
\end{aligned} \tag{11.8.6b}
$$

The coefficients $(r_2)_j$ and $(r_3)_j$ are given by

$$
\begin{aligned}
(r_2)_j = -2\Big\{ & h_j^{-1}\left[(bv)_j^{n-1/2,i} - (bv)_{j-1}^{n-1/2,i}\right] \\
& + v_{j-1/2}^{n-1/2,i}\left(c_7\theta_{j-1/2}^{n-1/2,i-3/2} + c_8\theta_{j-1/2}^{n-1/2,i-1/2}\right) \\
& + m_2^{n-1/2,i}\left[(u^2)_{j-1/2}^{n-1/2,i} - c_{j-1/2}^{n-1/2,i}(u_J^2)^{n-1/2,i}\right] \\
& + m_5^{n-1/2,i}\left[(uw)_{j-1/2}^{n-1/2,i} - c_{j-1/2}^{n-1/2,i} u_J^{n-1/2,i} w_J^{n-1/2,i}\right] \\
& + m_8^{n-1/2,i}\left[(w^2)_{j-1/2}^{n-1/2,i} - c_{j-1/2}^{n-1/2,i}(w_J^2)^{n-1/2,i}\right]
\end{aligned}
$$

$$
\begin{aligned}
&- m_{10}^{n-1/2,i}\left[\frac{(u^2)_{j-1/2}^{n,i} - (u^2)_{j-1/2}^{n-1,i}}{2k_n} - c_{j-1/2}^{n-1/2,i}\left(\frac{(u_J^2)^{n,i} - u_J^2)^{n-1,i}}{2k_n}\right)\right] \\
&- m_7^{n-1/2,i}\left[c_5 w_{j-1/2}^{n,i-1/2}\left(\frac{u_{j-1/2}^{n,i} - u_{j-1/2}^{n,i-1}}{r_i}\right)\right. \\
&- c_5 c_{j-1/2}^{n,i-1/2} w_J^{n,i-1/2}\left(\frac{u_J^{n,i} - u_J^{n,i-1}}{r_i}\right) \\
&+ c_6 c_j w_{j-1/2}^{n-1,i+1/2}\left(\frac{u_{j-1/2}^{n-1,i+1} - u_{j-1/2}^{n-1,i}}{r_{i+1}}\right) \\
&\left.\left. - c_6 c_{j-1/2}^{n-1,i+1/2} w_J^{n-1,i+1/2}\left(\frac{u_J^{n-1,i+1} - u_J^{n-1,i}}{r_{i+1}}\right)\right]\right\}
\end{aligned}
\tag{11.8.9a}
$$

$$
\begin{aligned}
(r_3)_j = -2\Big\{ & h_j^{-1}\left[(bt)_j^{n-1/2,i} - (bt)_{j-1}^{n-1/2,i}\right] \\
&+ t_{j-1/2}^{n-1/2,i}(c_7\theta_{j-1/2}^{n-1/2,i-3/2} + c_8\theta_{j-1/2}^{n-1/2,i-1/2}) \\
&+ m_3^{n-1/2,i}\left[(w^2)_{j-1/2}^{n-1/2,i} - c_{j-1/2}^{n-1/2,i}(w_J^2)^{n-1/2,i}\right] \\
&+ m_4^{n-1/2,i}\left[(uw)_{j-1/2}^{n-1/2,i} - c_{j-1/2}^{n-1/2,i} u_J^{n-1/2,i} w_J^{n-1/2,i}\right] \\
&+ m_9^{n-1/2,i}\left[(u^2)_{j-1/2}^{n-1/2,i} - c_{j-1/2}^{n-1/2,i}(u_J^2)^{n-1/2,i}\right] \\
&- m_{10}^{n-1/2,i}\left[u_{j-1/2}^{n-1/2,i}\frac{w_{j-1/2}^{n,i} - w_{j-1/2}^{n-1,i}}{k_n} - c_{j-1/2}^{n-1/2,i} u_J^{n-1/2,i}\frac{w_J^{n,i} - w_J^{n-1,i}}{k_n}\right] \\
&- m_7^{n-1/2,i}\left[c_5\left(\frac{(w^2)_{j-1/2}^{n,i} - (w^2)_{j-1/2}^{n,i-1}}{2r_i}\right)\right. \\
&- c_5 c_{j-1/2}^{n,i-1/2}\left(\frac{(w_J^2)^{n,i} - (w_J^2)^{n,i-1}}{2r_i}\right) \\
&+ c_6\left(\frac{(w^2)_{j-1/2}^{n-1,i+1} - (w^2)_{j-1/2}^{n-1,i}}{2r_{i+1}}\right) \\
&- c_6 c_{j-1}^{n-1,i+1/2}\left(\frac{(w_J^2)^{n-1,i+1} - (w_J^2)^{n-1,i}}{2r_{i+1}}\right)\Bigg\}
\end{aligned}
\tag{11.8.9b}
$$

The coefficients $(s_k)_j$ and $(\beta_k)_j$ $(k = 1, \ldots, 9)$ are

$$
(s_1)_j = h_j^{-1} b_j + \frac{1}{2}(c_7\theta_{j-1/2}^{n-1/2,i-3/2} + c_8\theta_{j-1/2})
$$

$$
(s_2)_j = -h_j b_{j-1} + \frac{1}{2}(c_3\theta_{j-1/2}^{n-1/2,i-3/2} + c_4\theta_{j-1/2})
$$

$$(s_3)_j = m_2^{n-1/2,i} u_j + \frac{1}{2} m_5^{n-1/2,i} w_j - m_{10}^{n-1/2,i} u_j k_n^{-1} - c_5 m_7^{n-1/2,i} w_{j-1/2}^{n,i-1/2} r_i^{-1}$$

$$\begin{aligned}(s_4)_j &= m_2^{n-1/2,i} u_{j-1} + \frac{1}{2} m_5^{n-1/2,i} w_{j-1} \\ &\quad - m_{10}^{n-1/2,i} u_{j-1} k_n^{-1} - c_5 m_7^{n-1/2,i} w_{j-1/2}^{n,i-1/2} r_i^{-1}\end{aligned}$$

$$(s_5)_j = \frac{1}{2} m_5^{n-1/2,i} u_j + m_8^{n-1/2,i} w_j - \frac{1}{2} c_5 m_7^{n-1/2,i} \frac{u_{j-1}^{n,i} - u_{j-1/2}^{n,i-1}}{r_i}$$

$$(s_6)_j = \frac{1}{2} m_5^{n-1/2,i} u_{j-1} + m_8^{n-1/2,i} w_{j-1} - \frac{1}{2} c_5 m_7^{n-1/2,i} \frac{u_{j-1/2}^{n,i} - u_{j-1/2}^{n,i-1}}{r_i}$$

$$(s_7)_j = c_8 v_{j-1/2}^{n-1/2,i}$$

$$(s_8)_j = (s_7)_j$$

$$\begin{aligned}(s_9)_j &= -2 m_2^{n-1/2,i} c_{j-1/2}^{n-1/2,i} u_J - m_5^{n-1/2,i} c_{j-1/2}^{n-1/2,i} (w_J)^{n-1/2,i} \\ &\quad + 2 m_{10}^{n-1/2,i} c_{j-1/2}^{n-1/2,i} u_J k_n^{-1} + 2 c_5 m_7^{n-1/2,i} c_{j-1/2}^{n,i-1/2} w_J^{n,i-1/2} r_i^{-1}\end{aligned}$$

$$(\beta_1)_j = h_j^{-1} b_j + \frac{1}{2}(c_7 \theta_{j-1/2}^{n-1/2,i-1/2} + c_8 \theta_{j-1/2})$$

$$(\beta_2)_j = -h_j^{-1} b_{j-1} + \frac{1}{2}(c_7 \theta_{j-1/2}^{n-1/2,i-1/2} + c_8 \theta_{j-1/2})$$

$$(\beta_3)_j = \frac{1}{2} m_4^{n-1/2,i} w_j + m_9^{n-1/2,i} u_j - \frac{1}{2} m_{10}^{n-1/2,i} \frac{w_{j-1/2}^{n,i} - w_{j-1/2}^{n-1,i}}{k_n}$$

$$(\beta_4)_j = \frac{1}{2} m_4^{n-1/2,i} w_{j-1} + m_9^{n-1/2,i} u_{j-1} - \frac{1}{2} m_{10}^{n-1/2,i} \frac{w_{j-1/2}^{n,i} - w_{j-1/2}^{n-1,i}}{k_n}$$

$$(\beta_5)_j = \frac{1}{2} m_4^{n-1/2,i} u_j + m_3^{n-1/2,i} w_j - \frac{1}{2} m_{10}^{n-1/2,i} u_{j-1/2}^{n-1/2,i} k_n^{-1} - c_5 m_7^{n-1/2,i} w_j r_i^{-1}$$

$$\begin{aligned}(\beta_6)_j &= \frac{1}{2} m_4^{n-1/2,i} u_{j-1} + m_3^{n-1/2,i} w_{j-1} \\ &\quad - \frac{1}{2} m_{10}^{n-1/2,i} u_{j-1/2}^{n-1/2,i} k_n^{-1} - c_5 m_7^{n-1/2,i} w_{j-1} r_i^{-1}\end{aligned}$$

$$(\beta_7)_j = c_8 t_{j-1/2}^{n-1/2,i}, \qquad (\beta_8)_j = (\beta_7)_j$$

$$\begin{aligned}(\beta_9)_j &= -m_4^{n-1/2,i} c_{j-1/2}^{n-1/2,i} w_J + 2 m_9^{n-1/2,i} c_{j-1/2}^{n-1/2,i} u_J \\ &\quad - m_{10}^{n-1/2,i} c_{j-1/2}^{n-1/2,i} \frac{w_J^{n,i} - w_J^{n-1,i}}{k_n}\end{aligned}$$

The boundary conditions are the same as those given by Eqs. (11.7.18a) to (11.7.18e). The solution of the linear system is again obtained with the bordering algorithm discussed in subsection 3.3.4 for two-dimensional flows. Further details can be obtained from the author.

11.8.2 Numerical Formulation of the Quasi-Three-Dimensional Boundary-Layer Equations

The numerical solution of the quasi-three-dimensional boundary layer equations given by Eqs. (11.6.20) to (11.6.23) is similar to the solution procedure used for two-dimensional flows discussed in Section 3.3. For the two momentum equations we now have five first-order equations rather than three for two-dimensional flows. The form of Eqs. (11.7.12a, b, c) and (11.8.6a, b) remain the same except that $(s_1)_j$ to $(s_9)_j$, $(\beta_1)_j$ to $(\beta_9)_j$, $(c_1)_j$, $(r_2)_j$, $(r_3)_j$ and $(r_4)_j$, with θ_j denoting $\theta_j^{n-1/2}$, are given by

$$(s_1)_j = h_j^{-1}b_j + \frac{1}{2}\theta_{j-1/2} + \frac{1}{2}m_{13}$$

$$(s_2)_j = -h_j^{-1}b_{j-1} + \frac{1}{2}\theta_{j-1/2} + \frac{1}{2}m_{13}$$

$$(s_3)_j = (m_2 - \alpha \cdot FLARE)u_j + \frac{1}{2}m_5 w_j$$

$$(s_4)_j = (m_2 - \alpha \cdot FLARE)u_{j-1} + \frac{1}{2}m_5 u_{j-1}$$

$$(s_5)_j = m_8 w_j + \frac{1}{2}m_5 u_j$$

$$(s_6)_j = m_8 w_{j-1} + \frac{1}{2}m_5 u_{j-1}$$

$$(s_7)_j = \frac{1}{2}(v_{j-1/2} + v_{j-1/2}^{n-1})$$

$$(s_8)_j = (s_7)_j$$

$$(s_9)_j = -c_{j-1/2}^{n-1/2}(m_2 - \alpha)u_J - \frac{1}{2}c_{j-1/2}^{n-1/2}m_5\bar{w}_e$$

$$\begin{aligned}(r_2)_j = & -L_{j-1/2}^{n-1} - \Big\{(bv)'_{j-1/2} + \theta_{j-1/2}(v_{j-1/2} + v_{j-1/2}^{n-1}) + m_2(u^2)_{j-1/2} \\ & + m_8(w^2)_{j-1/2} + m_5(uw)_{j-1/2} + c_{j-1/2}^{n-1}\bar{m}_{11} + m_{13}v_{j-1/2} \\ & - \alpha \cdot FLARE\big[(u^2)_{j-1/2} - (u^2)_{j-1/2}^{n-1}\big]\Big\}\end{aligned}$$

$$L_{j-1/2}^{n-1} = \{(bv)_{j-1/2} + m_2 u_{j-1/2}^2 + m_5(uw)_{j-1/2} + m_8(w^2)_{j-1/2} + m_3 v_{j-1/2}\}^{n-1}$$

$$\begin{aligned}\bar{m}_{11} = & -\{(m_2u_J^2 + m_5u_Jw_J + m_8w_J^2) \\ & + (m_2u_J^2 + m_5u_Jw_J + m_8w_J^2)^{n-1} - \alpha\big[(u_J^2) - (u_J^2)^{n-1}\big]\}\end{aligned}$$

$$(\beta_1)_j = (s_1)_j$$

$$(\beta_2)_j = (s_2)_j$$

$$(\beta_3)_j = -\frac{\alpha}{2} \cdot FLARE(w_{j-1/2} - w_{j-1/2}^{n-1}) + m_7u_j + \frac{1}{2}m_4w_j$$

$$(\beta_4)_j = -\frac{\alpha}{2} \cdot FLARE(w_{j-1/2} - w^{n-1}_{j-1/2}) + m_9 u_{j-1} + \frac{1}{2} m_4 w_{j-1}$$

$$(\beta_5)_j = -\frac{\alpha}{2} \cdot FLARE(u_{j-1/2} - u^{n-1}_{j-1/2}) + m_3 w_j + \frac{1}{2} m_4 u_j$$

$$(\beta_6)_j = -\frac{\alpha}{2} \cdot FLARE(u_{j-1/2} - u^{n-1}_{j-1/2}) + m_3 w_{j-1} + \frac{1}{2} m_4 u_{j-1}$$

$$(\beta_7)_j = \frac{1}{2}(t_{j-1/2} + t^{n-1}_{j-1/2})$$

$$(\beta_8)_j = (\beta_7)_j$$

$$(\beta_9)_j = \frac{\alpha}{2} c^{n-1/2}_{j-1/2} (\bar{w}_e - \bar{w}^{n-1}_e) - m_9 c^{n-1/2}_{j-1/2} u_J - \frac{1}{2} m_4 c^{n-1/2}_{j-1/2} \bar{w}_e$$

$$\begin{aligned}(r_3)_j &= -L^{n-1}_{j-1/2} - \{(bt)'_{j-1/2} + \theta_{j-1/2}(t_{j-1/2} + t^{n-1}_{j-1/2}) + m_3 (w^2)_{j-1/2} \\ &\quad + m_9 (u^2)_{j-1/2} + m_4 (uw)_{j-1/2} + c^{n-1/2}_{j-1/2} \bar{m}_{12} + m_{13} t_{j-1/2} \\ &\quad - \alpha \cdot FLARE[(u_{j-1/2} + u^{n-1}_{j-1/2})(w_{j-1/2} - w^{n-1}_{j-1/2})]\}\end{aligned}$$

$$L^{n-1}_{j-1/2} = \{(bt)'_{j-1/2} + m_3 w^2_{j-1/2} + m_9 u^2_{j-1/2} + m_4 (uw)_{j-1/2} + m_{13} t_{j-1/2}\}^{n-1}$$

$$\begin{aligned}\bar{m}_{12} &= -\{(m_3 w_J^2 + m_9 u_J^2 + m_4 u_J w_J) \\ &\quad + (m_3 w_J^2 + m_9 u_J^2 + m_4 u_J w_J)^{n-1} - \alpha(u_J + u_J^{n-1})(w_J - w_5^{n-1})\}\end{aligned}$$

and

$$(c_1)_j = -\frac{h_j}{4}[m_1 + \alpha(m_{10} + m^{n-1}_{10})]$$

$$(c_3)_j = -\frac{h_j}{4} m_6$$

$$\begin{aligned}(r_4)_j &= \frac{h_j}{2}\Big\{m_1 u_{j-1/2} + (m_1 u_{j-1/2})^{n-1} + m_6 w_{j-1/2} + (m_6 w_{j-1/2})^{n-1} \\ &\quad - \frac{\alpha}{2}(m_{10} + m^{n-1}_{10})(u_{j-1/2} - u^{n-1}_{j-1/2})\Big\} - (\theta_{j-1/2} - \theta^{n-1}_{j-1/2})\end{aligned}$$

11.9 Modelling of Viscous Effects in Three-Dimensional Inviscid Flows

Once a solution of the boundary layer equations is obtained for a wing or for a high lift configuration, the viscous effects into a three-dimensional invisicd method can be incorporated by using either the "solid" displacement surface concept or the "transpiration" model concept discussed in Section 1.2 for two-dimensional flows. The transpiration model again is more convenient to adopt and is used in the calculations presented in Chapter 13.

The expression for the blowing velocity again is obtained from the continuity equation by integrating Eq. (11.1.2) from the wall $y = 0$ to the edge of the boundary layer δ for the real flow so that

$$v(\delta) = \frac{-1}{h_1 h_2 \sin\theta}\left[\int_0^\delta \left(\frac{\partial}{\partial x}(uh_2 \sin\theta)\right) dy + \int_0^\delta \left(\frac{\partial}{\partial z}(wh_1 \sin\theta)\right) dy\right] (\equiv v_e) \tag{11.9.1}$$

Similarly, integrating Eq. (11.1.2) from δ to y for the fictitious flow (in which $u(y) = u_e$ and $w(y) = w_e$), assuming that the normal velocity component at δ is also v_e, yields

$$v^*(y) = v_e - \frac{(y-\delta)}{h_1 h_2 \sin\theta}\left[\frac{\partial}{\partial x}(u_e h_2 \sin\theta) + \frac{\partial}{\partial z}(w_e h_1 \sin\theta)\right] \tag{11.9.2}$$

where v^* is the normal velocity of the fictitious flow.

Substituting v_e in Eq. (11.9.2) into Eq. (11.9.1) and making use of the definitions of the displacement thickness δ_x^* and δ_z^* we obtain

$$\begin{aligned} v^*(y) = {} & \frac{1}{h_1 h_2 \sin\theta}\left[\frac{\partial}{\partial x}(u_e h_2 \sin\theta\delta_x^*) + \frac{\partial}{\partial z}(w_e h_1 \sin\theta\delta_z^*)\right] \\ & - \frac{y}{h_1 h_2 \sin\theta}\left[\frac{\partial}{\partial x}(u_e h_2 \sin\theta) + \frac{\partial}{\partial z}(w_e h_1 \sin\theta)\right] \end{aligned} \tag{11.9.3}$$

The value of $v^*(y)$ in the above equation is the boundary condition that must be applied to the fictitious inviscid flow at a distance y from the wall in order to account for viscous effects on the inviscid flow. In particular, at $y = 0$, the expression for the blowing velocity is obtained from

$$v_w = \frac{1}{h_1 h_2 \sin\theta}\left[\frac{\partial}{\partial x}(h_2 u_e \sin\theta\delta_x^*) + \frac{\partial}{\partial z}(h_1 w_e \sin\theta\delta_z^*)\right] \tag{11.9.4}$$

Similarly an equation for the displacement surface, Δ, in three-dimensional flows can be obtained by constructing a fictitious inviscid flow (Fig. 11.9b) which has the same mass flow as the real flow (Fig. 11.9a) through the faces $1, 2, 3, 4$ of the control volume. As in two-dimensional flows, the fictitious flow has constant properties along a normal to the wall below the line $y = \delta$.

Equating the mass flow for real and fictitious flows leads to an equation for displacement surface,

$$\frac{\partial}{\partial x}(h_2 u_e \sin\theta(\Delta - \delta_x^*)) + \frac{\partial}{\partial z}(h_1 w_e \sin\theta(\Delta - \delta_z^*)) = 0 \tag{11.9.5}$$

This equation can be solved by a method of characteristics, i.e. by integrating

$$\frac{dz}{dx} = \frac{w_e h_1}{u_e h_2}, \quad \frac{d\Delta}{dx} = \frac{f(x,z)}{u_e h_2 \sin\theta} \tag{11.9.6}$$

where

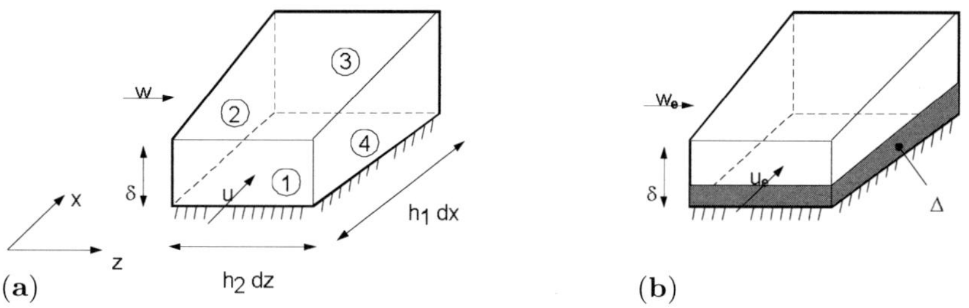

Fig. 11.9. Control volume for (**a**) the real flow and (**b**) the equivalent fictitious inviscid flow.

$$
\begin{aligned}
f(x,z) = {} & \frac{\partial}{\partial x}(h_2 u_e \sin\theta \delta_x^*) + \frac{\partial}{\partial z}(h_1 w_e \sin\theta \delta_z^*) \\
& - \Delta\left[\frac{\partial}{\partial x}(h_2 u_e \sin\theta) + \frac{\partial}{\partial z}(h_1 w_e \sin\theta)\right]
\end{aligned}
\tag{11.9.7}
$$

In addition to using the blowing velocity as a boundary condition in the inviscid flow, as in two-dimensional flows, it is also necessary to apply the Kutta condition (equal pressure for upper and lower surface at the trailing edge) at the displacement surface. Again, as in two-dimensional flows, the wake viscous effects are modeled by equivalent source distributions.

Appendix 11A Boundary-Layer Equations for Restricted Three-Dimensional Flows

A11.1 Boundary-Layer Equations with Infinite-Swept-Wing Approximation

For flows with yawed infinite cylinder or swept-wing approximation (Fig. A11.1), the spanwise flow is independent of the z-coordinate. Thus, Eqs. (11.1.2) to (11.1.5) can be simplified and conveniently expressed in a Cartesian coordinate system. The continuity, x-momentum (chordwise) and energy reduce to their two-dimensional form and with eddy viscosity and turbulent Prandtl number concepts [see Eq. (11.3.1)] can be written as

$$\frac{\partial}{\partial x}(\rho u) + \frac{\partial}{\partial y}(\overline{\rho v}) = 0 \tag{A11.1.1}$$

$$\rho u \frac{\partial u}{\partial x} + \overline{\rho v}\frac{\partial u}{\partial y} = \rho_e u_e \frac{du_e}{dx} + \frac{\partial}{\partial y}\left[\rho(\nu + \nu_t)\frac{\partial u}{\partial y}\right] \tag{A11.1.2}$$

$$\rho u \frac{\partial H}{\partial x} + \overline{\rho v}\frac{\partial H}{\partial y} = \frac{\partial}{\partial y}\left[\frac{\mu}{\mathrm{Pr}}\frac{\partial H}{\partial y} + \mu\left(1 - \frac{1}{\mathrm{Pr}}\right)\frac{\partial}{\partial y}\left(\frac{u_t^2}{2}\right) + \rho\frac{\nu_t}{\mathrm{Pr}_t}\frac{\partial H}{\partial y}\right] \tag{A11.1.3}$$

The z-momentum (spanwise) equation becomes

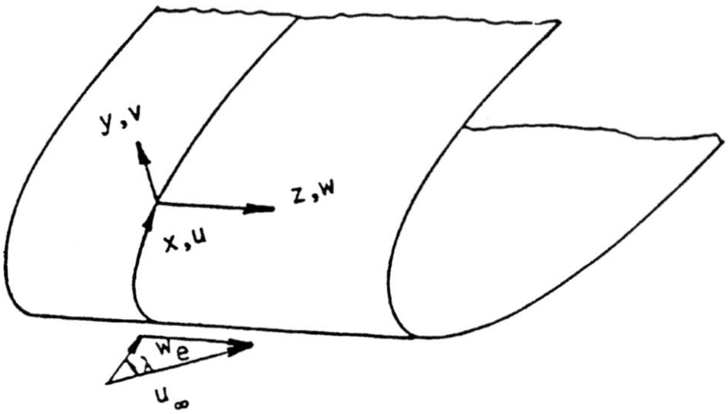

Fig. A11.1. Coordinate system for an infinite swept wing.

$$\rho u \frac{\partial w}{\partial x} + \overline{\rho v} \frac{\partial w}{\partial y} = \frac{\partial}{\partial y}\left[\rho(\nu + \nu_t)\frac{\partial w}{\partial y}\right] \tag{A11.1.4}$$

In terms of transformed variables defined by

$$d\eta = \left(\frac{u_e}{\rho_e \mu_e x}\right)^{\frac{1}{2}} \rho dy, \quad \psi = (\rho_e \mu_e u_e x)^{1/2} f(x, \eta) \tag{A11.1.5}$$

Eqs. (A11.1.2) to (A11.1.4) can be written as

$$(bf'')' + m_1 f f'' + m_2[c - (f')^2] = x\left(f'\frac{\partial f'}{\partial x} - f''\frac{\partial f}{\partial x}\right) \tag{A11.1.6}$$

$$(e_1 E')' + m_1 f E' + e_2' = x\left(f'\frac{\partial E}{\partial x} - E'\frac{\partial f}{\partial x}\right) \tag{A11.1.7}$$

$$(bg'')' + m_1 f g'' = x\left(f'\frac{\partial g'}{\partial x} - g''\frac{\partial f}{\partial x}\right) \tag{A11.1.8}$$

Here

$$b = C(1 + \nu_t^+), \quad C = \frac{\rho\mu}{\rho_e\mu_e}, \quad g' = \frac{w}{w_e}, \quad c = \frac{\rho_e}{\rho},$$

$$e_1 = \frac{C}{\Pr}\left(1 + \nu_t^+ \frac{\Pr}{\Pr_t}\right), \quad e_2 = \frac{C u_e^2}{H_e}\left(1 - \frac{1}{\Pr}\right)\left(f'f'' + \frac{w_e^2}{u_e^2} g'g''\right) \tag{A11.1.9}$$

$$m_1 = \frac{1 + m_2 + m_4}{2}, \quad m_2 = \frac{x}{u_e}\frac{du_e}{dx}, \quad m_4 = \frac{x}{\rho_e\mu_e}\frac{d}{dx}(\rho_e\mu_e)$$

As can be seen, for laminar flows, the solutions of Eqs. (A11.1.6) and (A11.1.7) are independent of the solutions of Eq. (A11.1.8). For turbulent flows, however, this is not the case and the solutions of Eqs. (A11.1.6) and (A11.1.7) are coupled to the solutions of the z-momentum equation, (A11.1.8), through the turbulence model discussed in Section 11.3. Obviously, for incompressible laminar flows, the solutions of the continuity and x-momentum equations, Eqs.

(A11.1.6), are also independent of the solutions of the energy equation, Eq. (A11.1.7).

In solving the compressible boundary-layer equations with infinite swept wing approximation, again two approaches can be pursued as in full three-dimensional flows. Even though the convergence rate of the solutions is better if Eqs. (A11.1.6) to (A11.1.8) are solved simultaneously, the second choice where Eqs. (A11.1.6) and (A11.1.7) are solved separately from Eq. (A11.1.8) may be preferred since this choice facilitates the solution procedure. It may even be simpler to solve Eq. (A11.1.6) first for assumed fluid properties, then solve the energy equation, Eq. (A11.1.7), and repeat the procedure until convergence.

A11.2 External Velocity Distribution for Infinite Swept Wings

The external velocity distribution for infinite swept wing boundary-layer equations can be obtained from a three dimensional inviscid flow field by using a strip theory approximation. To discuss this approximation, let us consider Fig. A11.2 which shows the application of the infinite swept-wing approximation locally to a three-dimensional (real) wing. With the sweep angle λ corresponding to the quarter chord sweep angle of the real wing, the coordinates of the airfoil cross section (x_n, y_n) which are normal to the wing leading edge can be determined from the coordinates of the streamwise airfoil section (x, y) by

$$x_n = x\cos\lambda \tag{A11.2.1a}$$

$$y_n = y \tag{A11.2.1b}$$

To calculate the external velocity distribution, $u_e(x)$ and w_e for the infinite swept wing, let us assume that the panel method of Chapter 10 is used to calculate the flow field of the real wing for two angles of attack, $(\alpha_{3\mathrm{D}})_1$ and $(\alpha_{3\mathrm{D}})_2$, producing two lift coefficients C_{L_1} and C_{L_2} at the spanwise location of interest. Since lift coefficient varies linearly with angle of attack, an equivalent local three-dimensional angle of attack can be calculated for any local lift coefficient. Using the notation in Fig. A11.1 we can write the following relations

$$\tan\alpha_{20} = \frac{\tan\alpha_{3\mathrm{D}}}{\cos\lambda} \tag{A11.2.2a}$$

$$\frac{u_e}{V_\infty} = \left(\frac{u_e}{u_\infty}\right)_{2\mathrm{D}}\frac{u_\infty}{V_\infty} = \left(\frac{u_e}{u_\infty}\right)_{2\mathrm{D}}\sqrt{1-\cos^2\alpha_{3\mathrm{D}}\sin^2\lambda} \tag{A11.2.2b}$$

$$\frac{w_e}{V_\infty} = \cos\alpha_{2\mathrm{D}}\sin\lambda \tag{A11.2.2c}$$

where the subscript 2D denotes the velocity calculated by a two-dimensional method about the airfoil defined by (x_n, y_n) at incidence $\alpha_{2\mathrm{D}}$. Given $\alpha_{3\mathrm{D}}$ and λ, $\alpha_{2\mathrm{D}}$ is determined from Eq. (A11.2.2a).

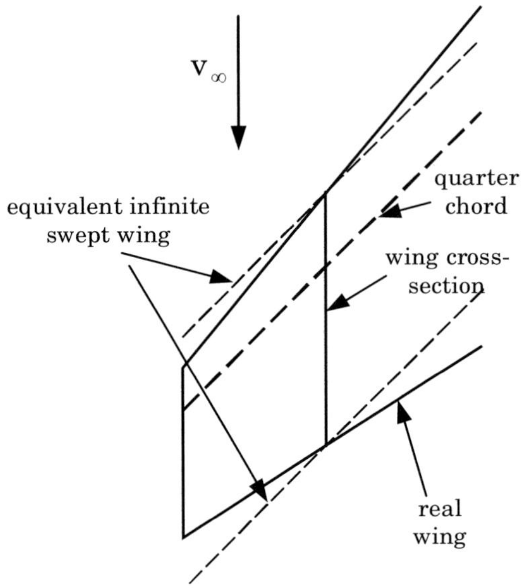

Fig. A11.2. Infinite yawed wing approximation

The two-dimensional panel method of Chapter 2 can be used to calculate $(u_e/u_\infty)_{2D}$, so that the streamwise external velocity distribution for an infinite swept wing (u_e/V_∞) can be calculated from Eq. (A11.2.2b). Finally the spanwise velocity component w_e/V_∞ is calculated from Eq. (A11.2.2c).

A11.3 Boundary-Layer Equations with Conical Flow Approximation

Pointed bodies such as circular cones are characterized by constant velocity along the rays (also called generators if on the surface) from the apex of the body and the flow is similar on all spherical surfaces of radius x centered at the apex (Fig. A11.4). Since the cone generators and spherical cuts intersect at right angles, the coordinate system on a conical body is orthogonal ($\theta = \pi/2$). Also a conical surface is a developable surface. This means that the surface can be rolled out on a flat plane (see Fig. A11.5).

As for infinite swept wings, the three-dimensional boundary-layer equations can be simplified for conical bodies. With x denoting the distance from the apex along each ray and z corresponding to the polar angle in the developed plane, with conical flow approximation, Eqs. (A11.1.1) to (A11.1.4) can be written as

$$\frac{\partial}{\partial x}(\rho x u) + \frac{\partial}{\partial z}(\rho w) + \frac{\partial}{\partial y}(\overline{\rho v} x) = 0 \qquad \text{(A11.3.1)}$$

$$\rho u \frac{\partial u}{\partial x} + \rho \frac{w}{x}\frac{\partial u}{\partial z} + \overline{\rho v}\frac{\partial u}{\partial y} - \rho \frac{w^2}{x} = \frac{\partial}{\partial y}\left(u \frac{\partial u}{\partial y} - \rho \overline{u'v'}\right) \qquad \text{(A11.3.2)}$$

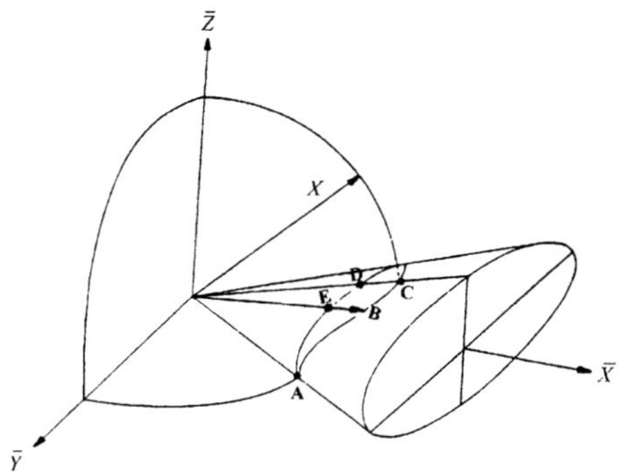

Fig. A11.3. Notation and geometry For a conical surface. ABC denotes the intersection between a sphere and a cone. AED denotes cone intersection with the plane z = constant.

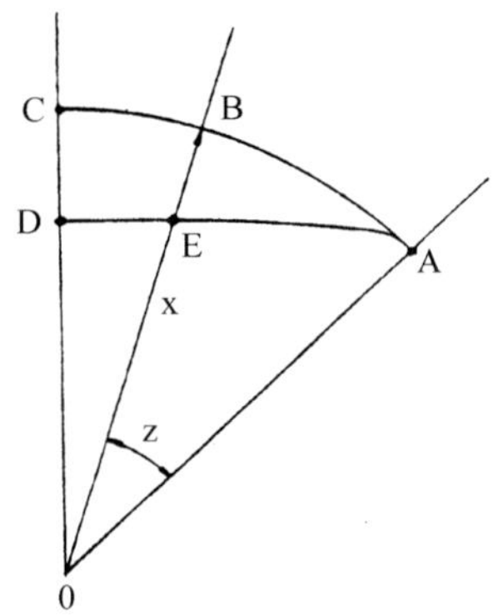

Fig. A11.4. The conical surface in the developed plane.

$$\rho u \frac{\partial w}{\partial x} + \rho \frac{w}{x}\frac{\partial w}{\partial z} + \overline{\rho v}\frac{\partial w}{\partial y} + \rho \frac{uw}{x} = -\frac{1}{x}\frac{\partial p}{\partial z} + \frac{\partial}{\partial y}\left(u \frac{\partial w}{\partial y} - \rho \overline{w'v'}\right) \tag{A11.3.3}$$

$$\rho u \frac{\partial H}{\partial x} + \rho \frac{w}{x}\frac{\partial H}{\partial z} + \overline{\rho v}\frac{\partial H}{\partial y} = \frac{\partial}{\partial y}\left[\frac{u}{\Pr}\frac{\partial H}{\partial y} + \mu\left(1 - \frac{1}{\Pr}\right)\frac{\partial}{\partial y}\left(\frac{u_t^2}{2}\right) - \rho \overline{v'H'}\right] \tag{A11.3.4}$$

On the attachment line, which is always aligned with a generator in conical flows, the velocity component w is identically zero. To avoid the singularity Eq. (A11.3.3) is differentiated with respect to z, and with $w_z \equiv \partial w/\partial z$ and $w = 0$, the attachment line equations on a conical body can be written as

$$\frac{\partial(\rho u x)}{\partial x} + \rho w_z + \frac{\partial}{\partial y}(\overline{\rho v} x) = 0 \tag{A11.3.5}$$

$$\rho u \frac{\partial u}{\partial x} + \overline{\rho v}\frac{\partial u}{\partial y} = \frac{\partial}{\partial y}\left(\mu \frac{\partial u}{\partial y} - \rho \overline{u'v'}\right) \tag{A11.3.6}$$

$$\rho u \frac{\partial w_z}{\partial x} + \frac{\rho}{x}(w_z)^2 + \overline{\rho v}\frac{\partial w_z}{\partial y} + \frac{\rho u w_z}{x} = -\frac{1}{x}\frac{\partial^2 p}{\partial z^2} + \frac{\partial}{\partial y}\left(\mu \frac{\partial w_z}{\partial y}\right) \quad \text{(A11.3.7)}$$

$$\rho u \frac{\partial H}{\partial x} + \overline{\rho v}\frac{\partial H}{\partial y} = \frac{\partial}{\partial y}\left[\frac{\mu}{\Pr}\frac{\partial H}{\partial y} + \mu\left(1 - \frac{1}{\Pr}\right)\frac{\partial(u_t^2/2)}{\partial y} - \overline{\rho v' H'}\right] \quad \text{(A11.3.8)}$$

For laminar-flows, the boundary-layer equations with conical flow approximation admit similarity. With the similarity variable η defined by

$$\eta = \sqrt{\frac{u_e}{\nu_e x}} \int_0^y \left(\frac{\rho}{\rho_e}\right) dy \quad \text{(A11.3.9)}$$

and the dimensional variables ψ and ϕ by

$$\psi = \sqrt{\rho_e \mu_e u_e}\, x^{3/2} f(z, \eta), \quad \phi = \sqrt{\rho_e \mu_e u_e}\, x^{3/2} \left(\frac{w_e}{u_e}\right) g(z, \eta) \quad \text{(A11.3.10)}$$

and with

$$\rho u x = \frac{\partial \psi}{\partial y}, \quad \rho w x = \frac{\partial \phi}{\partial y}, \quad \rho v x = -\frac{\partial \psi}{\partial x} - \frac{1}{x}\frac{\partial \phi}{\partial z} + (\rho v x)_w \quad \text{(A11.3.11)}$$

Equations (A11.3.1) to (A11.3.4) and their boundary conditions can be written as

$$(bf'')' + \theta f'' + m_5 f' g' + m_8 (g')^2 + m_{13} f'' = m_7 g' \frac{\partial f'}{\partial z} \quad \text{(A11.3.12)}$$

$$(bg'')' + \theta g'' + m_4 f' g' + m_3 (g')^2 + m_{12} c + m_{13} g'' = m_7 g' \frac{\partial g'}{\partial z} \quad \text{(A11.3.13)}$$

$$(e_1 E') + \theta E' + e_2' + m_{13} E' = m_7 g' \frac{\partial E}{\partial z} \quad \text{(A11.3.14)}$$

$$\theta' = m_1 f' + m_6 g' + m_7 \frac{\partial g'}{\partial z} \quad \text{(A11.3.15)}$$

$$\eta = 0, \quad \theta = f' = g' = 0, \quad E_w \text{ or } E_w' \text{ given} \quad \text{(A11.3.16a)}$$

$$\eta = \eta_0, \quad f' = g' = E = 1 \quad \text{(A11.3.16b)}$$

Here we have made use of the relation

$$du_e/dz = w_e$$

valid for a conical flow and again used primes to denote differentiation with respect to η and defined

$$f' = u/u_e, \quad g' = w/w_e, \quad E = H/H \quad \text{(A11.3.17a)}$$

$$b = C = \frac{\rho\mu}{\rho_e \mu_e}, \quad c = \frac{\rho_e}{\rho}, \quad e_1 = \frac{C}{\Pr}$$
$$e_2 = \frac{C u_e^2}{H_e}\left(1 - \frac{1}{\Pr}\right)(f' f'' + m_8 g' g'') \quad \text{(A11.3.17b)}$$

The coefficients m_1 to m_{13} are

$$m_1 = \frac{3}{2}, \quad m_3 = -\frac{1}{u_e}\frac{dw_e}{dz}, \quad m_4 = 1, \quad m_5 = -\left(\frac{w_e}{u_e}\right)^2,$$
$$m_6 = -m_3 + \frac{m_5}{2} + \frac{m_7}{2}\frac{1}{\rho_e\mu_e}\frac{d(\rho_e\mu_e)}{dz}, \quad m_7 = \frac{w_e}{u_e}, \tag{A11.3.18}$$
$$m_8 = -m_5, \quad m_{12} = -1 - m_3, \quad m_{13} = -\frac{\rho_w v_w}{\rho_e u_e}\sqrt{\frac{u_e x}{v_e}}$$

Similarly, with ψ and ϕ defined by

$$\psi = \sqrt{\rho_e\mu_e u_e}\, x^{3/2} f(\eta), \quad \phi = \sqrt{\rho_e\mu_e u_e}\, x\left(\frac{w_{ze}}{u_e}\right) g(\eta) \tag{A11.3.19}$$

where

$$\overline{\rho v}\, x = -\frac{\partial\psi}{\partial x} - \phi + (\rho v x)_w \tag{A11.3.20}$$

the attachment line equations given by Eqs. (A11.3.5) to (A11.3.8) become

$$(bf'')' + \theta f'' + m_{13} f'' = 0 \tag{A11.3.21}$$

$$(bg'')' + \theta g'' + m_4 f'g' + m_3(g')^2 + m_{12}c + m_{13}g'' = 0 \tag{A11.3.22}$$

$$(e_1E')' + \theta E' + e_2' + m_{13}E' = 0 \tag{A11.3.23}$$

$$\theta' = m_1 f' + m_6 g' \tag{A11.3.24}$$

The boundary conditions given by Eqs. (A11.3.16) remain the same. The definitions of f' and E', defined in Eq. (A11.3.17a) also remain the same except that now

$$g' = w_z/w_{ze}, \quad m_6 = -m_3$$

A11.3.1 Circular Cone

In general, the determination of the polar angle z in the developed plane is not simple because of the need to calculate the intersection of the conical body with a sphere. However, for a circular cone, the angle z is related to the cylindrical polar coordinates r and ϕ by (see Fig. A11.6)

$$xz = r\phi \tag{A11.3.26a}$$

Furthermore, the ratio r/x can be related to the cone half angle β by

$$\frac{r}{x} = \sin\beta \tag{A11.3.26b}$$

and ∂z can be replaced by $\sin\beta\,\partial\phi$ in all the equations given above.

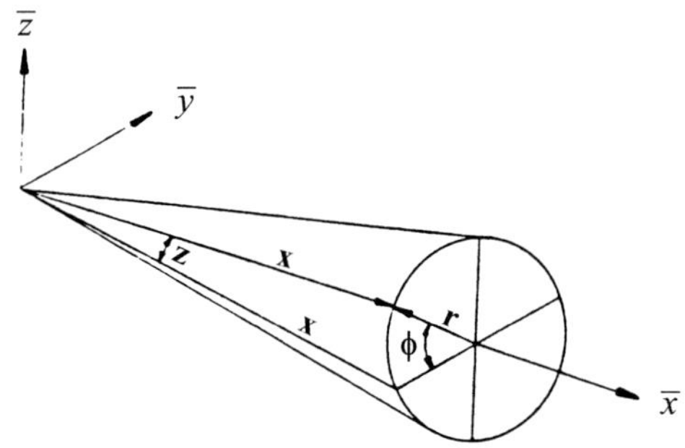

Fig. A11.5. Notation and geometry for a circular cone.

Sometimes it is more convenient to define ϕ in Eq. (A11.3.10) by

$$\phi = \sqrt{\rho_e \mu_e u_e}\, x^{3/2} \frac{u_0}{u_e} g(\phi, \eta) \tag{A11.3.27}$$

with u_0 representing a constant velocity in order to perform calculations on the windward plane of symmetry of a circular cone where $w_e = 0$. With the modifications introduced into the definitions and transformations defined by Eqs. (A11.3.9) to (A11.3.11), the transformed laminar boundary-layer equations for a circular cone can be written as

$$(bf'')' + \theta f'' + m_5 f'g' + m_8 (g')^2 + m_5 + m_{13} f'' = \frac{m_7}{\sin\beta} g' \frac{\partial f'}{\partial \phi} \tag{A11.3.28}$$

$$(bg'')' + \theta g'' + m_4 f'g' + cm_{12} + m_{13} g'' = \frac{m_7}{\sin\beta} g' \frac{\partial g'}{\partial \phi} \tag{A11.3.29}$$

$$(e_1 E')' + \theta E' + e_2' + m_{13} E' = \frac{m_7}{\sin\beta} g' \frac{\partial g'}{\partial \phi} \tag{A11.3.30}$$

$$\theta' = m_1 f' + m_6 g' + \frac{m_7}{\sin\beta} \frac{\partial g'}{\partial \phi} \tag{A11.3.31}$$

The boundary conditions become

$$\eta = 0, \quad \theta = f' = g' = 0, \quad E_w \text{ or } E_w' \text{ given} \tag{A11.3.32a}$$

$$\eta = \eta_e, \quad f' = E = 1, \quad g' = \frac{w_e}{u_0} \tag{A11.3.32b}$$

The definitions of the parameters C, c, e_1 and e_2 remain the same as those defined by Eq. (A11.3.17b); the definitions of the coefficients m_1 to m_{13} become

$$m_1 = \frac{3}{2}, \quad m_4 = -1, \quad m_5 = -\frac{u_0}{u_e}\frac{w_e}{u_e},$$
$$m_6 = \frac{m_5}{2} + \frac{m_7}{2\sin\beta\rho_e\mu_e}\frac{d(\rho_e,\mu_e)}{d\phi},$$
$$m_7 = \frac{u_0}{u_e}, \quad m_8 = m_7^2, \tag{A11.3.33}$$
$$m_{12} = \frac{w_e}{u_0}\left(1 + \frac{1}{u_e\sin\phi}\frac{dw_e}{d\phi}\right), \quad m_{13} = -\frac{(\rho v)_w}{\rho_e u_e}\sqrt{\frac{u_e x}{\nu_e}}$$

For the attachment-line equations, we define

$$g' = \frac{w_\phi}{u_0\sin\beta}, \quad \phi = \sqrt{\rho_e\mu_e u_e x}\frac{u_0}{u_e}g(\eta) \tag{A11.3.34}$$

The definitions of η and ψ remain unchanged, and the transformed attachment line equations and their boundary conditions for a circular cone become

$$(bf'')' + \theta f'' + m_{13}f'' = 0 \tag{A11.3.35}$$

$$(bg'') + \theta g'' + m_4 f'g' + m_3(g')^2 + m_{12}c + m_{13}g'' = 0 \tag{A11.3.36}$$

$$(e_1E')' + \theta E' + e_2' + m_{13}E' = 0 \tag{A11.3.37}$$

$$\theta' = m_1 f' + m_6 g' \tag{A11.3.38}$$

$$\eta = 0, \quad \theta = f' = g' = 0, \quad E_w \text{ or } E_w' \text{ given} \tag{A11.3.39a}$$

$$\eta = \eta_e, \quad f' = E = 1, \quad g' = \frac{w_{\phi e}}{u_0\sin\beta} \tag{A11.3.39b}$$

The parameters b, c, e_1 and e_2 with $m_8 = 0$ remain the same as those defined by Eq. (A11.3.17) but some of the definitions of the m-coefficients change,

$$m_1 = \frac{3}{2}, \quad m_3 = -\frac{u_0}{u_e}, \quad m_4 = -1, \quad m_6 = -m_3,$$
$$m_{12} = \left(1 + \frac{1}{u_e\sin\beta}\frac{dw_e}{d\phi}\right)\frac{1}{u_0\sin\beta}\frac{dw_e}{d\phi}, \tag{A11.3.40}$$
$$m_{13} = -\frac{(\rho v)_w}{\rho_e u_e}\sqrt{\frac{u_e x}{v_e}}$$

In all the above transformed equations based on conical flow approximation, the coefficient m_{13} represents the mass transfer parameter corresponding to either blowing or suction. The quantity $(\rho v)_w$ in the definition of m_{13} must vary as $x^{-1/2}$ for similarity.

A11.3.2 Conical Wing

The conical flow approximation can also be used for large aspect ratio tapered and swept wings; in a way it may be regarded as an extension of the infinite swept approximation used for nontapered wings to tapered wings. The approximation again excludes the root and tip regions of the wing. While it is usually associated with supersonic Mach numbers with the shock attached to the leading edge, it can also be applied to subsonic flows. In transonic flows wings are often designed to have nearly straight isobars at cruise conditions when the spanwise pressure gradient is negligible. If the wing is generated by straight-line development between the defining sections and the isobars run parallel to the generators, the conical flow assumption becomes a good approximation for flow over most of the wing. The wing does not have to fit the exact definition of a cone since the conical flow approximation is applied in a stripwise sense where a streamwise wing section is used to generate a cone with apex at the intersection of the extension of leading and trailing edges of the wing (see Fig. A11.6). The velocity components are calculated from the pressure distribution with the requirement that $du_e/dz = w_e$ which is also equal to

$$\sqrt{u_t^2 - u_e^2}$$

where u_t is the total velocity. The edge velocity $u_e(z)$ is obtained from the solutions of this equation.

For a conical wing, Eqs. (A11.3.1) to (A11.3.8) remain the same with x now denoting the distance from the fictitious apex and z the angle in the developed plane with $z = 0$ corresponding to the attachment line where $w_e \equiv 0$. Since the calculations are along a cut $x =$ const., the velocity component u_e becomes negative on swept-back wings. In order to use the transformations described in

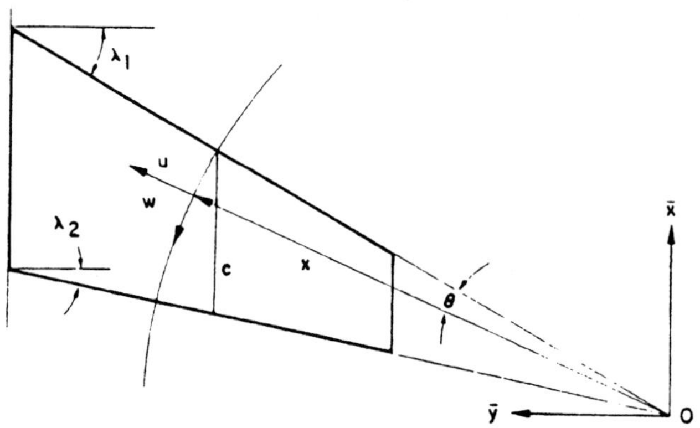

Fig. A11.6. Tapered wing and the polar coordinate system.

this subsection, it is necessary to replace u by $\bar{u}$ where $\bar{u} = -u$ and $\bar{u}_e = -u_e$ equations and transformations with the consequence that

$$f' = \bar{u}/u_e = u/u_e$$

For example, the term $\partial\psi/\partial x$ becomes positive in Eq. (A11.3.11) and there are sign changes in the m-coefficients of Eqs. (A11.3.28) to (A11.3.31): the modified m-coefficients for a conical wing are

$$\begin{gathered} m_1 = -\frac{3}{2}, \quad m_3 = -\frac{1}{\bar{u}_e}\frac{dw_e}{dz}, \quad m_4 = 1, \quad m_5 = \left(\frac{w_e}{u_e}\right)^2, \\ m_6 = -m_3 + \frac{m_5}{2} + \frac{m_7}{2}\frac{1}{\rho_e\mu_e}\frac{d(\rho_e\mu_e)}{dz}, \quad m_7 = \frac{w_e}{\bar{u}_e}, \\ m_8 = -m_5, \quad m_{12} = -m_3 - 1 \end{gathered} \tag{A11.3.41}$$

The definitions of b, C, c, e_1 and e_2 in Eq. (A11.3.17b) remain the same except that in the definition of e_2, $m_8 = m_5$.

The appropriate changes to the m-coefficients of the transformed attachment-line equations given by Eqs. (A11.3.35) to (A11.3.38) are

$$\begin{gathered} m_1 = -\frac{3}{2}, \quad m_3 = -\frac{1}{\bar{u}_e}\frac{dw_e}{dz}, \quad m_4 = 1, \\ m_6 = -m_3, \quad m_{12} = -m_3 - 1 \end{gathered} \tag{A11.3.42}$$

References

[1] Cebeci, T. and Cousteix, J., Modeling and Computation of Boundary-Layer Flows, Horizons Publishing, Long Beach and Springer-Verlag, Heidelberg 1998.

[2] Roget, C., "Structures Asymtotiques et calculs d'écoulements sur des obstacles bi et tridimensionels." These de l'Université Paul Sabatier.

[3] Isaacson, E. and Keller, H.B., Analysis of Numerical Methods, John Wiley, New York, 1966.

12 Three-Dimensional Flows: Transition Method

12.0 Introduction

The e^n-method discussed in Chapter 4 for predicting the onset of transition in two-dimensional incompressible flows can also be used for predicting the onset of transition in three-dimensional incompressible and compressible flows. In this chapter we shall describe its extension to three-dimensional incompressible flows (Sections 12.1 and 12.2) and to compressible flows (Section 12.3).

Again the e^n-method employs the solution of the linear stability equations, which for three-dimensional incompressible flows, can be written in the same form as two-dimensional flows,

$$\phi^{IV} - 2\gamma_2^2\phi'' + \gamma_2^4\phi - iR\gamma_1(\phi'' - \gamma_2^2\phi) + iR(\alpha u'' + \beta w'')\phi = 0 \tag{12.0.1}$$

where, with α and β denoting wave numbers in the x and z directions, and u and w the streamwise and spanwise velocity profiles, respectively,

$$\gamma_1 = \alpha u + \beta w - \omega, \quad \gamma_2^2 = \alpha^2 + \beta^2 \tag{12.0.2}$$

As discussed in [1], the boundary conditions for Eq. (12.0.1) can also be expressed in a form similar to those given for two-dimensional flows, Eqs. (4.0.5),

$$y = 0, \quad \phi = \phi' = 0 \tag{12.0.3a}$$

$$y \to \infty \quad (D^2 - \gamma_2^2)\phi + (\gamma_2 + \gamma_4)(D + \gamma_2)\phi = 0 \tag{12.0.3b}$$

$$(D + \gamma_4)(D^2 - \gamma_2^2)\phi = 0 \tag{12.0.3c}$$

where with $\mathrm{Re}(\gamma_2)$ and $\mathrm{Re}(\gamma_4) > 0$, γ_1 and γ_2^2 are given by Eq. (12.0.2) and γ_4^2 is given by

$$\gamma_4^2 = \gamma_2^2 + iR\gamma_1 \tag{12.0.4}$$

Equation (12.0.1) involves five scalars $\alpha, \beta, \omega_r, \omega_i$ and R in the temporal amplification theory and six scalars $\alpha_r, \alpha_i, \beta_r, \beta_i, \omega$ and R in the spatial amplification theory; its solution employing either a temporal amplification approach or spatial amplification approach, is considerably more difficult than the corresponding equation for two-dimensional flows. This is due to the eigenvalue problem. Unlike two-dimensional flows, the eigenvalue problem for three-dimensional flows requires a relationship between the two wave numbers α and β. The formulation of this relationship plays a crucial role in the eigenvalue problem since the onset of transition depends on the direction, magnitude and rate of the growth of the disturbances which propagate through the boundary-layer, and this information is represented by α and β which form a wave number vector, $\vec{k}$, at an angle ϕ to the flow direction.

There are several formulations developed to provide this relationship and these are discussed in the following section before the numerical solution of the three-dimensional linear stability equations employing the Cebeci-Stewartson eigenvalue formulation are described in Section 12.2 together with the calculation of the onset of transition with the e^n-method.

12.1 Eigenvalue Formulations for Three-Dimensional Flows

Malik [2], Mack [3], Arnal et al. [4] and Cebeci and Stewartson [5] have developed eigenvalue formulations for three-dimensional flows and used them in the solution of the linear stability equations. In the temporal amplification approach of Malik, the wave numbers α and β are related by the disturbance angle ϕ through

$$\phi = \tan^{-1}\left(\frac{\beta}{\alpha}\right) \tag{12.1.1}$$

with ϕ assumed and the dimensional frequency $\omega_r^*/2\pi$ specified, and α and β represent one unknown from Eq. (12.1.1); with R known, the second unknown is the growth rate of the disturbance ω_i. The stability equations are solved for initial estimates of α and ω_i for a given ω_r, ϕ and R. After convergence, α and ω_i are computed for different values of ϕ in order to find the maximum value of $\omega_i/|Vg|$, where Vg is the group velocity with components

$$Vg = \left[\left(\frac{\partial\omega}{\partial\alpha}\right)_{\beta,R}, \left(\frac{\partial\omega}{\partial\beta}\right)_{\alpha,R}\right] \tag{12.1.2}$$

Once the maximum value of $\omega_i/|Vg|$ is computed, the onset of transition is obtained from

$$n = \int_{x_0}^{x} \frac{\omega_i}{|Vg|} dx \tag{12.1.3}$$

Here x_0 corresponds to the x-location where the amplification rate is zero.

Mack [3] uses a spatial amplification theory and a different eigenvalue procedure that is restricted to infinite swept wings. In this formulation, he assumes the spanwise wave number β is real, thus allowing the wave to grow only in the chordwise direction. The value of β is obtained from the irrotationality condition applied to the complex wave number vector which, for an infinite swept wing, requires that the dimensional spanwise wave number β remains constant as the wave moves downstream. With these assumptions, the problem reduces to the calculation of the complex chordwise wave number α at each chordwise position for the specified dimensional values of β and ω. Arnal et al. [4] has improved this procedure by maximizing the amplification rate α_i with respect to β_r. More specifically, they solved the stability equations for initial estimate of α_r, α_i for a given β_r and specified $\omega^*/2\pi$ and R. After convergence, α_r and α_i are computed for different values of β_r in order to find the maximum value of α_i so that the onset of transition can be calculated from Eq. (4.2.4).

The eigenvalue formulation of Cebeci and Stewartson [5] is also based on the spatial amplification theory but differs from the approaches of Mack [3] and Arnal et al. [4] in that the relationship between the two wave numbers α and β is not assumed but is computed by making use of concepts based on group velocity using the saddle-point method discussed by Cebeci and Stewartson [5] and Nayfeh [6]. We assume the disturbance to be of the form

$$q'(x, y, z, t) = q(y)e^{i(\alpha x+\beta z-\omega t)}$$

use spatial amplification theory and consider a wave packet made of disturbances of all possible waves of given frequency. The wave packet disturbance is

$$Q = \int_c q(y)e^{i(\alpha x+\beta z-\omega t)}d\beta \qquad (12.1.4)$$

where c is a contour in the complex volume of β, extending to ∞ in either direction. Noting that at a given spatial location and for a given frequency

$$\alpha = \alpha(\beta)$$

we can write Eq. (12.1.4) as

$$Q = \int q(y)e^{ix[\alpha(\beta)+\beta\frac{z}{x}-\frac{\omega}{x}t]}d\beta$$

For large values of x (assuming that z/x is given), the saddle point method (or the stationary phase method) requires that the leading contribution to Q is obtained when the phase of the exponential term is stationary. This means that the derivative of the phase with respect to β is zero, that is,

$$\left(\frac{\partial\alpha}{\partial\beta}\right)_{\omega,R} + \frac{z}{x} = 0$$

or

$$\left(\frac{\partial \alpha}{\partial \beta}\right)_{\omega,R} = -\frac{z}{x} \tag{12.1.5}$$

Since x and z are real, the imaginary part of $(\partial\alpha/\partial\beta)_{\omega,R}$ must be zero to satisfy Eq. (12.1.5). According to the saddle point method, the real part of $(\partial\alpha/\partial\beta)_{\omega,R}$ is related to the disturbance angle ϕ

$$\left(\frac{\partial \alpha}{\partial \beta}\right)_{\omega,R} = -\tan\phi \tag{12.1.6}$$

The above relation provides the wave orientation and growth direction of the disturbance. This can be used to relate the two wave numbers α and β to each other as needed in the eigenvalue problem. The angle ϕ in Eq. (12.1.6) is also the direction of the group velocity $\vec{V}_g$, since

$$\left(\frac{\partial \alpha}{\partial \beta}\right)_{\omega,R} = -\frac{\left(\frac{\partial \omega}{\partial \beta}\right)_{\alpha,R}}{\left(\frac{\partial \omega}{\partial \alpha}\right)_{\beta,R}} \tag{12.1.7}$$

We observe that with α and β defined by Eq. (12.1.6), the magnitude of the disturbance at (x, z) (see Eq. (6.3.14)) is dominated by the exponential factor

$$\exp\left[-(\alpha_i x + \beta_i z)\right]$$

and, the disturbances are damped with increasing x if the amplification rate Γ defined by

$$\Gamma \equiv \alpha_i - \beta_i \left(\frac{\partial \alpha}{\partial \beta}\right)_{\omega,R} \tag{12.1.8}$$

is > 0, neutral if $\Gamma = 0$, and amplified when $\Gamma < 0$. In the eigenvalue procedure of Cebeci and Stewartson [5], once α and β are computed subject to the requirements of Eq. (12.1.6), additional calculations are then made for different values of $\partial\alpha/\partial\beta$ to determine the maximum amplification rate, Γ. In accordance with the e^n-method, the values of $\Gamma(x)$ are then integrated to find the value of n for the specified frequency. As in two-dimensional flows, this process is repeated for different dimensional frequencies to find the critical frequency that leads to the highest integrated amplification rate. In their use of this eigenvalue procedure for a rotating disk, Cebeci and Stewartson [5] fixed the direction of the disturbance angle ϕ as computed on the three-dimensional neutral stability curve, zarf, (Section 12.2) and did not maximize it during the calculations. However, as discussed in [7], and in Chapter 13, $\Gamma(x)$ can vary considerably with ϕ and must be maximized.

12.2 Numerical Method for Incompressible Flows

The numerical solution of Eqs. (12.0.1) and (12.0.3) is similar to the solution of Eqs. (4.1.3) and (4.1.5). If we let

$$\xi_1^2 = \gamma_2^2, \quad \xi_2 = \gamma_4, \quad \xi_3 = iR(\alpha u'' + \beta w'') \tag{12.2.1}$$

then Eq. (12.0.1) can again be expressed in the same compact form as Eq. (4.3.3). The difference approximations, Eqs. (4.3.6), boundary conditions, Eqs. (4.3.8b) and (4.3.10), remain the same, and the linear system can be expressed in the same form as Eq. (3.3.16) with relations given by Eqs. (4.3.12) and (4.3.12) remaining the same. The solution of Eq. (3.3.16) can again be obtained with the block-elimination method, with the solution now depending on four parameters α, β, ω and R rather than three parameters α, ω and R in two-dimensional flows. This time we can denote this dependence by rewriting Eq. (4.3.13) as

$$\vec{\vec{\delta}} = \vec{\vec{\delta}}(\alpha, \beta, \omega, R) \tag{12.2.2}$$

Recalling that α and β are complex and that ω is real in spatial amplification theory, the above relation implies that the solution of Eq. (3.3.16) depends upon six scalars $(\alpha_r, \alpha_i, \beta_r, \beta_i, \omega, R)$. With any four of these scalars fixed, the remaining two scalars can be computed in such a way that the missing boundary condition

$$f_0(\alpha, \beta, \omega, R) = 0 \tag{12.2.3}$$

is satisfied. Similarly, for temporal amplification theory, the solution of Eq. (3.3.16) depends upon five scalars $(\alpha, \beta, \omega_r, \omega_i, R)$. In either case, a relationship between the two complex wave numbers α and β is needed if ω and R are fixed as discussed in subsections 12.2.1 and 12.2.2 for predicting transition by the saddle-point method of Cebeci and Stewartson.

12.2.1 Calculation of Dimensional Frequencies: The Zarf

As discussed in Section 4.3.2, the dimensional frequencies needed in the e^n-method for two-dimensional flows are calculated by solving the eigenvalue problem on the lower branch of the neutral stability curve, where $\alpha_i = 0$. For three-dimensional flows, it is plausible to use a similar procedure by defining a neutral stability curve for three-dimensional flows. However, the definition of this curve requires some care, as discussed in [5] and except for the work of Cebeci and Stewartson, it has not been defined and used by others who employ the e^n-method for three-dimensional flows. Instead, the dimensional frequencies are specified in the stability calculations employing an eigenvalue formulation such as those discussed in Section 12.1 with the exception of the eigenvalue formulation of Cebeci and Stewartson. Amplification calculations are performed for several dimensional frequencies to determine the critical frequency which leads to the first and most amplified integrated amplification factor.

In the Cebeci and Stewartson procedure, the dimensional frequencies are computed by making use of an absolute neutral curve for three-dimensional flows. This curve, first introduced by Cebeci and Stewartson and called the zarf has a number of interesting properties as discussed in [5]. The first and most important of these is that along the zarf,

$$\alpha, \beta \quad \text{and} \quad \partial\alpha/\partial\beta \quad \text{are all real.} \tag{12.2.4}$$

Since α_i and β_i are zero on the zarf, Eq. (12.2.3) contains three unknowns $(\alpha_r, \beta_r, \omega)$ for a given R and an additional equation based on Eq. (12.1.6) is needed. This equation can be obtained by first expanding Eq. (12.2.3) with ω constant,

$$df = \left(\frac{\partial f}{\partial \alpha}\right)_\beta d\alpha + \left(\frac{\partial f}{\partial \beta}\right)_\alpha d\beta = 0 \tag{12.2.5}$$

Solving Eq. (12.2.5) for $d\alpha/d\beta$, we get

$$\left(\frac{d\alpha}{d\beta}\right)_\omega = -\frac{(\partial f/\partial \beta)}{(\partial f/\partial \alpha)} = e \tag{12.2.6}$$

Expanding Eq. (12.2.3) about α^ν, β^ν and ω^ν and retaining only the linear terms in the expansion, and after separating the real and imaginary parts of the resulting expression, we get

$$f_r^\nu + \left(\frac{\partial f}{\partial \alpha}\right)_r^\nu \delta\alpha + \left(\frac{\partial f}{\partial \beta}\right)_r^\nu \delta\beta + \left(\frac{\partial f}{\partial \omega}\right)_r^\nu \delta\omega = 0 \tag{12.2.7a}$$

$$f_i^\nu + \left(\frac{\partial f}{\partial \alpha}\right)_i^\nu \delta\alpha + \left(\frac{\partial f}{\partial \beta}\right)_i^\nu \delta\beta + \left(\frac{\partial f}{\partial \omega}\right)_i^\nu \delta\omega = 0 \tag{12.2.7b}$$

Similar to Eq. (12.2.3), we now expand Eq. (12.2.6) to obtain

$$e^\nu + \left(\frac{\partial e}{\partial \alpha}\right)^\nu \delta\alpha + \left(\frac{\partial e}{\partial \beta}\right)^\nu \delta\beta + \left(\frac{\partial e}{\partial \omega}\right)^\nu \delta\omega = \text{const.} \tag{12.2.8}$$

Since we are interested in the values of α, β and ω for which the imaginary part of $(d\alpha/d\beta)$ or e_i is zero, from Eq. (12.2.8) we can write

$$e_i^\nu + \left(\frac{\partial e_i}{\partial \alpha}\right)^\nu \delta\alpha + \left(\frac{\partial e_i}{\partial \beta}\right)^\nu \delta\beta + \left(\frac{\partial e_i}{\partial \omega}\right)^\nu \delta\omega = 0 \tag{12.2.9}$$

Solving Eqs. (12.2.7) and (12.2.9), we write $\partial\alpha$, $\partial\beta$ and $\partial\omega$ as:

$$\delta\alpha = \frac{1}{\Delta_1}[-f^r(f_\beta^i e_\omega^i - f_\omega^i e_\beta^i) + f^i(f_\beta^r e_\omega^i - e_\beta^i f_\omega^r) - e^i(f_\beta^r f_\omega^i - f_\omega^r f_\beta^i)] \tag{12.2.10a}$$

$$\delta\beta = \frac{1}{\Delta_1}[f_\alpha^r(-f^i e_\omega^i + e^i f_\omega^i) - f_\alpha^i(-f^r e_\omega^i + e^i f_\omega^r) + e_\alpha^i(-f^r f_\omega^i + f^i f_\omega^r)] \tag{12.2.10b}$$

$$\delta\omega = \frac{1}{\Delta_1}[f_\alpha^r(-f_\beta^i e^i + f^i e_\beta^i) - f_\alpha^i(-f_\beta^r e^i + f^r e_\beta^i) + e_\alpha^i(-f_\beta^r f^i + f^r f_\beta^i)] \quad (12.2.10c)$$

$$\frac{1}{\Delta_1} = e_i^\alpha(f_\beta^r f_\omega^i - f_\beta^i f_\omega^r) - e_\beta^i(f_\alpha^r f_\omega^i - f_\alpha^i f_\omega^r) + e_\omega^i(f_\alpha^r f_\beta^i - f_\alpha^i f_\beta^r) \quad (12.2.11)$$

For simplicity we have dropped the superscript ν and denoted the real and imaginary parts of f, $f_\alpha(\equiv \partial f/\partial\alpha)$, $f_\beta(\equiv \partial f/\partial\beta)$, $f_\omega(\equiv \partial f/\partial\omega)$, $e_\beta(\equiv \partial e/\partial\beta)$, $e_\alpha(\equiv \partial e/\partial\alpha)$, $e_\omega(\equiv \partial e/\partial\omega)$ by superscripts r and i. We note from Eq. (12.2.9) that if $\delta\alpha$, $\delta\beta$ and $\delta\omega$ are all zero, then e_i is zero and e is real.

To calculate the derivative of e with respect to α, β and ω we make use of Eq. (12.2.6), rewritten as

$$e = -\frac{f_\beta}{f_\alpha} \quad (12.2.12)$$

Differentiating with respect to β, we get

$$e_\beta = -\frac{f_{\beta\beta}}{f_\alpha} + \frac{f_\beta}{f_\alpha^2} f_{\alpha\beta} \quad (12.2.13a)$$

Similarly,

$$e_\alpha = -\frac{f_{\beta\alpha}}{f_\alpha} + \frac{f_\beta}{f_\alpha^2} f_{\alpha\alpha} \quad (12.2.13b)$$

$$e_\omega = -\frac{f_{\beta\omega}}{f_\alpha} + \frac{f_\beta}{f_\alpha^2} f_{\alpha\omega} \quad (12.2.13c)$$

Note from Eqs. (12.2.10) and (12.2.13) that the calculation of $\delta\alpha$, $\delta\beta$ and $\delta\omega$ requires the calculation of $f_{\alpha\beta}$, $f_{\alpha\alpha}$, $f_{\beta\omega}$, $f_{\beta\beta}$ and $f_{\alpha\omega}$, which can be obtained by differentiating Eq. (3.3.16) with respect to appropriate eigenvalues. For example, to find the eigenfunction $f_{\alpha\beta}$, we differentiate Eq. (3.3.16) first with respect to α then with respect to β.

$$A\left(\frac{\partial^2 \vec{\vec{\delta}}}{\partial\alpha\partial\beta}\right)^\nu = -\frac{\partial A}{\partial\beta}\left(\frac{\partial\vec{\vec{\delta}}}{\partial\alpha}\right)^\nu - \frac{\partial A}{\partial\alpha}\left(\frac{\partial\vec{\vec{\delta}}}{\partial\beta}\right)^\nu - \frac{\partial^2 A}{\partial\alpha\partial\beta}\vec{\vec{\delta}}^\nu \quad (12.2.14)$$

The above equation is the variational form of Eq. (3.3.16) with respect to α and β. Thus, to obtain the required derivative, Eq. (12.2.14) must be solved with the same coefficient matrix A already computed and factored for Eq. (3.3.16). The vectors on the right-hand side of Eq. (12.2.14) are determined from Eqs. (12.2.7), (12.2.12) and (12.2.9b). The eigenfunctions f_α and f_β are obtained from the solutions of the variational equations with respect to α and β, that is

$$A\left(\frac{\partial\vec{\vec{\delta}}}{\partial\alpha}\right)^\nu = -\left(\frac{\partial A}{\partial\alpha}\right)^\nu \vec{\vec{\delta}}^\nu, \quad A\left(\frac{\partial\vec{\vec{\delta}}}{\partial\beta}\right)^\nu = -\left(\frac{\partial A}{\partial\beta}\right)^\nu \vec{\vec{\delta}}^\nu \quad (12.2.15)$$

To find the variational equations with respect to $\alpha\alpha$, $\beta\beta$, $\alpha\omega$ and $\beta\omega$, a similar procedure is followed. For example, to find those for $\alpha\alpha$, Eq. (3.3.16) is differentiated twice with respect to α

$$A\left(\frac{\partial^2 \vec{\vec{\delta}}}{\partial \alpha^2}\right)^{\nu} = -2\frac{\partial A}{\partial \alpha}\left(\frac{\partial \vec{\vec{\delta}}}{\partial \alpha}\right)^{\nu} - \frac{\partial^2 A}{\partial \alpha^2}\vec{\vec{\delta}}^{\nu} \tag{12.2.16}$$

To summarize one step of the calculation of the eigenvalues of the zarf, we first solve Eq. (3.3.16) for initial estimates of α, β and ω (subsection 12.2.3). If the computed value of f_0 does not satisfy Eq. (12.2.3), then appropriate variational equations are solved in order to calculate the increments in α, β and ω from Eqs. (12.2.10) so that new estimates of α, β and ω can be obtained that will satisfy Eq. (12.2.3) and make e real. The procedure is repeated until convergence is achieved.

Since the eigenvalue formulation is based on Newton's method, the calculations usually indicate quadratic convergence provided that initial guesses are good estimates.

12.2.2 Calculation of Transition

With ω known from the specified dimensional frequency and R given, and with e_r fixed (equal to its value at zarf, say at R_1), we solve Eq. (3.3.16) for initial estimates of α and β to get a solution at $R = R_2$. To obtain the next values of α and β, we use Newton's method in which we expand f and e in Eqs. (12.2.3) and (12.2.6),

$$f_0 + \left(\frac{\partial f}{\partial \alpha}\right)_0 \delta\alpha + \left(\frac{\partial f}{\partial \beta}\right)_0 \delta\beta = 0 \tag{12.2.17a}$$

$$e_0 + \left(\frac{\partial e}{\partial \alpha}\right)_0 \delta\alpha + \left(\frac{\partial e}{\partial \beta}\right)_0 \delta\beta = e = \text{given} \tag{12.2.17b}$$

Solving for $\delta\alpha$ and $\delta\beta$, we get

$$\delta\alpha = -\frac{1}{\Delta_0}[f_0 e_\beta + f_\beta(e - e_0)] \tag{12.2.18a}$$

$$\delta\beta = \frac{1}{\Delta_0}[f_\alpha(e - e_0) + f_0 e_\alpha] \tag{12.2.18b}$$

where

$$\Delta_0 = f_\alpha e_\beta - f_\beta e_\alpha \tag{12.2.19}$$

As before, f_α, f_β, e_α and e_β are determined by solving the variational equations with respect to α, β, $\alpha\alpha$, $\alpha\beta$ and $\beta\beta$. This procedure is repeated until $\delta\alpha$ and $\delta\beta$ are less than a specified tolerance parameter. Upon convergence, the amplification parameter Γ is computed from Eq. (12.1.8). Since the calculation of Γ assumes that the value of e on the right-hand side of Eq. (12.2.17b) is fixed by the value it takes on the zarf, we compute a new value of e for which Γ is minimum, and this is achieved by assigning a new value of e. After convergence, we compute Γ from Eq. (12.1.8) and test to see whether Γ is increasing or decreasing. Since a positive Γ implies that disturbances are damped, we can increment e until a minimum value of Γ is computed.

12.2.3 Estimation of Eigenvalues

The solution of Eq. (3.3.16) requires initial estimates of the eigenvalues and, these are obtained by using the continuation method discussed in subsection 4.2.4 for two-dimensional flows.

We define

$$\begin{aligned}\bar{u} &= u_{\text{ref}} + n(u - u_{\text{ref}}) \\ \bar{w} &= w_{\text{ref}} + n(w - w_{\text{ref}}) \\ \bar{u}'' &= u''_{\text{ref}} + n(u'' - u''_{\text{ref}}) \\ \bar{w}'' &= w''_{\text{ref}} + n(w'' - w''_{\text{ref}})\end{aligned} \tag{12.2.20}$$

Here u and w correspond to the velocity profiles of the flow under consideration at a Reynolds number R, u_{ref} and w_{ref} denote reference velocity profiles with a Reynolds number R_{ref} and with eigenvalues of α_0, β_0 and ω_0. The parameter n is again a sequence of specified numbers ranging from 0 to 1. It follows from Eqs. (12.2.20) that the profiles $\bar{u}$, $\bar{w}$, and $\bar{u}''$, $\bar{w}''$ correspond to the reference profiles for $n = 0$ and to those for which eigenvalues α, β and ω are to be determined for specified R for $n = 1$.

12.3 Numerical Method for Compressible Flows

For compressible flows, the linear stability equations, with the parallel flow approximation, are given by the continuity, three momentum equations and one energy equation,

Continuity

$$i\gamma_1 \hat{\rho} + \rho\left[\frac{d\hat{v}}{dy} + i(\alpha\hat{u} + \beta\hat{w})\right] + \frac{d\rho}{dy}\hat{v} = 0 \tag{12.3.1}$$

x-Momentum

$$\begin{aligned}&\rho\left[i\gamma_1(\alpha\hat{u} + \beta\hat{w}) + \left(\alpha\frac{du}{dy} + \beta\frac{dw}{dy}\right)\hat{v}\right] \\ &= -i\gamma_2^2\frac{\hat{p}}{\gamma M_e^2} + \frac{\mu}{R}\left\{\left(\alpha\frac{d^2\hat{u}}{dy^2} + \beta\frac{d^2\hat{w}}{dy^2}\right) + \gamma_2^2\left[i\frac{d\hat{v}}{dy} - 2(\alpha\hat{u} + \beta\hat{w})\right]\right\} \\ &+ \frac{2}{3}\frac{\lambda - \mu}{R}\gamma_2^2\left[i\frac{d\hat{v}}{dy} - (\alpha\hat{u} + \beta\hat{w})\right] + \frac{1}{R}\frac{d\mu}{dT}\left(\alpha\frac{d^2u}{dy^2} + \beta\frac{d^2w}{dy^2}\right)\hat{T} \\ &+ \frac{1}{R}\left[\left(\frac{d\mu}{dT}\frac{d\hat{T}}{dy} + \frac{d^2\mu}{dT^2}\frac{d\hat{T}}{dy}\hat{T}\right)\left(\alpha\frac{du}{dy} + \beta\frac{dw}{dy}\right)\right] \\ &+ \frac{1}{R}\frac{d\mu}{dT}\frac{dT}{dy}\left[\left(\alpha\frac{d\hat{u}}{dy} + \beta\frac{d\hat{w}}{dy}\right) + i\gamma_2^2\hat{v}\right]\end{aligned} \tag{12.3.2}$$

z-Momentum

$$\rho\left[i\gamma_1(\alpha\hat{w}-\beta\hat{u})+\left(\alpha\frac{dw}{dy}-\beta\frac{du}{dy}\right)\hat{v}\right]$$
$$=\frac{\mu}{R}\left[\left(\alpha\frac{d^2\hat{w}}{dy^2}-\beta\frac{d^2\hat{u}}{dy^2}\right)+\gamma_2^2(\alpha\hat{w}-\beta\hat{u})\right]-\frac{1}{R}\left[\frac{d\mu}{dT}\frac{dT}{dy}\left(\alpha\frac{d\hat{w}}{dy}-\beta\frac{d\hat{u}}{dy}\right)\right.$$
$$\left.+\frac{d\mu}{dT}\left(\alpha\frac{d^2w}{dy^2}-\beta\frac{d^2u}{dy^2}\right)\hat{T}+\left(\frac{d\mu}{dT}\frac{d\hat{T}}{dy}+\frac{d^2\mu}{dT^2}\frac{dT}{dy}\hat{T}\right)\left(\alpha\frac{dw}{dy}-\beta\frac{du}{dy}\right)\right] \quad (12.3.3)$$

y-Momentum

$$i\rho\gamma_1\hat{v}=-\frac{1}{\gamma M_e^2}\frac{d\hat{p}}{dy}+\frac{\mu}{R}\left[2\frac{d^2\hat{v}}{dy^2}+i\left(\alpha\frac{d\hat{u}}{dy}+\beta\frac{d\hat{w}}{dy}\right)-\gamma_2^2\hat{v}\right]$$
$$+\frac{2}{3}\left(\frac{\lambda-\mu}{R}\right)\left[\frac{d^2\hat{v}}{dy^2}+i\left(\alpha\frac{d\hat{u}}{dy}+\beta\frac{d\hat{w}}{dy}\right)\right]$$
$$+\frac{1}{R}\left\{i\frac{d\mu}{dT}\left(\alpha\frac{du}{dy}+\beta\frac{dw}{dy}\right)\hat{T}+2\frac{d\mu}{dT}\frac{dT}{dy}\frac{d\hat{v}}{dy}\right. \quad (12.3.4)$$
$$\left.+\frac{2}{3}\frac{d}{dT}(\lambda-\mu)\frac{dT}{dy}\left[\frac{d\hat{v}}{dy}+i(\alpha\hat{u}+\beta\hat{w})\right]\right\}$$

Energy

$$\rho\left[i\gamma_1\hat{T}+\frac{dT}{dy}\hat{v}\right]=-(\gamma-1)\left[\frac{d\hat{v}}{dy}+i(\alpha\hat{u}+\beta\hat{w})\right]$$
$$+\frac{\gamma\mu}{\Pr R}\left\{\frac{d^2\hat{T}}{dy^2}-\gamma_2^2\hat{T}+\frac{1}{k}\left[\frac{dk}{dT}\frac{d^2T}{dy^2}+\frac{d^2k}{dT^2}\left(\frac{dT}{dy}\right)^2\right]\hat{T}+\frac{2}{k}\frac{dk}{dT}\frac{dT}{dy}\frac{d\hat{T}}{dy}\right\}$$
$$+\gamma\frac{(\gamma-1)}{R}M_e^2\left\{2i\mu\left(\alpha\frac{du}{dy}+\beta\frac{dw}{dy}\right)\hat{v}+\frac{d\mu}{dT}\left(\frac{d^2u}{dy^2}+\frac{d^2w}{dy^2}\right)\hat{T}\right.$$
$$+\frac{2\mu}{\gamma_2^2}\left[\left(\alpha\frac{du}{dy}+\beta\frac{dw}{dy}\right)\left(\alpha\frac{d\hat{u}}{dy}+\beta\frac{d\hat{w}}{dy}\right)\right.$$
$$\left.\left.+\left(\alpha\frac{dw}{dy}-\beta\frac{du}{dy}\right)\left(\alpha\frac{d\hat{w}}{dy}-\beta\frac{d\hat{u}}{dy}\right)\right]\right\} \quad (12.3.5)$$

where all the fluctuating and steady-state velocity components are normalized by u_e, and all lengths by a reference length L. The fluctuating and steady-state pressure, density, and temperature are normalized by p_e, ρ_e and T_e, respectively. Both dynamic viscosity μ and second viscosity λ and heat conductivity k are normalized by $c_p u_e$. The Reynolds number, R, and Mach number M_e are defined by

$$R=\frac{u_e L}{\nu_e}, \quad M_e=\frac{u_e}{\sqrt{\gamma\tilde{R}T_e}} \quad (12.3.6)$$

with $\gamma = c_p/c_v$ and $\tilde{R}$ denoting the universal gas constant. The hat ^ on each flow variable denotes the amplitude function of the fluctuating flow variable. The parameters α and β denote the wave numbers in the x- and z-directions, respectively, and ω represents the radian frequency. The parameters γ_1 and γ_2 are defined by Eq. (12.0.2).

Equations (12.3.1) to (12.3.5) are subject to the following boundary conditions

$$y = 0, \quad \hat{u} = \hat{v} = \hat{w} = \hat{T} = 0 \tag{12.3.7a}$$

$$y \to \infty, \quad \hat{u}, \hat{v}, \hat{w}, \hat{T} \to 0 \tag{12.3.7b}$$

The solution of the stability equations for three-dimensional compressible flows is similar to the solution of the flows discussed in Section 12.2 for incompressible flows. We again express Eqs. (12.3.1) to (12.3.5) and Eq. (12.3.7) as a first-order system by defining the following new variables with primes denoting differentiation with respect to y,

$$\begin{gathered} z_1 = \alpha\hat{u} + \beta\hat{w}, \quad z_2 = \alpha\hat{u}' + \beta\hat{w}', \quad z_3 = \hat{v}, \quad z_4 = \frac{\hat{p}}{\gamma M_e^2}, \\ z_5 = \hat{T}, \quad z_6 = \hat{T}', \quad z_7 = \alpha\hat{w} - \beta\hat{u}, \quad z_8 = \alpha\hat{w}' - \beta\hat{u}' \end{gathered} \tag{12.3.8}$$

In terms of these new variables, Eqs. (12.3.1) to (12.3.5) and the boundary conditions given by Eq. (12.3.7)

$$\begin{gathered} y = 0, \quad z_1 = z_3 = z_5 = z_7 = 0 \\ y \to \infty, \quad z_1, z_3, z_5, z_7 \to 0 \end{gathered} \tag{12.3.9}$$

can be written as

$$\tilde{z}' = B\tilde{z} \tag{12.3.10}$$

Here $\tilde{z} = (z_1, z_2, \ldots, z_8)$ and B is an 8×8 coefficient matrix defined by

$$B = \begin{vmatrix} 0 & 1 & 0 & 0 & 0 & 0 & 0 & 0 \\ b_{21} & b_{22} & b_{23} & b_{24} & b_{25} & b_{26} & 0 & 0 \\ b_{31} & 0 & b_{33} & b_{34} & b_{35} & 0 & 0 & 0 \\ b_{41} & b_{42} & b_{43} & b_{44} & b_{45} & b_{46} & 0 & 0 \\ 0 & 0 & 0 & 0 & 0 & 1 & 0 & 0 \\ 0 & b_{62} & b_{63} & b_{64} & b_{65} & b_{66} & 0 & b_{68} \\ 0 & 0 & 0 & 0 & 0 & 0 & 0 & 1 \\ 0 & 0 & b_{83} & 0 & b_{85} & b_{86} & b_{87} & b_{88} \end{vmatrix} \tag{12.3.11}$$

with its nonzero elements given below. We define

$$\gamma_3 = \alpha\frac{du}{dy} + \beta\frac{dw}{dy}, \quad \gamma_4 = \alpha\frac{dw}{dy} - \beta\frac{du}{dy}, \quad d = \frac{\lambda}{\mu}$$

and represent the elements of B by

$$b_{21} = \frac{iR}{\mu T}\gamma_1 + \gamma_2, \quad b_{22} = -\frac{1}{\mu}\frac{d\mu}{dT}\frac{dT}{dy}$$

$$b_{23} = \frac{R\gamma_3}{\mu T} - \gamma_2 \frac{i}{\mu}\frac{d\mu}{dT}\frac{dT}{dy} - \frac{1+2d}{3}\gamma_2 \frac{i}{T}\frac{dT}{dy}$$

$$b_{24} = \frac{iR}{\mu}\gamma_2 - \frac{1+2d}{3}\gamma M_e^2 \gamma_2 \gamma_1$$

$$b_{25} = \frac{1+2d}{3T}\gamma_1\gamma_2 - \frac{1}{\mu}\frac{d\mu}{dT}\left(\alpha\frac{d^2u}{dy^2} + \beta\frac{d^2w}{dy^2})\right) - \frac{1}{\mu}\frac{d^2\mu}{dT^2}\frac{dT}{dy}\gamma_3$$

$$b_{26} = -\frac{1}{\mu}\frac{d\mu}{dT}\gamma_3, \quad b_{31} = -i, \quad b_{33} = \frac{1}{T}\frac{dT}{dy}, \quad b_{34} = -i\gamma M_e^2\gamma_1$$

$$b_{35} = \frac{i}{T}\gamma_1, \quad E = \frac{R}{\mu} + \frac{2}{3}i(2+d)\gamma M_e^2 \gamma_1$$

$$b_{41} = -\frac{i}{E}\left[-2b_{22} + \frac{2}{3}(2+d)b_{33}\right], \quad b_{42} = -\frac{i}{E}$$

$$b_{43} = -\frac{i}{E}\left[-\gamma_2 + \frac{2(2+d)}{3\mu T}\frac{d\mu}{dT}\left(\frac{dT}{dy}\right)^2 + \frac{2(2+d)}{3T}\frac{d^2T}{dy^2} - \frac{iR}{\mu T}\gamma_1\right]$$

$$b_{44} = -\frac{2i}{3E}(2+d)\gamma M_e^2(-\gamma_1 b_{22} + \gamma_3 + b_{22}\gamma_1)$$

$$b_{45} = \frac{i}{E}\left[\frac{1}{\mu}\frac{d\mu}{dT}\gamma_3 + \frac{2(2+d)}{3T}(-b_{22}\gamma_1 + \gamma_3)\right]$$

$$b_{46} = \frac{2i}{3ET}(2+d)\gamma_1$$

$$b_{62} = -2(\gamma-1)\mathrm{Pr}\, M_e^2\left(\frac{\gamma_3}{\gamma_2}\right), \quad b_{63} = \frac{\mathrm{Pr}\, R}{\mu}b_{33} - 2i(\gamma-1)\mathrm{Pr}\, M_e^2\gamma_3$$

$$b_{64} = -\frac{i}{\mu}(\gamma-1)\mathrm{Pr}\, R M_e^2 \gamma_1$$

$$b_{65} = i\frac{\mathrm{Pr}\, R}{\mu T}\gamma_1 + \gamma_2 - \frac{1}{k}\frac{dk}{dT}\frac{d^2T}{dy^2} - \frac{1}{k}\frac{d^2k}{dT^2}\left(\frac{dT}{dy}\right)^2 - (\gamma-1)\mathrm{Pr}\, M_e^2 \frac{1}{\mu}\frac{d\mu}{dT}\left(\frac{d^2u}{dy^2} + \frac{d^2w}{dy^2}\right)$$

$$b_{66} = -\frac{2}{k}\frac{dk}{dT}\frac{dT}{dy}$$

$$b_{68} = -\frac{2(\gamma-1)}{\gamma_2}\mathrm{Pr}\, M_e^2\gamma_4$$

$$b_{83} = \frac{R}{\mu T}\gamma_4$$

$$b_{85} = -\frac{1}{\mu}\frac{d\mu}{dT}\left(\alpha\frac{d^2w}{dy^2} - \beta\frac{d^2u}{dy^2}\right) - \frac{1}{\mu}\frac{d^2\mu}{dT^2}\frac{dT}{dy}\gamma_4$$

$$b_{86} = -\frac{1}{\mu}\frac{d\mu}{dT}\gamma_4, \quad b_{87} = \frac{iR}{\mu T}\gamma_1 + \gamma_2, \quad b_{88} = b_{22}$$

The solution of the system given by Eqs. (12.3.9) and (12.3.10) is obtained with the numerical method described in Section 12.2.

References

[1] Cebeci, T. and Cousteix, J., Modeling and Computation of Boundary-Layer Flows, Horizons, Long Beach, and Springer-Verlag, Heidelberg, 1998

[2] Malik, M.R., COSAL – A Black-Box Compressible Stability Analysis Code for Transition Prediction in Three-Dimensional Boundary Layers, NASA CR 155952, 1982.

[3] Mack, L.M., Stability of Three-Dimensional Boundary Layers on Swept Wings at Transonic Speeds, IUTAM Symposium, Transsonicum III, Göttingen, Germany, 1988.

[4] Arnal, D., Casalis, G. and Julien, J.C., Experimental and Theoretical Analysis of Natural Transition on Infinite Swept Wing, IUTAM Symposium, Toulouse, France, 1989, Laminar-Turbulent Transition (eds. D. Arnal, R. Michnel) Springer-Verlag, 1989, pp. 311–326.

[5] Cebeci, T. and Stewartson, K., Stability and Transition in Three-Dimensional Flows. AIAA J., Vol. 18, 1980, p. 398.

[6] Nayfeh, A.H., Stability of Three-Dimensional Boundary Layers. AIAA J., Vol. 18, 1980, pp. 406–416.

[7] Cebeci, T., Chen, H.H., Arnal, D. and Huang, T.T., A Three-Dimensional Linear Stability Approach to Transition on Wings and Bodies of Revolution at Incidence. AIAA J., Vol. 29, No. 12, Dec. 1991, pp. 2077–2085.

13 Applications of the Calculation Method to Three-Dimensional Subsonic and Transonic Flows

13.0 Introduction

In this chapter we discuss the application of the calculation method of Chapters 10 to 12 to three-dimensional subsonic and transonic flows. In Section 13.1 we first demonstrate the accuracy of the e^n-method to three-dimensional flows by considering one subsonic and one transonic flow on infinite swept wings for which there is experimental data. In Section 13.2 we study the role of surface curvature on transition for an incompressible flow over an infinite sweep wing by varying the sweep angle and the chord Reynolds number in a systematic manner. This is followed by Sections 13.3 and 13.4 where the application of the calculation method is discussed for subsonic flows on wing alone and multi-wing configurations. Sections 13.5 and 13.6 address the prediction of ice shapes on lifting surfaces of an aircraft and the performance degradation of an aircraft in natural icing conditions. These two sections are similar to Sections 5.5 and 5.6 where the applications of the calculation method was directed to two-dimensional flows and not three-dimensional flows. The final section, 13.7, discusses the application of the calculation method to transonic flows over wing and wing/fuselage configurations.

13.1 Accuracy of the e^n-Method for Three-Dimensional Flows

Before we apply the calculation method of the previous two chapters to three-dimensional flows, it is useful to evaluate the accuracy of the e^n-method with limited available experimental data since the stability/transition method is a part of the calculation method. In this section this is done for both incompressible (subsection 13.1.1) and compressible (subsection 13.1.2) flows.

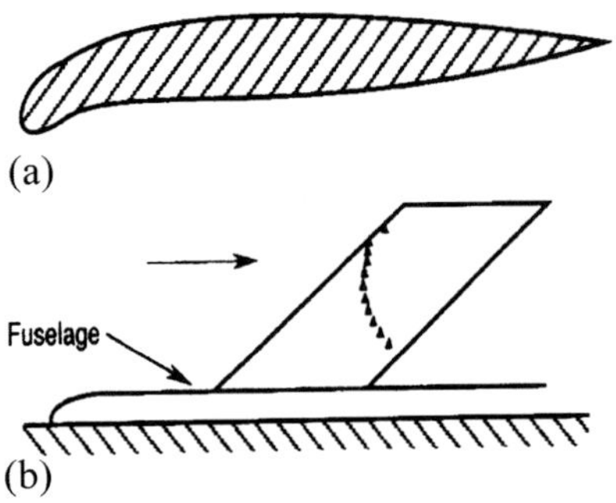

Fig. 13.1. ONERA-D airfoil with a cambered leading-edge: experimental setup.

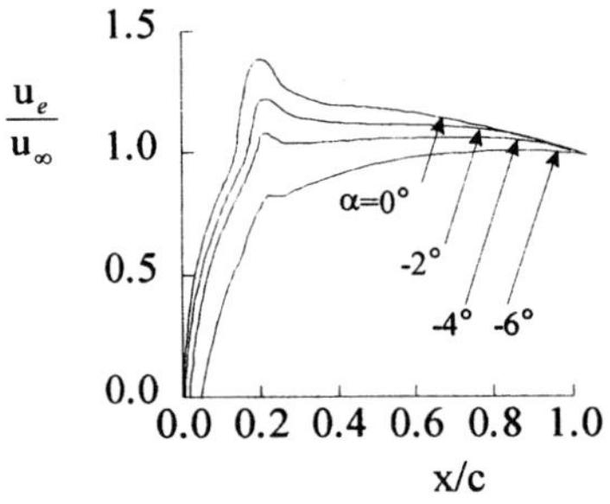

Fig. 13.2. Typical measured velocity distributions (suction side).

13.1.1 Incompressible Flows

The studies for an incompressible flow over an infinite swept wing were conducted in the F2 wind tunnel at Le Fauga-Mauzac Center by Arnal and Juillen [1]. The model had an ONERA-D airfoil section (subsection 5.2.2), symmetric between $x/c = 0.20$ and $x/c = 1$ and equipped with a cambered leading edge (Fig. 13.1a), so that the pressure distributions differed from those observed on the classical ONERA-D profile. The chord normal to the leading edge was 300 mm and the span 900 mm with zero angle of sweep. The wing and the half-fuselage were mounted on a turntable. The experimental arrangement is sketched in Fig. 13.1b. Three kinds of experiments were performed so that pressure distribution measurements, wall visualizations by sublimation technique, and hot-film measurements were obtained. Ten hot-films were glued on the model, from 32.5 to 86 percent of chord, and recorded simultaneously for more than one hundred combinations of the wind tunnel speed, angle of sweep, and angle of attack. The positions of the hot-films were indicated on Fig. 13.1b and were chosen to avoid interactions between probes.

These results were verified by comparing the transition positions obtained by the hot-films records and by the wall visualization (in the absence of hot-films): the results were similar. Three angles of sweep ($\lambda = 49$, 55 and 60°)

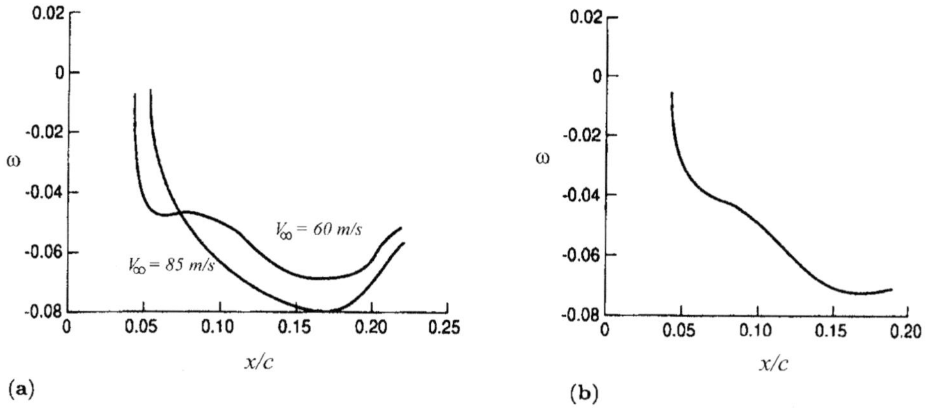

Fig. 13.3. Variation of ω on zarf near the leading edge; (a) $\alpha = 49°$, $V_\infty = 60$ ms^{-1} and 85 ms^{-1}, $\alpha = 0°$; (b) $\lambda = 49°$, $V_\infty = 60$ ms^{-1}, $\alpha = -2°$.

and four angles of attack ($\alpha = 0$, -2, -4 and $-6°$) were studied, and typical freestream distributions are plotted in Fig. 13.2; due to the presence of the cambered leading edge, the curves exhibit a complex evolution in the negative pressure gradient region.

Before we present calculated results obtained with the Cebeci-Stewartson eigenvalue formulation discussed in Section 12.1 [2], it is useful to discuss the procedure used to obtain the eigenvalues and the dimensional frequencies needed in the e^n-method. For given velocity profiles, the stability calculations begin on the zarf where the eigenvalue problem is solved subject to the requirements given in Eq. (12.2.4). Once the eigenvalues of the zarf are computed at a specified R, the eigenvalues of the zarf at the next Reynolds number are computed with the help of of the continuation method again, and the profiles of the previous station and their eigenvalues were assumed to correspond to the reference eigenvalues and profiles. Figure 13.3a shows the variation of zarf ω along the leading edge of the wing for $\lambda = 49°$, $V_\infty = 60$ ms^{-1} at $\alpha = 0°$, and Fig. 13.3b shows its variation for $\lambda = 49°$, $V_\infty = 60$ ms^{-1} at $\alpha = -2°$. As can be seen, the lower branch of the ω-zarf is negative with relatively flat values of ω away from the leading edge, whereas ω undergoes a very rapid variation near $x/c = 0.04$ and 0.05. Figures 13.4 shows that ω becomes positive and increases rapidly around $x/c = 0.04$ for $\alpha = 0°$, $V_\infty = 60$ ms^{-1}, and around $x/c = 0.05$ for $\alpha = -2°$, $V_\infty = 85$ ms^{-1}. This behavior of ω near the leading edge is very important since the e^n-method requires the calculation of the frequency that leads to the greatest amplification rate Γ. An accurate calculation of this frequency and its location is crucial to the accuracy of calculating the transition location.

Once the eigenvalues of the zarf are computed at a specified x/c-location, we proceed to the next x/c-station to solve the eigenvalue problem of Section 12.2 for a specified dimensional ω determined on the zarf with the assumption

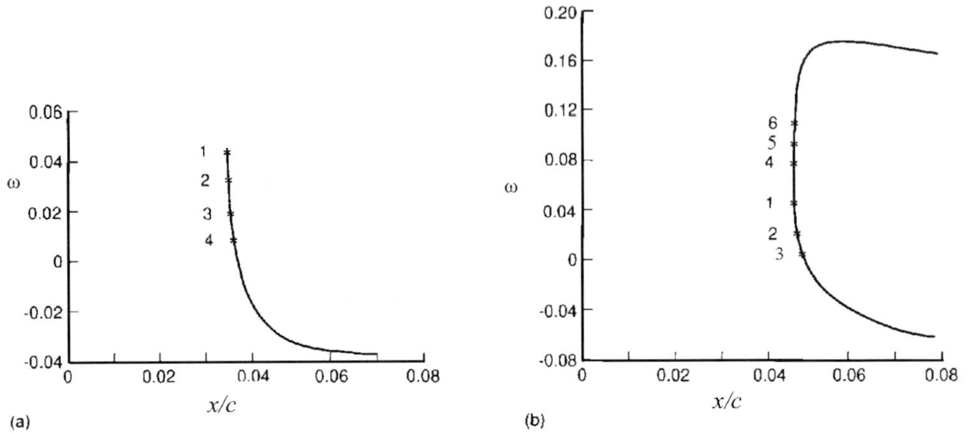

Fig. 13.4. Variation of ω on zarf very close to the leading edge for (a) $\alpha = 0°$, $V_\infty = 60\ \mathrm{ms}^{-1}$, and (b) $\alpha = -2°$. $V_\infty = 85\ \mathrm{ms}^{-1}$.

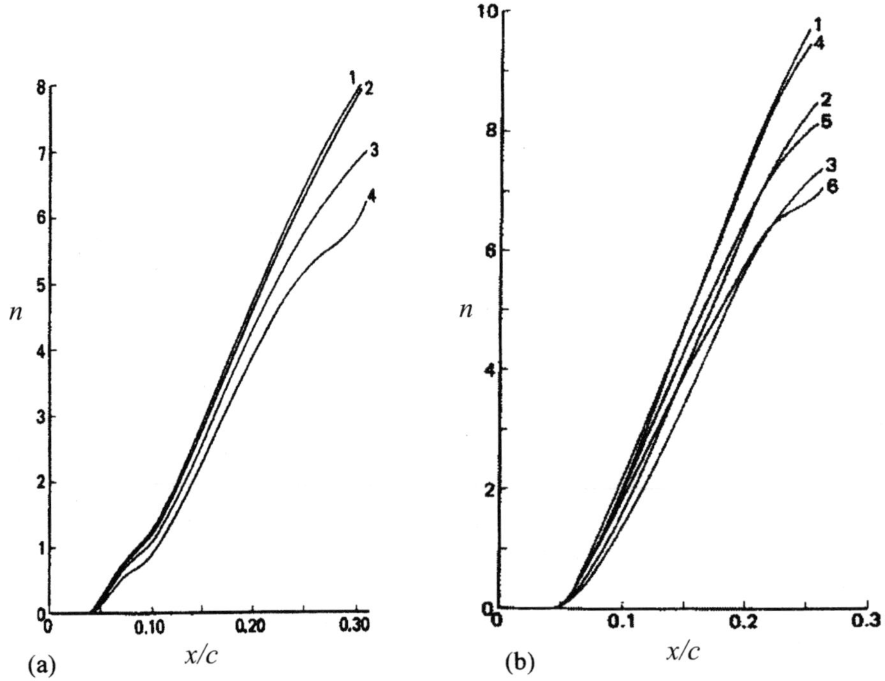

Fig. 13.5. Variation of the amplification factors for the frequencies of (a) Fig. 13.4a and (b) Fig. 13.4b.

that $\partial\alpha/\partial\beta$ is given by its value at the previous x/c-location, which amounts to specifying the disturbance angle ϕ. The amplification rate Γ, defined by Eq. (12.1.8) is calculated for different values of ϕ for which Γ is maximum. Once it is determined, we proceed to the next x/c-station and compute new values of α, to determine Γ and repeat the procedure to obtain its maximum value.

The above procedure is for one value of frequency ω chosen at a certain x/c-location. As in two-dimensional flows, it is repeated for different values of ω computed for a zarf at different x/c-locations. Since the frequency near the leading edge of the wing can vary drastically, it is important to choose these x/c-locations carefully. Figures 13.4a and 13.5a show the computed frequencies and the resulting amplification rates Γ for four different frequencies for $\lambda = 49°$, $\alpha = 0°$, $V_\infty = 60$ ms^{-1}. As can be seen, the computed amplification rates originate at nearly the same x/c-location and amplify differently depending on the choice of the frequency and may give different predictions of transition. Figures 13.4b and 13.5b show a similar behavior for $\lambda = 49°$, $\alpha = -2°$, $V_\infty = 85$ ms^{-1} and indicate clearly that the amplification rates computed with frequencies denoted by 1 and 4 lead to results which are substantially different from those computed with others.

The accuracy of the above procedure for computing transition was investigated for thirteen cases of which seven were for a sweep angle $\lambda = 49°$, three were for $\lambda = 55°$, and three were for $\lambda = 60°$. Calculated and experimental results are shown in Fig. 13.6 and summarized in Table 13.1, where x_0/c corresponds to the location of the critical frequency ω^*_{cr}. As can be seen, the predictions of the e^n-method with $n = 8$ are generally in agreement with the experimental data. Examination of the streamwise and crossflow velocity profiles of Fig. 13.7 for different speeds, angles of attack and sweep angles indicates, as expected, that the streamwise velocity profiles remain essentially the same but the crossflow velocity profiles undergo drastic changes. It is interesting to note from Fig. 13.7a, for example, that the velocity profile that leads to the criti-

Table 13.1. Comparison of Calculated and Experimental Results.

α	V_∞ [ms^{-1}]	x_0/c	w^*_{cr}	Exp	Computed
$\lambda = 49°$					
0	60	0.0375	0.03505 (2500 Hz)	$0.25 < \frac{x}{c} < 0.30$	0.295
0	85	0.0335	0.02567 (3000 Hz)	$0.15 < \frac{x}{c} < 0.20$	0.225
−2	30	0.27	0.04307 (700 Hz)	$\simeq 0.85$	0.81
−2	60	0.046	0.03296 (2000 Hz)	$0.27 < \frac{x}{c} < 0.31$	0.31
−2	85	0.445	0.02995 (3000 Hz)	$0.25 < \frac{x}{c} < 0.25$	0.22
−4	60	0.0565	0.07696 (4000 Hz)	$0.25 < \frac{x}{c} < 0.27$	0.25
−4	85	0.0520	0.06094 (5500 Hz)	$\simeq 0.15$	0.19
$\lambda = 55°$					
−2	60	0.053	0.04908 (3000 Hz)	$0.225 < \frac{x}{c} < 0.30$	0.26
−2	85	0.047	0.05461 (5500 Hz)	$0.15 < \frac{x}{c} < 0.20$	0.19

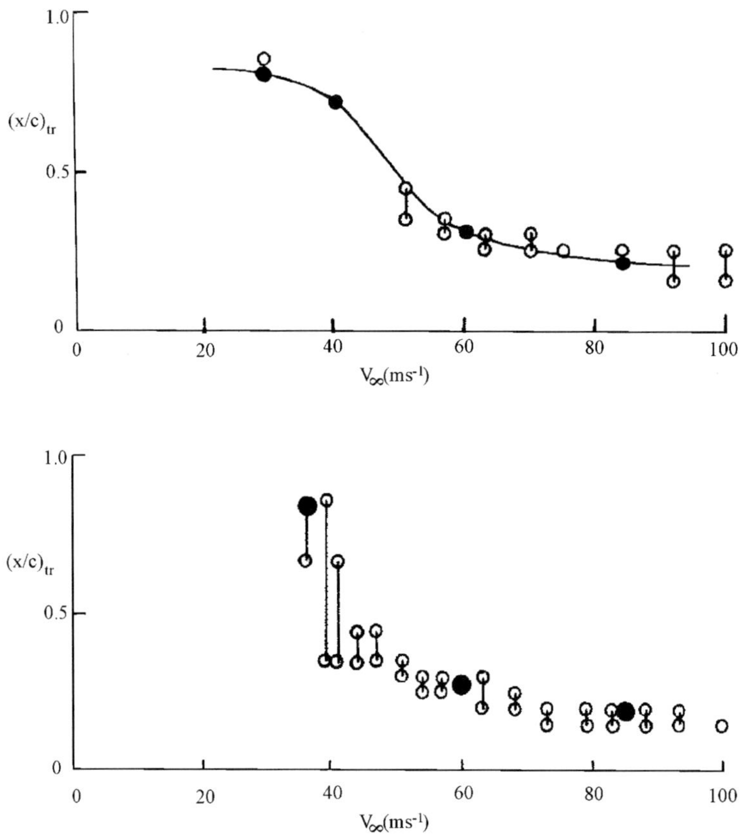

Fig. 13.6. Experimental (open symbols) and calculated (closed symbols) transition locations for (a) $\lambda = 49°$, $\alpha = -2°$, and (b) $\lambda = 55°$, $\alpha = -2°$.

cal frequency at $(x_0/c) = 0.0375$ is between the velocity profiles at $x/c = 0.025$ and 0.05, which resemble those at $x/c = 0.08$, yet does not lead to a critical frequency. The reason for this behavior is the rapid variation of the frequency with Reynolds number. Transition in this case occurs after the crossflow velocity profile develops a flow reversal which starts around $x/c = 0.26$.

The results in Fig. 13.7d show the effect of sweep on the behavior of the velocity profiles. Comparison between the velocity profiles for the same speed, Fig. 13.7b, indicates that the sweep angle reduces the flow reversal in the crossflow velocity profile and that the transition for $\lambda = 55°$ takes place with a crossflow velocity profile which shows almost no reversal.

13.1.2 Compressible Flows

The measurements for a compressible flow over a 15-degree swept tapered wing were conducted in the cryogenic wind tunnel T2 of ONERA/CERT at very

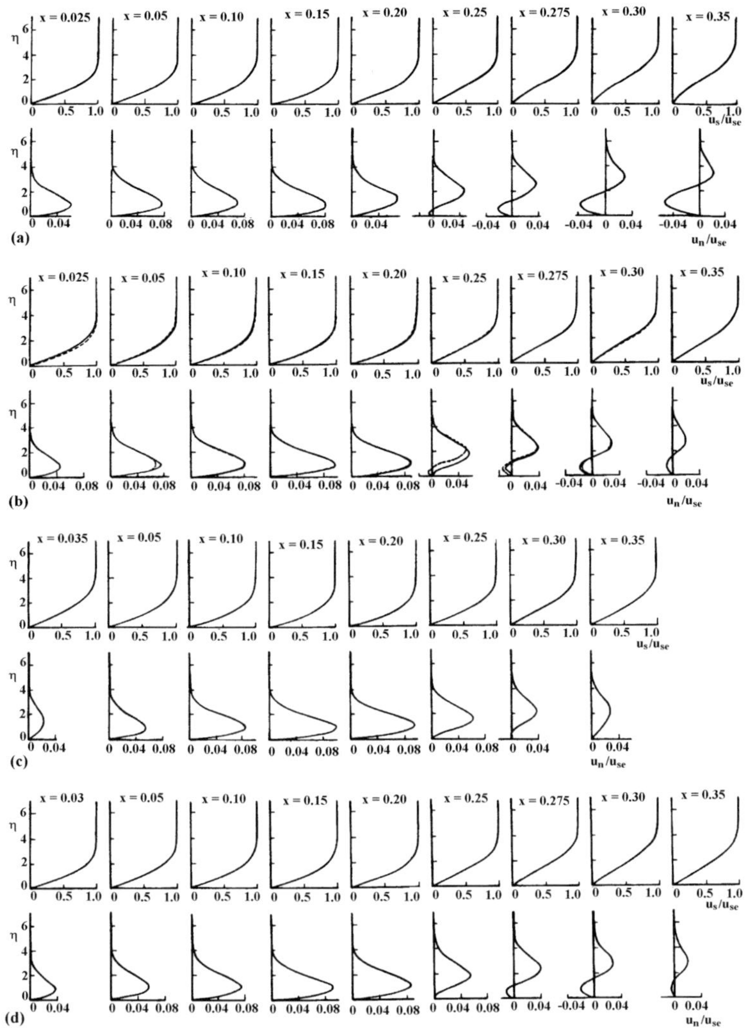

Fig. 13.7. Computed streamwise u_s/u_{se} (top) and crossflow u_n/u_{se} (bottom) velocity profiles for $V_\infty = 60\ \mathrm{ms}^{-1}$ at various x/c-stations starting from (right to left) $x/c = 0.025$, 0.05, 0.10, 0.15, 0.20, 0.25, 0.275, 0.30, (a) $\lambda = 49°$, $\alpha = 0°$ (b) $\lambda = 49°$, $\alpha = -2°$, (c) $\lambda = 49°$, $\alpha = -4°$, (d) $\lambda = 55°$, $\alpha = -2°$.

low stagnation temperatures [3, 4]. The wing, with a chord of 0.228 m at the root and 0.145 m at the tip, had a span of 0.39 m with an AS409 cross-section and a trailing-edge sweep angle of three degrees. It also had wavy surfaces less than 0.05 mm (from peak to valley) for a chordwise extent of about 2 cm. In order to avoid the need to perform fully three-dimensional stability/transition calculations, the measurements discussed in [3] and summarized in [4] were carried out under infinite swept wing conditions with the wing having a mean sweep angle of 13 degrees at an angle of attack of 0.3 degrees. The mean chord for computations was stated to be 0.186 m.

The accuracy of the e^n-method for this flow was investigated by using the ONERA/ CERT method using the eigenvalue formulation of Mack and Arnal and the saddle-point method of Cebeci and Stewartson discussed in Section 12.1. Here the results using the latter method are presented in order to demonstrate its accuracy for compressible flows [5]. Calculations were conducted for two different flow conditions referred to as Runs 42 and 79, with Run 42 having a stagnation temperature of 145 K and a stagnation pressure of 2.0 bar. Run 79 had higher temperature and pressure corresponding to 165K and 2.5 bar, respectively. In both cases the calculations were performed only for the upper surface of the wing.

The boundary-layer calculations for Run 42 were performed by solving the infinite-swept-wing equations for an adiabatic wall and specified wall temperature distributions at a Reynolds number of 13.8×10^6. The stability calculations were then performed for the velocity and temperature profiles determined by the boundary-layer method.

Figure 13.8 shows the zarfs for Run 42 for an adiabatic wall and measured wall temperature distributions. As can be seen, the frequencies originate at nearly the same location (on a vertical line) and vary drastically one from another. Figure 13.9 shows the computed n factors for the zarfs of Fig. 13.8. The results demonstrate that for adiabatic wall conditions, the maximum value of n for the experimental transition location is around 6.3; the corresponding value for the measured wall temperature is around 7. What is more important,

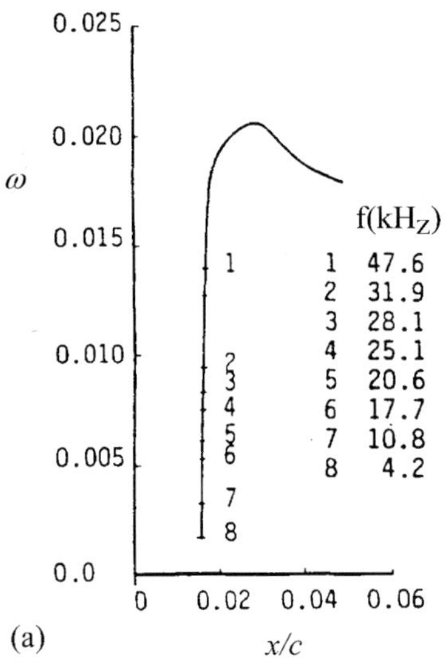

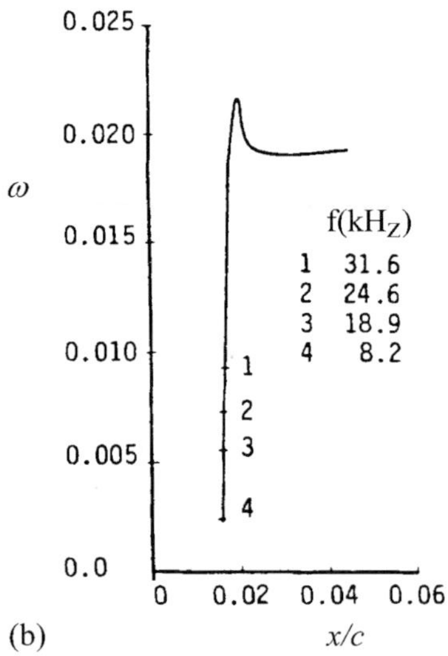

Fig. 13.8. Zarfs for Run 42 for (a) adiabatic, and (b) measured wall temperature distributions.

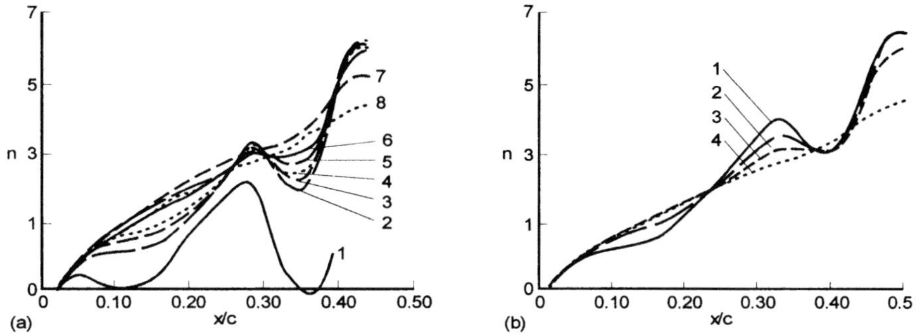

Fig. 13.9. Computed n-factors for Run 42 for (a) adiabatic, and (b) measured wall temperature distributions.

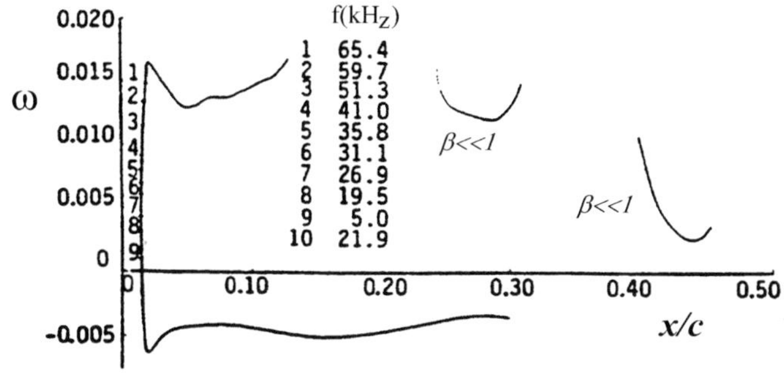

Fig. 13.10. Zarfs for Run 79 for adiabatic wall temperature distribution.

however, is the fact that in the latter case the computed transition location agrees very well with the observed location, considering that the n-value for transition in this tunnel lies between 7 and 8.

Figure 13.10 shows the zarfs for Run 79 with wall conditions corresponding to adiabatic temperature and Fig. 13.11 shows a comparison between the zarfs obtained under adiabatic and measured wall temperature conditions. Again the steep rise in frequencies close to a single x/c-location is noted. The zarf calculations were performed for a very fine x/c-grid, since most of the frequencies begin to exhibit amplification in an interval of $1/1000^{\text{th}}$ of chord around 0.015. Figure 13.10 also shows zarfs away from the leading edge. These zarfs have low values of β_r around 10^{-2}, occur in an almost zero pressure gradient region and do not lead to amplification rates that grow significantly.

Figure 13.12 shows the computed n-factors for Run 79 with stability calculations performed for the adiabatic and measured wall temperature conditions. For this flow, the n-value is much higher than those in Run 42. For an adiabatic wall, it reaches a maximum value of around 8 and a value of around 9 for the measured wall temperature case. If we take the n-value to be 7.5 (a mid-n value of the expected n-value range for this wind tunnel) then transition occurs at

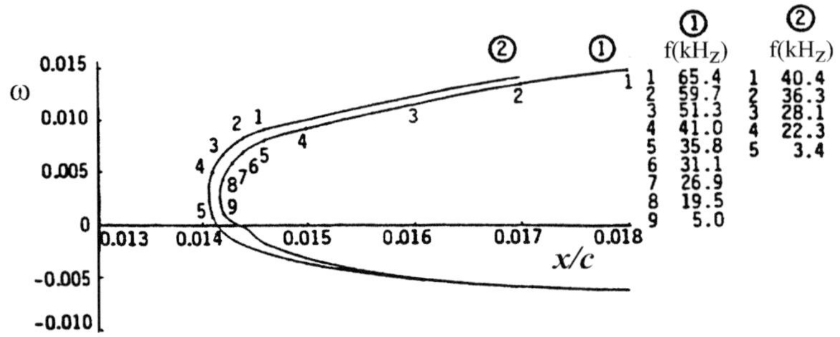

Fig. 13.11. Comparison between zarfs of Run 79 for (1) adiabatic, and (2) measured wall temperature distributions.

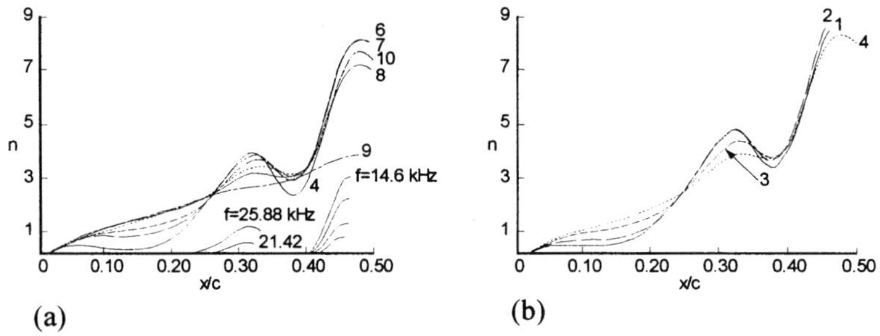

Fig. 13.12. Computed n-factors for Run 79 for (a) adiabatic and (b) measured wall temperature distributions.

$x/c \simeq 0.46$ for adiabatic wall conditions and $x/c = 0.47$ for measured wall temperature conditions. This compares well with the inferred transition location of $x/c = 0.47$.

13.2 Effect of Curvature on Transition

In this section we shall study the role of surface curvature on transition for an incompressible flow over an infinite-swept wing by varying the sweep angle and the chord Reynolds number in a systematic manner. This study is made by computing transition with the e^n-method by neglecting the effect of surface curvature on the velocity profiles and by only including the curvature terms in the stability equations. In this choice, the flowfield is obtained by solving the first order boundary-layer equations, and the stability equations are solved for three-dimensional flows with curvature terms.

The earliest work on including effects in three-dimensional flows was done by Malik and Poll [6] for incompressible flows and by Collier et al. [7] for compressible flows. The results of Malik and Poll with curvature effects and with parallel flow approximation differed significantly from the results obtained without curvature terms, and appeared to be in better agreement with the measurements of Poll [8] for example, the n-factor in the e^n-method, which is usually around 8 to 10, was found to be about 17 without curvature effects and about 11 with curvature effects. Calculations by Collier et al., conducted for the flight test on an F-14 aircraft, showed that the n-factor was about 16 without curvature effects and was about 9 with curvature effects. Their study concluded that flight test data on this aircraft can be correlated with the e^n-method employed in the COSAL code with n around 9, but only if curvature effects were included in the linear stability equations.

Cebeci, Chen and Kaups [9] questioned the claim of Malik and Poll that the stability calculations with curvature terms differ significantly from those without them. They solved the stability equations with and without curvature terms in the Cartesian and rotated curvilinear coordinate systems with eigenvalue formulations based on the temporal approach in order to study the effect of curvature on the onset of transition in three-dimensional incompressible laminar flows. They performed calculations for a yawed circular cylinder at Reynolds numbers of 1.68×10^6 (case 2) and 0.88×10^6 (case 4). The study showed that, in the absence of curvature effects, the results were in good agreement with the earlier calculations of Malik and Poll. Separate calculations, using the saddle point method of Cebeci and Stewartson also supported this observation. However, this was not the case when their results with curvature effects were compared with the results of Malik and Poll. At the lower Reynolds number their results indicated a small curvature effect that diminished at the higher Reynolds number, while the results of Malik and Poll indicated a strong curvature effect at both Reynolds numbers.

The claim of Cebeci, Chen and Kaups that the curvature effects were not important in influencing the stability/transition calculations performed for Poll's cylinder was later verified by Malik, who corrected an error in his method, which had led to inacurate conclusions regarding the role of curvature on transition.

In this section, we extend the study conducted for a yawed cylinder at relatively low Reynolds numbers to an infinite swept wing at low, moderate and high Reynolds numbers and study the effects of sweep angle and Reynolds number on transition in the presence of surface survature based on the parallel flow approximation, as described in subsection 13.2.1 using the saddle point method discussed in Section 12.1. A recent study reported in [10] shows that including the curvature effect in the boundary-layer equations as well as in the stability equations has practically no effect on the onset of transition location. For this reason, results are presented and discussed in subsection 13.2.2 with the effect of curvature neglected in the velocity profiles.

13.2.1 Incompressible Linear Stability Equations with Curvature Terms

For a Cartesian coordinate system, the linear stability equations including curvature terms with the parallel flow approximation and with appropriate boundary conditions are given in [9]; they can be expressed in the following dimensionless form similar to

$$i(\alpha\tilde{u} + \beta\tilde{w}) + \tilde{v}' + m_{12}\tilde{v} = 0 \tag{13.2.1}$$

$$\tilde{u}'' - \gamma_2^2\tilde{u} - iR\alpha\tilde{p} = R\gamma_1\tilde{u} + \bar{u}'R\tilde{v} + Rm_{12}\bar{u}\tilde{v} \tag{13.2.2}$$

$$\tilde{v}'' - \gamma_2^2\tilde{v} - R\tilde{p}' = R\gamma_1\tilde{v} - 2Rm_{12}\bar{u}\tilde{u} \tag{13.2.3}$$

$$\tilde{w}'' - \gamma_2^2\tilde{w} - iR\beta\tilde{p} = R\gamma_1\tilde{w} + R\tilde{w}'\tilde{v} \tag{13.2.4}$$

$$\eta = 0, \quad \tilde{u} = \tilde{v} = \tilde{w} = 0 \tag{13.2.5a}$$

$$\eta \to \infty, \quad \tilde{u}, \tilde{v}, \tilde{w} \to 0 \tag{13.2.5b}$$

Here the parameter m_{12} denotes the surface curvature of the coordinate lines $\eta =$ const.

$$m_{12} = \frac{1}{h_1}\frac{\partial h_1}{\partial \eta} \tag{13.2.6a}$$

and

$$\gamma_1 = i(\alpha\bar{u} + \beta\bar{w} - \omega), \quad \gamma_2^2 = \alpha^2 + \beta^2 \tag{13.2.6b}$$

with the metric coefficient h_1 given by

$$h_1 = 1 + k_w y \tag{13.2.6c}$$

and the surface curvature k_w calculated from

$$-k_w = \frac{d\theta}{ds} \tag{13.2.6d}$$

Here s denotes the arc length along the surface and θ is obtained from the slope of the surface in the plane normal to the leading edge,

$$\theta = \tan^{-1}\frac{d\bar{y}}{d\bar{x}}$$

where bars denote the Cartesian coordinate that defines the wing section.

Equations (13.2.1) to (13.2.5) are expressed as a first-order system by defining the following new variables

$$\begin{aligned} z_1 &= \alpha\tilde{u} + \beta\tilde{w}, \quad z_2 = \alpha\tilde{u}' + \beta\tilde{w}', \quad z_3 = \tilde{v} \\ z_4 &= \tilde{p}, \quad z_5 = \alpha\tilde{w} - \beta\tilde{u}, \quad z_6 = \alpha\tilde{w}' - \beta\tilde{u}' \end{aligned} \tag{13.2.7}$$

In terms of these new variables, Eqs. (13.2.1) to (13.2.5) become

$$\underset{\sim}{z}' = A\underset{\sim}{z} \tag{13.2.8}$$

Here $\underset{\sim}{z} = (z_1, z_2, \ldots, z_6)^T$, and A is a 6×6 coefficient matrix

$$A = \begin{vmatrix} 0 & 1 & 0 & 0 & 0 & 0 \\ a_{21} & 0 & a_{23} & a_{24} & 0 & 0 \\ a_{31} & 0 & a_{33} & 0 & 0 & 0 \\ a_{41} & a_{42} & a_{43} & 0 & a_{45} & 0 \\ 0 & 0 & 0 & 0 & 0 & 1 \\ 0 & 0 & a_{63} & 0 & a_{65} & 0 \end{vmatrix} \tag{13.2.9}$$

where the coefficients a_{ij} $(i, j = 1, 2, \ldots, 6)$ are given by

$$\begin{aligned} &a_{21} = \gamma_2^2 + R\gamma_1, \quad a_{23} = R(\alpha\bar{u}' + \beta\bar{w}' + m_{12}\bar{u}\alpha), \quad a_{24} = iR\gamma_2^2 \\ &a_{31} = -i, \quad a_{33} = -m_{12} \\ &a_{41} = \frac{1}{R}\frac{1}{\gamma_2^2}[2Rm_{12}\bar{u}\alpha + a_{31}a_{33}\gamma_2^2], \quad a_{42} = -\frac{i}{R} \\ &a_{43} = -\frac{1}{R}[m_{12}' + \gamma_2^2 + R\gamma_1 - a_{33}^2], \quad a_{45} = -\frac{2}{\gamma_2^2}m_{12}\bar{u}\beta \\ &a_{63} = R(\bar{w}'\alpha - m_{12}\bar{u}\beta - \bar{u}'\beta), \quad a_{65} = \gamma_2^2 + R\gamma_1 \end{aligned} \tag{13.2.10}$$

The solution of Eq. (13.2.8) subject to the boundary conditions given by

$$\eta = 0, \quad z_1 = z_3 = z_5 = 0 \tag{13.2.11a}$$

$$\eta \to \infty, \quad z_1, z_3, z_5 \to \infty \tag{13.2.11b}$$

is obtained with the numerical method described in Section 12.2.

13.2.2 Effects of Sweep Angle and Reynolds Number on Transition with Curvature Effect Included in the Stability Equations

Before we present results obtained with the calculation method of the previous section, it is useful to consider the differences between the geometries considered in Malik and Poll's study and the study conducted in [9].

Malik and Poll's geometry corresponds to a circular cylinder with a tall fairing. If c_n (see Fig. 13.13) denotes the chord length normal to the leading edge and r_0 denotes the radius of the circular cylinder, then this geometry can be regarded as an infinite swept wing with $c_n = 4r_0$. The curvature c_n/r_0, is 4, and extends over a chordwise distance of $s/c_n \sim 0.32$.

The study of [9] considers an infinite swept wing having a NACA 0012 airfoil section. The nose curvature of this airfoil is 63 and the curvature diminishes rapidly (see Fig. 13.14) as s/c increases. For example, at $s/c = 0.15$, $c_n/r \sim 1$.

Thus the curvature of Malik and Poll's geometry is considerably different than that of the study of [9] and is not a good representation of a geometry

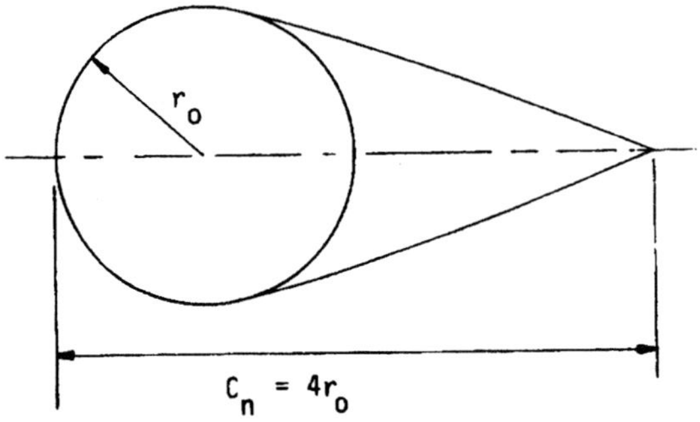

Fig. 13.13. Wing section normal to the leading edge used in [5].

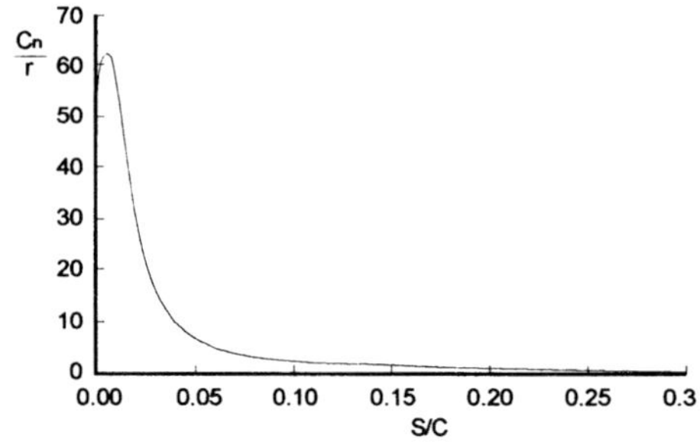

Fig. 13.14. Variation of surface curvature on a NACA 0012 airfoil.

to study the curvature effect on transition on wings: the effect of curvature is much weaker near the leading edge and extends for a longer distance.

The results for an infinite swept wing having the cross section of the NACA 0012 airfoil are presented below. The angle of attack is taken to be 2° and the curvature effect on transition is examined by taking the sweep angles to be 30°, 45° and 60°, and by taking the chord Reynolds numbers to be 1×10^6, 5×10^6, 10×10^6 and 20×10^6. Since for positive angle of attack the stagnation point is on the lower surface and away from the nose of the airfoil, the choice of $\alpha = 2°$ provides a longer extent of curvature effect than at $\alpha = 0°$.

The velocity profiles needed in the stability equations are obtained by solving the infinite swept wing equations without including the curvature effects in the boundary-layer equations. In all cases the external velocity distribution was obtained from a panel method.

Figures 13.15 to 13.18 show a sample of the results for a sweep angle of 30° with chord Reynolds number varying from 1×10^6 to 20×10^6. In all cases the dimensional frequencies needed in the e^n-method were obtained from the zarfs. Figure 13.15 shows the variation of the integrated amplification rates with and without curvature effects. As can be seen, the solid lines and dashed lines coincide and indicate no curvature effect at this rather low Reynolds number.

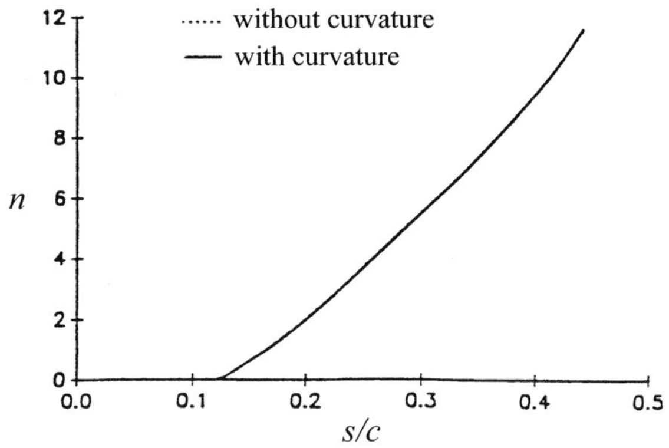

Fig. 13.15. Variation of the integrated amplification rates for $\lambda = 30°$, $R_c = 10^6$.

Figures 13.16 and 13.17 show the results for Reynolds numbers of 5×10^6 and 10×10^6, respectively. While the behavior of the integrated amplification factors for 5×10^6 are unaffected by the curvature, those for 10×10^6 show a different behavior. Here the curvature effect moves the location of the critical frequency away from the leading edge (see Fig. 13.17), and leads to lower amplification rates initially. As the curvature decreases with increasing surface distance, the magnitude of the n-factor computed without curvature effect, shows a reduced growth rate and eventually merges with the trend of the integrated amplification rate obtained with the curvature effect. The end result is that the location of transition remains unchanged and the curvature effect is again negligible. It is, however, interesting to note that the location of the critical frequency with curvature effect is different from the one without curvature effect. This indicates the advantage of using zarfs, which not only help to determine the magnitude of the critical frequencies but also their location.

Figure 13.18 shows the results for a Reynolds number of 20×10^6. At this high Reynolds number, the onset of transition occurs close to the stagnation point; it is at $s/c = 0.035$ without curvature effect. Since the curvature effect is very strong close to the leading edge, even though the locations of the critical frequencies remain the same, the integrated amplification rates are much lower with the curvature effect (Fig. 13.18b) and lead to transition at $s/c = 0.085$ with a 5% difference in transition location without and with curvature effect for $n = 9$.

Table 13.2 summarizes the effect of surface curvature on transition as the sweep angle varies from 30° to 60° and chord Reynolds number from 1×10^6 to 20×10^6.

The results at the lowest chord Reynolds number, R_c, of 10^6 show that surface curvature has no effect on transition at sweep angles varying from 30° and 60°. This may well be due to the initial disturbance growth which occurs at

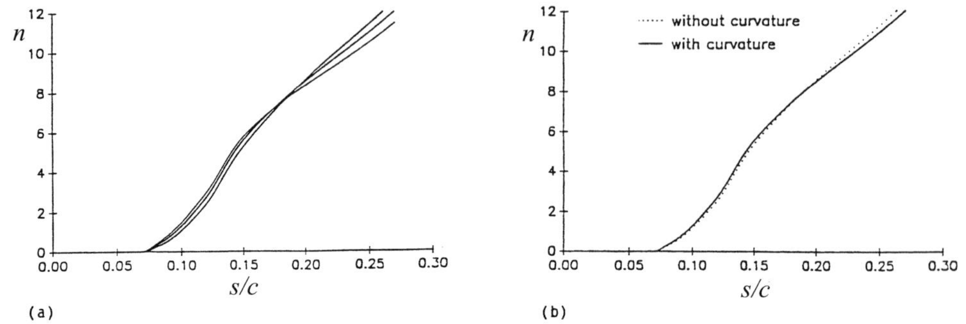

Fig. 13.16. Variation of the integrated amplification rates for $\lambda = 30°$, $R_c = 5 \times 10^6$ (a) with curvature effect, (b) with and without curvature effect.

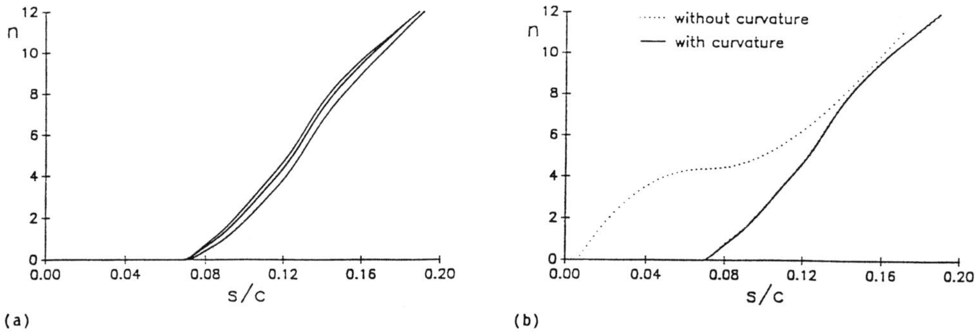

Fig. 13.17. Variation of the integrated amplification rates for $\lambda = 30°$, $R_c = 10 \times 10^6$ (a) with curvature effect, (b) with and without curvature effect.

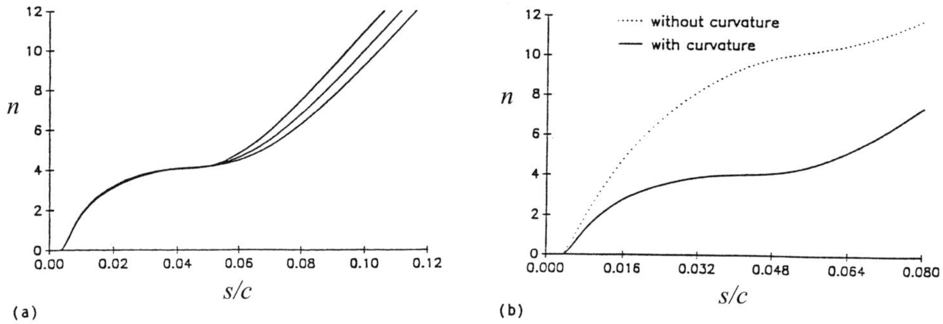

Fig. 13.18. Variation of the integrated amplification rates for $\lambda = 30°$, $R_c = 20 \times 10^6$ (a) with curvature effect, (b) with and without curvature effect.

Table 13.2. Computed transition locations $(s/c)_{tr}$ with and without curvature effect.

	$R_c \times 10^{-6}$	Without curvature $(s/c)_{tr}$	With curvature $(s/c)_{tr}$
$\lambda = 30°$			
	1	0.38	0.38
	5	0.19	0.19
	10	0.145	0.145
	20	0.038	0.085
$\lambda = 45°$			
	1	0.42	0.42
	5	0.19	0.23
	10	0.075	0.11
	20	0.019	0.026
$\lambda = 60°$			
	1	–	–
	5	0.20	0.29
	10	0.07	0.11
	20	0.021	0.027

a distance from the leading edge where the curvature effect is small. As the Reynolds number increases, the effect of surface curvature influences the location of the critical frequency along the zarf at some sweep angles. However, the different origins for disturbance growth do not necessarily result in different transition locations, as in the case for $\lambda = 30°$ and $R_c = 10 \times 10^6$, for example. Also, the calculations at the highest Reynolds number show that nearly identical zarf locations produce somewhat different transition location for the two calculations. There seems to be a sweep angle and Reynolds number combination where the curvature effect is maximum, However, it is clear that at higher sweep angles and high Reynolds numbers, the effect of surface curvature on the predicted transition location becomes negligible for all practical purposes, in spite of the fact that disturbances originate and grow in the region of high surface curvature. Thus, it may be said that the surface curvature has no effect on transition on swept wings at low Reynolds numbers and its effect is negligible at high Reynolds numbers and high sweep angles. Other sweep angle and Reynolds number combinations show some curvature effect which may be configuration dependent. However, the overall surface curvature effect on transition is much less than previously thought, and hence this arrangement should not be used to explain the failure of stability transition predictions to agree with experimental data.

13.3 Subsonic Flows: Wing Alone

In this section and the following two sections, the calculation method of Chapters 10–12 is applied to subsonic flows, with this section describing its application to two single wings. Section 13.4 describes its application to high lift configurations and Section 13.5 its application to the prediction of the aerodynamic performance degradation of an aircraft in natural icing conditions.

Before we discuss the application of the calculation method to two wings, it is useful to describe the calculation procedure in some detail. The calculations begin with the inviscid flow method of chapter 10 for a given wing geometry and freestream conditions (Fig. 13.19). Once the external velocity distribution is obtained, the input data to the interface program is rearranged so that at first the nonorthogonal coordinate system for the boundary layer program and the geometric parameters appearing in the boundary-layer equations are calculated with the procedure described in subsection 11.5. The velocity components $u_e(x, z)$ and $w_e(x, z)$ for the boundary-layer grid (x, z, η) are then calculated from the Cartesian components $V_{\bar{x}}, V_{\bar{y}}, V_{\bar{z}}$ with the procedure described in subsection 11.5.2. Upon the completion of these tasks, the upper and lower surfaces of the wing are identified in order to perform the boundary-layer calculations separately for each surface.

As discussed in Section 11.2, the boundary-layer calculations require initial conditions in the (z, y)-plane at $x = x_0$ and initial conditions in the (x, y)-plane at $z = z_0$. In the present program those in the (z, y)-plane are obtained by solving the attachment line equations given by Eqs. (11.2.5) to (11.2.8) subject to their boundary conditions given by Eq. (11.2.5) for laminar flows. The initial conditions in the (x, y)-plane are obtained by solving the quasi-three-dimensional boundary layer equations given by Eqs. (11.2.1) to (11.2.4) subject to their boundary conditions given by Eq. (11.1.9) and (11.1.11) for both laminar and turbulent flows at the first spanwise station near the wing-fuselage intersection. At the first cycle the onset of the transition location and the switch location from standard to inverse mode are estimated to be at an x-location before the pressure peak.

For all the calculations presented in this section and the following one as well those in Section 13.6, the boundary-layer calculations for $nz > nz_0$ were also performed by solving the quasi-three-dimensional boundary-layer equations. While the characteristic box method in the inverse mode can also be used for this purpose, this method has yet to be developed for flows with negative streamwise velocity component; presently it has only been tested for flows with negative spanwise velocity component.

Once the boundary-layer solutions are obtained for all spanwise and chordwise boundary-layer stations for either input or computed transition locations by making several sweeps in the chordwise direction at each spanwise location, the blowing velocity and displacement surface distributions are calculated for inviscid flow calculations as discussed in Section 11.9. The procedure is then repeated with inviscid flow calculations which now account for the viscous effects. The velocity components u_e and w_e and pressure coefficient c_p distributions are calculated at the displacement surface. Pressures are then integrated to ob-

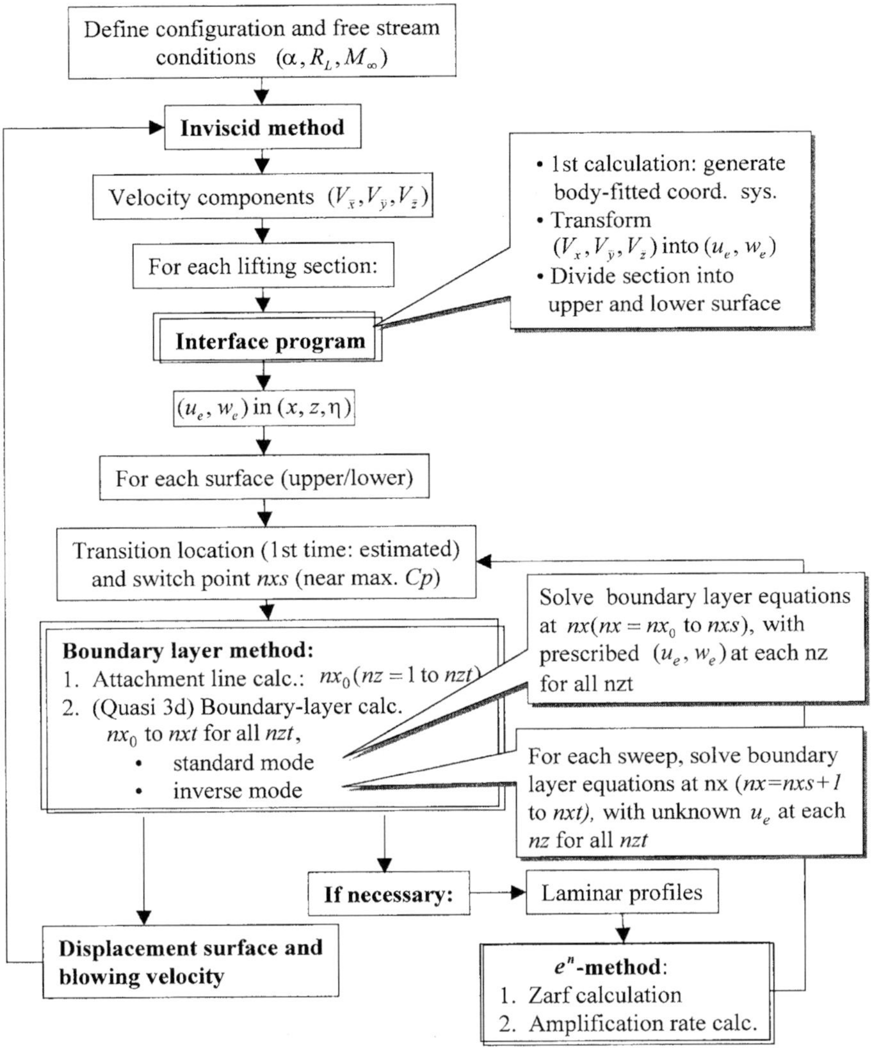

Fig. 13.19. Procedure for inviscid/viscous calculations.

tain the force coefficients, i.e., lift, moment coefficient and the induced drag coefficients discussed later in subsection 13.3.2.

In the next solution (cycle two) of the quasi-three-dimensional boundary-layer calculations, the onset of transition is calculated during the first sweep. It is not computed at each sweep as well as in each cycle to save computer time and effort. Since the linear stability equations do not contain partial derivatives with respect to x and z, the laminar boundary-layer profiles obtained by solving the quasi-three-dimensional boundary-layer equations in a nonorthogonal coordinate system must be expressed in an orthogonal coordinate system. With

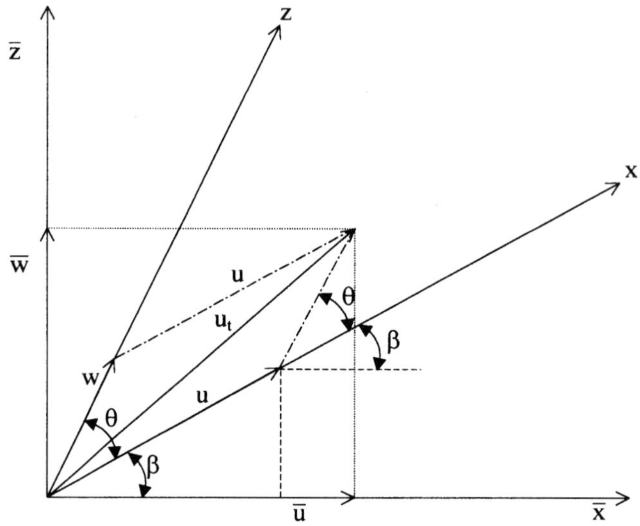

Fig. 13.20. The velocity profiles in (x, z) and $(\bar{x}, \bar{z})$ coordinate systems.

β denoting the positive angle between the $\bar{x}$ and the x axis (Fig. 13.20), the $\bar{u}$ and $\bar{w}$ profiles in the $(\bar{x}, \bar{z})$ orthogonal coordinate system are related to the u and w profiles in the (x, z) nonorthogonal coordinate system by

$$\bar{u} = u \cos \beta + w \cos(\theta + \beta) \tag{13.3.1a}$$

$$\bar{w} = u \sin \beta + w \sin(\theta + \beta) \tag{13.3.1b}$$

After the onset of transition is calculated, turbulent flow calculations are performed for one surface for all x-stations, including the wake. These laminar and turbulent flow calculations are then repeated for all spanwise stations, one by one until all the boundary-layer solutions on the wing and in the wake are obtained for the upper surface. An identical procedure is then repeated for the lower surface.

13.3.1 SAAB Wing

Bertelrud [11] tested the outer part of the wing of a SAAB 32 Lansen in a wind tunnel and measured the distribution of external velocity, pressure, momentum thickness, shape factor, and skin friction coefficient at various locations on the wing. The wing is symmetric, has an aspect ratio of 2.27, a taper ratio of 0.6 and a leading edge sweep angle of 39°.

Since the exact geometry of the tip of the wing was not precisely described, at first the calculations reported in [12] were performed with various tips to address the effect of the tip shape on the flowfield predictions where measurements were reported, Fig. 13.21.

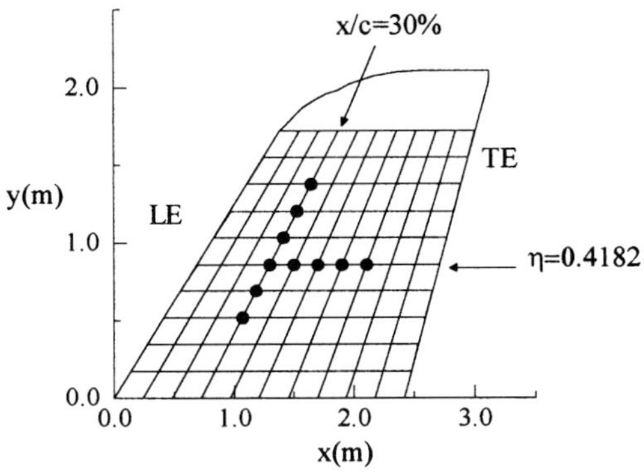

Fig. 13.21. Location of boundary layer measurements.

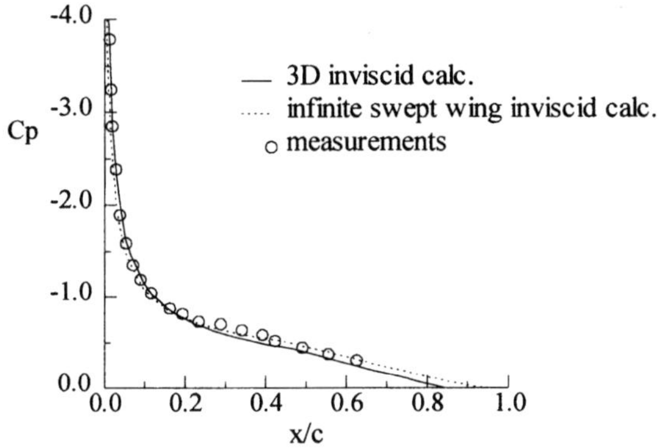

Fig. 13.22. Comparison of measured and calculated pressure distributions for the SAAB wing upper surface, $\eta_b = 0.40$, $\alpha = 14.4°$.

Inviscid calculations using the panel method of Chapter 10 were performed for three angles of attack (0°, 4.8° and 14.4°), for which experimental data was available. No noticeable differences in velocity and pressure coefficient distributions were observed at the locations where measurements were reported for the various tip shapes. Figure 13.22 shows a typical comparison between pressure coefficient measurements and inviscid calculations.

Measurements were made for several freestream velocities; $u_{\text{ref}} = 20$, 40 and 50 m/s. Since most of the data was obtained at $u_{\text{ref}} = 40$ m/s, all viscous/inviscid calculations reported in [12] were performed at that velocity, which corresponds to a Reynolds number of 5.6×10^6 based on the root chord. The transition calculations were initially made for two dimensional flows with the e^n-method, These calculations were then followed with the e^n-method for three-dimensional flows with velocity profiles obtained by solving the quasi three-dimensional boundary-layer equations.

Table 13.3 presents a comparison between the transition locations obtained by using the 2D (xctr2D) and 3D (xctr3D) e^n-methods at different spanwise locations and angles of attack. While the differences for the upper surface at $\alpha = 14.4°$ are minimal, they are relatively small at $\alpha = 0$, $4.8°$, and rather large at $\alpha = 14.4°$ for the lower surface due to the growth of crossflow disturbances in the accelerating flow region.

Table 13.3. Calculated transition locations on the SAAB wing; (a) upper surface and (b) lower surface.

(a)				(b)			
α (°)	η_b	$Xctr_{2D}$	$xctr_{3D}$	α (°)	η_b	$xctr_{2D}$	$Xctr_{3D}$
0	0.0843	0.57	0.56	4.8	0.0843	0.65	0.175
	0.2529	0.54	0.482		0.4216	0.64	0.134
	0.4216	0.52	0.441		0.6745	0.65	0.166
	0.6745	0.51	0.461		0.7588	0.65	0.187
	0.9148	0.50	0.445		0.8431	0.70	0.251
4.8	0.0843	0.0833	0.0320		0.9148	0.75	0.400
	0.2529	0.0612	0.0199	14.4	0.0843	0.80	0.109
	0.4216	0.0235	0.0172		0.2529	0.80	0.107
	0.6745	0.0156	0.0163		0.4216	0.80	0.118
	0.9148	0.0156	0.0181		0.6745	0.75	0.1647
14.4	0.0843	0.0015	0.0047		0.9148	0.75	0.186
	0.2529	0.0012	0.0028				
	0.4216	0.0010	0.0025				
	0.6745	0.0004	0.0027				
	0.9148	0.004	0.0029				

Figure 13.23 shows the variation of the calculated and measured streamwise displacement thickness δ^* along the span at $x/c = 0.3$ with transition specified at $x/c = 0.20$ arbitrarily and at the calculated transition locations ($xctr_{3D}$). Results demonstrate that the flow is indeed laminar at that chordwise station and that the determination of the transition location is important and should be included in any calculation method, including Navier-Stokes methods, to predict viscous effects accurately.

Figure 13.24 shows the variation of the streamwise momentum thickness, θ, distribution, along the chord at $\eta_b = 0.40$ with the specified transition location (xctr $= 0.20$), transition location determined from two- ($xctr_{2D} = 0.52$) and three-dimensional ($xctr_{3D} = 0.44$) calculations. As can be seen, the results obtained with the calculated transition locations agree better with experimental data, although they do not match the data very accurately.

Figure 13.25 shows the results for $\alpha = 4.8°$ at the same spanwise station. Both total skin friction and streamwise momentum thickness distributions are represented and the agreement between measurements and calculations is good.

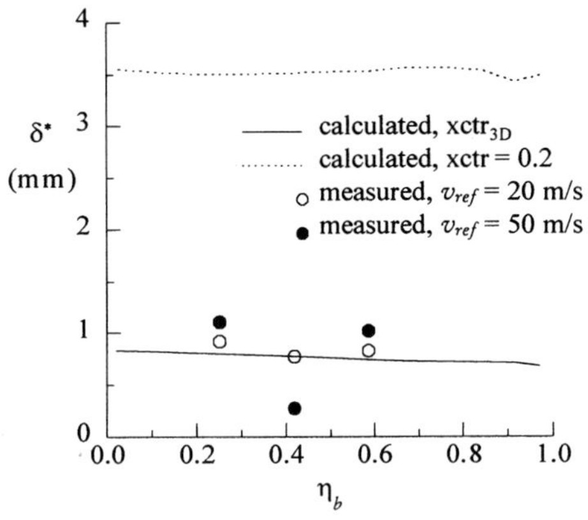

Fig. 13.23. Displacement thickness variation along the span at $\alpha = 0°$ and $x/c = 0.3$.

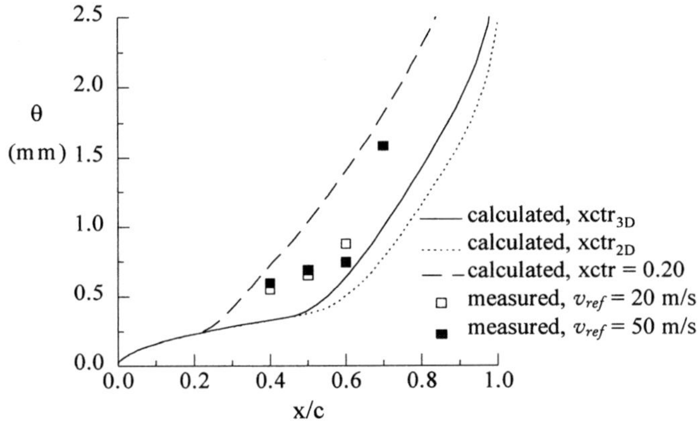

Fig. 13.24. Momentum thickness variation at $\alpha = 0°$ and $\eta_b = 0.40$.

The effects of transition are limited to the flow near the leading edge only, where no data was available for comparison.

Figure 13.26 shows the results for $\alpha = 14.4°$. Again, the agreement between measured and calculated skin friction and momentum thickness distributions is good. A slight improvement is observed with calculated transition locations (xctr_{3D}).

13.3.2 RAE Wing

The calculation method was also applied to a RAE wing tested by Lovell [13]. This wing has an airfoil section having a considerable rear loading with the maximum thickness of 11.7% occuring at 37.5% chord and the maximum camber occuring at 75% chord. It has no twist nor dihedral, but has a quarter-chord sweep angle of 28°, a taper ratio of 1/4 and an aspect ratio of 8.35. The ex-

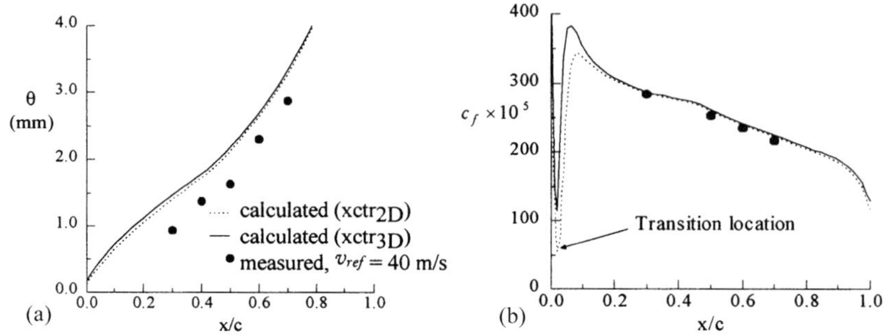

Fig. 13.25. (a) Momentum thickness and (b) skin friction coefficient distributions at $\alpha = 4.8°$ and $\eta_b = 0.40$.

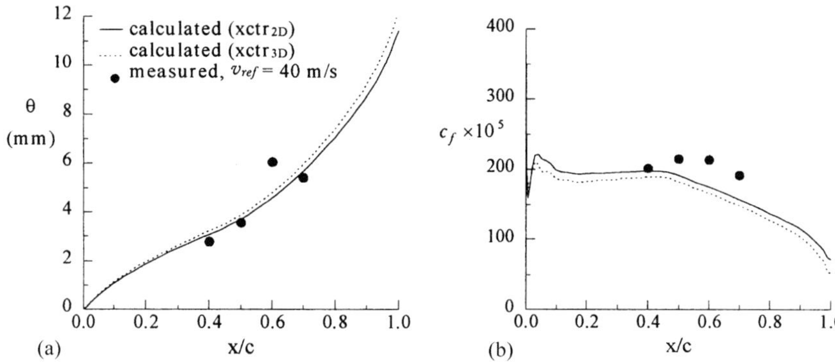

Fig. 13.26. (a) Momentum thickness and (b) skin friction coefficient distributions at $\alpha = 14.4°$ and $\eta_b = 0.40$.

periments were conducted at a test Reynolds number of 1.35×10^6 with one set of measurements corresponding to free transition and with another to fixed transition for a freestream Mach number of 0.223. The wing has a semispan of 1.07 m and a mean aerodynamic chord of 0.26 m.

In one set of Lovell's experiments, transition was set near the leading edge using a series of 0.5 mm diameter wires positioned streamwise around the leading edge from 10% chord on the upper surface to 10% chord on the lower surface. Ten wires were equally spaced across each half wing. Measurements showed that the maximum lift coefficient with fixed transition is higher than that with free transition. Lovell noted, however, that "a larger portion is probably caused by the tendency of the streamwise wires to act as vortex generators." Also, at post stall conditions, the flow became largely unsteady and only average measurements were reported. Therefore, near maximum lift, calculated results with fixed transition should be compared with free transition measurements. Both sets of experimental data are given in Fig. 13.27.

Effect of Turbulence Model on Lift and Drag Coefficients

Figure 13.27 shows the lift coefficient variation with angle of attack. Each calculated lift curve is obtained in approximately 15–20 minutes of CPU time on an SGI Octane workstation with the wing described by 10 cross-sections distributed along the span, each cross-section being defined by 63 points with a higher point concentration around the leading and the trailing edges. Calculations were performed with angle of attack increments of 2° until 10°. The increment in angle of attack is reduced to 1° beyond that point. Results show that with the original C.S. turbulence model, the lift coefficient keeps increasing past the measured stall angle (around 12° with free transition). With the modified C.S. turbulence model, on the other hand, the agreement between measured and calculated lift coefficient is excellent up to 14°. At 15°, the boundary layer calculations at $\eta_b = 0.7$ do not converge near the trailing edge due to the large separated flow region.

Figures 13.28 and 13.29 show the chordwise trailing edge displacement thickness and flow separation along the span at angles of attack, $\alpha = 13°$ and 14°. As can be seen, there is a significant increase in the growth of displacement thickness and the amount of flow separation with one degree increase in α. From $\alpha = 13°$ to 14° the maximum $(\delta_x^*)_{\text{te}}$ is multiplied by a factor of 1.6 indicating that the wing is at or very close to stall conditions. The results in Fig. 13.28 also indicate that for a swept non-twisted wing, stall initiates near the tip, at about $\eta_b = 0.7$–0.8.

Before we present the comparison between the measured and calculated drag coefficients, it is useful to discuss our approach to the drag calculations. As discussed in Chapter 10, the fundamental theorems of potential flow state that in three-dimensional cases a body cannot be lifting unless it sheds a vortex wake of the kind illustrated in Fig. 10.1. The wake represents energy imparted

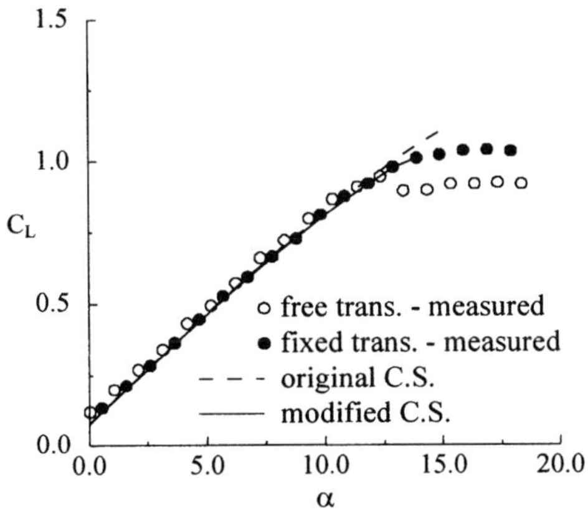

Fig. 13.27. Effect of turbulence model on the lift coefficient of the RAE wing.

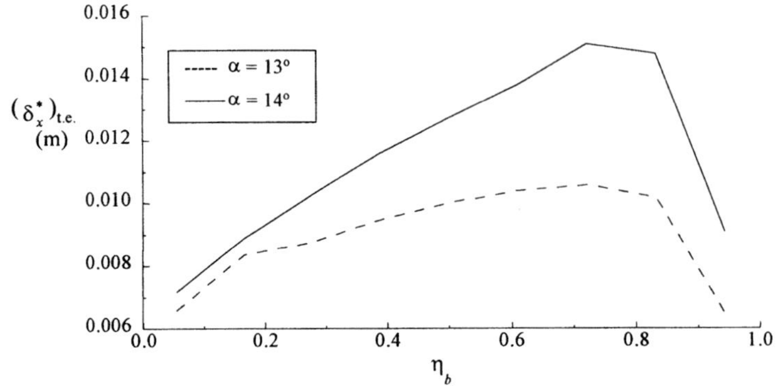

Fig. 13.28. Chordwise trailing edge displacement thickness distributions along the span.

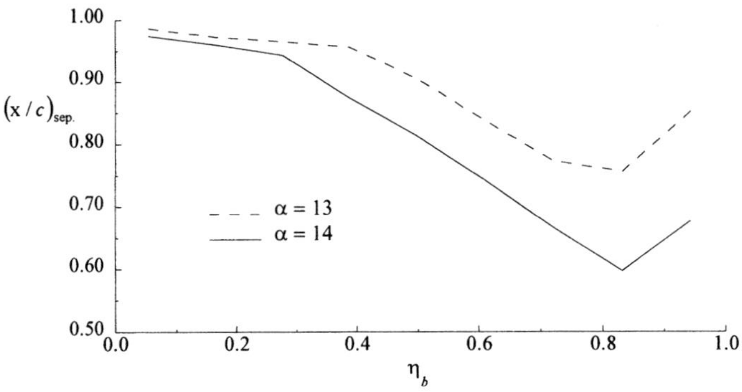

Fig. 13.29. Distribution of flow separation along the span at two angles of attack.

to the fluid, which implies a corresponding drag on the body, even in inviscid flow. Since this drag is present only because of the lift, it is often designated drag-due-to-lift or induced drag. Because of the presence of tip vortices, the flow incidence at a given spanwise location, the effective angle of attack, α_{eff}, differs from that of the overall geometry, α (existence of an induced angle, α_i, (see Fig. 13.30). The lift force (component normal to the incident flow) generated at that spanwise location is therefore not in the same direction as the lift force of the whole configuration. The offset constitutes the drag due to lift, or induced drag, and can be evaluated by calculating the energy in the wake by a far-wake procedure known as Treffz-plane analysis [14]. Alternatively, the calculated pressures may be integrated over the body panels to give the component of force in the freestream direction. In the limit of a large number of panels these two approaches converge to the same drag. It turns out that the Treffz-plane analysis approaches its asymptotic value more rapidly with panel number than the pressure integration. In past years, when panel numbers were restricted, the former approach was preferred. Now with more powerful computers, either approach may be used. The agreement between the drags computed by the two

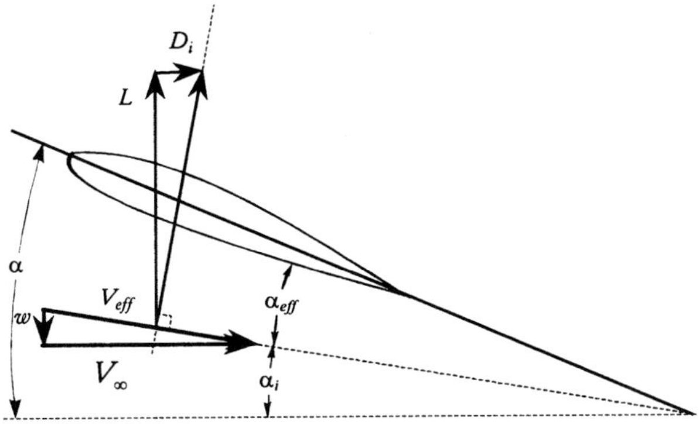

Fig. 13.30. Downwash and induced drag.

procedures is a measure of the accuracy of the calculations. In the calculations reported in [12], far field approach was used to calculate the induced drag.

For real flows, the viscous effects modify the pressure distribution, thus adding to the pressure drag, so that induced drag and pressure drag do not match any longer. For small to moderate angles of attack, i.e. relatively small induced angles, the drag coefficient for three-dimensional configurations can be calculated by adding the induced drag coefficient C_{D_i} (due to lift) to the profile drag coefficient, c_d, (due to viscosity), by

$$C_D = C_{D_i} + \frac{1}{S}\int_0^b c(y)c_d(y)dy \tag{13.3.2}$$

Here c is the local chord and S is the reference area used to non-dimensionalize drag. The profile drag coefficient c_d can be calculated from the Squire-Young formula [16].

$$c_d = 2\left(\frac{\theta}{c}\right)(\bar{u}_e)^{0.5(H+5)} \tag{13.3.3}$$

except that here θ, H and $\bar{u}_e$ ($\equiv \frac{u_e}{u_\infty}$) are calculated from the velocity profiles in the far wake expressed in an orthogonal coordinate system instead of using the trailing edge values. The boundary layer parameters in the streamwise direction are then calculated and used in Eq. (13.3.3).

Figure 13.31 shows a comparison between the measured and calculated drag coefficients obtained with the above procedure. Again, the agreement between data and calculations is quite satisfactory up to stall.

Strip-Theory Approximation

In some problems it may be useful to introduce another approximation, known as the strip-theory approximation, to the three-dimensional boundary-layer method or to the quasi-three-dimensional boundary-layer method. It can be

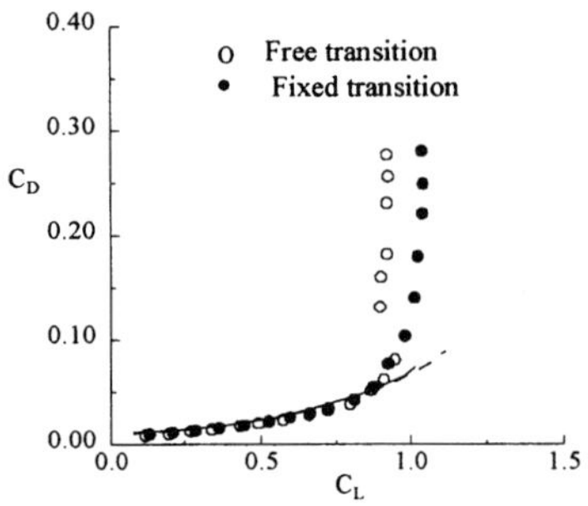

Fig. 13.31. Comparison of calculated and measured drag coefficients for the RAE wing.

used for flowfield calculations computed by either a panel method (Chapter 10) or an Euler method. The total velocity from the two velocity components, u_e and w_e, namely

$$v_e = \sqrt{u_e^2 + w_e^2}$$

is used in the solution of the two-dimensional boundary-layer equations applied to several streamwise airfoil sections at specified spanwise locations. The boundary-layer solutions are obtained for both laminar and turbulent flows with solutions initiating at the wing attachment line as a stagnation-point flow. The onset of transition is either specified or computed. In the latter case, it may again be computed with the e^n-method when the boundary-layer solutions are obtained by solving the quasi-three-dimensional or the two-dimensional boundary-layer equations. At each specified spanwise location, the boundary-layer calculations are carried out from the stagnation point up to the trailing edge and then into the wake along a given dividing streamline. As before, several sweeps are made at each specified spanwise location until convergence. The procedure is repeated at each spanwise location until the blowing velocities on the wing and in the wake dividing streamline together with the displacement thickness distributions are computed. A new inviscid flow solution is then obtained and the iterative procedure shown in Fig. 13.19 is repeated until the convergence of the solutions.

Figure 13.32 shows a comparison between the experimental data and two sets of calculations for the RAE wing, with one set of results obtained with viscous effects calculated by using the strip-theory approximation and the other using the solutions of the quasi-three-dimensional boundary-layer equations as in Fig. 13.27. In each case the onset of transition location was specified to be the same. There is practically no flow separation on the wing until $\alpha = 12°$. After that, flow separation takes place and increases in extent with increasing angle of attack.

The calculated results in Fig. 13.32 show that at low and moderate angles of incidence, there is very little difference between the two approaches. Near

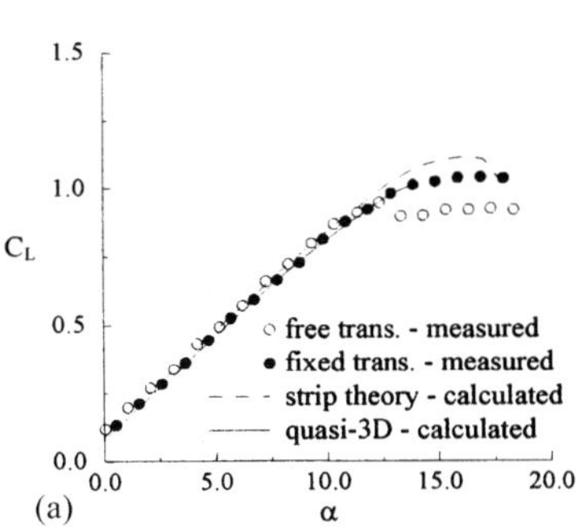

Fig. 13.32. Comparison between strip theory and quasi-three-dimensional approach.

stall, however, the quasi-three-dimensional approach provides results in better agreement with experimental data.

Effect of Transition Locations on Lift and Drag Coefficients

In subsection 13.3.1 it was shown that an accurate prediction of boundary layer parameters requires an accurate calculation of transition location, in particular on the lower surface of the wing. The importance of calculating transition location is investigated here for the purpose of determining the aerodynamic force coefficients for the RAE wing. Calculations are performed both with fixed and with calculated transition locations using the modified CS model again with the quasi-three-dimensional approximation [12].

Due to the combination of a fairly low Reynolds number (1.35×10^6) and moderate sweep angle (28°), cross-flow disturbances are small on the onset of transition and give rise to maximum amplification factors, n, less than 1. In all cases, transition is due to the growth of Tollmien-Schlichting waves in the adverse pressure gradient region. For low angles of attack, $\alpha = 0$ and 2°, transition on the upper surface is located downstream of the leading edge at $x/c = 0.75$ and 0.35 on the average, respectively. As the angle of attack increases, transition on the upper surface rapidly moves to a location very close to the leading edge. This rapid change in transition location is explained by the change in pressure distribution, as seen in in Fig. 13.33.

At $\alpha = 0°$, the pressure is almost constant up to $x/c = 0.75$ while at $\alpha = 4°$, a pressure peak already exists near the leading edge. Also, because of the absence of amplified cross-flow disturbances, natural transition on the lower surface occurs far downstream of the leading edge, around 45 percent of the chord on the average for $\alpha = 0°$ and moves towards the trailing edge as the angle of attack increases. For most spanwise stations on the upper surface, a very short separation bubble exists, with transition taking place within the bubble. Like for two-dimensional flows, setting transition at the separation location will not

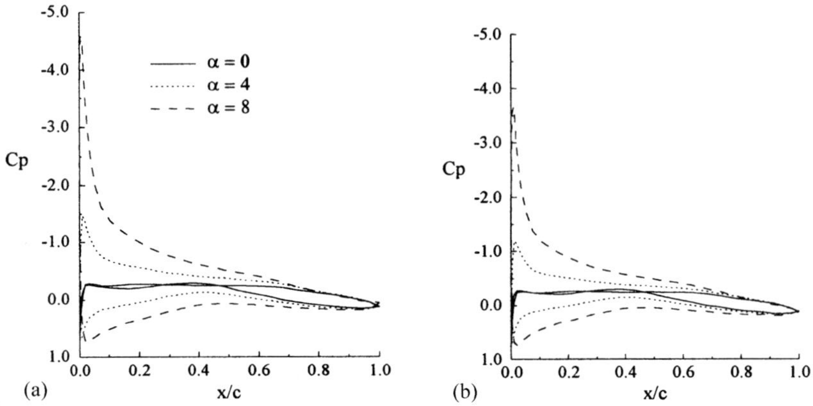

Fig. 13.33. Change of pressure distribution with angle of attack, (a) $\eta_b = 0.28$ and (b) $\eta_b = 0.77$.

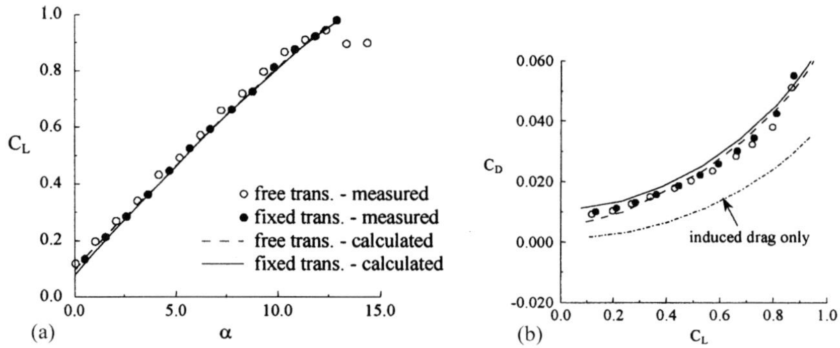

Fig. 13.34. Wing alone: Effect of transition location on (a) lift and (b) drag coefficients.

change the behavior of the flow, except possibly at stall conditions. In the latter case, however, the separation bubble is so short (less than 1 percent chord) that transition calculations cannot be performed adequately.

One should expect to have a lower drag for the test case with free transition than for the one with transition fixed near the leading edge. Results are shown and compared with experimental data in Fig. 13.34. The calculated increase in lift coefficient with natural versus fixed transition is less pronounced than the measured one. The opposite observation can be made for the drag coefficient. Also, as the angle of attack increases, the difference between the results with free and fixed transition gets smaller.

From the results presented here, transition can have an important effect for drag predictions, even if this effect is less pronounced than for two-dimensional flows since the viscous drag is only a part of the total drag. In fact, the need for calculating transition location depends on the goal set forth. For accurate boundary layer parameter predictions, transition should be calculated. Except near stall, at moderate and high Reynolds numbers, lift predictions are relatively insensitive to transition location. For the purpose of predicting the drag coefficient reasonably well, it might be sufficient to perform transition calculations for a limited number of angles of attack, focusing on the lower surface of the wing, in particular to determine if cross-flow disturbances cause transition. For the upper surface, at moderate to high angles of attack, transition would usually occur either some short distance upstream of the pressure peak (if due to cross-flow disturbances) or some short distance downstream of the pressure peak (if due to Tollmien-Schlichting waves), so that setting it at the pressure peak should yield satisfactory results.

For more accurate drag predictions, or for an accurate maximum lift prediction, transition should be calculated. Possible attachment line transition and/or relaminarization should also be included and could be incorporated into the present method

13.4 Subsonic Flows: Multielement Wings

The calculation method described in the previous section was also applied to several high lift configurations of the RAE wing for the purpose of predicting lift and drag coefficients. The full-span slat is either retracted or deployed at 25° and the full span flap is deployed at either 10° or 25°.

Each element is defined by 13 cross-sections, with each slat cross-section defined by 61 points, each main element cross-section defined by 81 points (slat retracted or deployed), and each flap cross-section defined by 71 points. For the inviscid method of Chapter 10, wake shapes need to be specified. These are generated with the two-dimensional panel method of Chapter 2 using an average angle of attack in the range of interest. The variation of the wake shape with angle of incidence and along the span is neglected here. This approximation could be the subject of another computational study. Figure 13.33 shows a typical panel distribution with every other panel ignored and wakes omitted for clarity.

In light of the results of the previous section, calculations are performed with transition on the upper surface fixed at the leading edge. Also, in order to be able to retract the various elements, surface irregularities (coves) are present on the lower surface of the slat (if deployed) and of the main element (if the flap is deployed). For the present calculations, these coves are faired and transition is assumed to take place at the location of the start of the fairing.

Results are first presented for wing-flap configurations, Fig. 13.36a, and are followed by results obtained for slat-wing-flap configurations, Fig. 13.36b. The deflection angles and relative positioning of the elements are described by Lovell [13].

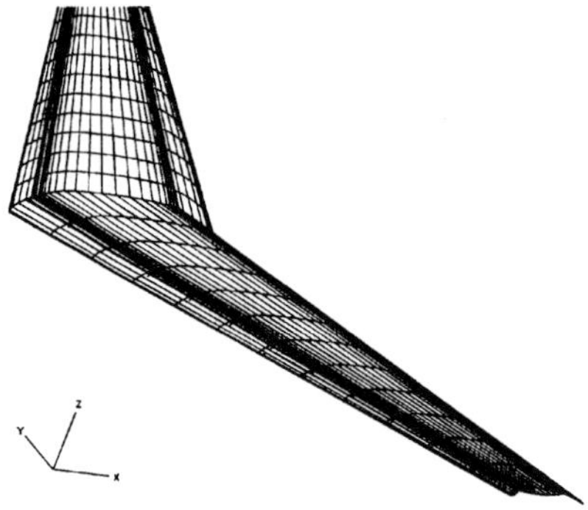

Fig. 13.35. Multi-element wing paneling with slat and flap deployed at 25° – wakes omitted.

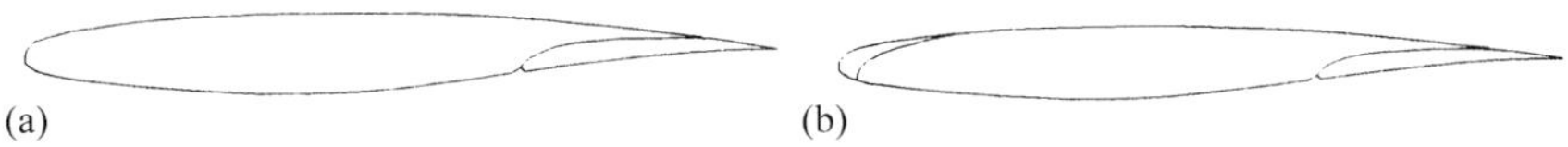

Fig. 13.36. Elements of the high lift system (a) main element and flap when slat is retracted and (b) slat, main element, and flap.

13.4.1 Wing-Flap Configurations

The cross-section of the wing-flap configuration with the flap deflected at 10° is shown in Fig. 13.37a. The lift coefficient variation with angle of attack and the drag polar are also shown in Figs. 13.37a and 13.37b. The inviscid lift coefficient is indicated to show how the introduction of the viscous effects allows for a better prediction of the lift curve. However, the lift coefficient is over-predicted while the drag coefficients, except near stall, are well predicted. From Fig. 13.37, it can be seen that the flap is positioned very close to the main element. Therefore, merging of the main element wake with the boundary layer developing on the flap upper surface is likely to occur.

This merging, observed in two-dimensional flows, may be the source of the discrepancy since it is not modeled in the present method. Also, several reasons can be advanced to explain why stall is not predicted by the calculation method. One of them is based on prior two-dimensional results. It was shown by Cebeci and Besnard [14] that even if the freestream Mach number is low (around 0.2), very high pressure peaks occur at the leading edge of the main element and can lead to locally supersonic flow. Also, a separation bubble is usually present near the pressure peak with transition taking place in that bubble. The stronger pressure gradient when the flap is deployed makes this separation bubble even shorter than for a single wing. The combination of these two phenomena – taking place on such a short distance – makes the calculations in that region a

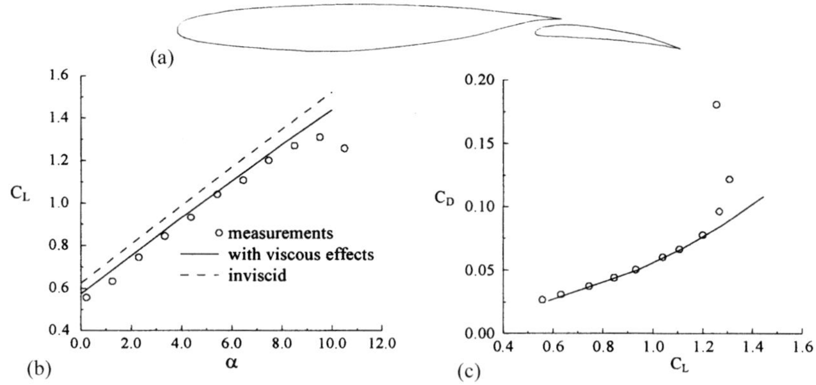

Fig. 13.37. RAE wing with slat retracted and flap deflected at 10°, (a) wing cross-section, (b) lift coefficient, and (c) drag coefficient.

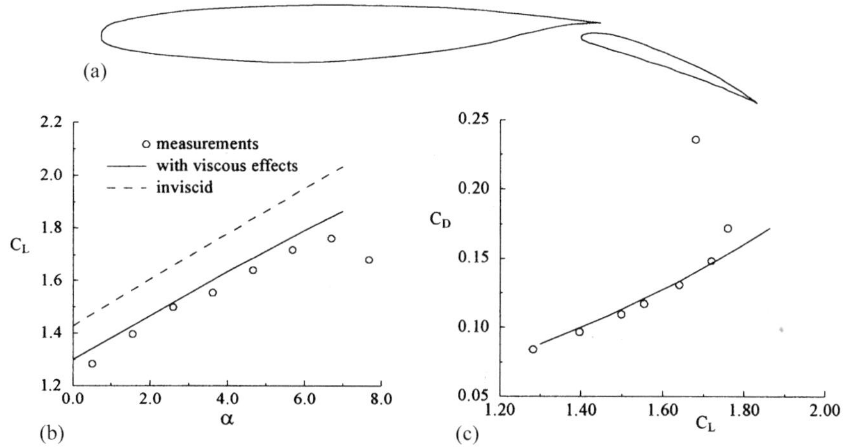

Fig. 13.38. RAE wing with slat retracted and flap deflected at 25°, (a) wing cross-section, (b) lift coefficient, and (c) drag coefficient.

challenging task. At the present time, there is no calculation method which can predict such flows accurately and reliably.

Figure 13.38 shows the results when the flap is deflected at 25°. Observations similar to those in Fig. 13.37 can be made. Agreement between measurements and calculations is satisfactory for both lift and drag coefficients. As expected, the larger flap deflection (25° instead of 10°) causes a larger difference between inviscid and viscous flow calculations.

13.4.2 Slat-Wing-Flap Configurations

Figures 13.39 and 13.40 present results for the configurations of the previous section with the slat deflected at 25°. and the flap deflected at 10° and 25°,

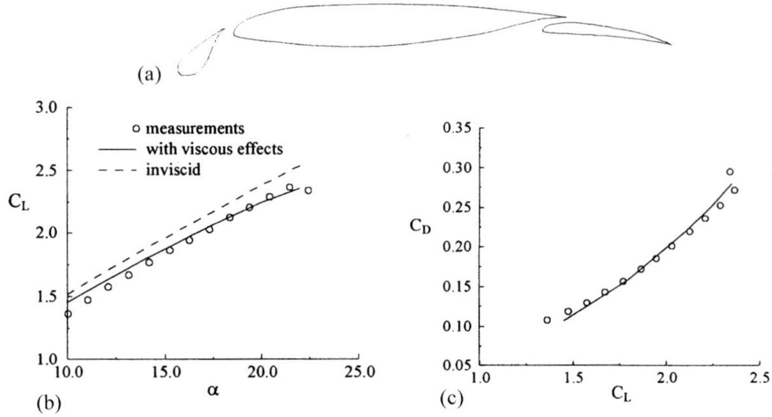

Fig. 13.39. RAE wing with slat deflected at 25° and flap deflected at 10°, (a) wing cross-section, (b) lift coefficient, and (c) drag coefficient.

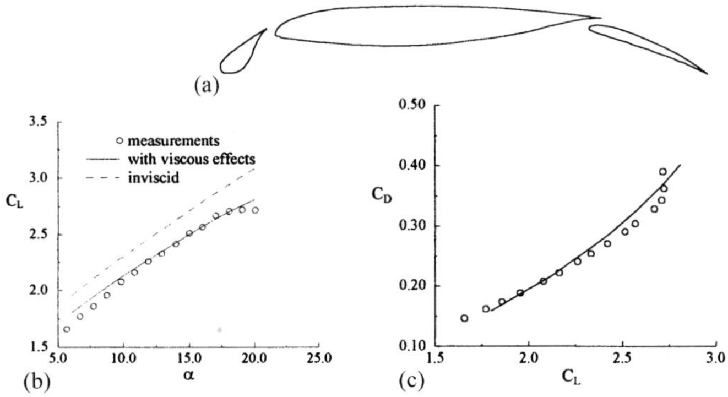

Fig. 13.40. RAE wing with both slat and flap deflected at 25°, (a) wing cross-section, (b) lift coefficient, and (c) drag coefficient.

respectively. Again, the inviscid lift coefficient is indicated to show how the introduction of the viscous effects allows obtaining reasonable predictions of lift and drag coefficients. The discrepancies can be partly explained by the merging of the shear layers which was not accounted for. In addition, the large recirculating flow region in the slat cove – larger at low angles of incidence – was removed with the fairing and may contribute to the disagreement at low angles of attack. Cove calculations, however, could be incorporated by extending the method of Cebeci and Besnard [14] to three-dimensional flows. The wake shape, prescribed in the inviscid method, may also be at the source of the discrepancy.

Stall is not captured for the configurations tested. However, it is worthwhile to note that, at the present time, the reliable prediction of stall for slat-wing-flap configurations still offers significant challenges for two-dimensional flows. Unlike for single element and wing-flap configurations, stall can occur without

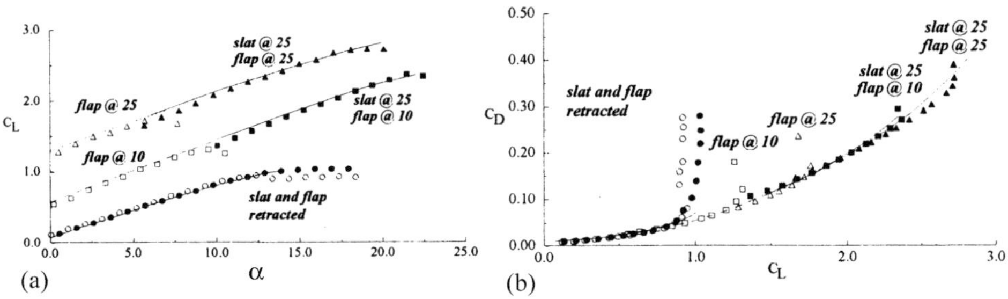

Fig. 13.41. Performance of high lift devices; (a) lift and (b) drag coefficients.

flow separation on the body but may be due instead to a sudden increase of the wake thickness thus reducing the circulation on the entire configuration (see Rogers et al., [15] for an illustration of this phenomenon). Therefore, the present results should be viewed as quite satisfactory.

13.4.3 Summary

Figure 13.41 shows the predicted effects of extending various high lift components and compares the results with measurements. Overall, except for stall conditions, the increments in lift and changes in drag are well predicted.

Further investigation is warranted to determine what influence the various approximations (quasi three-dimensional boundary layer, coves, shear layer merging, wake shape, compressibility, turbulence model) may have on the accuracy of solutions.

13.5 Prediction of the Aerodynamic Performance Degradation of an Aircraft in Natural Icing Conditions

In this section we discuss the prediction of the aerodynamic performance degradation of the Twin Otter icing research aircraft shown in Fig. 13.42. This requires the prediction of ice shapes on the lifting surfaces by an extension of the LEWICE code for airfoils (Section 5.4) to three-dimensional flows as discussed in subsection 13.5.1 and the prediction of the effects of icing on the aircraft lift and drag coefficients by the calculation method of Chapters 10 to 12, as discussed in subsection 13.5.2.

13.5.1 Prediction of Ice Shapes on Wings

Before we discuss the extension of LEWICE to three-dimensional flows, it is useful to point out that the heart of LEWICE is associated with the energy balance, which includes terms providing heat to or removing heat from the control

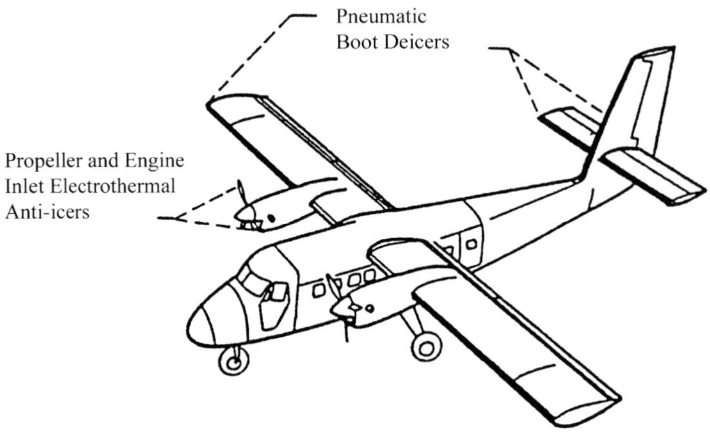

Fig. 13.42. NASA Lewis Twin Otter icing research aircraft.

volume. The dominant "heat removal" term is due to convective cooling and the dominant heat source term is due to the heat released by the freezing of water entering the control volume, which consists of: (1) water droplets impinging on the airfoil surface, and (2) water "running back" from an adjacent upstream control volume. This procedure for airfoils is remarkably good for predicting ice shapes (subsection 5.4.4) despite the very empirical nature of the expressions used in the heat balance as well as in the physical model of the ice accretion process. Studies conducted by Olsen and Walker [16] show that at aircraft speeds, there is no flow of liquid over the surface of the ice after a short initial period, even at subfreezing temperatures that are close to the freezing point. Instead, there are very large stationary drops on the ice surface that lose water from their bottoms by freezing as water is added to the droplets by the addition of microscopic cloud droplets. This observation disagrees with the existing physical model used in LEWICE, since it assumes that there is a thin liquid continuously flowing over the ice surface. While this model disagrees with the experimental evidence, especially for conditions leading to horn-shaped clear (glaze) ice, it does reasonably well for rime ice which forms when the air temperature is low enough to cause the cloud droplets to freeze almost immediately on impact.

The role that the liquid layer plays in the formation of glaze ice has been further elaborated on by Hansman and Turnock [17] and Bilanin [18], who point to the surface tension phenomenon as the most likely source for the observed deviations. There is convincing evidence from tests with variable surface tension that ice growth depends on it. Although a smooth water film may form at the stagnation point, it soon breaks up by the beading process and freezes into irregular rough lumps that enhance heat transfer and speed up the ice growth. The parameters controlling transition from a smooth film to beads and to rough ice formation are not well known at this time. For this reason, care is required to extend the ice accretion procedure from two- to three-dimensional flows, as discussed in this section.

In principle, the extension of the procedure used in LEWICE for airfoils to three-dimensional flows requires that the trajectory calculations be performed for the three components of the velocity obtained from a three-dimensional flowfield method and for a heat transfer balance which also includes the spanwise direction. While the former can be done with increased computer cost, the latter is difficult in view of the empirical nature of the heat transfer coefficient, h_c, used to represent the convective cooling given by Eq. (5.5.14). Since this equation is based mainly on data correlation with roughness parameters chosen so that the computed ice shapes fit experimental data, any attempt to add a spanwise heat transfer coefficient to the heat balance equation will be empirical and can only be done where experimental data exists. As a first step, however, it is prudent to leave the heat balance in its two-dimensional form and assume it to apply to a three-dimensional body which is expressed in an equivalent two-dimensional form. One approach is to perform trajectory calculations for a three-dimensional flowfield by determining the impingement patterns of the water droplets resulting from the three components of the velocity field. Another approach is to approximate the finite wing by an equivalent yawed infinite wing at each spanwise station and calculate the particle trajectories in a quasi-three-dimensional manner again for flowfield computed by a full three-dimensional flowfield method. A third approach, commonly called the strip-theory approach, is to apply the airfoil LEWICE computer program in its present form to each wing section for an equivalent angle of incidence determined from the local lift coefficient of the wing obtained by a three-dimensional flow method. Each approach has its merits and requires studies to see whether the use of full three-dimensional trajectory calculations are warranted in view of the very empirical nature of the heat balance equation, Eq. (5.5.9). Studies conducted by Cebeci et al. [19] cast doubts on the gains one may have using either the first or second approach. The calculations made using the three approaches show that there are essentially no gains from the first two approaches. It appears that the dominant effect lies with the heat transfer balance and that "more accurate" trajectory calculations do not have an appreciable effect on the ice shapes.

In Appendix 13A, we discuss the calculation of particle trajectories for three-dimensional flows since this is the appropriate approach to predict ice shapes if the energy balance were formulated for three-dimensional flows. Here in this subsection we use the strip theory approach for calculating ice shapes on the lifting surfaces of the Twin Otter aircraft.

Strip Theory Approach

In the strip-theory approach, at first an equivalent two-dimensional angle of attack, $\alpha_{2\mathrm{D}}$, which produces the same section lift coefficient, c_l, as the one

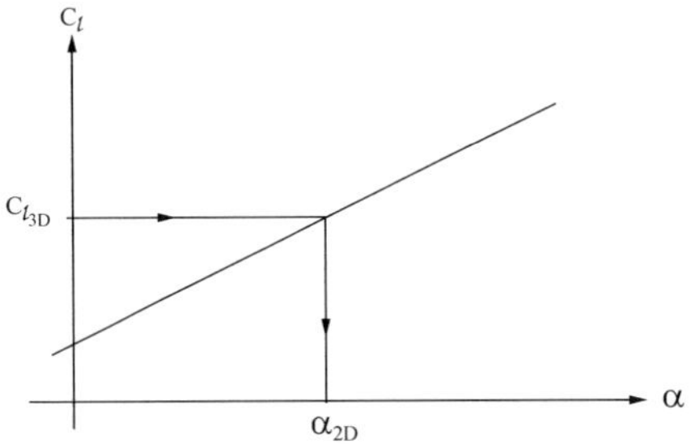

Fig. 13.43. Equivalent angle of attack determination.

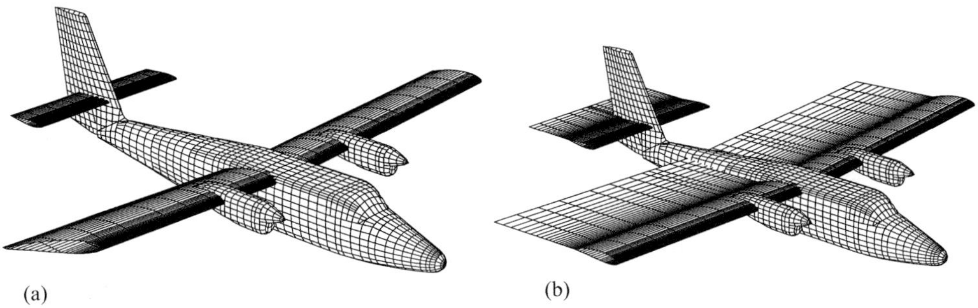

Fig. 13.44. Paneling of Twin Otter: (a) without wake, (b) with wake.

obtained by a three-dimensional flow calculation is determined. Knowing the section lift coefficient and the airfoil cross section, an equivalent angle of attack is determined from the solutions of a panel method for two-dimensional flows, as shown in Fig. 13.43.

Since the flow field is being calculated with the Hess panel method, the Twin Otter was paneled as shown in Fig. 13.44. A total of 11 and 5 lifting strips were taken on the wing and tail, respectively, with 72 and 67 grid points defining each airfoil section. The atmospheric conditions used in the calculations were taken from [20–22]. In each of these references, NASA identified a given flight with a number x-y, x being the year when they flew the Twin Otter and y being the flight number of the year x. For convenience, this identification number is kept the same here.

In calculating the ice shapes on the wing and on the tail with the strip theory approach, the following procedure is used:

1. Calculate the external velocity distribution on the airplane using the Hess panel method.
2. Use each section lift coefficient to determine an equivalent two-dimensional angle of attack.

3. Apply LEWICE to each strip and distribute the ice along the leading edge of the lifting surface.

This procedure was incorporated into a computer program that generated an iced aircraft for a given geometry and given atmospheric conditions.

The validity of the strip theory approximation as applied to the wing and tail configurations was studied in [23]. Calculations were first performed with the Hess panel method for the Twin Otter aircraft. For each of the lifting strips of the wing and tail, a pressure distribution and a section lift coefficient were obtained, a two-dimensional panel method was then used to compute the pressure distribution of the airfoil sections at angles of attack of 0.9 and 4.30 degrees. As expected, the differences between three-dimensional and "equivalent two-dimensional" velocity distributions for these two angles of attack were found to be very small for the sections when the three-dimensional flow is locally two-dimensional. The only sections where this assumption was not valid are those close to the fuselage for the tail and the wing where the flow is fully three-dimensional and the two-dimensional strip theory gave only an approximate pressure distribution.

However, as mentioned earlier, any improvement of the LEWICE code will have to rely on more accurate heat transfer computations rather than on better trajectory calculations. For this reason, and because our goal is to estimate how accurately the calculation method discussed in subsection 13.5.2 can predict the ice accretion, the strip theory approach was considered to be a good approximation in the study reported in [23].

The icing conditions considered in [23] are given in Table 13.4. The computed ice shapes correspond to the section where experimental results were available, which was at 69% of the semi-wing span. Some computed ice shapes for the tail are shown for 45% of the tail semi-span, even though no experimental ice shapes were available for comparison. With the present method, the computed ice shapes are not very different from one section to the other because the equivalent two-dimensional angle of attack is the only parameter which varies along the span.

Figures 13.45 and 13.46 show a sample of the ice shapes reported in [23]. Overall, the computed ice shapes agree well with the experimental data in the case of rime ice (flight 85-17). In the case of mixed ice – at slightly higher temperature – the method still gives a good prediction of the ice shape (flight 85-24a), although it does not predict the horns observed in flight (flight 85-24b and flight 84-34). No measured ice shapes were available for comparison under glaze ice conditions. However, it is expected that the generated ice shapes would have exhibited multiple horns which are difficult to predict.

Table 13.4. Atmospheric icing conditions for the Twin Otter Tests.

Flight No.	Type of ice	Pressure (pa)	Speed (m/s)	Duration of encounter (min)	Static Temp (K)	Liquid water content (g/m^3)	Medium particle diameter (μm)	Unit length roughness parameter
85-17	Rime	88150	59.72	65	261.5	0.22	13.4	5.764
83-11	Rime	84000	71.08	45	262.4	0.29	13.0	7.078
85-24a	Mixed	79600	71.38	15	258.3	0.45	19.5	7.814
85-24b	Mixed	79600	71.94	20	258.7	0.46	15.1	7.813
84-29	Mixed	79600	75.55	49	266.15	0.15	14.6	6.722
84-34	Mixed	82000	70.57	22	266.65	0.58	11.1	13.109
84-27	Glaze	73000	70.57	25	267.95	0.34	15.0	8.706
83-10	Glaze	85500	70.18	26	269.15	0.31	13.0	8.852

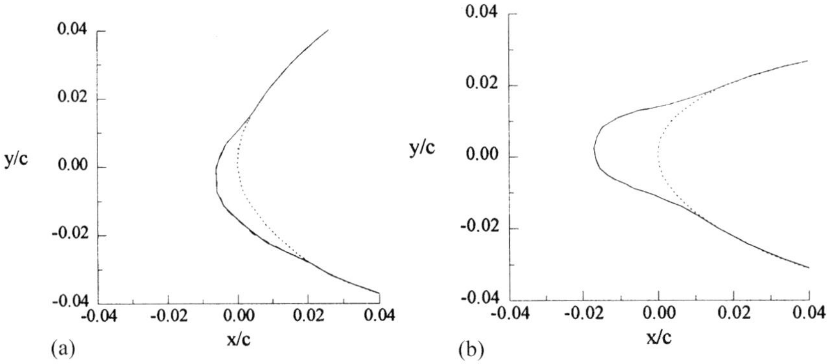

Fig. 13.45. Comparison of computed (solid lines) and experimental (dashed lines) ice shapes for rime ice, flight 85-17: (a) wing, and (b) tail.

13.5.2 Effect of Icing Conditions on Lift and Drag Coefficients

To determine the aerodynamic performance degradation of the aircraft due to ice, at first the studies in [23] were conducted for the "clean" aircraft with the inviscid flowfield of the aircraft again computed with the Hess panel method of Chapter 10. The viscous flow calculations are performed by applying the two-dimensional boundary-layer method of Chapter 3 to the lifting surfaces on the aircraft with a strip-theory approximation discussed in subsection 13.5.1. The Reynolds number is specified for cruise at an altitude of 7600 ft, at a speed of 70 ms^{-1}.

Table 13.6 shows the calculated aerodynamic characteristics of the clean aircraft with and without viscous effects, and Fig. 13.45 shows a comparison between the computed and experimental total lift, section lift and drag coefficients of the airplane. It can be seen that the introduction of the viscous effects on the lifting surfaces substantially lowers the lift. As expected, the lift curves are well predicted, but the computed section drag coefficient is too low, partly,

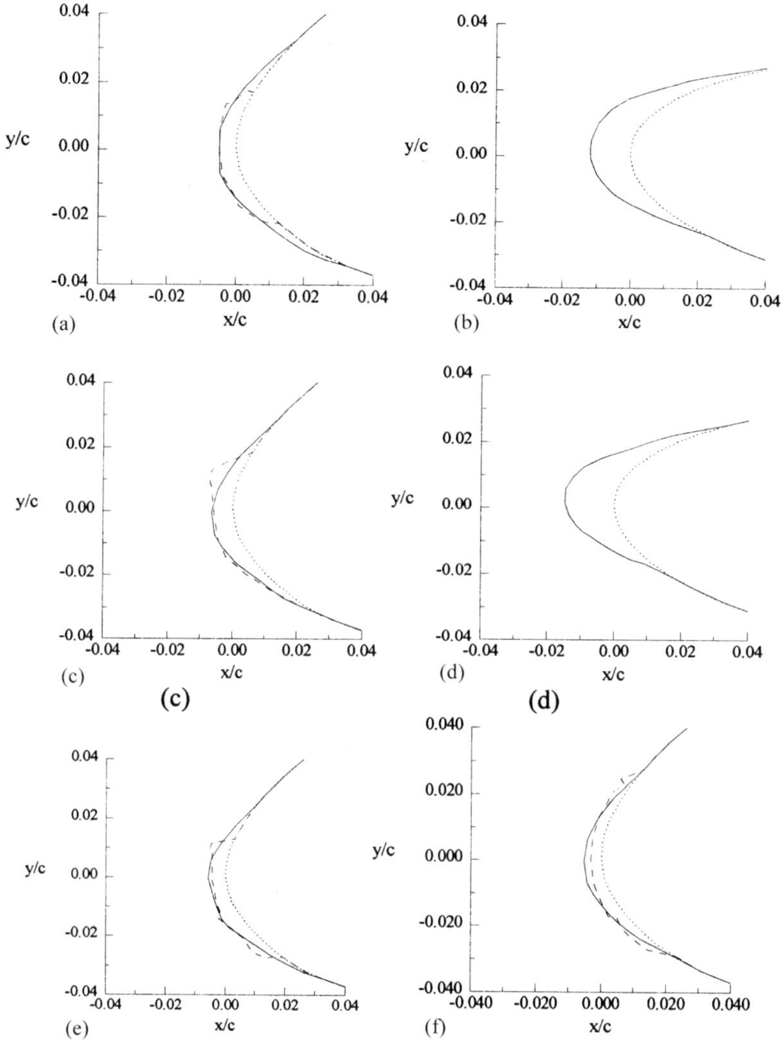

Fig. 13.46. Comparison of computed ice (solid lines) and experimental ice (dashed lines) shapes for mixed ice: (a) wing, flight 85-24a, (b) tail, flight 84-24a, (d) wing, flight 85-24b, (d) tail, flight 85-24b, (e) wing, flight 84-34, and (f) wing, flight 84-29.

at least, because the drag generated by the natural wing roughness (rivets, deicing boots, etc.) is not accounted for. The experimental data related to the lift was obtained from [21] and the drag data from [22]. The drag data from [21] were also reported, although they appear to be inconsistent.

Figure 13.48 shows a comparison between the total airplane lift coefficients with icing conditions corresponding to (a) rime ice of 45 minute accumulation, flight 83-11, (b) mixed ice of 22 minute accumulation, flight 84-34, and (c) glaze ice of 25 minute accumulation, flight 84-27.

Table 13.5. Aerodynamic characteristic of the clean Twin Otter.

Angle of attack (α)	Total aircraft lift coefficient		Section lift coefficient (69% of semi-span)		Section drag coefficient (69% of semi-span)
	Inviscid	Viscous	Inviscid	Viscous	
	0.4106	0.3516	0.4664	0.4334	0.00933
2°	0.6158	0.5756	0.6322	0.5910	0.00963
4°	0.8209	0.7675	0.7987	0.7457	0.01000
6°	1.0255	0.9509	0.9636	0.8909	0.01073
8°	1.2292	1.1309	1.1367	1.0238	0.01188
10°	1.4316	1.2405	1.2876	1.1504	0.01307
13°	1.6318	1.4388	1.4457	1.2719	0.01426

The comparisons also include the airplane lift coefficient with the clean wing and tail. The results in Fig. 13.48b also show the calculations for an iced airplane with the wings deiced. In this case (deiced wing), all the ice accumulation between the wing tip and the propeller was removed (deiced) but the wing ice between the propeller and the fuselage was not removed. The calculations for this case were performed for a clean wing between the wing tip and the propeller and for an iced wing between the propeller and fuselage. The tail calculations were performed for an all iced-tail conditions. Whereas the results are good at lower angles of attack, they are not so good at higher angles. This may be due to the ice shape with horns. Figure 13.49 shows a comparison between clean and iced section drag coefficients obtained at 69% wing semi-span. The calculated drag coefficients have been "corrected" such that the clean wing matches the experimental data. Figure 13.49a is for rime ice accumulation of 65 minutes and Fig. 13.49b is for mixed ice accumulation of 15 minutes. Both figures clearly show the drag increase due to ice on the wing.

13.5.3 Summary

The results and discussion in this section show that the approach and the method used to predict the performance degradation of the Twin Otter in natural icing conditions provide results which, in general, are in good agreement with measurements. The calculations indicate that LEWICE, with strip theory approximation as used in the study conducted in [23], produces ice shapes which are in good agreement with data. Of course, as in other studies, the predictions of glaze ice are not as good, but this is where the weakness of LEWICE lies, which is well known.

The calculation method in which the boundary-layer method is based on strip-theory approximation, predicts the viscous effects well and allows the determination of viscous drag and loss in lift. Even though the turbulence model is rather empirical, the results predict the icing and deicing effects remarkably

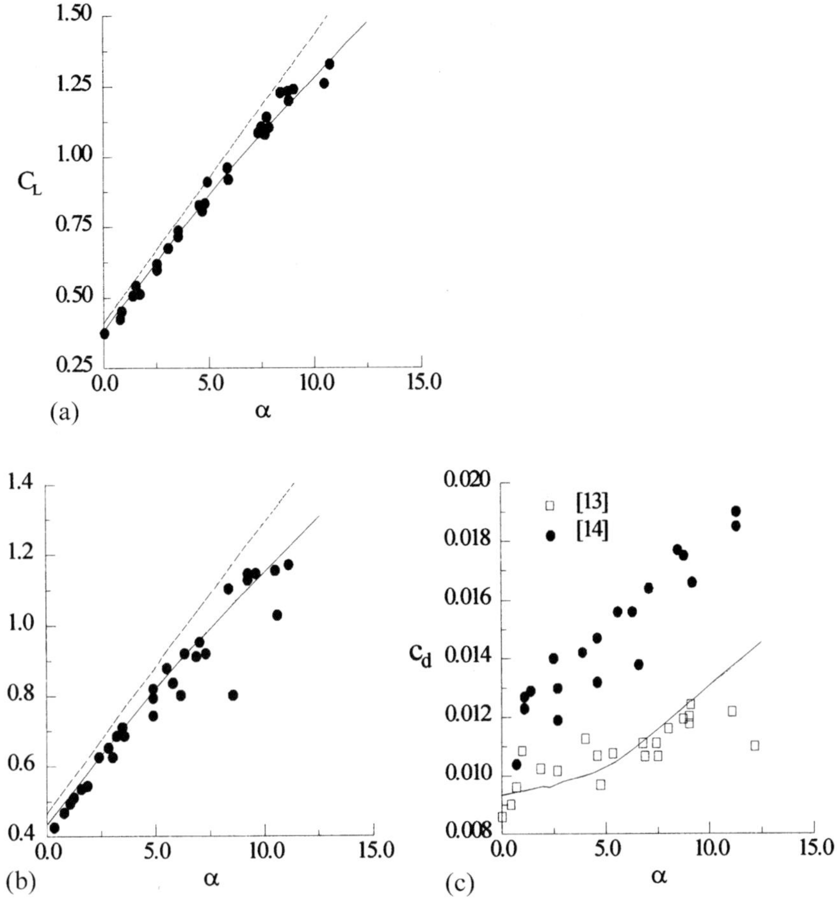

Fig. 13.47. Comparison of calculated (lines) and measured (circles) (a) total lift coefficient, (b) section lift coefficient, and (c) section drag coefficient. Dotted lines denote inviscid results and continuous lines results with viscous effects.

well. Additional studies, however, are needed before a firm conclusion can be reached.

In general, the study shows that the goal of cost-effective prediction of the aerodynamic performance degradation of an aircraft due to natural icing conditions can be obtained with this approach.

13.6 Transonic Flows

In this section we extend the applications of the calculation method of Chapters 10 and 12 to wing and wing/body configurations in transonic flows. The main difference in the calculation method is the replacement of the panel method

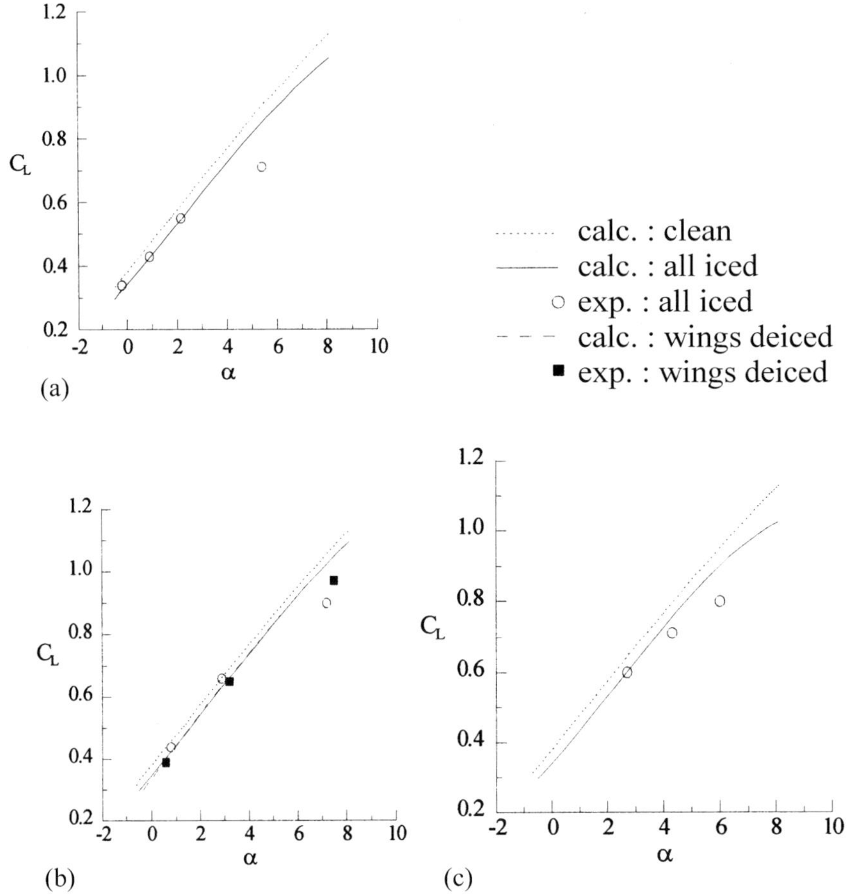

Fig. 13.48. Comparison of calculated and measured airplane lift coefficient for (a) rime ice (flight (83-11), (b) mixed ice (flight 84-34), and (c) glaze ice (flight 84-27).

with an Euler method and use of the compressible version of the boundary-layer method discussed in Chapter 11.

There are several Euler methods that can be used to compute the inviscid flow and these are discussed in several references, see for example, [24–28]. Popular and versatile Navier-Stokes codes such as OVERFLOW [29], CFL3D [30,31] and TLNS3D [32] can also be used for this purpose since they all have options for inviscid methods.

The accuracy of the calculation method of Chapters 10 and 12, as in two-dimensional flows, not only depends on the solutions of the boundary-layer equations and transition prediction, but also on the solutions of the Euler equations, most of which employ either cell-centered schemes or cell-vertex schemes. In [33] Koduma discusses a cell-centered finite-volume upwind scheme with global conservation (meaning that the conservation property is satisfied everywhere in

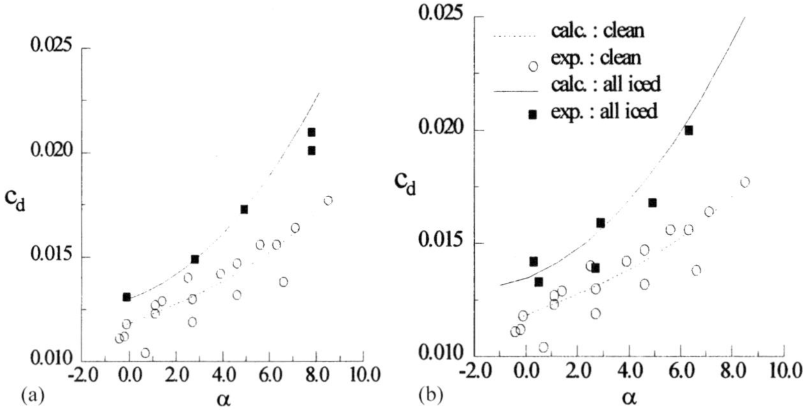

Fig. 13.49. Increase in section drag coefficient due to (a) rime ice (flight 85-17), and (b) mixed ice (flight 85-24a).

the computed flow domain, all the way down to boundaries) and shows how one can remove the "flaws" existing in conventional methods (cell-vertex scheme) that prevent global conservation by adopting a cell-centered scheme that solves for the flowfield at the nodal points. In both schemes the fluxes across cell surfaces and the residuals of the Euler equations at cell centers are calculated at each time step during iterations.

In the cell-centered scheme, the fluxes across cell surfaces are calculated by making use of the cell-centered solutions of the two neighboring cells adjacent to the cell surface, while in the cell vertex scheme the fluxes across the cell surfaces are calculated by making use of the four nodal solutions of the cell surface. After knowing the fluxes at each cell surface in both schemes, the residuals of the Euler equations are calculated at the centers of each cell. In the cell vertex scheme, the residuals at nodal points are obtained by averaging the cell-centered values of residuals of the surrounding cells. For regular cells, both schemes are equally good in terms of the solution accuracy. For cells with relatively large aspect ratios, the cell vertex scheme seems to introduce smaller truncation errors because of more accurate summation of the fluxes through the surface. However, for surface cells with a high degree of distortion, the cell-centered scheme seems to be more stable, probably because of the avoidance of direct solving for solutions at nodal points on the surface.

To introduce the viscous effects into an Euler method via the blowing velocity approach, it is necessary to modify its surface boundary condition. In the cell-centered scheme, Fig. 13.50a, ρ, ρu, ρv, ρw and ρe are computed at the centers of mesh cells during each time step. Imaginary cells inside the wing surface are used to formulate the flux balance in the surface normal direction and also the dissipation terms near the surface. With flow variables ρ, ρu, ρv, ρw and ρe

(a) (b)

Fig. 13.50. Control volume with (a) cell-centered and (b) cell-vertex schemes. Symbols • and × denote grid and flow variable nodes, respectively.

given at the centers of mesh cells next to the wing surface, and a prescribed blowing velocity distribution on the surface, the flow variables at the centers of imaginary cells can be determined by using a simple linear extrapolation. The normal pressure gradient term on the wing surface is computed according to the prescribed blowing velocity distribution and is then used to evaluate the pressure distribution on the wing surface.

In the cell-vertex scheme, Fig. 13.50b, the blowing velocity distribution is prescribed on the surface, the residual is calculated at the centers of the cells adjacent to the surface, and the pressure distribution on the surface is evaluated directly from the cell-vertex solutions obtained on the surface.

It is clear that the choice of the grid system used in the solution of the Euler equations plays a crucial role, since the grid generation method not only must be able to model the geometry of the configuration accurately, but also be able to resolve the detailed flowfield in high pressure gradient regions, such as along the leading and trailing edges of the body, along the wing/body juncture, and near wing tip and fuselage nose.

The application considered in this section deals with wing/body configurations as well as with the wing alone. An appropriate grid generation method for such a configuration is a *C-C-C* grid, which is referred to as the QGRID by Chen and Bui [34]. In this method, based on a hybrid mapping/numerical grid generation scheme, the grid-point distributions near the fuselage surface and in the wing cross-sectional surface are similar to that in the *C-C-H* grid, discussed in [26], but the mesh lines in the wing tip region form a *C*-mesh, as shown in Fig. 13.51, resulting in a concentration of grid points near the wing tip. A global view of the QGRID is given in Fig. 13.52, which shows the grid distributions on the body surface, the symmetry plane, and the far-field boundary. With this grid, there is no need to add a fictitious wing extending from the wing tip to the spanwise farfield, as in some other grid systems.

In the following two subsections, we consider two finite-volume Euler methods called FLO-58 and FLO-67, both developed by Jameson [24,26] using the QGRID discussed above. While FLO-58 uses a cell-centered scheme, FLO-067 uses a cell-vertex scheme, allowing us to evaluate the accuracy of each scheme against data on wing/fuselage and wing-alone configurations in conjunction with

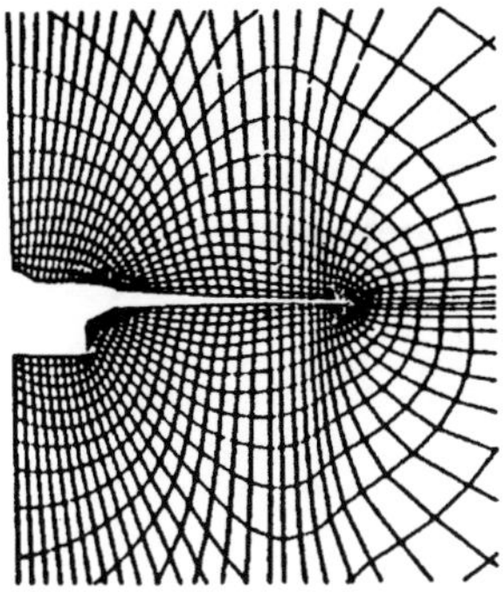

Fig. 13.51. A cross-section view of the QGRID.

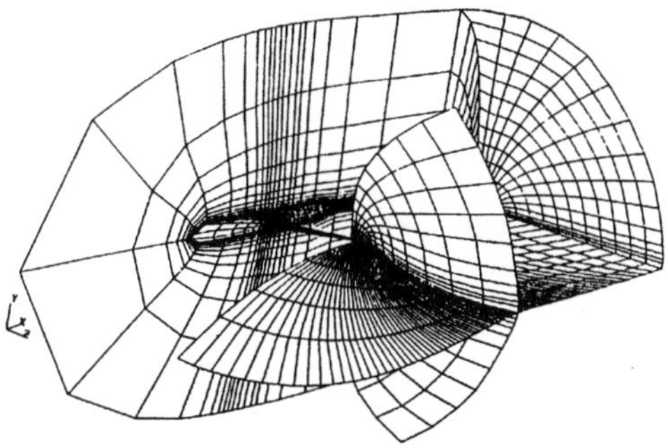

Fig. 13.52. A global view of the QGRID.

the boundary-layer method discussed in Chapter 11. We assume that the distribution of the onset of transition along the wing span is known.

13.6.1 ONERA-M6 Wing

Figure 13.53 shows a comparison between calculated and measured chordwise pressure distributions at four spanwise stations for the ONERA-M6 wing for $\alpha = 3.06^\circ$, $M_\infty = 0.84$. The inviscid flow is computed from FLO-58 and FLO-67. This wing has a 30° leading-edge sweep and a 15.8° trailing-edge sweep. Its wing tip is closed by a half body of revolution. The test data was obtained with free transition, and pressure data was obtained using seven rows of pressure taps located from 20% to 99% semispan. The Euler calculations with the QGRID used a total of $193 \times 25 \times 33$ grid points. The location of transition was fixed at $x/c = 0.03$ on the upper surface. On the lower surface its location varied, being at $x/c = 0.66$ at the root and 0.50 at the tip. As can be seen, the interactive solutions in which the inviscid flow solutions obtained with two different Euler methods are essentially the same and agree well with measured values.

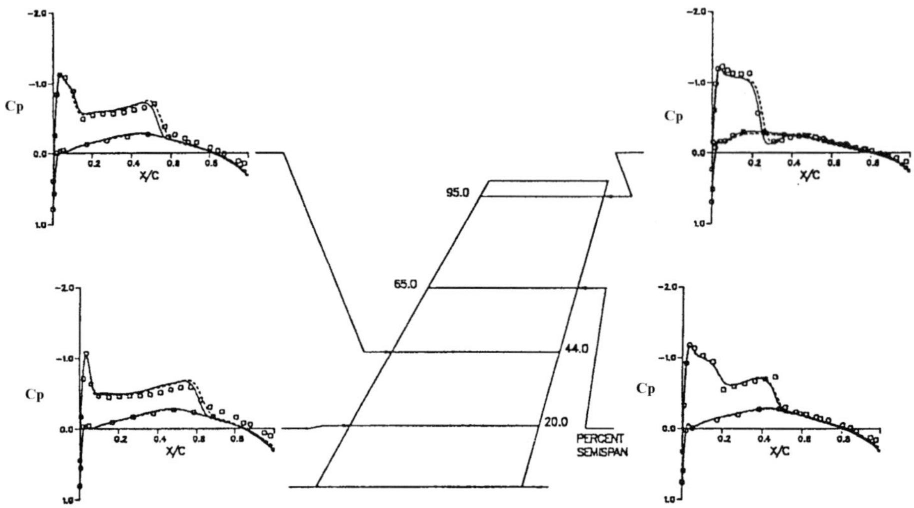

Fig. 13.53. Comparison of interactive boundary-layer solutions with the experimental data for the ONERA-M6 wing for $\alpha = 3.06°$, $M_\infty = 0.84$. Symbols denote data and solid and dashed lines denote the interactive flow solutions in which the inviscid flow solutions are obtained from FLO-58 and FLO-67 codes, respectively.

13.6.2 Douglas Wing/Fuselage Configuration

Figure 13.54 shows a comparison between the solutions of the calculation method and experimental data for a Douglas wind-tunnel model LB 488 for $M_\infty = 0.80$, $\alpha = 2.30$, $Re = 5.4 \times 10^6$. Again the inviscid flow calculations used the QGRID and two Euler methods, as in the wing-alone case, for a total of $193 \times 25 \times 25$ grid points, and the distribution of transition locations was specified. Unlike the wing-alone case, however, the calculated solutions for the inviscid flow computed with two Euler codes using different numerical schemes differ from each other. Those computed with FLO-67 (cell-vertex scheme) predict a weaker shock near the wing root and are not in good agreement with experimental data in the inboard region of the wing, but are in good agreement with data in the outboard region. The calculations performed with FLO-58 (cell-centered scheme) agree well with measurements in the inboard region but predict a stronger shock in the outboard region. Additional calculations performed for wing/body configurations show that even though the FLO-67 solutions are better for predicting the correct shock, the predictions of FLO-67 are consistently poor in the wing-root region, and overall Euler solutions using the cell-centered scheme produce results which, when used in the calculation method, yield much better agreement with data, including the outboard region.

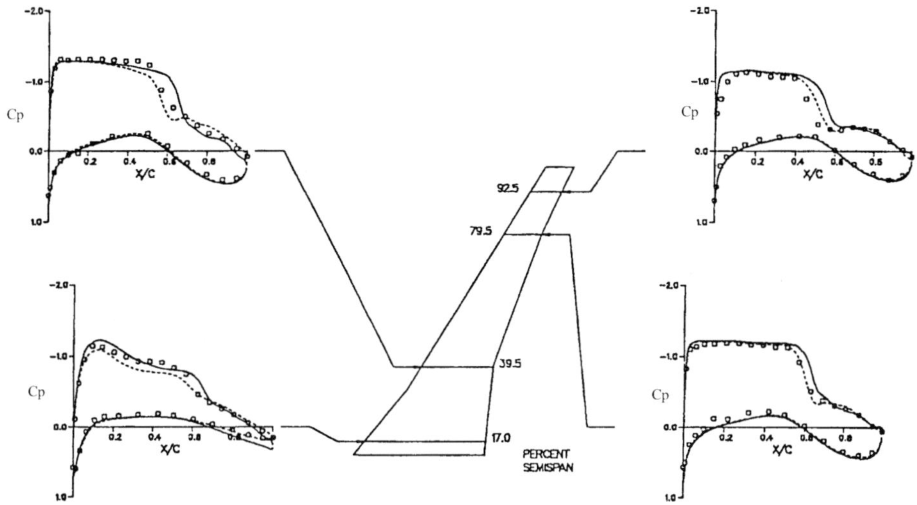

Fig. 13.54. Comparison of interactive boundary-layer solutions with the experimental data for the LB-488 wing/body configurations for $\alpha = 2.30°$, $M_\infty = 0.80$, $Re = 5.4 \times 10^6$. Symbols denote data and solid and dashed lines denote the interactive flow solutions in which the inviscid flow solutions are obtained from FLO-58 and FLO-67 codes, respectively.

Appendix 13A Calculation of Particle Trajectories

The calculation of particle trajectories requires the calculation of the flowfield. For three-dimensional flows, we use the Hess surface-source panel method discussed in Chapter 10. In this method, for a wing defined by streamwise sections, the inviscid velocity components $V_{\bar{x}}, V_{\bar{y}}, V_{\bar{z}}$ are computed in a Cartesian coordinate system $\bar{x}, \bar{y}, \bar{z}$ where $\bar{x}$ is along the fuselage axis, $\bar{y}$ is spanwise and $\bar{z}$ normal to the $\bar{x}, \bar{y}$ plane, for a given incidence angle α measured from the $\bar{x}$-axis. If we denote the unit vectors in this coordinate system by $(\hat{\imath}, \hat{\jmath}, \hat{k})$, the velocity vector can be written as

$$\vec{V} = V_{\bar{x}}\hat{\imath} + V_{\bar{y}}\hat{\jmath} + V_{\bar{z}}\hat{k} \tag{A13.1}$$

The particle trajectory equations are obtained from Newton's second law of motion by assuming that the forces acting on the water droplet or particle consist of an aerodynamic drag force $\hat{F}_a$ in the direction of the relative velocity vector $\hat{V}_r$ and a gravity force $\hat{F}_g$ in the vertical plane and with m representing the mass of a water particle,

$$m\frac{d^2\vec{r}_p}{dt^2} = \vec{F}_a + \vec{F}_g \tag{A13.2}$$

Here $\vec{r}_p$ denotes the position vector of the particle defined by $\vec{r}_p = x_p\hat{\imath} + y_p\hat{\jmath} + z_p\hat{k}$. We next assume level flight so that both the angle of attack α and the gravity force vector are in the $\bar{x}\bar{z}$-plane. The relative air velocity vector $\hat{V}_r$ that the particle experiences is given by

$$\vec{V}_r = \vec{V} - \vec{V}_p \tag{A13.3}$$

where $\vec{V}_p$ is the particle speed with Cartesian components $V_{p\bar{x}}$, $V_{p\bar{y}}$, and $V_{p\bar{z}}$. In terms of the velocity components, the relative air velocity with respect to the moving particle can be expressed as

$$\vec{V}_r = (V_{\bar{x}} - V_{p\bar{x}})\hat{\imath} + (V_{\bar{y}} - V_{p\bar{y}})\hat{\jmath} + (V_{\bar{z}} - V_{p\bar{z}})\hat{k} \tag{A13.4}$$

so that the unit vector, $\hat{e}_r$, in the $\vec{V}_r$ direction can be computed from

$$\hat{e}_r = \frac{\vec{V}_r}{|\vec{V}_r|} = \frac{V_{\bar{x}} - V_{p\bar{x}}}{V_r}\hat{\imath} + \frac{V_{\bar{y}} - V_{p\bar{y}}}{V_r}\hat{\jmath} + \frac{V_{\bar{z}} - V_{p\bar{z}}}{V_r}\hat{k} \tag{A13.5}$$

where

$$V_r = [(V_{\bar{x}} - V_{p\bar{x}})^2 + (V_{\bar{y}} - V_{p\bar{y}})^2 + (V_{\bar{z}} - V_{p\bar{z}})^2]^{1/2} \tag{A13.6}$$

The drag force $\vec{F}_a$ and the gravitational force $\vec{F}_g$ are

$$\vec{F}_a = \hat{e}_r C_d \rho \frac{\vec{V}_r \cdot \vec{V}_r}{2} A_p \tag{A13.7a}$$

$$\vec{F}_g = mg(\hat{\imath} \sin\alpha - \hat{k}\cos\alpha) \tag{A13.7b}$$

Here C_d is the drag coefficient of the particle, ρ the density of air, A_p is the particle reference area and g the acceleration of gravity.

Resolving Eq. (A13.2) into its components and using the relations in Eqs. (A13.7) and dropping the bars on $\bar{x}$, $\bar{y}$ and $\bar{z}$ we can write

$$\frac{d^2 x_p}{dt^2} = KV_r(V_x - V_{px}) + g\sin\alpha \tag{A13.8a}$$

$$\frac{d^2 y_p}{dt^2} = KV_r(V_y - V_{py}) \tag{A13.8b}$$

$$\frac{d^2 z_p}{dt^2} = KV_r(V_z - V_{pz}) + g\cos\alpha \tag{A13.8c}$$

where $K = \rho C_d A_p/2m$. Since the aerodynamic force is assumed to act along the relative velocity vector direction, the flow around the particle must be axially symmetric. The simplest configuration conforming to this requirement is a sphere which is also a good approximation to the water droplet shape. For a spherical particle of radius r, $A_p = \pi r^2$ and the mass is $(4/3)\pi\rho_w r^3$ so that

$$K = \frac{3}{4}\frac{C_d}{d}\left(\frac{\rho}{\rho_w}\right) \tag{A13.9}$$

where ρ_w is the density of liquid water, and d is the particle diameter.

Noting that

$$\frac{dx_p}{dt} = V_{px}, \quad \frac{dy_p}{dt} = V_{py}, \quad \frac{dz_p}{dt} = V_{pz} \tag{A13.10}$$

Eqs. (A13.8) can be written as a system of six first-order equations in dimensionless form by normalizing all velocities by freestream velocity V_∞, all lengths by L and time by L/V_∞. The resulting equations in addition to those in Eq. (A13.10) with F denoting the Froude number, $\sqrt{V^2/Lg}$, are

$$\frac{dV_{px}}{dt} = KV_r(V_x - V_{px}) + \frac{\sin\alpha}{F^2} \tag{A13.11a}$$

$$\frac{dV_{py}}{dt} = KV_r(V_y - V_{py}) \tag{A13.11b}$$

$$\frac{dV_{pz}}{dt} = KV_r(V_z - V_{pz}) - \frac{\cos\alpha}{F^2} \tag{A13.11c}$$

The quantities V_x, V_y, and V_z on the right-hand side of Eqs. (A13.10) and (A13.11) are functions of x_p, y_p and z_p, and are all evaluated at each calculation step from the inviscid flowfield solution. The drag coefficient C_d for a sphere in incompressible flow is given as a function of Reynolds number ($R \equiv \frac{V_r d}{\nu}$) for $10\,\mu\text{m} < d < 500\,\mu\text{m}$ by

$$C_d = \begin{cases} \dfrac{24}{R} & 0 < R < 0.05 \\ \dfrac{1}{R}(24.167 + 3.254R - 0.23564R^2) & 0.05 < R < 3.04726 \\ \dfrac{1}{R^2}(-28.339 + 38.969R + 0.73204R^2 \\ \quad - 0.0005608R^3) & 3.04726 < R < 82.230 \\ \dfrac{1}{R^2}\exp\{LR[5.3220062 - 1.937238(LR) & R > 82.230 \\ \quad + 0.4372856(LR)^2 + 0.001926421(LR)^4]\} & \end{cases} \tag{A13.12}$$

where $LR = \ln(R)$.

For $d < 10\,\mu\text{m}$, the drag coefficient from the above relations is corrected for slip-flow effects. The correction factor CF is given in tabular form in Table A13.1 as a function of the particle diameter d. The corrected drag coefficient is obtained from C_d/CF.

The system given by Eqs. (A13.10) and (A13.11) constitutes an initial value problem so that the values of the unknowns must be specified at the starting point of the calculations. It is common to start them at points where $|\vec{V} - \vec{V}_\infty| < \varepsilon$, where ε is a small number. This ensures that particles, which are motionless relative to the still air, initially move with nearly freestream speed with respect to the body-fixed coordinate system. It is sufficient to set $V_{px} = V_x$, $V_{py} = V_y$ and $V_{pz} = V_z$ and proceed with the numerical integration of the equations. In this initial value problem, it is desirable to use a predictor-corrector method with a variable step size in conjunction with error control in order to obtain accurate solutions. Gear's method [35] used in LEWICE is a good one for this purpose since it incorporates automatic step size and error control to a specified tolerance.

Table 13.6. Cunningham correction factor for standard air.

$d(\mu)$	CF	$d(\mu)$	CF	$d(\mu)$	CF	$d(\mu)$	CF
0.001	221.600	0.010	22.680	0.1	2.911	1.0	1.168
0.002	112.100	0.020	12.650	0.2	1.890	2.0	1.084
0.003	74.250	0.030	7.978	0.3	1.574	3.0	1.056
0.004	55.830	0.040	6.151	0.4	1.424	4.0	1.042
0.005	44.780	0.050	5.060	0.5	1.337	5.0	1.034
0.006	37.410	0.060	4.337	0.6	1.280	6.0	1.028
0.007	32.150	0.070	3.823	0.7	1.240	7.0	1.024
0.008	28.200	0.080	3.441	0.8	1.210	8.0	1.021
0.009	24.140	0.090	3.145	0.9	1.186	9.0	1.019
						11.0	1.017

References

[1] Arnal, D. and Juillen, J.C.: "Three-Dimensional Transition Studies at ONERA/CERT, AIAA Paper No. 87-1335, 1987.

[2] Cebeci, T., Chen, H.H., Arnal, D. and Huang, T.T.: "Three-Dimensional Linear Stability Approach to Transition on Wings and Bodies of Revolution at Incidence," AIAA Journal, Vol. 29, No. 12, pp. 2077–2085, 1991.

[3] Séraudie, A., Archambaud, J.P. and Payry, M.J., "Etude de la laminarité sur l'aile AS409 jusqu'a des nombres de Reynolds de l'ordre de 14 millions dans la soufflerie T2," ONERA/CERT Internal Report No. 33/5006-29, June 1989.

[4] Niethammer, R., "Boundary-layer stability computations related to laminar flow experiments at low temperatures," Project de fin D'etudes. Department d'etudes et de recherches en aerothermodynamique, Toulouse Cedex, July 1991.

[5] Cebeci, T., Chen, H.H. and Arnal, D., "Natural transition in compressible flows on wings: Spatial theory and experiment," AIAA Paper No. 94-0824, Jan. 1994.

[6] Malik, M.R. and Poll, D.I.A., "Effect of curvature on three-dimensional boundary-layer stability," AIAA Journal, Vol. 23, 1985, pp. 1352–1369.

[7] Collier, F.S., Bartlett, D.W., Wagner, R.D., Tat, V.V. and Anderson, B.T., "Correlation of boundary-layer stability analysis with flight transition data," Proceedings of IUTAM, Third Symposium on Laminar-Turbulent Transition, Sept. 11–15, 1985, Toulouse, France.

[8] Poll, D.I.A., "Some observations of the transition process on the windward face of a long yawed cylinder," J. Fluid Mech., Vol. 150, 1985, pp. 329–356.

[9] Cebeci, T., Chen, H.H. and Kaups, K., "Further consideration of the effect of curvature on the stability of three-dimensional flows," Computers and Fluids, Vol. 21, No. 4, pp. 491–502, 1992.

[10] Schmitz, A., "Effect of Longitudinal Curvature on Stability and Transition," Aerospace Eng. Dept., California State University, Long Beach, California, 1994.

[11] Bertelrud, A., "Measurements on a Three-Dimensional Swept Wing at Low Speeds," The Aeronautical Research Institute of Sweden, Rept No. 131, Stockholm, 1977.

[12] Besnard, E., "Prediction of High Lift Flows with Separation", Ph.D. Dissertation, California State University, Long Beach, 1998.

[13] Lovel, D.A., "A Wind-Tunnel Investigation of the Effects of Flap Span and Deflection Angle, Wing Planform and a Body on the High-Lift Performance of a 28° Swept Wing". RAE CP 1372, 1977.

[14] Cebeci, T. and Besnard, E., "An Efficient and Accurate Approach for Analysis and Design of High Lift Configurations," Canadian Aeronautics and Space J., Vol. 44, pp. 212–222, 1998.

[15] Rogers, S., Menter, F., Durbin, P., and Mansour, N., "A Comparison of Turbulence Models in Computing Multi-Element Airfoil Flows," AIAA Paper No. 94-0291, 1994.

[16] Olsen, W. and Walter, E., "Experimental Evidence for Modifying the Current Physical Model for Ice Accretion on Aircraft Structures," NASA TM 87184, 1986.
[17] Hansman, R.J. and Turnock, S.R., "Investigation of Surface Water Behavior During Glaze Ice Accretion," J. of Aircraft, Vol. 26, No. 2, p. 140, 1989.
[18] Bilanin, A.J., "Proposed Modification to Ice Accretion/Ice Scaling Theory," J. of Aircraft, Vol. 28, No. 6, p. 353, 1991.
[19] Cebeci, T., Chen, H.H., Kaups, K. and Schimke, S., "Analysis of Iced Wings," AIAA Paper 92-0416, 1992.
[20] Ramando, R.J., Mikkelsen, K.L. and McKnight, R.C., "Performance Degradation of a Typical Twin Engine Commuter Type Aircraft in Measured Natural Icing Conditions," NASA TM 83564, 1984.
[21] Mikkelsen, K.L., McKnight, R.C., Ramando, R.J. and Perkins, P.J., "Icing Flight Research: Aerodynamic Effects of Ice and Ice Shape Documentation with Stereo Photography," NASA TM 86906, AIAA Paper 85-0468, 1985.
[22] Mikkelsen, K.L. Juhasz, N. Ramando, R., McKnight, R.C., Freedman, R. and Greishing, J., "In-Flight Measurements of Wing Ice Shapes and Wing Section Drag Increase Caused by Natural Icing Conditions," NASA TM 87301, 1986.
[23] Cebeci, T. and Besnard, E., "Progress toward the prediction of the aerodynamic performance degradation of an aircraft in normal icing conditions", California State University, Long Beach, Aerospace Eng. Dept. Report No. AE-93-6, 1993.
[24] Jameson, A., Schmidt, W. and Turkel, E., "Numerical Solutions of the Euler Equations by Finite-Volume Methods Using Runge-Kutta Time-Stepping Schemes," AIAA Paper No. 81-1259, June 1981.
[25] Rizzi, Arthur, "Damped Euler-Equation Method to Compute Transonic Flow Around Wing-Body Combinations," AIAA J., Vol. 20, pp. 1321–1328, Oct. 1982.
[26] Jameson, A. and Baker, T.J., "Multigrid Solution of the Euler Equations for Aircraft Configurations," AIAA Paper No. 84-0093, Jan. 1984.
[27] Jameson, A. and Yoon, S., "Multigrid Solution of the Euler Equations Using Implicit Schemes," AIAA J. Vol. 24, pp. 1737–1743, Nov. 1986.
[28] Roe, P.L., "Characteristic-Based Schemes for the Euler Equations," Annual Review of Fluid Mechanics, Vol. 18, pp. 337–365, 1986.
[29] Buning, P.G. et al., "OVERFLOW User's Manuel," Version 1.7r, NASA Ames Research Center, Moffett Field, CA., 1996.
[30] Thomas, J.L., Taylor, S.L. and Anderson, W.K., "Navier-Stokes Computations of Vortical Flows Over Low Aspect Ratio Wings," AIAA Paper No. 87-0207, Jan. 1987.
[31] Krist, S.L., Biedron, R.T. and Rumsey, C.L., "CFL3D User's Manual," NASA Langley Research Center, Hampton, VA, 1996.
[32] Vatsa, V.N. and Wedan, B.W., "Development of an Efficient Multigrid Code for 3-D Navier-Stokes Equations," AIAA Paper No. 89-1791, June 1989.
[33] Kodama, Y., "A Cell-Centered Finite-Volume Upwind Scheme with Global Conservation," Paper presented at the Autumn Meeting of the Society of Naval Architects of Japan, Nov. 1990.
[34] Chen, L.T. and Bui, M.N., "An Interactive Scheme for Transonic Wing/Body Flows Based on Euler and Inverse Boundary-Layer Equations," AIAA Paper No. 90-1586, June 1990.
[35] Hoffman, J.D., Numerical Methods for Engineers and Scientists. McGraw-Hill, N.Y., 1992.

Subject Index

Computer to plate: Mercedes Druck, Berlin
Binding: Buchbinderei Lüderitz & Bauer, Berlin